Franz Reuleaux (1829–1905)

THE KINEMATICS OF MACHINERY

Outlines of a Theory of Machines

Franz Reuleaux

Translated and Edited by
ALEXANDER B. W. KENNEDY, C.E.

and With an Introduction by
EUGENE S. FERGUSON

DOVER PUBLICATIONS, INC.
MINEOLA, NEW YORK

INTRODUCTION TO DOVER EDITION

The Book

OUR modern way of thinking about kinematics of mechanisms was established by this book. The study of kinematics, which is concerned with the discovery and application of principles underlying motions occurring in mechanisms, is based upon the assumptions and propositions set forth in these pages. The author's work has been elaborated upon, added to, and superseded in many details, but his conviction that this is the way we must look at mechanisms if they are to be understood has withstood the test of nearly a century of time.

Written by Franz Reuleaux (1829–1905) while he was professor of kinematics and director of the Royal Industrial Academy in Berlin, the treatise was published serially in the *Transactions* of the Society for the Advancement of Industry in Prussia, from 1871 through 1874. In 1875 the entire work was published in book form under the title *Theoretische Kinematik: Grundzüge einer Theorie des Maschinenwesens*; and in 1876 the present English translation by Alexander Blackie William Kennedy (1847–1928) was published in London.

Many of the ideas and concepts introduced in this book have become so familiar to us that we are likely to underestimate Reuleaux's originality and consider him merely a recorder of the obvious. We may feel that there is indeed no other way of approaching kinematics. But that is perhaps the hallmark of genius: to state a new idea in such convincing and uncompromising terms that it becomes immediately obvious and soon a truism.

Many men, before and after Reuleaux, have contributed to the development of kinematics of mechanisms. By any standards, however, the original and incisive ideas of Reuleaux, contained in this book, overshadow the contributions of others in the field.

Reuleaux's *Kinematics of Machinery* is not merely of interest historically, however. In it the reader has the opportunity of observing at work a brilliant mind and of sharing with the author some of the penetrating insights that enabled him to discover and identify the permanent truths of his subject. Reuleaux's sections on the process of innovation, for example (Chapter VI and parts of the Introduction and Chapter I) will be of interest to anybody who has ever seriously wondered how we go about doing anything that has not been done before.

The fundamental and original ideas that were first stated clearly by Reuleaux can be summarized briefly. While the concepts are few and simple, it is instructive to note that they establish the point of view from which we contemplate mechanisms today.

The author observed, for one thing, that machine elements are never employed singly, but always in pairs. He was thus able to identify the " kinematic pair." The idea of expansion of pairs led to the recognition of kinematic similarities between mechanisms that physically are quite dissimilar in form and arrangement.

Reuleaux was first to recognize, moreover, that the fixed link of a mechanism is kinematically the same as any of the moving links. From this discovery followed the powerful concept of inversion of linkages, in which different links are fixed successively to change the function of a mechanism.

His chapter on rotary engines and pumps (Chapter IX) demonstrates clearly the extent to which designers may be confused and deluded in their ignorance of the notions of expansion and inversion of linkages.

Reuleaux's list, furthermore, of the six classes of mechanical components from which all mechanisms are built, which appears near the end of the book (p. 480), has provided chapter or section headings for many if not most subsequent books on kinematics and machine design.

Finally, the possibilities of kinematic synthesis—that is, a systematic approach to the design of a mechanism to perform a given function—were first explored in this book. The author's chapter on synthesis (Chapter XIII), which draws upon all the rest of the book, stresses a point that modern students of kinematics will heed with profit : that kinematic synthesis will be successful in direct proportion to the designer's understanding and appreciation of analysis. Reuleaux's objective was to show the designer " the essential simplicity of the means with which we have to work " and to remind him of the analytical principles that must be observed in any successful new design.

Reuleaux contributed little to velocity or acceleration analysis. Kennedy, on the other hand, in his *Mechanics of Machinery*, published in 1886, made extensive use of instant centers in velocity analysis. Kennedy receives proper credit for his law of three centers, which he discovered independently. It should be noted, however, that the law had been stated in 1872 by Siegfried Aronhold in an article that appeared alongside one written by Reuleaux in the journal that Reuleaux edited. The fact that Reuleaux did not include Aronhold's law in his book can be taken as evidence of his failure to appreciate its significance.

When we meet a subject at the undergraduate level we seldom have the background or the maturity to learn (or even to be concerned about) the underlying concepts upon which the subject is based. It is true that the technical axioms of the present book can be, and have been, boiled down to a relatively few pages in current textbooks. It is also true that subsequent additions by Reuleaux's successors have provided materials for extensive manipulation and problem solving. We can successfully use the techniques of kinematics without troubling ourselves to inquire into their origins. Nevertheless, unless we know about such things and can appreciate the organic nature of the subject's growth and development, it is unlikely that we shall have any clear or useful vision as to the directions in which we may guide the future development of kinematics.

The Author

Franz Reuleaux was recognized within his own lifetime as a chief authority in mechanical engineering subjects associated with machine design. He wrote in this field many dozens of articles and three important pioneering books, of which the present book was second to appear. He was also for forty years a gifted teacher, influencing a long generation of engineers directly as well as through his writings.

Reuleaux was born in Eschweiler, near Aachen (Aix-la-Chapelle), on September 30, 1829, the fourth son of Johann Josef Reuleaux, who had established one of the early machine shops in the Rhineland. At the age of twenty-one, having acquired in addition to traditional schooling a background of practical experience in machine building, Franz entered the polytechnic school in Karlsruhe, drawn thither by the reputation of Ferdinand Redtenbacher (1809–1863), professor of mechanical engineering. Redtenbacher was an able and perceptive teacher who conveyed to others his deep impatience with the existing traditional order of machine design and construction, and who was in his writings groping toward true principles of machine design. If the young Reuleaux had not yet settled upon a career when he arrived in Karlsruhe, it is clear that he had, after his two years with Redtenbacher, a sense of direction and dedication that stayed with him throughout his life. He went on to universities in Berlin and Bonn, where he pursued studies covering a wide variety of interests in the physical sciences and in philosophy. In 1856, at the age of twenty-seven, he was appointed professor of machine design in the polytechnic school in Zurich, where he remained for ten years and where his revolutionary ideas on the kinematics of mechanisms took shape.

Reuleaux, in collaboration with a classmate named Moll, had published two books before he was twenty-five years old. The first was a brief work on the strength of materials ; the second was a long (983 pp.) practical treatise called *Principles of Machine Design.* Three years later he published a short treatise on springs, which had not before been treated analytically. The first of his three major books was on the theory of machine design. Gaining wide acceptance after its publication in 1861, *Der Konstrukteur* reached its fourth and final edition in 1889. This book was translated into English by the American engineer Henry H. Suplee. It was also translated into French and Russian.

In 1864, while he was yet in Zurich, Reuleaux had begun to lecture on the new kinematics. In 1865, he accepted a call to the Royal Industrial Academy in Berlin, of which he was made director in 1868. In 1879 he was a moving spirit in the founding of the Royal

Institute of Technology in Berlin, where he served as professor until he retired in 1896.

Theoretische Kinematik, Reuleaux's second major work, was, as noted above, published in 1875. His third major work, an extension of and enlargement of the author's ideas of kinematics, appeared in 1900 as *Die Praktischen Beziehungen der Kinematik zur Geometrie und Mechanik.* This final book, although not translated into English, has also made a lasting impression upon scholars of kinematics.

In addition to the books that he wrote from time to time, Reuleaux produced a constant stream of technical papers. His writing was loose and often prolix, showing little evidence of rewriting or revision, but it had a practical flavor and clarity of explanation that moved the editor of *Engineering* (London) to write in his review of the present book : " The subject is treated theoretically, but with a recognition of the claims of practice such as Englishmen do not generally associate with the writings of a German scientific professor." The reviewer found in Reuleaux's introductory comments on theory and practice (p. 54) the " promise that the theoretical treatment will not be too abstract to be useful " ; he went on to conclude that " this promise is amply fulfilled."

Another side of Reuleaux's career was concerned with the promotion and advancement of German industry. He made significant contributions through his work in connection with international exhibitions. He was a judge of awards for the London Exhibition of 1862 and the Paris Exhibition of 1867 ; he was a commissioner to the Vienna Exhibition in 1873 and the Centennial Exhibition of 1876 in Philadelphia. He managed the German departments of the Sydney and Melbourne Exhibitions of 1879 and 1881, respectively, and he visited and wrote about the Columbian Exhibition of 1893 in Chicago. He wrote also a short history of the development of exhibitions.

It was after the summer of 1876, when he had served as a German commissioner to the Centennial Exhibition in Philadelphia, that Reuleaux became known throughout Germany as the author of the severely critical " Letters from Philadelphia," in which he characterized the industrial products of Germany as *billig und schlecht* (cheap and ugly). Publication of these letters helped steer technical effort toward the manufacturing excellence that has long been associated with German industry.

Reuleaux found time also to concern himself with the relationships between technology and the culture in which it exists. He wrote a number of papers on this subject, one of which, " Kultur und Technik," was translated into English and published in the Smithsonian *Annual Report* for 1890. While he had little that was original to say about culture, he undoubtedly made a positive contribution by showing other engineers that he considered such questions worth investigating.

In addition to his other projects, Reuleaux found diversion in such activities as translating into German Longfellow's *Hiawatha,* buying paintings and sculpture for the Royal Museum of Art, and encouraging the German Language Association to arrest the deterioration of the German language. According to one of his translators, who had known him personally, he was an accomplished linguist, speaking and writing freely in French, English, and Italian. However, he wrote for publication only in German.

A man of commanding personality, Reuleaux was one of those rare individuals who combine a brilliance of penetration and a clarity of original thought with methodical and untiring industry to produce a significant effect upon the direction in which his contemporaries and successors will go.

He died at his home in Berlin on August 20, 1905, at the age of 75.

EUGENE S. FERGUSON

Ames, Iowa
April, 1963

SELECTED BIBLIOGRAPHY

HISTORY OF KINEMATICS

De Jonge, A. E. Richard, "A Brief Account of Modern Kinematics,"
A.S.M.E. *Transactions*, vol. 65 (1943), pp. 663–683.
Ferguson, Eugene S., "Kinematics of Mechanisms from the Time of
Watt," U.S. National Museum (Smithsonian Institution) Bulletin
228, paper 27 (1962), pp. 185–230.
Freudenstein, Ferdinand, "Trends in the Kinematics of Mechanisms,"
Applied Mechanics Reviews, vol. 12 (1959), pp. 587–590.
Hartenberg, Richard S., and Denavit, Jacques, *Kinematic Synthesis
of Linkages* (New York: McGraw-Hill Book Company, Inc., in prep.),
chapter I: "An Outline of Kinematics to 1900."

FRANZ REULEAUX—BIOGRAPHY

Suplee, Henry H., Obituary notice in A.S.M.E. *Transactions*, vol. 26
(1905), pp. 813–817.
Weihe, Carl, "Franz Reuleaux und die Grundlagen seiner Kinematik,"
Deutsches Museum (Munich) *Abhandlung und Berichte*, 1942.
Zopke, Hans, "Professor Franz Reuleaux," *Cassier's Magazine*, vol. 11
(December, 1896), pp. 133–139.

FRANZ REULEAUX—BIBLIOGRAPHY

Poggendorff, J. C., *Biographisch-literarisches Handwörterbuch*, vol. 3
(1858–1883), vol. 4 (1883–1904).

PREFACE TO FIRST ENGLISH EDITION

THE greater part of the *Theoretische Kinematik* of Prof. Reuleaux, which I have now the pleasure of presenting to English and American readers, was originally published in chapters in the *Berliner Verhandlungen,* under the title of *Kinematische Mittheilungen.* These papers, revised and enlarged, and with the addition of a chapter on Kinematic Synthesis, were published collectively in 1874–5 in the work of which the present is a translation. They have attracted considerable attention in Germany, and the principles laid down in them have already made their way into Polytechnic School instruction, not only in that country but also in Russia and Italy.

The book addresses itself to somewhat different classes of readers, or rather to readers who have had very different training, on the Continent and here. Its readers there are to a great extent the past or present students of the Polytechnic Schools, or at least those who are acquainted with polytechnic teaching. They are familiar with a regularly systematised system of machine instruction and its somewhat extended literature. Here, on the other hand, neither systematised instruction nor extended literature exists. The book addresses itself greatly to practical engineers and mechanicians, men who have often enough worked out their knowledge of the subject for themselves to a far greater extent than they have acquired it from books or lectures. To these readers some sections of the book may appear unnecessary, as referring to opinions or combating conclusions of which they have scarcely heard, and the erroneousness of which they are perfectly ready to admit. No doubt had the work originally been written for its English readers these passages might have been omitted or changed; as it is, I must merely remind those readers of the fact I have just mentioned. Here and there I have made small alterations in the text on this account, otherwise the sections referred to remain as in the original. The conclusions arrived at in them are not the less interesting that they might have been reached here, sometimes, in a more direct manner.

It may be well for me to mention here some of the leading characteristics of Prof. Reuleaux's treatment of his subject, and to point out in what respects it differs from that of his predecessors. In the oldest books upon machinery each machine was taken up as a whole, to be described and treated by itself from beginning to end. Gradually it became recognised that similar parts occurred again and again in different machines, and these parts received the name of mechanisms. They sometimes appear in a more or less abstract form in text-books of Elementary Mechanics, and have received more complete treatment in separate works. With the growth of clear ideas in physical science it became possible to separate the ideas of force, time and motion, and to consider the latter merely for its own sake without reference to the other two. Prof. Willis adopted this treatment unreservedly in his *Principles of Mechanism*—a work too well known to need any characterisation here—calling the study thus marked out the "Science of Pure Mechanism." Here, however, the matter stopped, later writers have been content to follow upon Willis's lines, not carrying the analytic process further, and contenting themselves with the examination of mechanisms as a whole in the forms in which they are presented to us by tradition or invention, without attempting to analyse them, or to investigate their mode of formation.

It is unquestionably true that by the aid of mathematics this treatment of mechanisms has given us many most valuable results, but it is equally true that the method itself is defective, and was only used for want of a better. This better method Prof. Reuleaux has attempted, and I think with great success, to indicate. Starting with the idea of motion as change of position only—and limiting himself to cases where such changes are absolutely determinate at every instant,—as always in the machine—he points out that they are conditioned simply by the geometric form of the moving bodies. Two bodies, such for instance as a screw and nut, having such forms that at any instant there is only one possible motion for each relatively to the other, form the simplest combination available for machinal purposes—such bodies he calls a pair of elements. Two or more elements from as many different pairs can be combined into a link, and such links united into kinematic chains, and it is by fixing, that is, preventing the motion of, some one link of such a chain that a mechanism is obtained. Stated thus in a few words the analysis is simple and obvious enough; like many other simple things, however, it leads to most important consequences. As one illustration merely of this, I may point to the collection of "rotary" engines and pumps examined in Chap. IX. Here will be found, among others, over thirty forms of "rotary" engines of which the kinematic chain used in the driving mechanism is absolutely identical with that of the common direct-acting engine! Their constructive forms differ most widely, and have of course too often misled their inventors, but the application of what Prof. Reuleaux calls "kinematic analysis" shows at once both their identity as kinematic chains and their relation as mechanisms. In Fig. 3, Pl. XX., for instance, is shown a rotary engine which has been patented every few years since 1805 in one or another form, and in which no doubt some of my readers will recognise an old friend "schemed" in the days of their apprenticeship. Its driving mechanism is absolutely the same as that of the direct-acting engine, but with the crank fixed and the frame allowed to move round it.

In order to utilise the kinematic analysis Prof. Reuleaux has devised and elaborated the notation which is explained in Chap. VII. and used in the later part of the book. That this notation is both exceedingly simple and of practical use will be admitted by all readers of Chap. IX., but its full advantages will only be realised by those who use it for themselves. The way in which it aids the resolution of apparently complex mechanisms into quite familiar forms is often most remarkable. Use will no doubt suggest modifications and improvements in its details, but Prof. Reuleaux is very anxious that its essential features, and especially the symbols for the elements (which have been so chosen as to be as suitable as possible for the principal European languages) should not be altered.

I may mention here only one other feature in Reuleaux's work, namely, his treatment of fluids when they occur in mechanisms or machines (Chap. IV. &c.). It has long been customary, of course, to treat cords, chains, belts &c., as organs which could legitimately form part of machines, but fluids have been universally (so far as I know) excluded from consideration in this way. Reuleaux points out that fluids—"pressure-organs" —are simply contrapositives of the "tension-organs" just mentioned, and that if one be included in the study of "pure mechanism" there can be no reason for excluding the other. He gives also many instances of the way in which engineers use the one or the other as the column of fluid or the cord best suits their purpose. In examining mechanisms we consider the motions of each body as a whole, ignoring altogether its molecular condition, or more strictly assuming that it is so arranged that its molecular stability is not disturbed during the motion. This pre-supposition is made tacitly in the case of "rigid" bodies, where molecular stability is independent of the application of external force. It is made also in the case of ropes, belts, &c., for when these occur in machines it is always assumed that they are kept in tension by some force external to themselves, in any other case their motions would be quite indeterminate. With fluids it is not necessary to make any other assumption than this, but the external force must be a pressure instead of a pull, and must be supplied in directions other than that in which motion takes place. § 126 shows some of the interesting results to which this treatment of fluid organs may lead.

My own work in connection with Prof. Reuleaux's book has been chiefly, of course, that of translation; but a comparison of this edition with the German one will show several not unimportant improvements. Some of these have been suggested by the author; in all cases where they involved more than the changing of a few words they have been submitted to him. I may take this opportunity of acknowledging the assistance I have received from him and the interest he has taken in the progress of this English edition of his work, (which has been already published in Italian, and is now being translated into

French). I have also great pleasure in acknowledging the help I have received on many occasions from my friend and colleague Prof. Henrici, F.R.S.

The references given in footnotes are mostly those of the original edition; in Chapters IX and X I have added English references where I was able to do so. The longer footnotes I am responsible for, except in cases where I have placed "*R*" after them. Of the notes at the end of the book I have added those which are placed in square brackets. A few of the notes in the original, which referred to matters with which English readers would probably be unacquainted, or to passages which have been altered in the text, have been shortened or omitted.

The names which Prof. Reuleaux gives to the various mechanisms have in most cases been invented by himself, and in several other instances he has had to coin words to express ideas to which individual distinctness has now first been given. Such names and words I have not tried to translate, but only to replace by equally good English ones, with what success I must leave my readers to judge. I shall be happy to receive suggestions for improvements in this matter. The names have, however, been very carefully considered, and so arranged as to fit in with each other—I venture to hope, therefore, that those who use them for instruction will not alter them without good reason. For the word "centroid," for which I anticipate great usefulness, I have to thank Prof. Clifford.

Prof. Reuleaux uses the word Kinematics in a limited sense, for the Science of constrained motion—that is, motion as it occurs in machines—without reference to the ideas of either time or force (p. 40, &c.), and has therefore called his book *Theoretische Kinematik. Grundzüge einer Theorie des Maschinenwesens.* Whether this use of the word be advisable or not, it was obviously impossible to adopt it in this country, where it has obtained firm hold in a much more extended, but quite legitimate, sense. While retaining the second part of Prof. Reuleaux's title I have therefore been compelled to change the first part to *Kinematics of Machinery.* It is very unfortunate that we have as yet no word for the study of motion as change of position merely. Phoronomy, which is used in Germany very nearly in this sense, is very unprepossessing;—I would suggest Metastatics for the purpose, unless a better word can be found. It has at least the merit of expressing the idea clearly, and with both philologic and scientific accuracy.

ALEX. B. W. KENNEDY.

I may add that Prof. Reuleaux has sent over some three hundred models, a portion of his Kinematic collection at the Berlin Gewerbe-Akademie, to the Exhibition of Scientific Apparatus now (May 1876) being held at South Kensington, where they will remain throughout the summer. Among these models are a great number of the mechanisms described in the following pages, along with many others. They will well repay a visit, or more than one, and as close examination as circumstances permit. Herr Kirchner, of the Berlin Akademie, has charge of the models, and will be as well pleased as he is well able to give explanations about them.

CONTENTS.

INTRODUCTION.

THE aim of the following chapters is to determine the conditions which are common to all machines, in order to decide what it is, among its great variety of forms, that essentially constitutes a machine;—they are therefore called Outlines of a Theory of Machines. The whole study of the constitution of machines—the Kinematics of Machinery—naturally divides itself into two parts, the one comprehending the theoretical, and the other the applied or practical side of the subject; of these the former alone forms the subject of this work. It deals chiefly with the establishment of those ideas which form the foundation of the applied part of the science, and in its treatment of these its method differs in great part essentially from those hitherto employed.

As I have here to do chiefly with theoretical questions, it might seem that I could hardly expect to interest others than those concerned only with the theoretical side of this special study. But Theory and Practice are not antagonistic, as is so often tacitly assumed. Theory is not necessarily unpractical, nor Practice unscientific, although both of these things may occur. Indeed in any department thoroughly elucidated by Science the truly practical coincides with the theoretical, if the theory be right. The popular antithesis should rather be between Theory and Empiricism. This will always remain, and the more Theory is extended the greater will be the drawback of the empirical, as compared with the theoretical methods. The latter can never be indifferent, therefore, to any who are able to use them, even if their work be entirely "practical," and although they may be able for a while longer to get on without them. The theoretical questions, however, which are here to be treated, are of so deep-reaching a nature that I entertain the hope that those who are practically, as well as those who are theoretically concerned with the subject, may obtain help from the new method of treating them. I am thus obliged to lay before both equally the grounds upon which I have given up the customary ideas upon the subject and put others in their place.

In attempting to place the theory of the constitution of the machine upon a new basis I do so with the conviction that my trouble will be repaid only if it prove of some actual advantage in the right understanding of the machine. I venture to promise such a result with confidence. He who best understands the machine, who is best acquainted with its essential nature, will be able to accomplish the most by its means. It is not a matter of merely setting forth in a new form and order what is already well known, or of substituting a new classification and nomenclature for the old. Possibly with such improvements the subject might be more conveniently and elegantly taught, but for practical purposes the old forms might be used for a long time to come. On the contrary, if the new theory is to lay claim to general interest, it must be capable of producing something new; it must make problems solvable which before could not be solved in any systematic way. This may certainly be said to be the case if it succeeds in making Machine-Kinematics, down to its simplest problems, truly scientific.

This subject has indeed, in a certain sense, been scientifically treated hitherto, in so far, namely, as particular portions of it admit of mathematical treatment. But this concerns a part only of the subject, and not that part which is peculiarly its own; so far as the treatment has been scientific, too, it has been mathematical or mechanical, and not kinematic. This last in its essence, in the ideas belonging specially to it, has been left indistinct, or made clear accidentally at a few single points only. It is like a tree which has grown up in a dark tower, and thrown out its branches wherever it could find an outlet; these, being able to enjoy the air and light, are green and blooming, but the parent stem can only show a few stunted twigs and isolated leaf-buds.

The mathematical investigations referred to bring the whole apparatus of a great science to the examination of the properties of a given mechanism, and have accumulated in this direction rich material, of enduring and increasing value. What is left unexamined is however the other, immensely deeper part of the problem, the question : How did the mechanism, or the elements of which it is composed, originate ? What laws govern its building up ? Is it indeed formed according to any laws whatever ? Or have we simply to accept as data what invention gives us, the analysis of what is thus obtained being the only scientific problem left—as in the case of natural history ?

It may be said that the last method has been hitherto followed exclusively, for only traces appear now and then of penetration behind these data. The peculiar condition consequently presents itself throughout the whole region of investigation into the nature of the machine that the most perfect means have been employed to work upon the results of human invention—that is of human thought—without anything being known of the processes of thought which have furnished these results. Terms have been made with this inconsistency, which would not readily be submitted to in any other of the exact sciences, by considering Invention either avowedly or tacitly as a kind of revelation, as the consequence of some higher inspiration. It forms the foundation of the kind of special respect with which any man has been regarded of whom it could be said that he had invented this or that machine. To become acquainted with the thing invented we leap over the train of thought in which it originated, and plunge at once, designedly, *in medias res.*

If, for instance, we consider the well-known parallel motion which Watt invented for his steam-engine, or that of Evans, or that of Reichenbach, according to the method hitherto adopted, we find that after we have classified them we have nothing further to do than to ascertain the laws of motion by which they are governed as mechanisms, to fix upon the constructive forms most suitable for them—and, if it be required to go further, to elucidate their more intimate mutual relationships. *How*, however, their inventors arrived at them we leave unentered upon, except in so far as our personal feeling of interest in this point is concerned Now and then we are glad to overhear the Genius in his thought-

workshop, but more from curiosity than in any spirit of investigation. And yet it would appear from what we have said that here
there is a great step further to be made. Let us try.

Watt has left behind for us in a letter some indications of the
line of thought which led him directly to the mechanism just
alluded to. "The idea," he writes to his son in November 1808,
"originated in this manner. On finding double chains, or racks
and sectors, very inconvenient for communicating the motion of
the piston-rod to the angular motion of the working beam, I set to
work to try if I could not contrive some means of performing the
same from motions turning upon centres, and after some time it
occurred to me that $A B$ and $C D$ being two equal radii revolving

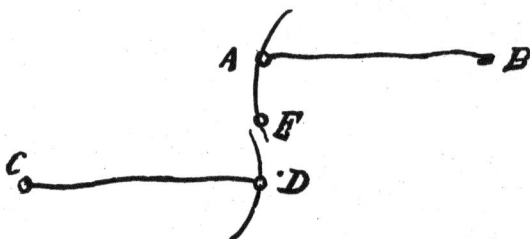

Fig. 1.*

on the centres B and C, and connected together by a rod $A D$, in
moving through arches of certain lengths, the variation from the
straight line would be nearly equal and opposite, and that the
point E would describe a line nearly straight, and that if for
convenience the radius $C D$ was only half of $A B$, by moving the
point E nearer to D the same would take place, and from this the
construction, afterwards called the parallel motion, was derived.
. . . . Though I am not over anxious after fame, yet I am more
proud of the parallel motion than of any other invention I have
ever made."

Interesting as this letter is, a closer examination of it reveals a
deficiency which perhaps the questioner also may have discovered.
We quite appreciate the motives as well as some of the final
results of Watt's exertions, but we obtain no indication of a
methodical train of ideas leading up to them. Moreover it must
be remembered that the description is written twenty-four years

---

* Facsimile from Watt's letter. See Muirhead's *Mechanical Inventions of James
Watt*, vol. ii. p. 88.

after the invention, that therefore reflection and recollection have had time to work upon each other. Watt expressed himself much more directly and distinctly in a letter written in 1784 to Boulton, giving him the first idea of the invention :—

" I have started a new hare," he writes ; " I have got a glimpse of a method of causing a piston-rod to move up and down perpendicularly by only fixing it to a piece of iron upon the beam, without chains, or perpendicular guides, or untowardly frictions, arch-heads, or other pieces of clumsiness, by which contrivance, if it answers fully to expectation, about 5 feet in the height of the [engine] house may be saved in 8 feet strokes, which I look upon as a capital saving ; and it will answer for double engines as well as for single ones. I have only tried it in a slight model yet, so cannot build upon it, though I think it a very probable thing to succeed, and one of the most ingenious simple pieces of mechanism I have contrived, but I beg nothing may be said on it till I specify."* [1]

If we examine the specification referred to we find no less than six methods of guiding described, and among them the very " perpendicular guides " and " arch-heads " found fault with above ; two of these methods take the form which our mechanism can assume. One of the six is specially notable,—it is exactly the parallel motion of Reichenbach, and seems to lead up to the motion more generally known by Watt's name. Watt has evidently not recognised this,—and later on it has completely escaped him, at which one cannot wonder, considering the uncouth garb of timber beams and hammered rods in which the elegant mechanism was at that time disguised.

We see that even a thinker like Watt was at fault in the essential elucidation of the matter. Yet we note at the same time that each thought in the inventor's stream of ideas is developed out of another,—that the ideas form a ladder up which he presses step by step,—through labour and exertion,—to his goal. His eventual success gains from us the more respect that he did not find the end of his exertions close at hand. But the inspiration,—the instantaneous illumination,—cannot be detected ; —he says above "and after some time it occurred to me," which

* See Muirhead as above, vol. ii. p. 93;—where also the Specification is given at full length with the necessary engravings.

only points to previous uninterrupted search, continuous following-out of thoughts. "By continuous thinking about it," answered Newton to the question how he had discovered the law of gravitation. Göthe also gives us the same idea in his sentence, "What is Invention? It is the end of seeking." The links which connect isolated thoughts seem indeed to be almost entirely destroyed,—we have to reconstruct them. We see the whole before us only like a faintly outlined or half-washed-out picture, and the painter himself can hardly furnish us with any better explanation of it than we can discover for ourselves. Indeed the comparison holds good in more than one point. In each new region of intellectual creation the inventor works as does the artist. His genius steps lightly over the airy masonry of reasoning which it has thrown across to the new standpoint. It is useless to demand from either artist or inventor an account of his steps.

Observations similar to those made in this single case can be made throughout the whole history of invention, wherever the genius of past generations has busied itself in bringing forward new things. The invention of the steam-engine, for example, stretches back through a whole century,[*] without ever making a step in advance of the natural development going on in other departments of knowledge.

In the school of Galileo,—where his experiments on falling bodies first threw a ray of light through the scholastic cloud which had veiled all knowledge,—there began early in the seventeenth century those experiments in physical science with the growth of which the invention of the steam-engine is inseparably bound up. It is no mere chance that the place is distinguished also as the centre of great artistic development;—art and science flourish together on a rich soil. It is as if the proud citizens of Pisa had made their marble tower to lean expressly for Galileo's experiments. In Florence (1643) Galileo's disciple Torricelli, still in the freshness of youth, made his discovery of the heaviness of the air,—on which there followed directly alarming outcries for the preservation of the "horror vacui" and the whole threatened belongings of the wiseacres of the time. The centre of disputation and investigation passed from Tuscany to

---

[*] See Prof. Reuleaux's *Geschichte der Dampfmaschine*, Brunswick, 1864.

France when Pascal took up the question in 1646,—he, after a while, came quite over to the new ideas. He caused the memorable first barometric measurement to be made on the Puy-de-Dôme in 1648. It was conclusive, and the joyful bells of Münster and Osnabruck rang in the triumph of the young science.

The line in which the centre of this newly-awakened activity moved now turned to the north-east, towards Germany, into the country of the great Electors. Tilly had not been able to destroy the intellectual life of Magdeburg. There in 1650 Otto von Guerike brought a new idea into what was the question of the day in physical science,—that namely of the possibility of utilising the atmospheric pressure as a force. He showed this both popularly and scientifically with the air-pump and other experimental apparatus. The search for means of utilising the power stored in the atmosphere by some simple vacuum arrangement now began everywhere. For a long time no satisfactory results were obtained, but at last, in 1696, Papin at Marburg discovered the true solution—the condensation of steam in a cylinder fitted with a piston. The steam-engine was invented. Papin, an experimenter really worthy of respect, who attempted to solve the problem in the most various ways, and who has made a whole series of other remarkable inventions, is its true inventor. His arrangement, however, was as yet very incomplete and unpractical; the great ideas required still to be carried beyond learned circles and Latin treatises into practical life. In this Papin did not succeed; he never got beyond the beginning. His first large steam cylinder still stands incomplete as a monument in the court of the Museum at Cassel, but his ideas crossed the English Channel and found their way directly to practical men. The mechanics Newcomen and Cawley produced in 1705 a really useful pumping-engine, which soon found actual employment in mining operations.

The spirit of invention now rested for a while, as if exhausted with the exertions of the last few years. The pause was unavoidable, for the necessary means for making progress did not exist. It was necessary that more should first be known of heat, which could not yet even be measured. The thermometer had first to be perfected, the whole theory of heat had to be greatly advanced. Then came Watt (circa 1763), whose far-reaching genius furnished.

a whole series of mechanical and kinematic inventions, and brought the machine in a short time to a high degree of completeness, enriching at the same time all the kindred realms of knowledge. From that time onward the rapidly extending employment of the machine and its continual development, improvements from a thousand heads and a thousand hands, have brought it to its present perfection, and made it absolutely the common property of all.

I have entirely omitted in this hurried sketch what Humboldt called the "horrible wrangling about priority." From such a summary review we might almost be led to believe in the entirely spontaneous unfolding of ideas, were it not that each separate energetic step forward shows us the grasp of more distinguished endowments, and convinces us afresh of the importance of genius in the further growth of the race. Throughout the whole, however, we can discern the one idea developing itself out of the other, like the leaf from the bud or the fruit from the blossom; just as throughout nature everywhere each new creation is developed from those which have preceded it.

I believe I have shown in the preceding paragraphs that a more or less logical process of thought is included in every invention. The less visible this is from outside, the higher stands our admiration of the inventor,—who earns also the more recognition the less the aiding and connecting links of thought have been worked out ready to his hand. To-day, when scientific aids are so abundant in every branch of technical work, progressions of the greatest importance are frequently made without receiving any such high recognition as in former times. Everything lies so clearly and simply before us, that it can be reached and comprehended by commonplace intellects. At the same time the relative difference between the work of the commonplace and that of the highly cultivated intellect is even greater than before, and by this may be explained the apparently almost feverish progress made in the regions of technical work. It is not a consequence of any increased capacity for intellectual action in the race, but only of the perfecting and extending of the tools with which the intellect works. These have increased in number just like those in the modern mechanical workshop;—the men who work them remain the same.

Let us return now to our special subject, and examine in a more strictly historical way what has been done hitherto for theoretical Kinematics. The reader need not fear that I shall disturb thr dust on old parchments in order to build up from dry dates the foundations of a science. We look now for the beginnings of the idea of our subject, and this delicate material can be taken from the ancient volumes without disturbing the moths.

In earlier times men considered every machine as a separate whole, consisting of parts peculiar to it; they missed entirely or saw but seldom the separate groups of parts which we call mechanisms. A mill was a mill, a stamp a stamp and nothing else, and thus we find the older books describing each machine separately from beginning to end. So for example Ramelli (1588), in speaking of various pumps driven by water-wheels, describes each afresh from the wheel, or even from the water driving it, to the delivery pipe of the pump. The concept "waterwheel" certainly seems tolerably familiar to him, such wheels were continually to be met with, only the idea "pump"—and therefore also the word for it — seems to be absolutely wanting.[2] Thought upon any subject has made considerable progress when general identity is seen through the special variety;—this is the first point of divergence between popular and scientific modes of thinking. Leupold (1724) seems to be the first writer who separates single mechanisms from machines, but he examines these for their own sakes, and only accidentally in reference to their manifold applications. The idea was certainly not yet very much developed. This is explained by the fact that so far machinery had not been formed into a separate subject of study, but was included, generally, under physics in its wider sense. So soon, however, as the first Polytechnic School was founded, in Paris in 1794, we see the separation between the study of mechanisms and the general study of machinery, for which the way had thus been prepared, systematically carried out.

The completion of this separation connects itself with the honoured names of Monge and Carnot. The new line of study appeared first as a subdivision of descriptive geometry, from which it has only by degrees released itself. It fell to Hâchette to give instruction in it, and he, working upon outlines given by Monge, constructed (1806) a programme of which Lanz and

Bétancourt afterwards filled up the details in their "Essai sur la composition des machines" (1808). Monge had entitled the subject "Elements of Machines," which he intended to be equivalent to means for altering the direction of motion. He understood by these "means" mechanisms, and based on this idea the arrangement of mechanisms according to the possible combinations of four principal kinds of motion, viz.: continuous and reciprocating rectilinear, and continuous and reciprocating circular. Omitting repetitions, these give ten classes of mechanisms, corresponding to the changes of motion between

Continuous rectilinear and
$\begin{cases} \text{continuous rectilinear} \\ \text{reciprocating } \text{ ,,} \\ \text{continuous circular} \\ \text{reciprocating } \text{ ,,} \end{cases}$

Continuous circular and
$\begin{cases} \text{reciprocating rectilinear} \\ \text{continuous circular} \\ \text{reciprocating } \text{ ,,} \end{cases}$

Reciprocating rectilinear and
$\begin{cases} \text{reciprocating rectilinear} \\ \text{,,} \quad \text{circular} \end{cases}$

Reciprocating circular and
,,   ,,

This scheme,—or "system,"—was capable of extension,—and in the second edition (1819) it was extended by the addition of other fundamental motions,—continuous and reciprocating motions in curved lines,—increasing the number of classes from 10 to 21, but not altering the principle of classification. This indeed has remained with unimportant alterations in tolerably general use until the present time, and has thus acquired the sanction of very general recognition. Hâchette himself, who assisted in the production of Lanz' work,* adopted it unconditionally in his Traité élémentaire des machines, published in 1811. Borgnis, in his Traité complet de mécanique (1818) departed from it to a certain extent; he considered the problem more generally than his predecessors, and divided the parts of machines into six classes, which he called récepteurs, communicateurs, modificateurs, supports, régulateurs, and opérateurs respectively.

---

* A third edition of this appeared in 1840 : it is a repetition of the second, slightly enlarged and got up in a better form.

He did not take the change of motion as a leading principle, but used it only in determining the subdivisions. His system has not been accepted as necessarily antagonistic to Monge's, his method of division being taken as more or less suitable rather for the general study of machinery than for Machine-Kinematics. One leading idea at least of Borgnis' scheme has since become universally familiar;—his division of machinery into the parts receiving effort, the parts transmitting it, and the working parts. Through the brilliant works of Coriolis [3] and Poncelet [4] these have become supporting pillars, one might almost say articles of belief, in the modern study of machines. At the risk of being considered a heretic, I must say here that these fundamental notions require essential modification. The honoured Nestor of applied mechanics must pardon me for the sceptical saying : *Amicus Plato, sed magis amica veritas.** We shall further on obtain the means of putting Borgnis' ideas to the proof,—but it is clear that his principles play too important a part in reference to the motions of the different parts of machinery to be altogether foreign to the study of Mechanisms. Borgnis' work itself is to-day quite out of reckoning,—his classification of machines and their parts has borne little fruit;—it serves for the most part as little more than a somewhat systematic exercise of the reader's memory. Nevertheless we shall find later on that more lies behind some of his thoughts than has been commonly supposed.

The year 1830 saw a notable change in the position of the study of Mechanisms, through the critical examination of its principles by the great physicist Ampère in his Essai sur la Philosophie des Sciences. In his system of sciences Ampère ranked the study created by Monge and Carnot as one of the third order, and attempted to lay down its exact limits. He considered in connection with it Lanz's treatise, and said, among other things—" It (this science) must therefore not define a machine, as has usually been done, as an instrument by the help of which the direction and intensity of a given force can be altered, but as an instrument by the help of which the direction and velocity of a given motion can be altered." He completely excludes forces from the investigations proper to the science, and says further, " To this science, in which motions

---

* This was written before the death of Poncelet.

are considered by themselves as observed in the bodies surrounding us, and specially in those systems of apparatus which we call machines, I have given the name Kinematics (Cinématique), from κινημα, motion." He further on encouraged the treatment of the science in text-books, for which he foresaw great use,— but he did not enter into any further details regarding it.

The seed thus sown by Ampère has borne rich fruit,—the Science of Kinematics was soon taken up (in France first of all) as a separate study, and a literature for it came rapidly into existence. The proposed name met with the most ready acceptance in France, and has since become more or less familiar in many other places.[5] In the scientific limitation of the nature and aims of the study, however, the clearness so much to be desired has by no means been attained.

The next important original work is the "Principles of Mechanism" of the late Professor Willis (1841), a remarkable book, full of valuable illustrations from applied Kinematics, and of thoughts in relation to their real connection with each other. In system Willis differs from Monge. He considers that the scheme of Lanz, "notwithstanding its apparent simplicity," must be considered "a merely popular arrangement." He finds further in Lanz and Bétancourt a contradiction of Ampère's definition, in their inclusion of waterwheels, windmills and so on, among mechanisms, and will only allow that those mechanisms are pure which consist entirely of rigid bodies. With these mechanisms he lays special stress upon the important characteristic that they do not determine the actual motions in direction and velocity, as Monge says, but only the relations in direction and velocity between the motions occurring in the machine. According to whether these relations in any mechanism are constant or varying, he placed it in one or other of three classes, each of which has subdivisions corresponding to the means used for transmitting the motion (rolling-contact, —sliding contact, &c.)

Willis' observations bear throughout the mark of careful and earnest investigation, but while there is much in them that is true, there are also some things that are incorrect, as especially the exclusion of hydraulic and other such machines. To this I will return further on. It is worthy of note, however, that Willis' classification has never taken root in his own country, but that much more commonly people have been content to follow the well-trodden path marked out by Lanz.*

In Italy, Ampère's seed has also taken root. In the Cinematica applicata alle arti, a text-book for technical schools, which first appeared in 1847 under a somewhat different title, Giulio has left to his country a valuable gift. This book unites admirably Kinematics and Mechanics; it follows Willis pretty faithfully in essentials, but not without an attempt,—incompletely successful,— to replace the hydraulic machines which Willis had struck out. A delicate intellectual inspiration breathes through the whole work, and is the more notable that it was written for pupils not having more than an elementary acquaintance with mathematics. The concise mathematical expressions, which speak for themselves, have thus had to be replaced by explanations in words,— a method which presupposes a deeper understanding on the part of the author than is shown in many books bristling with formulæ.

In 1849, Laboulaye, again in accordance with Ampère's suggestion, attempted in his Cinématique to set forth the science of Mechanisms in a complete form. He also discards Willis' limitation of mechanisms to those constructed of rigid bodies, and points out also that Ampère required something impossible in wishing absolutely to exclude the consideration of force from Kinematics. Besides this he attempts to determine a new theoretical method of a general character. This consists in dividing the whole "machine-elements" into three classes, which he calls système levier, système tour, and système plan, corresponding respectively to the making fixed (inébranlables) for the time, one, two, or three and more points of the moving body. These "systems," however, do not really cover the problem, of which we shall find the proof in its proper place. Even their originator has not made any real use of them, feeling, doubtless, that there was not much to be gained by doing so. So far as applied Kinematics is concerned, he seems rather to return to the system of Lanz, with suitable subdivisions. Indeed, he goes so far in this direction as to construct Monge's system a priori, thus showing that it forms, in

* Since the first publication of these remarks the second edition of Prof. Willis' work has appeared. (London 1870.) It is considerably in advance of the first, but in all essential points the principles originally laid down are unchanged. It so far confirms my view that the author, while retaining his original divisions and subdivisions, has inverted their order. R.

reality, the groundwork of his theory. Laboulaye has done no service to the science of Kinematics by this philosophical experiment, for by the apparently convincing form of his proof he has prevented those familiar with the subject from making further investigations. His *a priori* construction is based entirely upon the motion of a point, and is for that very reason inapplicable to the motions of a body, or system of points. In other matters, Laboulaye's book is valuable, and it has without question been the means of widely spreading much useful information ; it relies in the practical part confessedly very much upon the infinitely industrious Willis, with whom it sometimes even shares errors.

Morin also, in a little book (1851) intended for elementary instruction, has made a collection of the principles of Kinematics, called, in the later editions, Notions géométriques sur les mouvements. It is unpretending, and written in a very intelligent manner, and contains some capital leading thoughts, but in essentials it follows Monge's scheme.

In Germany it may almost be said that nothing was done during the period under consideration for the development of theoretical Kinematics. Weisbach, in his article " Abänderung der Bewegung," (Alteration of Motion), in Hülsse's Encyclopædia (1841), adhered altogether to the system of Lanz,—his own scientific work however had admittedly quite a different direction. Something new might have been expected from Redtenbacher, whose work was so continually connected with mechanisms. His highly philosophic brain perceived strongly the insufficiency of Monge's system ; but drawn away from the subject, first in the development of scientific machine construction, and afterwards by his work in mechanical physics, he abandoned it, but did not bring anything new into its place. This was probably his reason for holding that no true system of the study of mechanisms was possible, that they could be arranged only according to their practical usefulness, and for the rest must be treated mathematically. This nihilism may be read between the lines of his valuable work, Die Bewegungsmechanismen (1857), in which he describes and treats theoretically the mechanisms of the collection of models at Karlsruhe. That this book, systemless as it is, has had no inconsiderable circulation, shows that our technical public feel the lively necessity which exists for the theoretical exposition of the subject.

In France, meanwhile, progress which has been of great importance to Kinematics has taken place in the region of Geometry. The geometrical method of treating the motion of rigid bodies developed by Euler in the last century was again taken up, and soon further extended, by Chasles and (especially) by Poinsot. The works of the latter,—Théorie de la rotation des corps, and Théorie des cônes circulaires roulants, gave a great impulse also to the employment of geometrical methods of representation in the study of mechanisms. The propositions of Euler, which had hitherto possessed no more than a purely theoretical and abstract interest, were now formed by the French kinematists into fundamental doctrines. They breathed fresh life into what had become a somewhat dull study. Under their influence appeared Girault's Elémens de géom. appl. à la transformation du mouvement 1858,—Belanger's Cinématique 1864,—and Haton's Traité des mécanismes 1864; the two first specially rich in geometrical, *i.e.* in theoretical parts, the third bearing rather on the application of theory to the mechanisms themselves. All three books, however, valuable and important though they be, fall into the old difficulties as to classification as soon as they enter upon the applications of the science. They all differ from Monge, for the inadequacy of the "*ancien système*" became too evident to escape detection in the light of the new ideas,—nevertheless they all remained, for better or worse, partly involved in it. They differ too among themselves, and each seems to hesitate in his own particular way between Monge and Willis. Girault and Belanger take their principal divisions from the changes of motion,—but in entirely different ways,—using the various methods of transmitting the motion as subdivisions, Belanger with the addition of Willis' relative velocities. Haton recognises the want of the old system, and points out, for example, that to arrange toothed wheels according to it would require in all some 21 different classes;—he himself takes his principal divisions from the methods of transmission; these he divides into nine classes, of which Rollers, Guide-bars, Eccentrics, Toothed-wheels, Connecting Rods, and Cords form the first six, the three last bearing in common the fatal designation *appareils accessoires.* One whole third of the subject therefore has been cut off,—placed as it were like a note below the text.

Nor has this new geometrical development, into the further

growth of which I need not enter, been able to bring about any common method of treatment for applied Kinematics. Something still remains doubtful, and the results obtained have correspondingly been uncertain. Even those who wish to employ pure scientific methods only seem to be convinced that such methods cannot be used in the applied part of the science. They fall into Redtenbacher's nihilism, and cut "Cinématique pure" away from "Cinématique appliquée." Résals' Cinématique pure (1861) is an example of this,—it shows that the evaporation, as it were, of kinematic problems into those of pure mechanics can scarcely, with such a method of treatment, be avoided.

Moreover, as a further fruit of this uncertainty there has been an attempt to construct another special study which demands mention. This is the so-called Automatics, the study of the realization in mechanism of motions either supposed or given by mathematical expressions. For this further attempt at separation we have to thank the engineer E. Stamm, who wishes again to subdivide his subject into pure and applied parts, as fully described in his Essai sur l'automatique pure, 1863. Stamm has earned considerable distinction in connection with the self-acting spinning-machine, —that is to say in a special case of applied Kinematics,—which those technically concerned know how to appreciate. His separation of Automatics from the science to which it belongs must, however, be considered unpractical; it cannot exist by itself,—it is only a portion of a synthetic method based upon the fundamental laws of Kinematics, to which science it therefore belongs inseparably.

We have reached the end of our literary review.* We have found, on the one hand, a most unsatisfactory confusion of attempts to find a form for one and the same circle of ideas. As many systems as authors, no resting point reached, always new trying and seeking,—and as a conclusion we have Ampère's well-defined science split into two, indeed into four sciences, as if it were one of those Infusoria which are propagated by division. On the other hand, we may make the consolatory remark that the observations have grown more and more in exactness and delicacy, as also that both the methods of examination and the mechanisms examined

---

* The work of Prof. Rankine in connection with Machine-Kinematics, which was in some respects remarkable, I have mentioned in a note at the end of the Introduction.

have by degrees greatly increased in number. The two sides of the question, the theoretical and the applied, have thus carried on a separate existence side by side; but their union has not yet been accomplished. The cause of this must be sought in the systems alone, for the applications, the mechanisms themselves, have been quietly developed in practical machine-design, by invention and improvement, regardless of whether or not they were accorded any direct and proper theoretical recognition. Indeed the theories used hitherto have contributed to this development only in regard to the form of execution in certain cases, as *e.g.* in the methods of drawing wheel-teeth, &c.; they have however furnished no new mechanisms. This circumstance is very remarkable; it explains the conservative tenacity with which practical mechanists, where they have adopted any theory whatever, have always fallen back upon the old and apparently natural ideas of Monge in spite of the promises held out by the more recent theories; this the technical journals everywhere sufficiently show.

I believe I have proved the insufficiency of the theoretical Kinematics hitherto taught, and the necessity for some reform. The question now comes, wherein exactly does the error of the existing methods lie?

Monge's classification, however natural it appears, does not in the first place correspond to the real nature of the matter. Did it really do so—did it resemble, for instance, the classifications of Linnæus and Cuvier in organic nature—it would, like them, be able to make its footing firm; and the obstinate conservatism above referred to may perhaps be explained by the existence of a dim feeling that some analogous relation must also exist here. But this is not the case. Even if it were supposed that the function of the science did not extend beyond arranging the mechanisms in classes, still their division cannot be made according to the changes of motion, for this would necessitate endless repetitions. Almost all mechanisms might be looked for, and might also be placed, in at least two such classes, and most in from four to six or even ten to fifteen, for they can be used, and in practice are actually employed, for just so many different kinds of change. Willis, who strictly investigated this very point, and had a strong perception of the necessity for logical order, showed a distrust of the elasticity of the basis of his own system, so that his classification does not carry conviction with it. He wishes to adhere consistently to his fundamental principle of relative motions, and so finds it necessary to treat together very various kinds of mechanisms; and as a single mechanism frequently contains in itself several kinds of relative motions, he is compelled repeatedly to enter upon repetitions of an extended kind. Other objections might be raised against the modified classifications of Laboulaye, Girault, Belanger, Haton and the others, generally as well as individually, for no true science can be moulded at will in six or eight different ways.

The real cause of the insufficiency of the system is not, however, the classification itself; it must be looked for deeper. It lies, as I have already pointed out, in the circumstance that the investigations have never been carried back far enough,—back to the rise of the ideas; that classification has been attempted without any real comprehension being obtained of the objects to be classified. The formation of a science cannot be entered upon *in medias res;* it requires that we should start, as in mathematics, from the very simplest elements, the axiomatic beginnings. Without the determination of these the goal can never be reached. A single trial of the method commonly employed shows this very clearly.

In the old classification a commencement was made very commonly with the changing of one rectilinear motion into another; but no one asked whence the first rectilinear motion came, why it existed, how it had been created. To take a special case, Hâchette and Lauz choose for their first mechanism the so-called "fixed pulley." In this case it is the rectilinear motion of the cord as it runs off the pulley which is changed into another such motion in the part of the cord running on in the opposite direction. Why, however, the first motion is rectilinear we do not understand. Indeed it is not necessarily rectilinear, for the cord running off may be pulled to one or the other side, if it only be kept stretched, without in any way altering the mechanism. Then also the motion of the cord originates in the circular motion of the points of the pulley or drum; only after this motion do we have that of the cord itself. Thus the very first problem takes us beyond the limits of the class. The same indeterminateness which has been pointed out in the motion of the one cord belongs also to the other. We see therefore that, even in the very first example, inexactness in a number of ways makes its appearance. It may be noticed too that this first problem of the fixed pulley, so far as concerns the theory of the motions involved in it, is already a very complicated one, as I shall show in the text later on.

The ideas by which alone the nature of a simple mechanism can be arrived at may be very complex, or in certain circumstances may be quite the reverse. But equally whether these ideas be simple or complex, if it be desired to examine and understand the simple apparatus scientifically, it is necessary to work through the whole succession of them, passing from each one to the next higher from which it was developed, until really general principles are arrived at. However difficult this may be, and however little use it may appear to have, it must be done; its omission by kinematists hitherto has caused the wreck of all their theories. What they have done is what I have already indicated as incorrect,—they have accepted the simple or apparently simple mechanism as it came from the hand of the inventor, whether he were a well-known person, or nameless in the traditions of the dawn of national history.

An examination of these hazy traditions gives the kinematist much subject for remark. Denying myself here the prosecution of this attractive subject, upon which I shall enter in some detail further on, I must dismiss it with one remark. Following back machines to the earliest forms in which they have been used in historic times, we find in different places contrivances of various kinds in use,—from somewhat complicated machines to the very simplest arrangements,—all of which must be called "machine." We are not yet in a position to discuss the criterion of the comparative difficulty of their invention; we require here only to note that they appear in various places independently of each other. The rollers upon which the Assyrian as well as the Egyptian builders moved their enormous stones are among these primitive machines; carriages of wood and metal, both for war and for transport, were possessed by the Egyptians, Babylonians and Indians in the remotest antiquity; water-wheels were in use in old Mesopotamia and in Egypt, as well as in China, India and Central Asia; toothed wheels were known to the Greeks, as were also the screw, the pulley, certain systems of levers, &c. Some of these arrangements have come down to us unaltered; others, however

in the way already enlarged upon, formed only steps leading up to
their present successors. All have been thought out by human
brains,—now and then by brains of special capacity, and then praised
as God-sent gifts,—but in all cases thought out, produced
by a mental process which has contained more or less
well-defined gradations.

And to-day just as formerly they must still be arrived at by
a mental process; and this forms the problem which it must be
the chief aim of theoretical Kinematics to solve. So long as it
could not reach the elements and mechanisms of machines without
the aid of invention, present or past, it could not pretend to the
character of a science, it was strictly speaking mere empiricism—
(sometimes even of a very primitive kind),—appearing in garments
borrowed from other sciences. When however its investigations
enable it to furnish the means of producing any required
kind of motion, it will begin to deserve the name of Science.
It will then itself point out the true classification of its
own material. It can put before itself the question as to the
change of one motion into another, and decide as its own director
whether really and to what extent a division founded upon this is
important. As a genuine science, moreover, it will find its laws in
itself, and require no Lycurgus to deliver them from without.

Here we reach another weighty and notable consequence. If
the processes of thought by which the existing mechanisms have
been built up are known, it must be possible to continue the use
of these processes for the same purpose ;—they must furnish the
means for arriving at new mechanisms—must, that is to say, take
up the position hitherto assigned to invention. I hope not to
fall under the suspicion of saying anything so absurd as that the
new method will make it possible for the commonplace head to
become oracularly inventive like that of the genius. The case is
rather that it will become possible to introduce into machine-
problems those intellectual operations with which science everywhere else pursues her investigations. I have attempted already to
show that Invention, in those cases especially where it succeeds, is
Thought; if we then have the means of systematizing the latter,
so far as our subject goes, we shall have prepared the way for the
former.

Göthe,—who had so great an interest in the inner nature of

everything which could enlarge the circle of our ideas,—expresses himself in the following noteworthy sentence : " Everything that we call Invention, discovery in the higher sense, is the ultimate outcome of the original perception of some truth, which, long perfected in quiet, leads at length suddenly and unexpectedly to productive recognition." Schopenhauer too, whose thoughts not unfrequently seem to take the same direction as those of Göthe, says upon a very similar question :—" Our best, most able and deepest thoughts often seem to enter our consciousness like an inspiration,—sometimes directly in the form of a weighty sentence. Evidently, however, they are the result of long and unconscious meditation, and numberless long past and often entirely forgotten thoughts and conclusions. The whole process of thought and conclusion seldom lies on the surface,—seldom takes, that is, the form of a chain of reasoning clearly thought out, although we may endeavour to attain this in order to be able to give ourselves and others an account of what has occurred : commonly, however, the rumination by which the material received from without is converted into thoughts, occurs as it were in darkness,—taking place almost as unconsciously as the change of the nourishment into the fluid and substance of the body. Thus it comes about that we can often give no account of the origination of our deepest thoughts ;—they are born from our most secret being. Out of its depths arise unexpectedly Opinions, Ideas, and Conclusions."[6]

There is nothing, however, impossible in the ideas necessary for the origination of a mechanism being " clearly thought out," and they can then lead to what is sought just as in mathematics the clearly-reasoned and well-connected ideas lead up to the result. In other words, the invention of a mechanism will be to the scientific kinematist a synthetic problem,—which he can solve by the use of systematic, if also difficult, methods. The clever man, supplied with such powerful instruments, will leave the less clever behind in future as hitherto,—just as the mathematical genius leaves behind the mere algebraist who works only with operations learnt by rote.

The thorough understanding of old mechanisms, however, is even more important than the creation of new ones. It is indeed astonishing to how small a depth the methods hitherto used have penetrated into their real nature, and how incompletely known therefore are most of the mechanisms in common use. To the thorough, thoughtful mechanician who looks at his work as a serious matter, a scientific investigation of the Kinematics of Machinery will in this respect be specially valuable. It will relieve him of the minute and often worrying search after solutions of his problems by rendering it possible for him to work systematically. The technologist too, who hitherto has scarcely made any use of Kinematics, will find in it an important assistance in understanding old machines and devising new ones. The deepening of the comprehension which must occur in such cases as these renders it certain that the remodelled science will take its share in the real end before us,—the progressive development of the machine.

If we look back over the representation I have tried to give of the way in which the subject has hitherto been treated, and of the Ideal to be aimed at, the old method appears to have no inner unity, although the scientific methods of investigation which it employs have prevented this from being generally recognised. We have, however, shown that these are only of secondary importance compared with the establishment of the special ideas and principles peculiar to the subject. This question, moreover, must at once be seen to be one which actually concerns a department of investigation belonging distinctly to the exact sciences. This being recognised, the former method must be considered insufficient and not permanently tenable, for it permits of deductions only to a limited extent, and does not make it possible to give reasons for existing phenomena.

The remodelling which has become necessary requires undisturbed adherence to clear, simple, logical principles. What, however, is to be drawn from our criticism of the system hitherto used,—what I have endeavoured to illustrate and develope by single instances,— what the philosophical sentences I have quoted bring before us in a condensed form,—we may contract into one word. So far as our special problem is concerned, the question is to make the science of machinery deductive. The study must be so formed that it rests upon a few fundamental truths peculiar to itself. The whole fabric must be reducible to their strictness and simplicity, and from them again we must be able, conversely, to develope it. Here again is a point from which the weakness of the method hitherto employed can be surveyed at a glance. Its difference from the ideal method is not that it employs the inductive instead of the deductive method ; that would indeed be no advantage, but it might still be defensible. No, it has been entirely unmethodical. It has chosen no fixed method of investigation, or rather, it has not found any in spite of zealous search ; indeed it has so often cried *Eureka* that it now rests quietly in the impression that some such fixed standpoint has really been found.

In the development of every exact science, its substance having grown sufficiently to make generalisation possible, there is a time when a series of changes brings it into clearness This time has most certainly arrived for the science of Kinematics. The number of mechanisms has grown almost out of measure, and the number of ways in which they are applied no less. It has become absolutely impossible still to hold the thread which can lead in any way through this labyrinth by the existing methods.

It cannot be denied that the difficulties in the way of remodelling the science are great. We often do not know ourselves how closely wedged-in our ideas are by the boundaries which education and study have drawn around us. If new ideas therefore are to be substituted for the common ones, it can only be by a wrench sufficiently powerful to overcome the cohesion of established notions and prejudices. There are the traditional courses of instruction in the schools,—the widely extended and important technical literature,—the force of habit, acquired with difficulty, and on that very account firmly rooted; there is also the real difficulty that the new study requires to be grasped as a whole, and not taken up partially and occasionally,—all these pile up mighty hindrances. I cannot therefore shorten the way, although the truths to which it leads are of great simplicity. The careful removal of preconceived ideas, the slow seeking of the right path among those inviting us, prevents rapid motion. The following chapters are therefore intended not so much to add to the positive knowledge of the mechanician as to increase his understanding of what he already knows, so that it may become more thoroughly his own property. For, to conclude in the words of Göthe, " What is not understood is not possessed."

NOTE.—Had Prof. Reuleaux been acquainted with Prof. Rankine's "Machinery and Millwork," the first edition of which was published in 1869, he would no doubt have mentioned it here. Some 300 pages of it are devoted to Machine-Kinematics, or as Prof. Rankine calls it, the Geometry of Machinery,—and this subject is treated in a way which has some points in common with that now adopted by Reuleaux, although greatly differing from it.

Although neither Rankine's nomenclature nor his classification (given first partly in his Applied Mechanics) is now likely to be followed—it may be interesting to compare them with those of Reuleaux, the superiority of which I believe Rankine would have been the first to recognise had he lived to know them. Rankine considers a machine to be made up of a "frame" and "moving pieces," the latter being "primary" and "secondary." The frame is the fixed link of the mechanism (in the language of Reuleaux), the primary moving pieces are links the nature of whose motions are determined solely by their connection with the frame,—the secondary moving pieces are all other links. He considers (erroneously) that the primary pieces can have no other motions than those he calls shifting, turning and helical, and that they must be connected to the frame by one of the lower pairs of elements. He then goes on to examine the simpler conditions of these three kinds of motion, which he does by the ordinary geometrical methods. The general nature of the motion of secondary moving pieces is treated by itself, principally by the method of instantaneous centres and axes, which Rankine afterwards uses freely and with great advantage throughout the work. In the next chapters, unfortunately, an entirely different set of ideas is introduced, and the comparative distinctness, as to system, of the earlier part of the work is lost. The idea of "elementary combinations" is brought in, and under this head are treated an immense number of kinematic chains of the most various descriptions, as well as the delineation of the profiles of several higher elements. The "mechanical powers" are placed as a subdivision under "elementary combinations." The lever and wheel and axle are taken together as cases of motion about a point, *i.e.* of turning,—and the inclined plane, wedge and s c r e w as cases of sliding,—the "powers" are therefore not considered to be connected with the three simple motions of the first chapter. The place of the pulley among them is left unexplained. Some compound chains and some simple ones which in no way differ from "elementary combinations" are treated in two further chapters on "aggregate combinations" and "adjustments."

Notwithstanding the excellence of Prof. Rankine's book—and the value of some parts of it will be increased rather than diminished when read in the fresh light of Reuleaux's investigations,—it must be confessed that it contains neither a general theory of machines nor a s y s t e m a t i c treatment of their motions. Most of its solutions are special rather than general, fresh methods being adopted for each new class of mechanisms.

The real points of connection and of difference between them are thus lost, and the subject presents itself to the student as a series—or rather as many series—of interesting problems, which have only a very indistinct relation to each other. He is not led to any standpoint from which he can take a general survey of the whole. For this purpose the apparently simple, but in their results most wonderful ideas of the nature, pairing and inversion of machine-elements and kinematic chains were essential. Rankine unfortunately had no opportunity of making use of these; he does not seem either to have recognised the use of centroids, or the possibility of constrained motion in the higher pairs. He does not even point out that two of the motions of his "primary" pieces are in reality only special cases of the third, the twist,—or that the method of instantaneous centres is as applicable to these motions as to more complicated ones. If it had not been for my own feeling as to the value of Prof. Rankine's work, I should not have alluded at such length to what appear to me to be defects or omissions in it. He has done great and lasting service in connection with the scientific study of machinery in this country, and his "Machinery and Millwork" is so familiar to engineers that it appeared to me impossible to leave unnoticed the bearing upon it of Reuleaux's Theory of Machines.

CHAPTER I.

G E N E R A L   O U T L I N E S.

§ 1.

Nature of the Machine-Problem.

WHILE in appearance a machine differs greatly from any of the force- or motion-distributors of nature, yet for the theoretical or pure mechanician no such difference exists,—or rather it completely disappears on analysis, so that to him the problems of machinery fall into the same class as those of the mechanical phenomena of nature. He sees in both forces and motions existing, and subject to the same great laws which, developed in their most general form, govern and must govern every single case. In pure Mechanics machines are now treated only as illustrations; they no longer receive their complete development as they did when many of their problems were still new and strange, and apparently stood opposed to those of Mechanics. This present subdivision is quite correct, so far as the question is one of scientific comprehension only. As, however, the actual machinery itself, deriving its existence from various sources, and having its own characteristic features and methods of classification and treatment, forms a quite distinct and special subject, a separation of its scientific mechanical problems from those of Mechanics in general is possible, and indeed has already been made.

27–29

It must be admitted, too, that the sense of the reality of this
separation has been felt not only by Engineers or others actually
engaged in machine design, but also by those theoretical writers
who have had any practical knowledge of machinery, in spite
of the increasing tendency in the treatment of mechanical science
to thin away machine-problems into those of pure mechanics.

There are good reasons for this feeling. Such a treatment of
machine-problems is first of all greatly to be deprecated because
it would place the scientific part of machine-construction upon
a base too indefinite and widely extended. The fundamental
notions of force and motion themselves are subject to uncertain
interpretation. In the attempt to define ideas standing on the
boundary line between Physics and Metaphysics an uncertainty
which demands the closest mathematical and philosophical investigation makes itself felt. This uncertainty or indistinctness,
by holding open a perspective of ideas entirely beyond all purpose of the study concerned, exercises a disturbing influence on it.
It affects every definition, every explanation intended to be exhaustive; it compels the teacher who desires to express himself
with scientific accuracy either to use generalisations of which he
feels the unpractical nature, or to employ illogical limitations such
as " common practice," " usual arrangements," and so on. He who
knows laws only is fain to content himself with rules where he
would far sooner employ strict scientific methods. Not every
generalisation, that is to say, is practical, nor from a certain
point of view indeed, even correct. This point of view is
that from which Geometry separates itself from Mathematics
in general, Descriptive Geometry from Geometry in general, still
more from which Kosmical Physics, Hydraulics, Aerostatics, branch
away from Natural Philosophy,—in other words, the point of view
from which special sciences are seen to separate themselves from
the more general sciences to which they are subordinate.

Such a separation becomes at once possible and advisable if any
complete circle of ideas lie at the base of the region separated.
In the case of machine-problems their separation from those of
general Mechanics can be demonstrated. A distinct line of demar-
cation, although in certain examples less distinct than in others,
shows itself between them. To find the real nature of this difference let us endeavour to look at the whole question from outside,

examining, without regard to any existing machine-theories, one and the same motion as it appears in Nature and in the Machine.

Let us take the case of a circular motion, which shall be supposed to occur first as the motion of a satellite about its planet; and then as the revolution of a wheel.

Suppose that from any cause the satellite T (Fig. 2) so move about the planet P that its centre describes a circle about the centre of P in a plane passing through that point. So long as the conditions remain unaltered the motion continues the same. So soon however as any external disturbing force Q_1 (shown here perpendicular to the plane of motion), begins to act upon one side of T, T alters its path. If this is to be prevented, another external force Q_2, equal and opposite to Q_1, must be brought simultaneously into action. If Q_1 be a pound, Q_2 must be also a pound; if Q_1 increase to a ton, Q_2 must increase to a ton also; the absolute value of Q does not further enter into the question, which is one solely of the equilibrium of the external disturbing forces acting upon T. The existence of such an equilibrium in nature would presuppose the existence always of equally divided causes of force; it probably does not once occur in the case of celestial bodies. We are however at liberty to assume its possibility in order to simplify the matter.

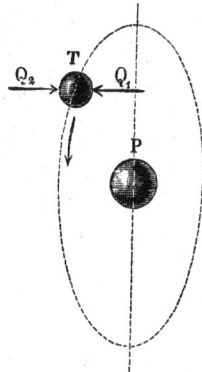

Fig. 2.

In the machine the case is quite different, and much more simple. In order that points of the wheel R, Fig. 3, may move in circles, let it be fixed upon a rigid shaft of which the ends at A and B are turned down and fitted in holes in fixed and rigid supports $L L$, the whole system having a common geometrical axis. If now the wheel be set in motion by some suitable handle, every point in it lying beyond its geometrical axis describes a circle about some point in that axis. If any disturbing force Q act sideways upon the wheel, then (if we suppose the material of the wheel, shaft, and bearings to be completely rigid) no alteration of the circular motion occurs; and this is true equally whether Q be great or small, continuous or intermittent, constant or changing in direction. There is nevertheless continuous equilibrium here, but in another way than before. So soon as the force Q begins to act it calls forth in the interior of the wheel, the shaft and the supports, internal molecular forces, opposite in direction and exactly equal to it. The action of the forces therefore, considered by themselves, is here exactly the same as, or exactly corresponding to, their action in the case of the satellite. There exists however the difference that there all external forces are independent of each other, while here the action of an external force becomes at once the cause of the opposite action of a molecular force.

In actual machines we do not employ absolutely rigid material, for no such material exists; we use however only those materials which when of suitable dimensions alter their form under the

Fɪɢ. 3.

action of external forces very little, so little that the corresponding variations from the original form may be neglected. The choice of suitable dimensions and forms is the work of the machine designer. If we disregard the very small variations of form which actually occur, it appears that the solution of the problem by the machine exists, and also that it differs essentially from that occurring in nature.

Whilst in the first system, which we may call kosmical, the external measurable mechanical forces are opposed by similar external forces, in the second, the machine system, there are opposed to all external forces others concealed in the interior of the bodies forming the system, and appearing there,—and acting in exactly the required manner,—in consequence of the action of the external forces. One might apply to these forces,—with a very small alteration which I hope the reader will permit,— Schiller's riddle about the spark:—

> "Sleeping, yet ready for the expected foe,
> I lie concealed within my iron walls ;
> He comes, he feels my iron weapon's blow,
> We fight ; I sleep again,—for soon he falls."

The force is challenged, and immediately it appears ;—the external challenging force ceases, and immediately its opponent, which has so energetically defended the form of its dwelling, also disappears. Nothing is to be seen of the inner force so long as it is not awakened by an outer one. It is as it were concealed in the interior of the body. We shall not be carrying the analogy with Thermal Physics too far if we call these molecular forces, which in their hiding-places guard the stability of the material world, latent forces, as opposed to the directly measurable sensible forces which externally influence bodies through gravitation and other causes. The difference between the two systems is therefore that sensible forces are in the one case opposed by other and independent sensible forces, and in the other case by dependent latent forces.

We have considered both systems in a form of special simplicity which, it may be thought, does not permit sufficiently general deductions to be made. Then let us suppose the kosmical system to be enlarged into a solar system with sun, planets, and satellites moving in their circular or elliptic orbits, and let us add to our wheel other wheels and shafts connected with the first as spur-gearing or in any other way, so that rotation occurs throughout the whole system, and a machine suitable for any particular purpose is formed. We shall then note that in the kosmical system the mutual motions of the bodies, both as to their paths and their velocities, are entirely dependent on the influence of sensible forces, while in the machine system the paths of motion are absolutely determined, and at the same time no point can alter its velocity without the velocities of all other points being correspondingly altered ; that in the latter case, therefore, disturbing sensible forces are without influence,—they are everywhere balanced by the latent forces. Not less is it these last which carry the moving forces from body to body. The difference we found above, then, is general, so far as it relates to the nature of the forces coming into action, and is in no way limited to the simple case supposed.

Both the cases chosen as illustrations are extreme,—in general the kosmical and machine systems do not differ so widely, but approach each other mutually more or less. The plant, for instance, so far resembles a machine system that the motion of its sap takes place in tolerably rigid channels or tubes, and in definite prescribed directions. The correspondence is not, however, exact, for the leaf-stalks, twigs and boughs undergo alterations of form both small and great from kosmical forces. The nearest approach to our machine system in the vegetable kingdom is perhaps the circulation of the sap in the tissue of a firm, strong tree-stem, for here only are the alterations of form small enough to be neglected. In a few existing machines also actions occur which must be classed as kosmical, as for instance the motion of the water in the ancient water-wheels (*Straube-räder*) used sometimes in mountainous districts to drive saw-mills, upon which the stream dashes almost like a waterfall. Thus the two systems are not divided by a hair-line, but still their differences are always notable, and become the more distinct the more decidedly each belongs to its own class. The more perfectly the water-wheel is made, the more completely do the freely-playing streams of water disappear; the rude wheel becomes the smooth and quietly running turbine, where the foaming and splashing of the water is reduced to the smallest limits. From the huge swinging lever, by the help of which the Walloon brickmaker or the Hindoo builder lowered his empty bucket into the brook and raised it again full, has grown the beam-engine, with its quiet and regularly working pumps. The kosmical freedom of natural forces is brought in the machine under order and law, which no ordinary external force can shake. On the other hand, latent forces also act with those of nature, as in the waterfall, hurling the rebounding streams of water upwards from the rocky channel; or as in the meteoric stone, diverting it from its original path by atmospheric resistance. The balancing of sensible by latent forces is therefore not solely a distinctive mark of the machine but we have in it a principal characteristic of the machine-like or machinal as distinguished from the kosmical, and it must be kept distinctly in view in endeavouring to understand the exact idea conveyed by the word machine.

The prevention of disturbing motions by latent forces is then a principle in the machine. Its application is connected with various objects. When a machine is constructed it is meant to be an arrangement for carrying on some definite mechanical work—it may be the moving of some body, or the alteration of its form, or both together. For such a purpose we require that so soon as motion is caused by any effort in any part of the machine that motion shall be of an absolutely defined nature. Thus our wheel in Fig. 3 might be used for lifting weights if we made the disc R a drum and passed a cord over it,—or it might serve as a grindstone if the disc were made of suitable material, and so on. Every motion then which varies from the one intended will be a disturbing motion, and we therefore give beforehand to the parts which bear the latent forces—the bodies, that is, of which the machine is constructed,—such arrangement, form and rigidity that they permit each moving part to have one motion only, the required one. This having been done, so soon as the external natural forces which it is intended to employ are allowed to act, the desired motion occurs. Our procedure is therefore two-fold ; negative first—the exclusion of the possibility of any other than the wished-for motion ; and then positive—the introduction of motion. The result is that the natural force when applied accomplishes the required mechanical work.

A machine may be perfect, or may contain more or fewer imperfections ; it approaches perfection just in proportion as it corresponds to what we have recognised as its special object,— the special end for which it has been constructed. After the insight we have now obtained into its nature it is possible for us to frame a definition of the machine. It is as follows :—

A machine is a combination of resistant bodies so arranged that by their means the mechanical forces of nature can be compelled to do work accompanied by certain determinate motions.[7]

This shows within what distinct limits machine-problems lie, and that they allow themselves to be readily separated from the general problems of Mechanics, as we have already maintained.

While the science of Mechanics examines motion caused in the most general cases by the action of mechanical forces, Machine-mechanics occupies itself with certain special cases only, with motions produced by a limited circle of means. It draws its first laws from the same fountain as Mechanics, to which, as the more comprehensive science, it is subordinate. But the region which it specially concerns can be separated from the science as a whole, and its function is to create systematic order within this separate region, and to investigate the laws specially belonging to it. Here is work enough, challenging some one to undertake it. It is greatly to be wished that those who are familiar with machine-design should not leave these investigations entirely to others, as has of late years often been the case with us, and still more in France. It is this that has caused what I have already alluded to as the volatilisation, the thinning away, of the problem,—a method of treatment from which practical mechanists, upon whom the machine depends for its further development, and for whose benefit specially the investigations have been undertaken, turn away dissatisfied. They have the right to demand, within certain limits, complete concentration upon their special problems, and will not allow the question to be carried off into another region, where the solid ground seems to them to disappear without any counterbalancing advantages being gained.

§ 2.

The Science of Machines.

The scientific carrying-out in practice of the requirements covered by our definition of the machine has caused the rise of an extended apparatus of sciences in connection with the progressing development of Polytechnic instruction. From the foundation sciences of Mathematics and Physics three or four other sciences specially concerning the machine have separated themselves. Their common object is the elucidation of the causal-connection of machine phenomena. Together they have been happily enough called Practical Mechanics. I speak of them here as sciences without pretending to insist on their absolute right to the title; they may be called sciences of the second or third order, or by their usual names; they employ scientific methods, and treat by their means special regions of investigation; within these they have reached by degrees an independence which has made necessary their separation from the more general sciences.

First comes the study of machines in general, looked at in connection with the work they have to do. This is known in Germany under a number of somewhat vague titles, as general or descriptive, special and theoretical " *Maschinenlehre.*" In its general form it deals, descriptively, with the whole of existing machines,—it teaches what machines exist and how they are constituted, and thus affords us a glance at their manner of growth. It proceeds teleologically in the fullest sense of the word, seeking always to refer everything to the special object for which the machine was constructed. Its methods of classification are made as general as possible. At present a complete descriptive, or really general treatment of machinery in this way, is hardly possible on account of the enormous number of existing machines. To be really general only classes and types can be treated of. Quietly adapting itself to the every-day wants of the learner the study thus becomes specialised,—single classes are taken up and treated singly in detail. Along with the construction of each special machine its theory is also, for the most part, considered,—that is, the nature of the sensible forces which come into action and the motions to which they give rise are examined, and deductions are drawn in regard to the most suitable way for turning these forces to account. This method of treatment is therefore based also on existing machines, but differs from the former in not only describing their arrangement and purpose, but in examining also how they can best be arranged in order to carry out the given purpose. In Germany at present it is for the most part rightly grasped and comprehended, the machine itself being taken as both the end and the beginning of the problem. The French, however, have not always freed themselves from the idea that the machine occurs merely as an illustration—an example—in Applied Mechanics; if this idea were right, however, it is clear that all other applications of Mechanics should be treated in the same way. If,—coming somewhat nearer to the heart of the matter,—the applications of

mechanics "in machinery" be classed by themselves, as is done
by Poncelet,—still the principle is not carried sufficiently far, for
under this title all machines of every kind must be treated, which,
however, has not been the case. Redtenbacher first removed this
stigma of indistinctness from the matter, and thereby laid the
foundation of the freshness and power which the German system
of machine-instruction shows as contrasted with the French.
Redtenbacher's most lasting services, which have not always been
understood by his successors, lie in this direction,—in the separation of the questions connected with machinery into separate
sciences or branches of science. It was on this ground that his
influence was, I may say, so electric, and brought to him so quickly
in his time the engineering students of Germany.

The existing treatment of the theory of machines (*theoretische
Maschinenlehre*) confines itself principally to prime-movers,—
Steam-engines, Water-wheels, Turbines, Windmills, and so on,—
or in terms of our definition, it concerns itself with the nature
of the various arrangements by means of which natural
forces can be best applied in machinery. Yet it does also
consider machines in general (other than prime-movers), and obviously these all belong to its province. To the general examination
of the theory of these machines the name mechanical technology
is often given. This is not universal, nor indeed is it correct, for
mechanical technology must include all mechanical processes of
manufacture, and in a multitude of cases machines are not employed in these. It possesses therefore a domain of its own, and
must be treated in its own proper way. From its own point of view
it also examines the machine, but in a way entirely differing from
that in which it is examined for its own sake in the studies of
which we are speaking. While therefore it can easily be understood how both studies should set up claims to the same object
of instruction, it is on that very account important that they
should not be confused with each other.

The special part of technology here coming into question,—
or what may be called the technological part of the study of
special machines,—concerns itself with the action of the natural
forces, through their various applications in the machine, on the
bodies to be worked upon. It examines, in other words, by what
special arrangement of the parts of the machine the

required action can be best obtained. As a whole, there-fore, the specialised study of machines (*specielle Maschinenlehre*) considers both the application of the natural forces to a given machine and their action in it.

The third science is that of Machine-design. It also has been freed by Redtenbacher from its incorrect treatment under Applied Mechanics, and placed by him on an independent footing. Its province is to teach how to give to the bodies constituting the machine the capacity for resisting alteration of form mentioned in our definition. In order to determine this pro-perty fully it must be considered in reference not only to sensible but also to latent forces.

The first it accepts as found by the aid of the Science just ex-amined, in the shape for example of the steam pressure upon a piston, the water pressure in a turbine, and so on ; these determine the strength of the bodies. The latter, the latent forces, carry the force-action from body to body,—*e.g.* from piston-rod to con-necting-rod, from spur-wheel to spur-wheel, and so on ; and cause therefore necessarily friction and wear. The problems of machine-design extend in both the directions thus pointed out. In solving these problems in such a way as to conform to the technological conditions of each special case, machine-design forms itself into a really technical science. Its twofold nature, as concerning itself both with sensible and with latent forces, which hitherto has been recognised in fact without being known to theory, I wish to raise into the position of a leading principle ; its reality has been clearly proved from the general development of fundamental propositions.

Now, lastly, our definition covers a fourth characteristic of the machine which has not been a leading idea in either of the thre studies we have considered. This is the arrangement of the means for insuring that only certain determined motions shall occur in the machine. So far certainly as the motions are conditioned by forces, and are regarded solely in connection with force-actions, they have been considered in studying the theory of machines in the way already described. But that study simply takes the mo-tions looked at as changes of position as given. Hence another series of investigations remain, their subject being the nature of the mutual dependence of the changes of position of parts of the machine. If the problems here presenting themselves be treated separately, those of the three former studies being supposed to be solved, they form a province of investigation which can be worked in by means of Applied Mathematics and Mechanics. The systematised study of the solutions of these problems forms the science with which we have to do, Kinematics, the "Science of Pure Mechanism." * It is, as follows from what we have said, the study of those arrangements of the machine by which the mutual motions of its parts, considered as changes of position, are determined.

The difference between this definition of Kinematics and that which Ampère indicated rather than gave fully (see Introduction, p. 11) requires to be noticed. It is principally this, that here Kinematics is made to belong essentially not to Mechanics, as with Ampère, but to the Science of Machines, as has been done more or less, but without any distinct admission of it, by most of Ampère's followers. Its objects and methods subordinate themselves therefore to the chief laws which affect the machinal as distinguished from the kosmical, and must at the same time fit in with the methods of treatment received by the machine in the three different studies already described. So far, that is to say, Kinematics is not an absolutely isolated science, as it would be under Ampère's definition, but works in consciousness of the neighbourhood of other systems of investigation having a common object with it. On the other hand we have in our own way arrived at the same conclusion with Ampère, that Kinematics observes changes of position only. Only we do not thereby shut out the actions of forces, as Ampère does; we take the problems connected with them as solved in every case, and consider the conditions imposed by them, which is a real and important difference. The indistinctness remaining with Ampère upon this point has been the cause of the unavoidable introduction by his followers of fragments of three other studies, with which they could not dispense : thus, for example, Haton gives an abstract of the strength of materials, Laboulaye this and the study of friction also, and so on.†

---

* This was the name used by Prof. Willis.

† I have given in the Preface my reasons for thinking that, in this country at least, it is now too late to make the limitation proposed by Reuleaux, in the meaning

Summarising this section, we see that " Practical Mechanics " has been subdivided into—

The study of Machinery in general.
The special or theoretical study of Machinery.
The study of Machine-design.
The study of pure Mechanism.

For the understanding of the nature of machines the last-named science is evidently as important as the three first; indeed in many respects it must stand first and prepare the way for them, and on this account single sections of Kinematics are often included in all three. The union of the three last sciences is necessary that the machine may be completely understood, the first having pointed out its existence and treated it teleologically All four interact continually ; only as a whole do they furnish the practical mechanist with complete solutions to all the problems of his work.

§ 3.

General Solution of the Machine-Problem.

We must now proceed to establish the general principles of kinematic procedure, in order to gain a standpoint from which to survey generally the method of solving our problem. The ideas above developed concerning the essential nature of machine-systems will serve as an introduction to this. Those parts of a machine transmitting the forces by which the moving points are caused to limit their motions in the definite and required manner must be bodies of suitable resistant capacity, the moving points themselves must belong also to similar bodies. In the machine, consequently, the moving bodies are prevented, by bodies in contact with them, from making any other than the required motions. This contact also, if the problem is to be entirely solved, must take place continually, which presupposes the possession of certain properties by the bodies in contact. In proceeding to examine these properties more closely, we shall

of the word k i n e m a t i c s, and for substituting m a c h i n e - k i n e m a t i c s in the title of his book.

assume in the first instance that the bodies possess complete rigidity, and shall pay no attention to their size,—in other words we take all questions belonging to what we have called the special study of machinery and the study of machine-design as solved, so that only geometrical properties remain for us to consider.

Now in order that any moving body A, of given form, may remain continually in contact with a stationary one B, we must give to the latter a special form. This can be found if the body A be caused to take up consecutively the series of positions which it is intended to occupy relatively to B, and the figure which envelopes all the positions of the outline of the body A determined. If for example A be a parallelopiped (Fig. 4), of which one surface remains during its motion in a plane, the figure B will become a curved channel. The geometrical form thus found for B is called the **envelope** of the moving figure A. A has to B also the same relation as B to A, that is, A is the envelope to B, or

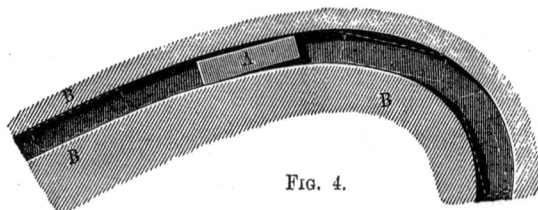

Fig. 4.

at least those points in the figure A with which B comes into contact form the envelope of the body A in respect to B. The relation, therefore, is reciprocal.

Many such reciprocal envelopes can be actually constructed. If a moving body be surrounded by stationary ones containing its envelope in such a way that at no instant is more than one motion of the body possible, then its motion must necessarily be such as belongs to the envelopes, and is determined by their form.

We see at once that at least one other body is necessary for the envelopment of a moving form. If it be found necessary to use several—perhaps because the one first found, while actually forming an envelope, does not exclude all motions but the one required,—then these can be united with the first into one body. Thus for instance we can suppose the upper and lower half brasses of a plummer block joined together. We find, that is to say, that in all cases at least two bodies correspond in being reciprocally envelopes each of the other. A machine consists solely of bodies which thus correspond, pair-wise, reciprocally. These form the kinematic or mechanismal elements of the machine.

The shaft and the bearing, the screw and the nut, are examples of such pairs of elements. We see here that the kinematic elements of a machine are not employed singly, but always in pairs; or in other words, that the machine cannot so well be said to consist of elements as of pairs of elements. This particular manner of constitution forms a distinguishing characteristic of the machine.

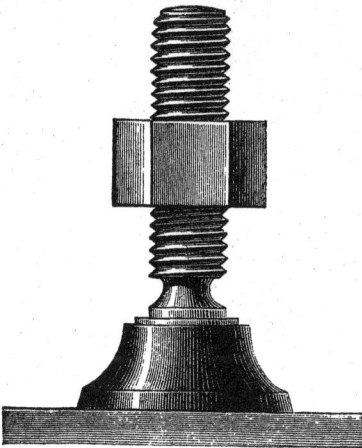 

Fig. 5.   Fig. 6.

If a kinematic pair of elements be given, a definite motion can be obtained by means of them if one of the two be held fast or fixed in position,—that is, be brought to rest relatively to a given portion of space chosen with reference to the motion which is to be observed. The other element is then free to be moved, but only in the one particular way allowed by the constitution of the pair. Its motion relatively to its companion element is under these conditions the same as its absolute motion in the given portion of space. Thus, for example, in the pair of elements shown in Fig. 5,— a screw and nut,—if the former be fitted with a suitable foot or its equivalent, so as to prevent its motion, then each point in the nut, if it be moved, will describe a helix of determinate magnitude.

In the pair shown in Fig. 6, which consists of a solid prism enclosed by a corresponding hollow one, all points of the latter describe straight lines of equal length if it be set in motion after the former has been fixed.

A large number of motions can be obtained in this way simply by pairs of elements, as we shall have occasion further on to see more in detail, while the complete development of their properties affords the means of multiplying indefinitely the motions obtain-

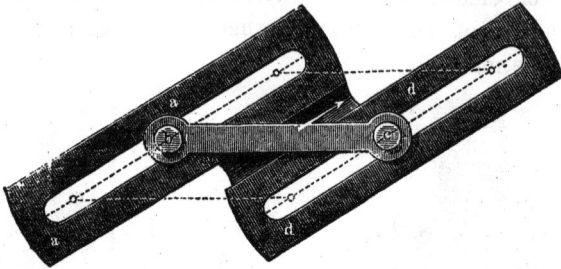

FIG. 7.

able by single pairs. This can be done by the combination of pairs.

Let it be desired to combine two pairs of elements, *a b* and *c d*;—this must take place so that each of the elements of one pair be

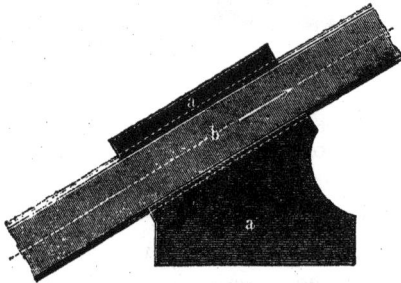

FIG. 8.

combined with, that is made part of the same solid body as, one of the elements of the other pair. This, moreover, may occur so that the mutual relation of the parts is not altered, and no new motion obtained. If the element *b* be joined to *c*, then *d* must be combined with *a* ; or if *b* be combined with *d*, then *a* must be joined to *c*. We may illustrate this by an example, Fig. 7. Suppose that *a b* and *c d* be two similar pairs, *b* and *c* being cylinders, *a* and *d* prismatic slots in bars, having such a form as to prevent either sideway or endlong motion of the cylinders. Let *b* and *c* be so joined that their axes are parallel, and let the two slotted bars be also placed parallel to each other and joined. Then evidently, if *a d* be fixed, every point in *b c* must move parallel to the centre line of the slots, as shown by the arrow. All such points therefore describe equal straight lines. Thus the motion takes place exactly as it would if *b c* were a solid prism enclosed by a hollow one, like the pair *a b* in Fig. 8. Thus by combining two pairs we have obtained nothing but what we could have got by means of a single one. So far the experiment has led to no result.

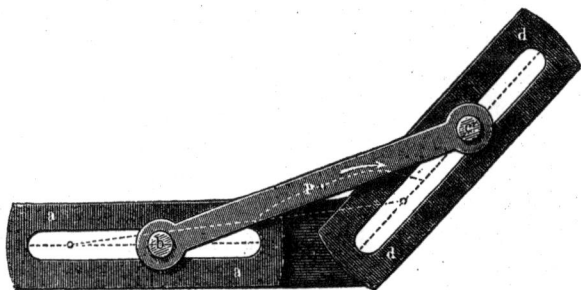

Fig. 9.

But if we do not place *a* and *d* parallel to each other, but set them obliquely as shown in Fig. 9, the case is entirely altered. The centres of *b* and *c* no longer move in similar directions in the slots, and consequently the various points in *b c* no longer have similar paths,—the point *p*, for example, describes a curve. The motion is thus quite different from what it was before.

Nevertheless in these two different cases, Figs. 7 and 9, the same relation obtains; in both, *b c* and *a d* form rigid bodies, or what may be considered as such,—that is to say, we have in the end one pair of elements only, by combining two pairs of bodies. With different methods of combination different results are obtained, but in every case there results only one pair.

Accordingly the reciprocal combination of the elements of two pairs gives us again a pair of elements, which may differ from either of the single pairs of which it is composed. Thus we obtain already an important result, and one having many consequences.

We may now proceed to the combination of three or four pairs of elements. Suppose the pairs

$$a\,b \qquad c\,d \qquad e\,f \qquad g\,h$$

to be given. Let each element of each of these be joined to one element of another pair,—then every pair keeps its own peculiarity and at the same time has another added to it. The combination may take place in a number of different ways, for example in the same order as above.

$$b \longrightarrow c\,d \longrightarrow e\,f \longrightarrow g\,h \longrightarrow a\,;$$

or in the order

$$b \longrightarrow d\,c \longrightarrow e\,f \longrightarrow h\,g \longrightarrow a,$$

and so on. The whole now forms a linkage returning upon itself, like an endless chain, consisting simply of single links connected together. A combination of pairs of elements in this way we shall call a chain, or more fully a kinematic chain. The body which is formed by the junction of the elements of two different pairs is then a link of the kinematic chain. Every link of the above-mentioned chain consists of two elements, so that the chain here has as many links as it contains pairs.

In the chain every two adjacent links have a definite relative motion, that namely which belongs to the pair of elements connecting them. But two links which are connected by a third do not possess definite reciprocal motions except under certain conditions. Such motions can occur only if the chain be so arranged that every alteration in the position of a link relatively to the one next to it be accompanied by an alteration in the position of every other link relatively to the first. In a kinematic chain which possesses this peculiarity, each link has only one relative motion to each other link; if, that is to say, any relative motion occur in the chain, all the links are constrained to execute determinate relative motions. Such a kinematic chain I call a constrained closed—or simply a closed—chain.

We may take as an illustration the simple chain shown in figure 10. It consists of four similar pairs $a\,b$, $c\,d$, $e\,f$, $g\,h$, each being a cylindrical pin fitting a corresponding eye, the axes of all being parallel. Here each link has only motion in a circle relatively to the one next to it. Every turning of $h\,a$ relatively to $g\,f$ must necessarily be accompanied by alterations in the positions of $b\,c$ and of $d\,e$—the chain is therefore closed.

In itself a closed chain does not postulate any definite absolute motion. In order to obtain this a similar method must be adopted to the one employed above with pairs of elements,—namely, to hold fast or fix in position one link of the chain relatively to the portion of surrounding space assumed to be stationary. The relative motions of the links then become absolute. A closed kinematic chain, of which one link is thus made stationary, is called a mechanism or train.

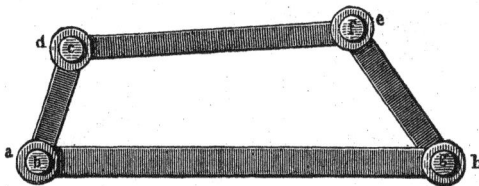

FIG. 10.

The above chain can be made a mechanism in four different ways, as shown in the following table, in which the stationary link is underlined in each case :—

1. b —— c d —— e f —— g h —— a
2. b —— c d —— e f —— g h —— a
3. b —— c d —— e f —— g h —— a
4. b —— c d —— e f —— g h —— a

In general, therefore, a constrained closed kinematic chain can be formed into a mechanism in as many ways as it has links.[8]

In order that a link may be made stationary it must be provided with suitably formed fastenings or carriers.

To make the demonstration complete, let us suppose that we employ a sufficiently rigid pedestal, such as that shown in Fig. 11, as a support to which one link of our chain, $a\,h$ for example, can be clamped, so that kinematically it may form one piece with $a\,h$. The motion which can now take place in the chain is indicated by the dotted lines, and will be at once recognised as that of the beam and crank of a steam-engine.

The form of the pedestal or equivalent body is of course, so far as the motion is concerned, indifferent. Yet it will be noticed at

once that as a rule there is a certain inclination to treat it as a
piece of architecture, with which it certainly has in common the
property of rest or immovability, and in the neighbourhood of
which also it is often placed. The stationary parts of mechanisms
have often attracted the attention of theorists. We have already
seen this in the Introduction (p. 10) in Borgnis' division of the

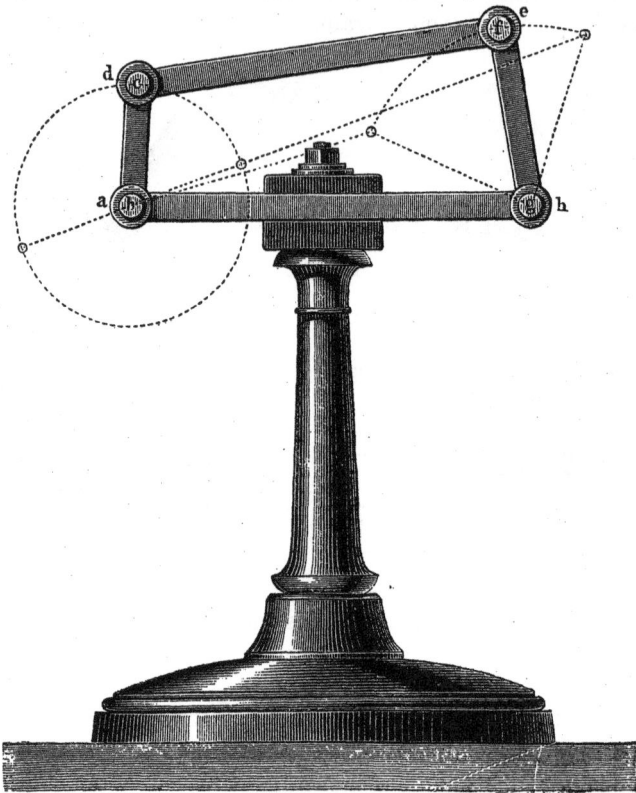

Fig. 11.

parts of machines, where the " supports " figure as a class by
themselves. Another indication of the same feeling is to be found
in the division which so often occurs of the parts of machines into
" active " and " passive." The latter are nothing else than the elements connected with those links of a kinematic chain which are,
for the time, fixed. No absolute distinction exists however between

these and the "active" parts, for in the various mechanisms constructed out of the same chain, the same part of the machine may sometimes be fixed and sometimes movable.

In the mechanisms which can be constructed out of a chain of the above-described arrangement, the motion of a link next the fixed link is determined by the nature of the element by means of which it is paired to that link; this one pair of elements alone influences its motion. With the link upon the further side of this one the case is different, its motion depends upon the motion of its neighbour-elements as well as upon the motions due to the elements at its points of attachment; in our illustration, for instance, it is influenced by four pairs of elements.

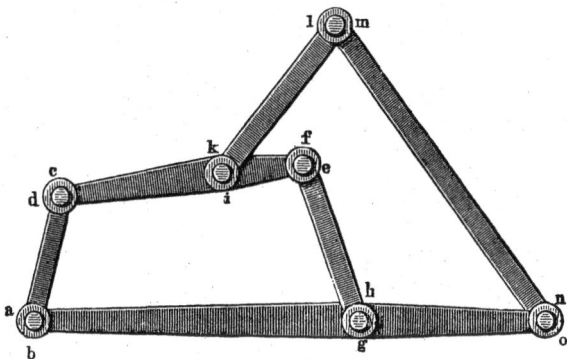

FIG. 12.

Its motion relatively to the fixed link is however as determinate as if the two were connected by one pair of elements only. Hence we can again use the method by which in the first instance we obtained the chain,—we can, that is, combine an element of a new pair with it and so further extend the chain. In order to obtain at the same time the requisite closure, this extended chain must be brought back again into connection with the link at which it started. We obtain in this way a compound kinematic chain, as distinct from which we may call the one described above a simple chain. Fig. 12 shows such a compound chain, formed of six links of exactly the same description as those used before. Two of the links now contain three elements:—

$$d \longrightarrow i \longrightarrow e$$
$$\text{and } a \longrightarrow h \longrightarrow o$$

If we suppose a —— h (that is a —— h —— o) again to be fixed, k —— l will have a yet more complex motion than d —— e ; the method of building up the chain further allows the possibility of obtaining motions according to more and more complex laws, and so to serve as the means of procuring a great, indeed an infinite number of different forms of motion. It holds equally good with the compound as with the simple chain that it can be set in motion after fixing any one of its links; in as many ways, that is to say, as it has links.

Closed mechanisms also can again combine, and so unite into higher forms; we may however allow these compound mechanisms to class with those built up from compound chains.

We have now before us a general view of the method of construction of Mechanisms :—

The mechanism is a closed kinematic chain; the kinematic chain is compound or simple, and consists of kinematic pairs of elements; these carry the envelopes required for the motion which the bodies in contact must have, and by these all motions other than those desired in the mechanism are prevented.

A kinematic mechanism is moved if a mechanical force or effort be applied to one of its movable links in such a way as to alter its position. The effort thus applied performs mechanical work which is accompanied by determinate motions; the whole, that is to say, is a **Machine.**

The arrangement by which the natural force is thus brought into action must correspond to the purpose for which the machine is intended. If for example the natural force act continuously, the machine receives a continuous motion, as in water-wheels, turbines, and so on. If the part acted upon by the force comes after a time into such a position that the latter exercises no further influence upon it, then, if the motion is to be continued, artificial means must be provided for restoring it to a position where such influence can again be exerted; as for instance in the clock. In many machines the action is limited to very small alterations of position of the moving parts, as in the balance, after which alteration they must be restored to their original position. This much by way of illustration only, later on we shall have to consider these questions systematically.

The title Machine has not hitherto been used logically. Commonly it is applied only in those cases where force or motion appears continuously or to some large extent. Many would not call the balance which we have just mentioned a machine, in consquence of the narrowness of the limits within which its motion is confined ; but force and motion are employed in it in exactly the same way as in other machines ; it certainly ought therefore to receive the same name. We may much rather say that the Engineer's measuring instruments, the theodolite, level and so on, are not machines. Here indeed mechanisms, in exactly the above described meaning, are used, and forces must be applied to these in just the way we have supposed in order that they may be used. The forces however are very small, and the mechanisms are only used at intervals, so that the name Instrument may properly be preferred for them. But the title Machine is even here not incorrect, as one may convince oneself by looking at the English giant telescope with its massive foundation and all the appliances for working it. With all these it differs in degree only, and not in kind, from a little pocket telescope. To such machines as occur naturally, also, the name is denied by many. Two blocks of stone which, like " dog-knee " levers, grip a third between them, may form kinematically the same combination as the train shown in Fig. 11 ; the so-called Rocking-stones, which have been weathered into existence in many places, are formed like balance beams ; the Geysers of Iceland act in a way to a certain extent resembling the steam-engine, forcing the water through distinct vertical tubes formed by stalactitic deposits ; from all these we cannot withhold the name of Machine. I mention these things, however, merely to show the availability of the word for our purpose, for the strictly scientific meaning of the name employed cannot be a matter of indifference to us. It is far from my intention to urge the employment of the name in cases in which its use is of no importance. But the examples just given come as well within our definition of a machine as within the above demonstration of its general nature. They show also that, in spite of the non-employment of the name, it is yet perfectly correct in the circumstances we have supposed ; it serves, that is to say, to indicate that they jointly possess the characteristics summed up in our definition.

We have already seen how a mechanism becomes a machine.

In its complete form the machine consists of one or more mechanisms, which can, in the way we have already pointed out, be separated into kinematic chains, and these again into pairs of elements. This separation is the analysis of the machine, the investigation of its kinematic contents, arranged in mechanisms, kinematic chains, and pairs of elements. The reverse of this operation is synthesis, the placing together of the kinematic elements, chains and mechanisms, from which a machine can be built up so as to fulfil its required function.

There is a large region among the exact sciences in which analysis and synthesis can exist without each other, where at least important results can be obtained by the use merely of deduction from fixed general laws. In our case, however, the two intellectual operations cannot go on separately, because the machine never, or scarcely ever, comes to us as a ready-made production of nature, but as something which we ourselves have made,—because, that is, it has been created by us essentially by a synthetic method. The induction by which we have arrived at it has often been very indistinct, and hence deduction and analysis are or must become means enabling us to reach it by an induction or synthesis which is conscious and definite.

The synthesis is here, as in most cases, by far the more difficult of the two processes. On this account it has scarcely ever been undertaken other than empirically. Its province is simply that which is assigned, in common language, to invention, and about which we spoke at length in the Introduction. Essentially, invention is nothing other than induction, a continual setting down and thereafter analysing of the possible solutions which present themselves by analogy. The process continues until some more or less distinct goal is reached,—a goal which generally seems itself to be indefinite on account of the haziness which envelopes the whole procedure. In this way a result lying close to the starting point is too often reached only after traversing a whole labyrinth of solutions each one depending upon the one before it, and each thrown away as soon as it has been found. I do not doubt that many of my readers, who have spent hours and days poring over mechanical problems, have found, after many laborious trials, that they have been travelling through a circle of experiment only to reach some well known, but unfortunately not so well recognised, problem.

The chief cause of all this trouble is that the mechanisms are not seen, or not recognised, because their proper nature, the kinematic linkwork with its laws, has not been present in the thoughts of the mechanician. The acquaintance with this would, in nine cases out of ten, have shown him any near-lying result immediately, and would have greatly shortened the way to results further off. For the scientific theory of mechanisms, if it give a complete mastery over analysis, sweeps entirely away a great portion of the difficulties, and entirely alters the nature of those which remain. While the empirical method is only a groping in the dark in the hope that by good luck we may lay hold of the solution, we come here to the application of an inductive method, based upon a well-understood analysis. The difficulties now consist only in the increased demands upon the capacity for induction. In this itself, however, Kinematics follows its own strict laws, like all other sciences. There will be further frequent opportunity for showing how great the difference is between this method and the old one,—at this point I can only place it before the reader in general propositions.

We see now the Machine-problem theoretically solved, or in other words, we have the general features of the method of solution sketched in an abstract form before us ; these point out the direction in which we must work. The general propositions laid down as to pairs of elements, chains and mechanisms, are, as it were, only the titles of volumes as yet unopened, the contents of which we must now commence to study page by page ; for it is necessary, in order that our solution of the problem may be brought down from general first principles to their detailed applications, that the latter should be carefully examined. This study we shall begin in the following section.

It can be readily understood that such an investigation is neither simple nor easy, to me at least it does not appear possible to pass quickly over such wide-reaching questions. Whoever attentively examines the nature of the machine, discovers in it so many phenomena having mutual relations difficult to understand that he cannot penetrate to the deep under-lying laws which connect them, and he comprehends how it has often taken the whole power of single men to carry forward even one step some of the problems which present themselves. When we consider the spinning-machine, for example, which has been gradually developing for three generations into its present form, (one still capable of further improvement), notwithstanding that the best mechanics have worked at it;—or look at the changes through which the sewing-machine has passed, and examine each step by step, we can form some conception of the difficulties which the theory has to overcome. In addition to this, the propositions to be developed are completely new. They therefore require that numerous details should be carefully entered into, of which some may appear to the engineer to be already well understood, although in reality the laws upon which they depend have not yet been investigated, and in the light of these they may be seen in many new aspects. It will therefore be some time before we arrive at such propositions as are adapted for direct application.

When, however, we have gone so far as to have demonstrated these existing laws and their mutual relations, we shall have reached the limits up to which theory can be our guide.

For the right application of these laws demands certain special qualities in the designer of a machine besides a mere knowledge of its theory, if his work is to be what is called " practical,"—by which is meant that the required object is to be fully and permanently attained, without too great an expenditure of means. This art of making practical work can be but very partially communicated by teaching, it can only be made quite clear by example. The scientific abstraction only serves to show the possibility of the machine, it affords no means whatever of judging between " practical " and " unpractical." This is often cited as an essential imperfection of theory, a notion which only arises from an obstinate ignoring of its real province. We have separated the department of practice from that of abstract theory in order to see more clearly the complicated course of our subject. Every time, however, that we have to choose between the useful and the useless, we are compelled to return from the abstract to the concrete. In the school, therefore, kinematic science must frequently be connected with its practical applications—it has not only to show what theoretical solution applies to problems already solved empirically, but in most cases to construct the theory as well as to find it. It is remarkable that there is scarcely any kinematic problem, scarcely any turning, however bold, in the theoretical propositions, for which we cannot find an example in practice. It must not, however, be considered that theory has only and always to limp behind practice, as is too often the case; it may rather be said to comprehend in itself all the mutual relations of the laws which in their application constitute this practice; it raises a clear flame out of each spark of truth, and so renders possible new and various roads to its higher development. The attitude of Theory and Practice to each other, in connection with the Machine, must be one of mutual respect.

CHAPTER II.

PHORONOMIC PROPOSITIONS.

§ 4.

Preliminary Remarks.

THE science of Mechanism is a derived science, and has—as we have already said — its foundations in Applied Mechanics and Mathematics. It chooses from the first a special part, in considering chiefly those motions which occur through latent forces, and are therefore conditioned by the geometric properties of the pieces transmitting these forces. Its problems therefore lie, for the most part, in a well defined portion of Mechanics—the geometrical. This department does not always receive the same name in scientific books; it is best called Phoronomy. Frequently it is simply called Kinematics, but this seems to be a misunderstanding of the word. Ampère at least, who invented the name, did not intend it to be used in this sense. It is at the same time unnecessary so to employ it, for Phoronomy is quite sufficient, and is besides more distinctive than Kinematics. Phoronomy is, therefore, the study of the measurement of the motions of bodies of every kind, and has become specially the study of the geometric representation of motions. We therefore retain the designation Phoronomy, and as we shall frequently have to employ its propositions, especially so far as they concern rigid bodies, it will be well here to review a portion of them.

The purely mathematical part of the phoronomic propositions coming thus into consideration forms what Professor Aronhold has called "Kinematic Geometry." It is, however, really a part of Phoronomy, and there does not appear any reason why it should not receive that name. If along with this the masses of the moving point-systems be also taken into consideration, we have "Phoronomic Mechanics." This is unquestionably what the later French writers (see p. 16) mean by "pure Kinematics." I shall be glad if my view of the matter be generally accepted, so that some sort of common understanding may be reached as to the general directions of these studies. It is greatly to be wished that some end could be brought to this multitude of new names.

Until recently only the lower and simpler part of Phoronomy, that namely which relates to the motion of a point, has been systematically taught in our German Polytechnic Schools; and this has been so frequently treated in text-books as to be familiar to all practical mechanicians who take any interest in the theory of their subject. Problems connected with point-systems have been only occasionally treated—in the text-books familiar to practical men they appear but seldom, and then rather as interesting corollaries than as important problems. These problems, however, are of the highest importance in Kinematics, and it is only to them that we need turn our attention here. I must therefore suppose that the following propositions, here taking their place in our investigations, will be in great part new to my Engineer readers. One of the most important characteristics of the method of treatment we shall employ is that it enables us to make the progressive changes of position visible in form to the imagination ; I have tried, wherever it has appeared possible, to develope this conception still more fully than has hitherto been done.

§ 5.

Relative Motion in a Plane.

We are unable to grasp with our senses the absolute motion of a point; we observe only the relation of its successive positions to other points or bodies in our neighbourhood. This relation, when known for every instant, is called the relative motion of the point—or if instead of a point we have a body, the relative motion of the body,—to the portion of space surrounding us. If this were itself motionless, the absolute and relative motions of the point would be identical, but if not, they must differ. Absolute motion in the universe being of no importance in our investigations, we may limit the meaning of the term absolute motion, and understand by it only motion relative to the portion of space in which our observations are made,—to the earth, for instance, or to a ship or a train.

We shall examine first the case in which this portion of space is a plane merely, the motions to be considered being thus motions in a plane.

Prop. I.—The motion of a point P relatively to another point Q in the plane $P Q$ takes place along the line $P Q$ which joins those

$$\underline{P \qquad\qquad\qquad Q}$$

Fig. 13.

points, no matter what motion the point may have relatively to the plane itself. The motion of P relatively to Q and of Q to P is known when the distance $P Q$ is known for every instant. This first proposition is not limited to motion in a plane, but is entirely applicable to the general motion of two points in space.

Example.—The motion of the centre (P) of a planet relatively to that (Q) of a body around which it revolves in any orbit, is an oscillation along the line $P Q$ joining their centres.

Prop. II.—The motion of a point P relatively to a plane in which it moves* is known if its motion relatively to two fixed points A and B in the plane of motion be given.

The path of motion is then the locus of the vertex P of the triangle $A P B$, which takes, for instance, the position $A P' B$ (Fig. 14).

Example.—The motion of any point P in an ordinary connecting-rod relatively to the plane in which it swings, is a curve which can be

* In Fig. 14 the plane of the paper.

determined by distances measured from any two points in the section of the frame traversed by that plane.

Prop. III.—The motion of a plane figure relatively to a plane in which it moves is known if the motions of any two of its points P and Q (Fig. 15) relatively to two fixed points A and B in the plane of motion be given. For when the positions of P and Q are known, the positions of every other point in the moving figure

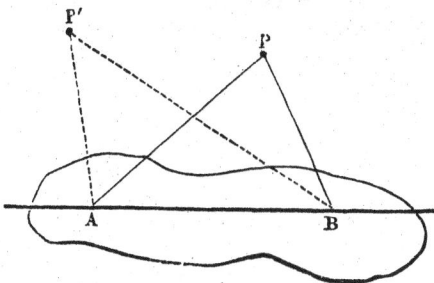

FIG. 14.

can be found by considering it as the vertex of a triangle of which the position of the base and the lengths of all three sides are known. The motion of any plane figure may therefore be expressed by that of any line in it. The motion of the line PQ relatively to the line AB is therefore the same as that of the figure PQ to the figure AB.

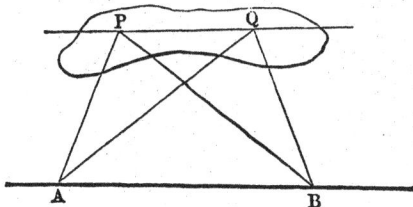

FIG. 15.

Example.—In order to determine the motion of a plane equatorial section of any planet relatively to the plane of the solar system (assuming the section to move in that plane), the motions of at least two points in that section relatively to the plane of motion must be known. Visible inequalities on the surface of a star, or a spot on the sun, may be supposed to furnish points relatively to which the motion of the section can be observed.

Prop. IV.—The motion of a plane figure $P\,Q$ relatively to one point A in its plane of motion is expressed by that of the points P and Q relatively to A, and this, as we have seen (Prop. I), takes place along the lines joining P and Q with A. But the position of these lines in the plane remains indeterminate, so that complex motions in the plane may occur without any alteration in the lines defining the motion relatively to the given point. [Thus in Fig. 16 the triangle $P\,Q\,A$ is similar and equal to $P'\,Q'\,A$, so that the position of $P\,Q$ relatively to A is the same in both cases; its position in the plane is, however, very different. The same is true of $P^2\,Q^2$ and $P^3\,Q^3$.]

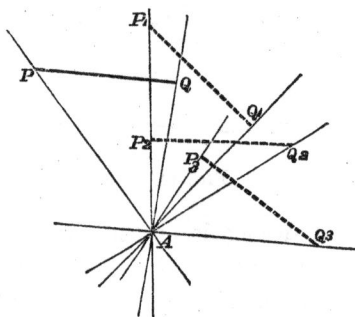

Fig 16.

Example.—A kinematic chain adapted for motion in a plane, but having one point only fixed in that plane, gives no determinate absolute motion, although the motion of every point in it relatively to the chain may be absolutely fixed, the chain being closed. The apparent contradiction that certain recent parallel motions require only one fixed point arises from a misapprehension of their nature. If only one point, or more strictly one axis, of such mechanisms be fixed, no "constrained" motion of the other parts occurs.

§ 6.

Temporary Centre; the Central Polygon.

The foregoing four propositions constitute the groundwork of the Phoronomics of point-systems, and are in certain respects exhaustive. They give, however, no distinct idea of the way in which the relative positions assumed by the moving point or body follow one another. We must now examine this more closely.

If for any plane figure two positions PQ and $P_1 Q_1$ in the same plane be given, the figure can in every case be moved from the one position to the other by turning about some point O in the plane, which can be determined by joining PP_1 and QQ_1 and finding the intersection of perpendiculars drawn from the middle points of these two lines. This intersection, O, is the required point, because the two triangles OPQ and $OP_1 Q_1$ are similar and equal, OP being equal to OP_1 and OQ to OQ_1. The point O is called the temporary centre for the given change of position.

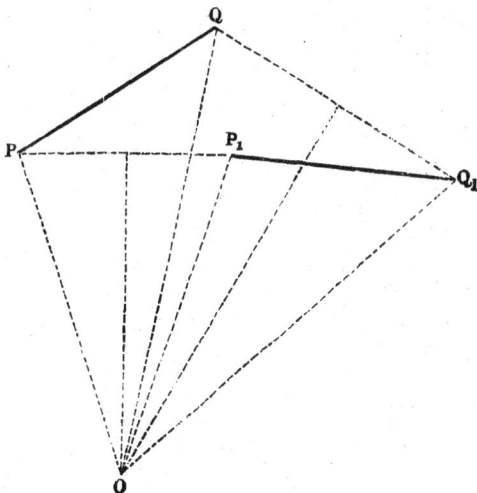

FIG. 17.

If the temporary centres for further changes of position,—from $P_1 Q_1$ to $P_2 Q_2$, $P_3 Q_3$ and so on, be determined in the same way, we obtain a series of points $O \, O_1 \, O_2 \, O_3$ etc., which may be joined together by straight lines. We thus obtain a polygon which has the centres for its corners, and which we may therefore call a central polygon. If the figure PQ return into its original place after a series of changes of position, the polygon is closed, otherwise it is open. The figure itself in every case makes a series of turnings about the temporary centres,—its points, that is, move always in arcs of circles; these are completely determined if the

angle through which each single turning takes place be given. This
angle we must therefore look at more closely.

Rotation about the centre O extends through the angle $P O P_1 =$
ϕ_1. It will help us in our examination of the matter if we suppose
the line $M M_1$,—which is equal to $O_1 O$, and which is so placed
that $\angle O_1 O M_1 = \phi_1$, the point M coinciding with O,—to be
rigidly connected with $P Q$. Then in the first turning the line

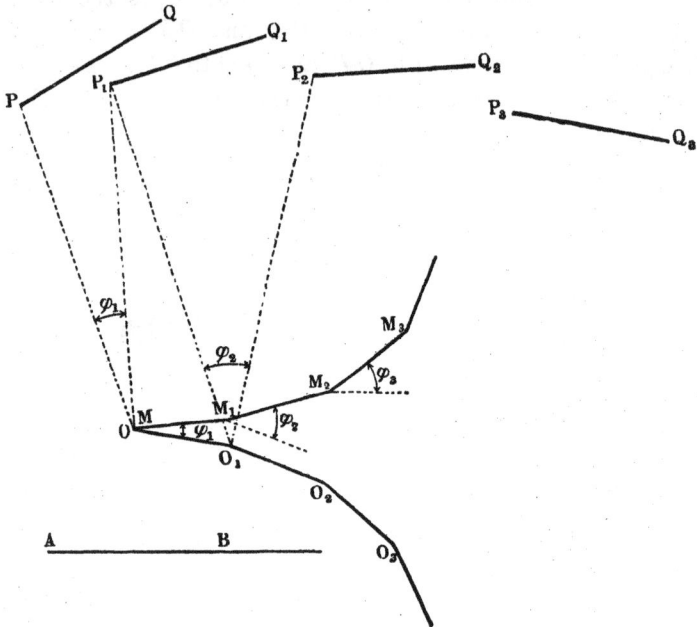

Fig. 18.

$M M_1$, turning about O, will take the position $O O_1$, and at the
same time $P Q$ (with which it is rigidly connected) will be moved
as before into the position $P_1 Q_1$, so that so far as the determination
of the motion is concerned, $M M_1$ may replace $P Q$. If we repeat
the whole process for the rotation about O_1, by joining the line
$M_1 M_2 (= O_1 O_2)$ to $M M_1$ in such a way that when M_1 coincides
with O_1, the angle $O_2 O_1 M_2$ is equal to ϕ_2, we can again replace
the figure $P Q$ by $M_1 M_2$, or rather by the polygon $M M_1 M_2$.
In this way a second polygon, $M M_1 M_2 M_3$, can be found, which
by the consecutive turnings of its corners about the corresponding

corners of the first polygon will give to $P \, Q$ the required changes of position relative to the fixed plane, or to any stationary figure, as $A \, B$, lying in it.

If we examine the relation of the two polygons to each other we notice the special and important peculiarity, that each has the same properties in reference to the other, that is, that they are reciprocal. Thus in any one of the positions in which two corresponding sides coincide, the polygons show not only the position of the figure supposed to be movable relatively to the fixed figure, but also conversely the position of the fixed relatively to the movable figure. (Prop. III. of § 5.) We can thus determine as many relative positions of the two figures by means of the central polygons as the latter have pairs of corresponding sides.

§ 7.

Centroids; Cylindric Rolling.

The method above described affords us the means of representing a succession of given separate positions of two figures. It leaves, however, the actual changes of position undetermined, or substitutes for them a series of rotations about isolated points. But if we suppose the assumed positions $P Q$, $P_1 \, Q_1$, $P_2 \, Q_2$, etc., taken nearer and nearer together until at last the intervals between them disappear, we shall have a complete representation of the whole motion. The corners of both the central polygons will at the same time have approached each other until each is removed from its neighbour by only an indefinitely small distance, and thus the two polygons become curves, of which infinitely small parts of equal length continually fall together after infinitely small rotations about their end points,—which, that is to say, turn or roll upon each other during the continuous alteration in the relative positions of the two figures. The turning which takes place about each point in the curves is not now, as before, temporary, but in general for an instant only, and each point is therefore called an instantaneous centre. The curves into which the polygons are transformed both pass through the whole

series of instantaneous centres point by point, and on this
account we may call them the centroids of the moving figures.
If these be known for any given pair of figures, their relative
motions for a series of positions infinitely near together are also
known,—their changes of position are completely determined,
and can be found by rolling one of the centroids upon the other.

It will be evident from the foregoing that in general the
relative motions of two plane figures to each other are not alike,
for none of the conditions of the problem necessitate the similarity
of the centroids;—whenever the centroids are similar, however,
the relative motions become the same.

Example 1.—The construction of trochoids illustrates the relative motion
of plane figures of which the centroids are known. If a circular cylinder
roll upon a plane, the normal sections of both figures move in a common
plane, and therefore come within the conditions of our problem. The
circle *P Q* and the straight line *A B* (the forms of these sections) are the
centroids both of the two figures and of all figures or points connected
with them. All points of *P Q* describe linear trochoids [13] relatively to *A B;*
these being common, curtate, or prolate, according to whether the point lies
upon, without, or within the circle. All points connected with *A B*
describe involutes relatively to *PQ;* these again being common, curtate,
or prolate, according as the describing point lies upon, beyond, or within
the straight line.

Example 2.—Two equal circles rolling upon one another have the
same relative motions ; points in both at equal distances from their centres
describe equal epicycloids.

Our examination applies generally to the relative motions of
plane figures in a common plane, or, as we shall in future call
them shortly, con-plane figures, and the result of it may be
summed up as follows:—

All relative motions of con-plane figures may be
considered to be rolling motions, and the motion of
any points in them can be determined so soon as the
centroids of the figures are known.

If solid bodies be laid through the supposed figures *P Q* and
A B, and rigidly connected with them, then every pair of sections
of such bodies which (like the pair of figures) lie parallel to

the plane of motion have a pair of centroids identical with theirs. The series of centroids, which we may suppose in this way to be lying closely one behind the other, form together two cylinders (in general non-circular), which always touch along one line, and turn or roll upon one another. Each line in which the cylinders come in contact for an instant is for that instant the axis of rotation, and is called therefore the instantaneous axis of the motion. The motion itself in such a case is called a cylindric rolling. We may extend the law just enunciated for plane figures equally to this relative motion of solids. The characteristic of a series of reciprocal positions of a body undergoing cylindric rolling is that its sections normal to the instantaneous axis are figures which remain always con-plane during the motion. We can therefore say :—Every relative motion of two con-plane bodies may be considered to be a cylindric rolling, and the motions of any points in them may be determined so soon as their cylinders of instantaneous axes are known.

§ 8.

The Determination of Centroids.

With the transition from irregular to continuous motion the perpendiculars (see Fig. 17) upon the lines joining pairs of consecutive positions of the points P and P_1, Q and Q_1, become normals to the curve-elements in which at the given instant the points P and Q are moving. In order therefore that the centroid for the motion of a figure $P Q$ relatively to another $A B$ may be known, there must be known for every position of $P Q$ the directions in which at least two of its points are moving, that is the position of the tangents to their paths. The normals to such tangents for any number of points in the moving figure all intersect in the same point, from which it follows that only one pair of centroids is possible for any relative motion of con-plane figures.

Centroids can always be found by determining separately a sufficient number of points in them, and often by a general

investigation into the nature of the curves to which they correspond. We shall examine briefly both methods of determination.

Let the relative motion of any two con-plane figures PQ and
AB be known. In order to find the corresponding centroids we
must first convert this given relative motion into an absolute
one (in the limited sense in which we use the word), by supposing the system as a whole to receive such a motion that
one of the figures, *e.g. A B.* (Fig. 19), comes to rest in reference
to ourselves. We can then find the paths or curves in which

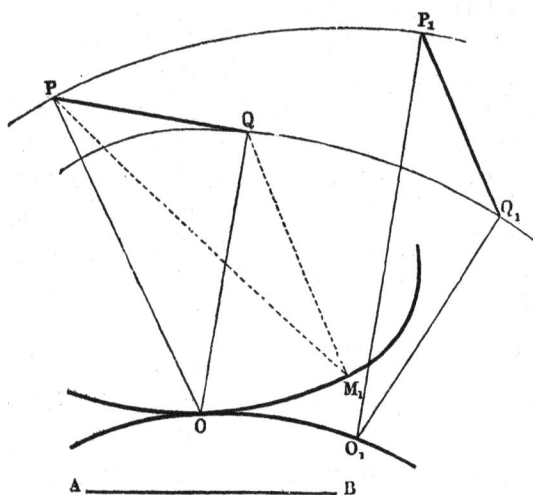

FIG. 19.

two points P and Q move, and draw normals to those paths at
P and at Q; their intersection gives a point O in the centroid
belonging to AB. Another pair of normals, drawn from P_1 and Q_1
give another point O_1 of the same centroid, and so on. The
second centroid, OM_1 . . . . . can be found in a similar way by
bringing PQ to rest and making AB the moving figure,—but it may
be determined more easily as follows. The instantaneous centre O_1
is a point in both curves when the moving figure is in the position
P_1Q_1, its distance from the point P in that figure is O_1P_1 and from
Q, O_1Q_1;—it therefore is necessary only to describe arcs of circles
about P and Q with the radii $= O_1P_1$ and O_1Q_1 in order to find

the point M_1 in the second centroid which corresponds to O_1, and which must be at their intersection.

Example 1.—The nature of the kinematic chain (Fig 20) discussed in § 3 allows none but con-plane motions; it may therefore serve us here for an illustration. Let it be required to find its centroids. As each of its links can have a motion relatively to all the three others, there are in all six pairs of centroids belonging to the chain, four for the motions of adjacent and two for those of opposite links. The four first are very simple, each curve being a point only; the two others are not so readily found. We will here examine the pair of centroids belonging to the links a —— h and d —— e. For this purpose we first bring the link a —— h to rest (we may suppose it connected with a fixed pedestal, as shown in Fig. 21); then a —— d rotates about a, while e —— h swings to and fro in circular arcs about h. The centres of the elements c and f

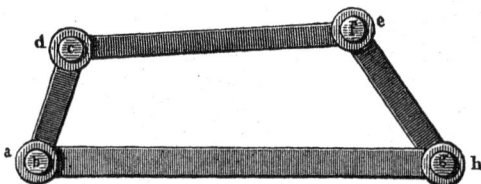

Fig. 20.

describe therefore paths to which the normals are always radii passing through the centres of a and h. By producing these radii until they intersect we can consequently obtain any number of points in the centroid of the fixed link a —— h. The curve found in this way is shown in Fig. 22. Q or M is the instantaneous centre for the original position a —— d —— e —— h obtained by producing a —— d and h —— e until they cut each other. The whole figure O O_1 .... O_4 is not simple in form. It contains four infinitely distant points, corresponding to the two parallel positions of a —— d and h —— e. The second centroid, that of the link d —— e, drawn in the way above described, is shown in M M_1 .... M_4; it also contains, necessarily, four infinitely distant points. The two centroids which in the engraving touch at O M, roll upon one another as the mechanism moves (O O_1 .... O_4 remaining stationary), and supply completely the means of examining the whole complicated motion of the link d —— e. As regards easy comprehension this geometric representation still leaves something to be wished,—the infinitely distant points impair its clearness not a little.

But the principal question here is not whether the solution be easy or difficult to comprehend, but whether it be a real and complete solution of the problem. We shall have occasion to point out, further on, the way in which very intricate cases may often be made very simple and easy to realise.

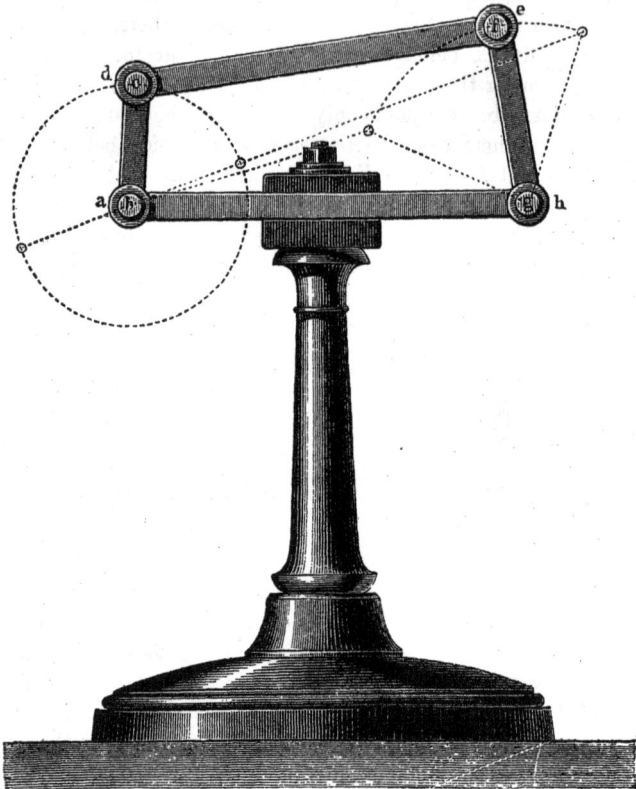

FIG. 21.

Example 2.—A very simple illustration of finding centroids by general investigation occurs in common spur-gearing, or generally in any two bodies which turn about parallel axes at a fixed distance apart with a uniform velocity-ratio. If *a* and *b* (Fig. 23) be two con-plane sections of such bodies, and *c* and *d* the two centres about which they revolve, then if we consider *a* as fixed, the point *d* must move in a circle round *c*, the distance *d c* being unalterable, and at the same time *b* must be turning about the as yet unknown instantaneous centre. But the normal to the path of *d* must always coincide with the line of centres *d c*; the instantaneous centre must therefore lie in that line or its prolongation. Assume, provisionally, that O is the instantaneous centre, and suppose the line $c\,d$ to be fixed, so that both a and b may revolve as they did at first, then the two centroids, both moving, roll upon one another in O, and have therefore

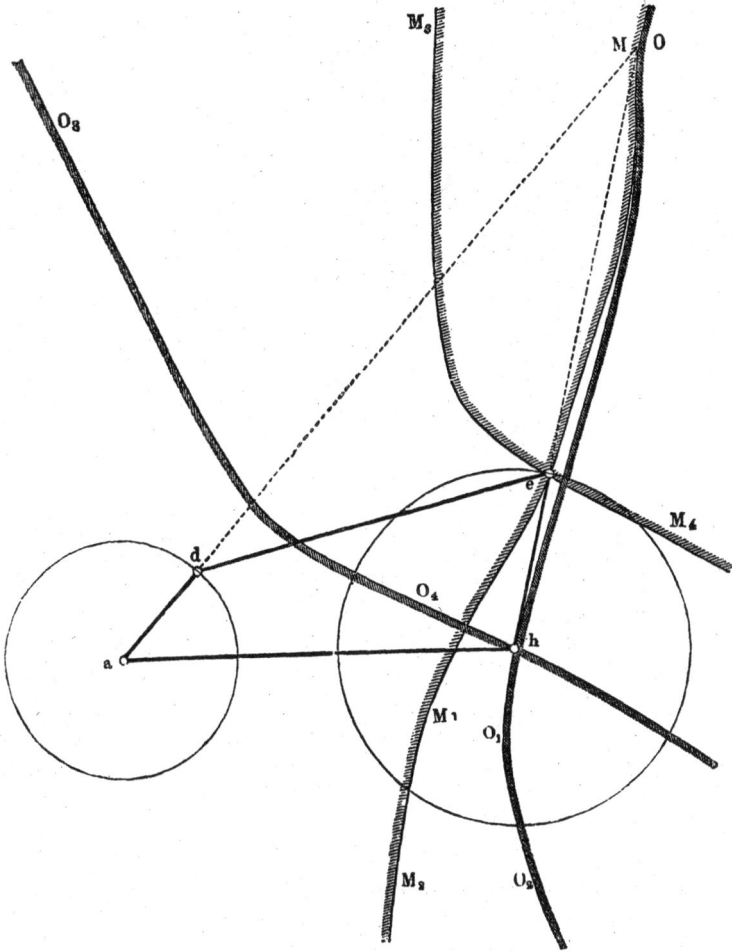

Fig. 22.

at that point the same peripheral velocities. Their angular velocities will have to each other the ratio $dO : cO$. But by hypothesis this ratio is constant, therefore also dO and cO themselves must be constant; and the centroids become circles described about c and d with radii which are in inverse proportion to the angular velocities of the bodies. The pitch circles of spur-wheels are thus simply the centroids of their normal

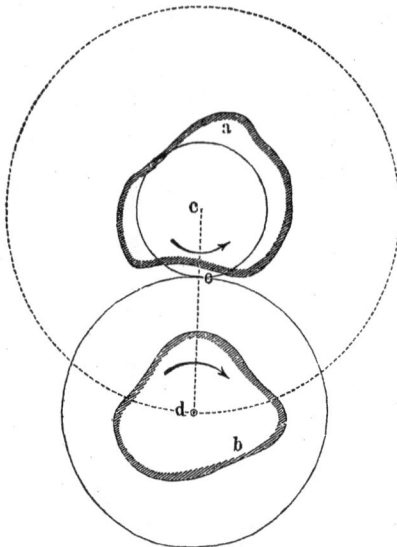

FIG. 23.

sections. We touch here, consequently, upon a case in which the centroids are of peculiar simplicity, and which moreover has the most extended practical application in the construction of machines.

§ 9.

Reduction of Centroids.

The pair of centroids which we found in the first example of § 8 completely determines the relative motion between the links d —— e and a —— h of the given kinematic chain, and must therefore equally express that motion, whichever link of the chain be fixed or in whatever way it be set in motion. Let us suppose the chain to be arranged in the first of the four methods given on page 47, that is by making b —— c the fixed link; we can obtain in this way another important mechanism, one in which the links, a —— h and d —— e both describe circles, connected with each other always by the link $f \text{——} g$. The mechanism is that known as a drag-link coupling. The arms or cranks $d \text{——} e$ and $a \text{——} h$ revolve with a varying velocity ratio which can be

Fig. 24.

ascertained for each particular position from the corresponding radii of the centroids. If we imagine, for instance, the centroids shown in Fig. 22 to be fixed to the arms to which they belong, we can see that they will both turn about the fixed points a and d, and so roll upon each other like the pitch lines of non-circular spur-wheels.[9] It is very difficult, however, to realize

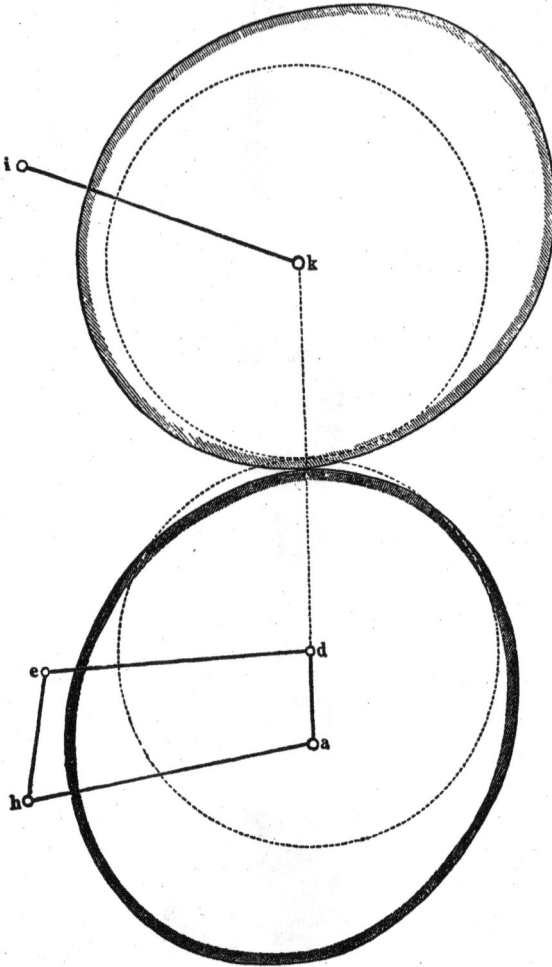

FIG. 25.

the matter from this point of view on account of the peculiar form of the centroids, and some method which can be more easily comprehended, for showing at least the velocity-ratio of the arms, is required. For this purpose let us suppose a cylindric spur-wheel fixed to the axis of one of the arms, as d —— e, and gearing with another equal and similar wheel upon a third axis k This axis then turns with precisely the same velocity as d —— e but in the opposite direction. If now the centroids be found for the motion of an arm $i\,k$ upon the axis k relatively to the first arm a —— h, these can evidently,—so far as the velocity-ratio between a —— h and d —— e goes,—take the place of the less easily comprehended centroids of Fig. 22. We transform, as it were, the first two centroids into two new ones. Fig. 25 shows these transformed or reduced centroids for this special case. If these be considered as the pitch lines of non-circular spur-

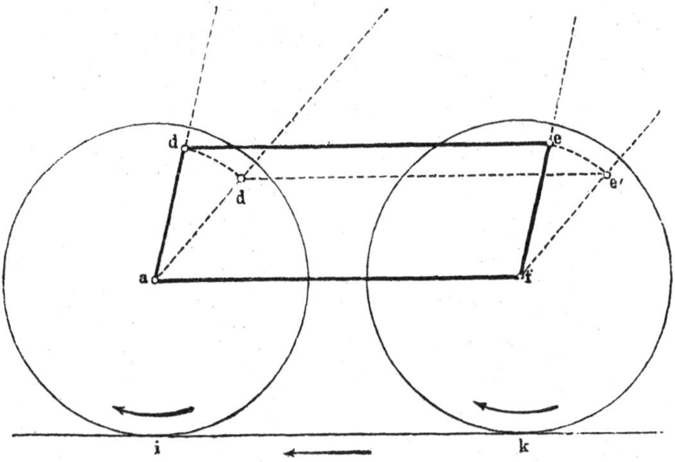

FIG. 26.

wheels, we have at once an easily understood representation of the communication of rotation between a —— h and d —— e. We shall return later on to the methods of drawing such reduced curves. It is sufficient just now to point out that here the **sum** of the instantaneous radii is constant (being $= a\,k$), while with the original centroids—Fig. 22—their difference was constant (being $= Oa\text{-}Od = a\,d$). The infinitely distant points have disappeared, as will be seen, and the whole representation is very simple and can very frequently be employed.

The infinitely distant parts of centroids may under some circumstances be even more troublesome than in the case we have supposed, where they can to some extent be used through their asymptotes. If for example the opposite links of the mechanism were made of equal length (Fig. 26), the centre lines of the four arms would form a parallelogram, and their intersections would always be at an infinite distance; so that both centroids become infinite and therefore cannot be drawn. The method of reduction which we have just used, if applied to this case, gives us two rolling circles of equal diameter. We can here, however, obtain the same result in a still clearer manner. For this purpose let equal circles i and k, Fig. 26, be described about a and f, with radii less than half as long as a ——f, and let a straight line $i\,k$ touch them both externally. If now this line move without sliding upon the circles, they will turn about their centres a and f in the same way as they would if they were connected with the arms of the

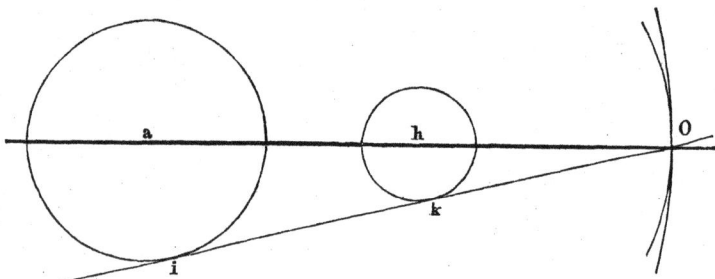

FIG. 27.

parallelogram,—that is, with a uniform angular velocity-ratio equal, in this case, to unity. Thus the three centroids, to which we have here reduced the original infinitely large curves, completely express the relative motion which it was desired to represent. Further, we are able to imagine them existing and moving simultaneously with the original curves. They second, as it were, the movement which has been set up, for which reason we may call them secondary centroids. It will be noticed that in this case we have not two but three connected figures, which is noteworthy; for we already know (§ 8 above) that only a pair of (primary) centroids accompanies any relative motion of con-plane figures. That more than a pair of mutually rolling figures should result from the secondary representation of such a motion is not peculiar to this particular case, but is general.

Secondary centroids are of service to us also in many cases where it is possible, and even easy, to draw the primary curves, but where they would be inconveniently large; as for instance in the case of a pair of bodies which revolve in the same direction with a uniform angular velocity-ratio not equal to unity. Such bodies would have as centroids a pair of circles of which one would touch the other internally. Their secondary centroids would be circles whose diameters bear the same ratio to each other as those of the original curves,* by which means the position of the tangent

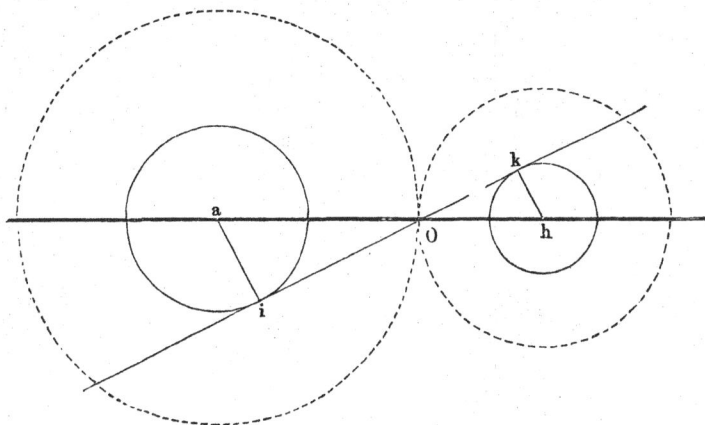

FIG. 28.

i k can be determined, as in Fig. 27. These centroids are also suitable for cases in which rotation with constant-velocity-ratio takes place in opposite directions, as in Fig. 28. In the various methods for drawing the forms of the teeth of spur-wheels,[10] these secondary centroids often play an important and acknowledged part. They have thus already made their way into practice, and enable us to reduce a much used practical method to general phoronomic principles.

---

* The secondary circles being drawn, the point O in which their tangent cuts the line of centres is always the point of contact of the primary centroids; and from this, conversely, the secondary circles can be drawn as circles touching any line *i k* passing through it. It must be remembered that the point of contact of the centroids is always the instantaneous centre for the motion actually occurring.

§ 10.

Rotation about a Point.

Having now investigated the general methods of representing relative motion in a plane, we must proceed to consider the more difficult problem of relative motion in space; in the first place, however, with the limitation that one point of the body which we are considering be supposed not to alter its position in space in reference to us. When a body moves in such a way that each of its points remains always at a constant distance from some one fixed point, it is said to turn about that point. In order to find the motion which thus takes place relatively to a stationary body rigidly connected with a fixed point, let us describe about that point A, Fig. 29, a sphere of such a size that it shall pass through

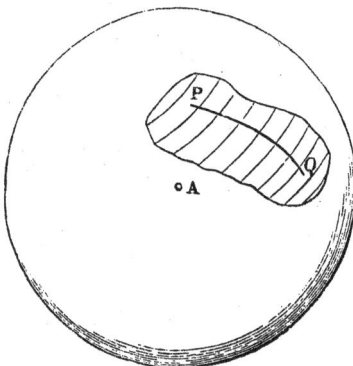

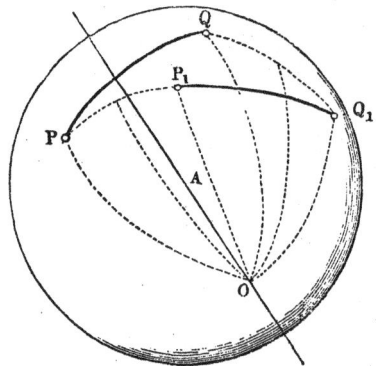

Fig. 29.   Fig. 30.

the moving body, giving us a spheric section of it, PQ. If we then know the motion of such a sectional figure upon the sphere, it is evident that we shall know the motion of the body itself. But the motion of the figure PQ is known if all the positions of two of its points, as P and Q, or of the great circles connected with them, be known. For from the position of this curved line the positions of all other points in the figure may be found by considering them as vertices of spherical triangles, of which the position of the base (PQ) and the magnitudes of all three sides are known. Thus on investigation we see that the motion of the figure about a fixed point reduces itself simply to that of a circular arc, PQ, upon a sphere, and, generally, that we may determine the motion of any spheric figure by that of any arc lying within it, just as we have already found that in con-plane motions we could replace a figure by a line.

Every spheric figure, as PQ in Fig. 30, which moves upon the surface of a sphere, can be moved from one position PQ into another $P_1 Q_1$ through spheric turning about some point O on the surface of the sphere; and this point can be determined by finding the intersection of two great circles passing through the centres of the lines PP_1 and $Q Q_1$, and perpendicular to those lines. The point of intersection is then the required instantaneous centre O, because the spherical triangles $O P Q$ and $O P_1 Q_1$ are similar and equal-sided. The point O is the temporary centre for the supposed spherical turning. The two great circles cut each other twice, once at O, and once at the other end of the diameter passing through O.

But by hypothesis the distance of the figure PQ from the fixed point A is constant, and therefore the diameter passing through O and A is stationary during the turning relatively to the figure, and so becomes the temporary axis of the assumed motion.

A new turning supplies a second pole O_1, another, a third pole O_2, and so on, and by joining these with arcs of great circles we have a spheric central polygon. A second spheric central polygon, rigidly connected with the moving figure, corresponds to the first. If a series of straight lines be drawn passing through the corners of these polygons and the fixed point A,—that is, a series of diameters of the sphere passing through these corners,—we obtain two pyramids, about the angles of which the separate turnings take place.

§ 11.

Conic Rolling.

It will be seen that the method we have here used bears the most complete analogy to that employed in the consideration of motions in a plane. If we continue the process further by supposing the consecutive positions of PQ to be infinitely near together, the spheric central polygons become spheric centroids, the temporary axes become instantaneous, and the pyramids become cones (in general non-circular), which have a common vertex at A, and which roll or turn upon one another. The cones are c o n e s of i n s t a n t a n e o u s axes, and the whole motion is called c o n i c r o l l i n g. We arrive therefore at the following law, connecting together the phenomena we have been considering : All relative motions of two bodies which have during their motion a common point, may be considered as conic rolling, and the motions of any points in the bodies are known so soon as the corresponding cones of instantaneous axes are determined.

It is evident that our former examination of the methods of de-termining centroids and reducing them applies equally to conic and to cylindric rolling, so that it is not necessary to reconsider these matters here.

§ 12.

Most general Form of the Relative Motion of Rigid Bodies.

If the positions of three points in a body be known, the positions of all other points may be found by making them the vertices of triangular pyramids of which the magnitude and position of the base are fixed and the length of the edges known. We can there-fore determine the relative motion of any two rigid bodies by means of two fixed triangles, PQR and ABC, in them. Let the body ABC be brought to rest, so that only PQR moves relatively to us, and let the latter move into any position, as $P_1 Q_1 R_1$, Fig. 31. This change of position may take place in many ways. If, for example, we join P and P_1 by a straight line, and cause PQR to undergo translation parallel to it until P falls upon P_1 and the figure takes the position $P_1 Q' R'$,—we have only further to turn PQR about an axis SS (which may be found in every case), passing through P_1, in order that it may assume the required position $P_1 Q_1 R_1$. Thus in the most general case any motion of

P Q R relatively to *A B C* may be obtained by combining a simple translation and a simple rotation about an axis; and this in an endless number of ways. It is in no way essential that the translation should be parallel to the line joining the two positions of one of the points (as *P P₁* above). Among the infinite number of possible methods just mentioned one is of special simplicity, that, namely, in which the translation takes place

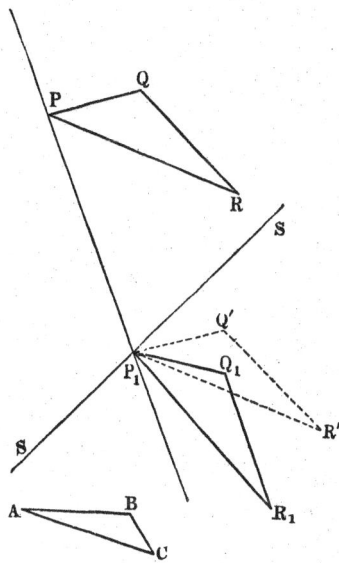

FIG. 31.

parallel to the axis of rotation. Here the motion resolves itself into an endlong sliding upon and rotation about one and the same axis; and if in such a case the changes of position of *P Q R* in reference to *A B C* be taken indefinitely small, the instantaneous axes of rotation, along which also simultaneous sliding takes place, become infinitely near together.

§ 13.

Twisting and Rolling of Ruled Surfaces.

Many attempts have been made to render easy the comprehension of the kind of motion just described, but the task is far from being a simple one. Poinsot proposed that the fixed body (or body assumed to be stationary relatively to the observer) should be imagined to be a screw, and the moving body a nut, in which case the sliding would take place along, and the rotation about, the axis of the screw, as above supposed. But as the motion takes place with variable velocity, both as regards the sliding and the rotation, the angle of the thread in the screw and the nut must be imagined to be continually changing. It is difficult to realize this distinctly: a form so irregular is no longer a body, it cannot be realized even with the most determined effort,—indeed, things of so varying a nature as are these screws and nuts can scarcely be grasped better than the abstract idea of rotation and translation in space.[11]

Belanger makes two proposals. The first is to imagine a pair of bodies having conic rolling (as above, § 11), in which both cones have a motion of translation in space. The a rotation then takes place through the conic rolling, and the sliding through the translation of the pair of bodies. By this means the motion may certainly be realized, but only by using three bodies to find the relative motions of two, a thing which is in certain cases advisable, and even necessary (§ 9), but which can only be justified when no simpler method is equally satisfactory.

Belanger's second proposition is to consider the consecutive positions of the axis as forming a pair of ruled surfaces, one for each body, so that the motion is reduced to a rolling of the two ruled surfaces upon each other, with a simultaneous endlong sliding upon each other of the generators which are in contact. Other later investigations have attached themselves to this method of representation. Indeed it follows, as a direct conclusion from what we found above, that the consecutive positions of the instantaneous axes of turning and sliding in each of the two bodies enclose such ruled-surface forms as solids of instantaneous axes.[12]

The special motion in which translation or sliding along a straight line and turning about it take place simultaneously is called twisting. We may now also, as we have arrived at the most general standpoint, indicate with a common name the bodies we have found, which, by their motions upon each other, determine the relative motions of the bodies with which they are connected. As the surfaces of these solids are always the loci of a series of axes, they may be called axoids. What we have

found may then be collectively expressed in the following law :—
All relative motions of two bodies may be considered
as the twisting or rolling of ruled surfaces or axoids.

The laws already found separately must follow from this general
proposition by a diminution and simplification of its conditions.
There is indeed no difficulty in realizing the transformation of a
ruled surface into a cone or a cylinder (of any form) where instead
of the twisting motion only rolling occurs. Nevertheless the conclusion must not be drawn from this that a pure rolling motion

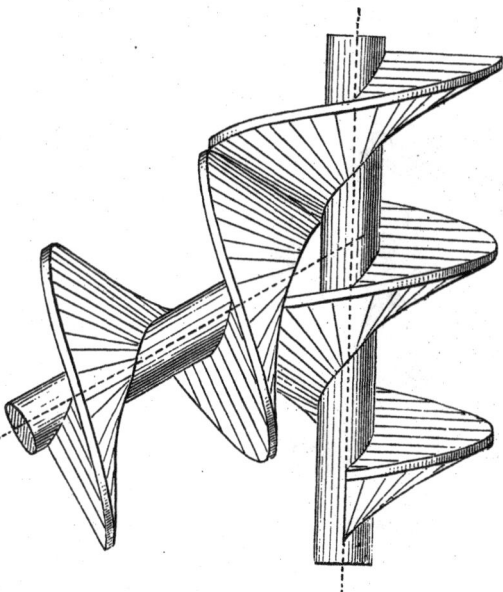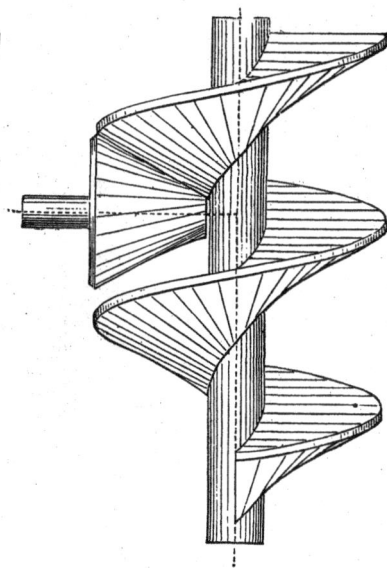

Fig. 32.   Fig. 33.

can be ascribed only to these two cases (as has been generally done
so far as I know). The condition determining the absence of the
sliding of the edges or generators is neither that all the axes
should intersect in one point, as in the cone, nor that they should
be parallel, as in the cylinder, but the higher condition that the
two ruled surfaces should be so formed that their infinitely near
generators in homologous positions include surfaces of the same
figure, or, in geometrical language, that they can be developed
upon one another.

The surface of any cone or cylinder can be developed upon that
of any other cone or cylinder (respectively), because the portions
of surface included between infinitely near generators in corresponding or homologous positions vary similarly. For the same
reason many other ruled surfaces can roll upon one another, their
forms being such as can be developed upon each other. Thus, for
example, two screw surfaces may be mutually developable, and if
placed suitably, as in Fig. 32, will roll upon each other; and
similarly, a screw surface and a hyperboloid, Fig. 33. Forms very
closely related to these are in actual use in machines;*—the consideration of them is therefore of practical as well as theoretical
interest.

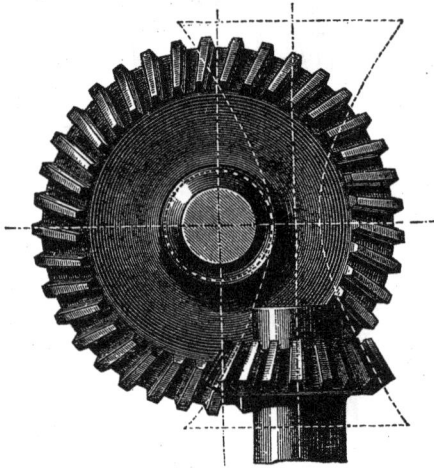

Fig. 34.

Ruled surfaces which twist upon one another are also employed
in machine construction; the axoids of hyperboloidal spur-wheels,
as in Fig 34, belong to this class. These axoids are not themselves
constructed, although in the case before us their nature is indicated
by the form and position of the teeth upon the wheel. This however is not the case in the pair of elements shown in Fig. 35, the
worm wheel and endless screw. The axoids representing the
motions of these elements are again hyperboloids,— but this is not

* See for example Johnson's *Imperial Cyclopædia,* "Steam Engine," Moison's
regulator, page 49.

recognisable in any way from their constructive form. Even the ratio between the pitch radii R and R_1 does not give us the relative diameters of the throats of the axoids. The wheels which I have elsewhere described* as hyperboloidal face gear, form a case in which the twisting motion becomes very visible. Here one axoid is a circular cone, the other a plane hyperboloid, or one in which the generator moves always in a plane normal to the axis of rotation (Fig. 36). The sliding of the instantaneous axes as they pass the line of contact is

Fig. 35.

here very distinct, especially near the vertex of the cone.

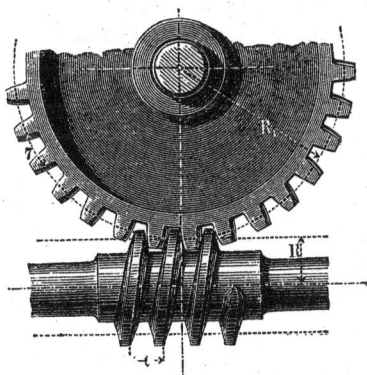

We have seen that all axoids belong necessarily to one particular class of geometric forms,— viz. ruled surfaces,— that pairs of such surfaces can, therefore, express all possible motions. We are therefore justified in considering these, in preference to any of the other geometric forms above mentioned (p. 80), as the general representatives of the motions occurring in machines. The most general characteristic of the relative motions of the axoids is their rolling. This exists even in the special case where of the two motions constituting the twist the

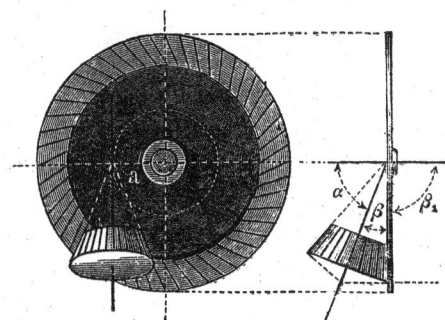

Fig. 36.

turning becomes infinitely small and the sliding only remains, for the latter may itself be considered as a turning about an infinitely distant axis,† and is therefore only a particular case of cylindric rolling.

* *Der Constructeur* ; 3rd Edition, p. 451.

† This is the special case of the proposition given on p. 61, where the two positions PQ and $P_1 Q_1$ are parallel, and where therefore the normals to PP_1 and QQ_1 intersect only at infinity.

Every motion which occurs in the machine thus connects itself with one leading idea, of which the single propositions considered contain special applications. Just as the old philosopher compared the constant gradual alteration of things to a flowing, and condensed it into the sentence: "Everything flows;" so we may express the numberless motions in that wonderful production of the human brain which we call a machine in one word, " Everything rolls." Through the whole machine, hidden or apparent, the same fundamental law of rolling applies to the mutual motions of the parts. The same idea, as we have already seen occasionally, can be extended to all the phenomena of kosmic motion, for our investigations do not merely cover the movements of the different parts of a machine, but are applicable generally to all moving bodies.

But the rolling geometrical figures which we can imagine to connect themselves with kosmic bodies are not in themselves constant. They are parts of the universal "flowing;" they alter themselves incessantly as the motions vary, now disappearing into nothing, now remodelling themselves into other continually changing forms, exactly determinable at every instant only in the rolling point itself. In the planetary motions also, that regularity which is capable of exact representation by axoids exists only approximately. In the machine, on the other hand, the rolling figures are rendered constant by artificial limitations of the motions,—this end at least is sought by all possible means, and practically attained,—so that, considered in the abstract, we are entitled to say that here this constancy exists.

Here these figures pass periodically through their mutual changes of position unnumbered times; they rest when the machine stops, and commence their play anew exactly as before so soon as the driving force again infuses life into the whole; one part only remains stationary, that which serves as a connecting piece between the machine itself and the unmoving space surrounding it.

For the practical mechanician, who has made himself familiar with modern Phoronomy, and still more for the theorist, the machine becomes instinct with a life of its own through the rolling geometrical forms everywhere connected with it. Some of these stand out in bodily form, as in belt-pullies or friction-wheels (as for example those of a railway carriage); others, such as spur-wheels, are slightly disguised in a very transparent veil; others still are closely drawn together in the interior of solid bodies which in their exterior forms scarcely give any indication of them, as in the case of the curved discs which we shall shortly have to examine more closely; and others, lastly, such as those of the link-work mechanisms, are widely extended, encircling the bodies to which they belong at a great distance, their branches indeed stretching to infinity, their outward forms quite undiscernible. These all carry on, partly before the bodily eye of the student and partly before the eye of his imagination, the same never tiring play. In the midst of the distracting noise of their material representatives they carry on their noiseless life-work of rolling. They are as it were the soul of the machine, ruling its utterances—the bodily motions themselves—and giving them intelligible expression. They form the geometrical abstraction of the machine, and confer upon it, besides its outer meaning, an inner one, which gives it an intellectual interest to us far greater than any it could otherwise possess.

CHAPTER III.

PAIRS OF ELEMENTS.

§ 14.

Different Forms of Pairs of Elements.

WE have already found, in the general solution of the machine problem (p. 35 *et seq.*), that the elementary—or what may be called the elementary—parts of a machine are not single, but occur always in pairs,—so that the machine, from a kinematic point of view, must be divided rather into pairs of elements than into single elements. It is the geometrical form of these pairs with which we must first of all make ourselves acquainted.

We shall in the first instance limit our investigations to rigid bodies,—that is, to such as possess approximately complete rigidity;—the problem before us in the construction of pairs of elements will then be the determination of a given or required motion by means of two such bodies or elements only. As we have found in the last chapter, these elements must satisfy the following conditions:—

1. That one element be fixed relatively to the surrounding portion of space,—itself assumed to be stationary;

2. That this element be so formed as to carry upon or within itself the envelope of the second and moving element, which

3. must be so arranged as to prevent every motion of the second element except the one which is required.

The stationary element then holds the moving one as it were imprisoned,—preventing every motion except a single one,—forcing every point in it, when it has begun to move, to travel in a determinate path, on which account such a pair of bodies may be called constrained. We referred in the last article to the immense number of forms which the relative motion of two bodies might take; remembering this, it is easily seen that pairs of constrained bodies may have very many geometrical forms. All pairs of such forms, however, which conform to the two last of the above conditions, have this in common, that they are envelopes, and indeed reciprocal envelopes, for the given motion,—which can be represented through their axoids. Hence they may, like their axoids, be more or less simple. We can imagine a case fulfilling both the conditions, and in which at the same time the one element not merely forms an envelope for the other, but encloses it,—in which, that is to say,· the forms of the elements are geometrically identical, the one being solid or full and the other hollow or open. Such a pair of bodies may be called a closed pair.

It is evident that, in their simplicity, closed pairs differ notably from pairs in which the elements are not identical in form. We shall on this account consider them separately and in the first instance.

§ 15.

The Determination of Closed Pairs.

The geometrical properties of the bodies from which closed pairs can be constructed are so well defined that we do not require to look first for these pairs in existing machinery, but may attempt to discover them by *a priori* reasoning.

Two bodies forming a closed pair cover each other with their surfaces; on these we may imagine any number of pairs of coincident curves, and among these some may be supposed to be such that the single motion possible for the time being occurs along them,—such in other words, as slide on one another. If a pair of these curves,—(one belonging to the one and the other to the other element),—be isolated, they can therefore be caused to slide upon one another without destroying their coincidence. Thus if in each of two points A and A' of both curves (Fig. 37), an

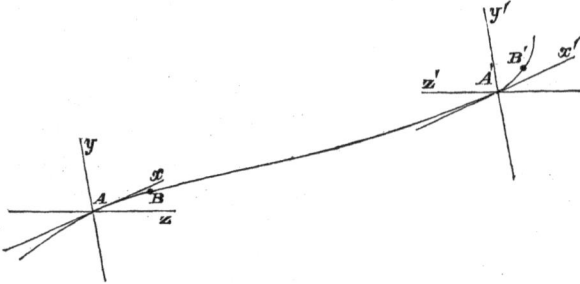

FIG. 37.

osculating plane be laid, and if further by its means the homologous systems of coordinates $x\,y\,z$, $x'\,y'\,z'$ be drawn, then if $A\,B$ and $A'\,B'$ be portions of equal length, and A be brought to A' and x to x', B must come to B', and the whole line $A\,B$ must coincide with $A'\,B'$. Among the lines which fulfil this condition we have first, where extension alone is concerned, the straight line ;— among plane curves, that is curves of two dimensions, we have the circle only, and among general curves of three dimensions, the cylindric helix only. The two first may, however, be considered as special cases of the last, so that we may say that the only curve fulfilling the required condition is the cylindric helix. The common screw and nut therefore form a closed pair (Fig. 38).

FIG. 38.

The form of the screw is not however in this case completely indifferent, on account of the third condition that only a single motion shall be possible, which here must be along the helix itself. Screw and nut must therefore be so shaped that each and every motion normal to the screw line must be impossible. With this object various forms are given in the common screw to the section perpendicular to the direction of sliding,—all being formed so that the sectional profile is double sided, as in the various forms shown in the annexed figure.

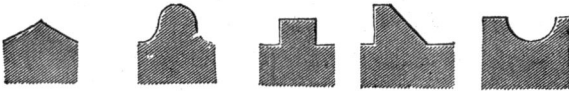

Fig. 39.

In a closed pair having a profile such as the foregoing every motion oblique to the helix is rendered impossible,—so that only motion along the helix itself can take place. Everyone is familiar with the use of these profiles in the ordinary screws employed in construction. If a straight line parallel to the axis and longer than the pitch of the screw be used to generate a screw surface, the surface obtained coincides with that of a circular cylinder, so that in a pair of bodies formed of screw surfaces of this kind only relative radial motion would be prevented,—they would not be constrained so far as motion in any other direction is concerned.

We may now proceed to examine more closely the effect of altering the two essential dimensions of the closed pair which we have found, a screw and nut having a thread of suitable profile. Any such screw owes its special properties to two quantities, its radius or circumference and its pitch-angle, (or angle which has for its tangent the ratio pitch : circumference).

The alteration of the radius alone, (the pitch-angle remaining constant), gives us no new form,—all the properties of the original screw remain unchanged.

With the angle of pitch it is otherwise. If this be gradually diminished,—the diameter remaining constant,—the pitch becomes less and less, until at last if the angle be reduced to zero the pitch disappears altogether ; the profile has a motion of rotation only, and describes simply a solid of revolution. The profile-section, however, in a plane at right angles to the direction of sliding remains unaltered, and has become the form by which the solid is generated, and we have as the result a closed pair consisting of two solids of revolution, of which the profiles are such as to prevent all motion in the direction of the axis. Fig. 40 is an illustration of such a pair,—in which the nut has become an eye (shown in section); the only motion which is possible for it being rotation.

If we now suppose the pitch-angle to increase instead of diminishing, the screw becomes steeper and steeper. If we make the angle = 90°, the screw lines become parallel to the axis,—

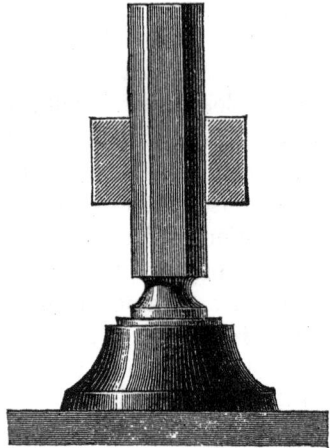

Fig. 40.   Fig. 41.

the screw becomes a prism, and the nut a corresponding hollow prism. The normal profile of the screw thread is now at right angles to the axis, and so is transformed into the normal section of a prism, which always retains a profile preventing cross-motions,—that is, a non-circular profile. As a result we have a closed pair consisting of two prisms having such profiles as to prevent any turning about their axes (Fig. 41). The single possible motion of the open prism is here sliding in the direction of the edges of the full one.

Further alteration of the pitch-angle gives no new result—if we make it > 90° the screw passes from right to left-handed; it remains, however, always a screw; and we have found it unnecessary to make any distinction between a right and left-handed thread. The problem of the closed pair has thus been exhausted by our investigation. It will be well to consider the

three forms which we have found as distinct from each other,—
although they might all be considered as modifications of the
screw,—and we have thus three closed pairs to distinguish.
These are,—to recapitulate in a few words :—

1. The common screw and nut (twisting pair);
2. The solid and hollow solid of revolution, which for the sake
   of brevity we shall call full and open revolutes respectively (turning pair);
3. The full and open prism, (sliding pair);

all three having normal profiles which prevent what we have
called cross-motion; they are adapted for the production of

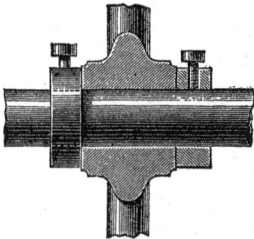 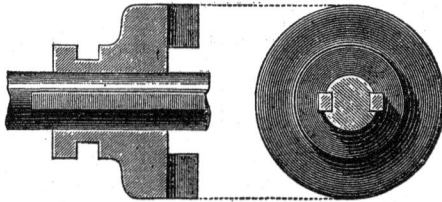

FIG. 42.   FIG. 43.

three kinds of constrained motion,—viz. motion (*a*) in helical
paths; (*b*) in circular paths; and (*c*) in rectilinear paths.
  All three are well known in machine construction,—the screw-
pair both in fastenings and in moving pieces ; the pair of revolutes
in journals, bearings, &c. and the prism-pair in guides of all
sorts. The use of such normal profiles as prevent any motion of
a pair except in the one required direction, is also very familiar.
The rings, collars, or flanges of the bearings of shafts and spindles,
for instance, carry these profiles. If it be desired to use a
cylindrical shaft, which for convenience has been made without
collars, as one element of a turning-pair,—the well-known "loose
ring" (Fig. 42) is used to close the pair. The ease with which
cylinders can be formed in the lathe causes them often to be
employed in places where they have to become elements in
sliding-pairs. To make a cylinder into a prism for this purpose
is the object of the common arrangement of feather and groove

shown in Fig. 43. The same process of converting a cylinder into a prism is often used in cases where two bodies have to be so connected that they may resist all forces tending to move one upon the other;—here keys, cutters, and so on are employed. In short, the machine-maker is accustomed to fulfil the above described condition in most numerous ways in his practical work.

§ 16.

Motion in Closed Pairs.

We have found in the last section that there are three pairs of elements which fulfil the conditions necessary for the complete and continuous enclosure of the bodies of which they consist. It is specially notable that there are only three,—in itself a remarkable result of the investigation,—for judging from the immense variety of cases which occur in machinery we might have been inclined beforehand to assume the existence of a very much larger number. These three single cases are, however, still further characteristic, on account of the nature of the constrained motions which can be carried out by their means.

In the screw-pair all points in the nut describe helices,—and equal helices if the describing points lie at equal distances from the axis. These motions are compounded of a rotation about an axis and a sliding along it, and this axis is always that of the screw-spindle itself. The axoid of the screw-spindle (see § 13) is hence a straight line coinciding with the axis of the screw. We can find the axoid of the nut at once, by supposing it fixed and causing the spindle to move: all points of the spindle then describe helices relatively to the nut, and equal helices if the points are equally distant from the axis,—exactly the same motion, that is, as that of the screw. The axoid of the nut is therefore likewise a straight line coinciding with its geometrical axis. This axoid slides endlong upon the first and at the same time revolves about it, the angular motion bearing always a constant relation to the sliding. We have then before us in this pair of elements,—the screw and nut,—the most general case of the twisting of axoids, reduced at the same time to the most simple imaginable form, where both axoids are concentrated in the twisting axes themselves.

With the turning-pair we can observe something very similar to this. Here all points in the moving eye, the open element, describe circles about points in the geometrical axis of the stationary cylinder, these circles being equal for points at equal distances from the axis. The axoid of the fixed body is thus again a straight line coinciding with its geometrical axis, and we find the axoid of the eye to be the same, if we fix it and cause the cylinder to move. Thus for the axoids of this pair of elements,— the full and open revolute,—we have again two coincident axes turning about each other, forming the simplest case of cylindric rolling which we can conceive, one in which both the cylinders of instantaneous axes have become merely straight lines.

With the prism-pair all rotation ceases; the twisting of the instantaneous axes becomes a simple sliding of them one along the other. The geometrical axes of both prisms may be considered to be their axoids, but the notion of the geometrical axis is not so determinate in the prism as in the cylinder or screw; and we can consider any given pair of coincident edges or parallels to edges to be axoids.

Here therefore the other extreme of the most general case of twisting is realized—that in which the sliding alone remains.

We may now advance a small but important step. We have above laid down as the first condition for the attainment of a given motion by one pair of bodies that one of the elements must be rigidly connected with the portion of space which we have considered as stationary. We may now release ourselves from this condition. For if two elements which have been rightly paired be both set in motion, there still continues between the one element and its partner the former absolute motion, (or motion which we have agreed to consider absolute,) but it has now become the relative motion of the element to its partner. We may therefore arrange the pairs of elements which we have found in kinematic chains, where then the relative motion of the paired elements becomes that also of the links which they connect.

From all this we see that the three closed pairs represent the three typical cases of the most general form of relative motion described in §§ 12 and 13; viz. (reversing their order), simple sliding or translation, simple rotation, and simple sliding combined with simple rotation proportional to the sliding.

This is one of the characteristics of the closed pairs. Another very notable one we have already noticed, but without enlarging

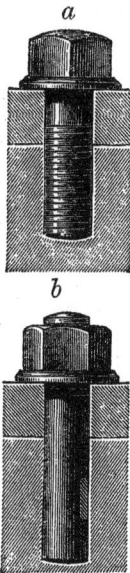

a

b

Fig. 44.　　　　Fig. 45.　　　　Fig. 46.

upon it. It is this, that the exchange of the fixed element with the movable one causes no alteration in the resulting absolute motion. The equality of the axoids proves this generally. It is however extremely important, and for the purposes of machine construction extraordinarily valuable. The exchange of one element of a pair with the other,—or, as we can say, the exchange, in respect to its fixedness or movability, of an element with its partner,—we shall call the inversion of the pair:—the inversion of closed pairs causes no alteration in the motion belonging to them.

Continual use is made of this principle in machine construction. Where, for example, a screw with a head (Fig. 44 *a*) is employed

instead of one with a nut (Fig. 44 *b*), it is simply an inversion of
the closed pair screw and nut. In the common waggon wheel
the axle is fixed to the waggon body, the wheel with the open
revolute moving upon it; in a railway carriage the latter is
attached to the frame, the solid body or shaft being connected with the wheel, and movable. For a guide for rectilinear
motions either of the arrangements shown in Figs. 45 and 46 is
used, as may be more convenient; in the first of these a solid prism
A slides in a prismatic slot, or open prism, *B*, while in the
other an open prism, *A*, is moved backwards and forwards
upon a straight bar *B*.

The familiar and easy realization of this invertibility of the
elements of closed pairs is in many cases of the greatest value

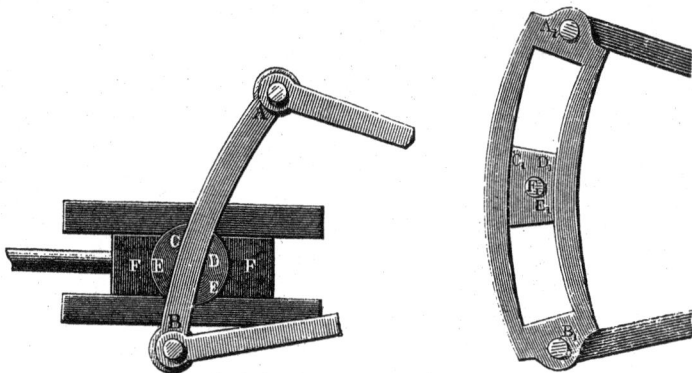

Fig. 47. Fig. 48.

to the constructor; in using such pairs he at once recognises
the possibility of employing either the one or the other element
as the hollow body, or making the contact of each with its partner
partly internal and partly external. The recognition of this
principle sometimes removes differences between constructions
which in their external appearance differ more or less widely, or
at least gives a simple expression to what was before an indistinct
sense of relationship between them. There may be mentioned,
for instance, the exchange, which has of late been frequently
seen, of the cylinder with the piston, which (*e.g.*) distinguishes
Condie's Steam-hammer from Naismyth's. What the constructor
has here carried out is the inversion of a prism-pair, whilst in
other matters the functions of the different mechanisms remain

unaltered; the changed arrangement of the ports presented some difficulties, the slide-valve also required to be somewhat differently arranged for convenience' sake, but in essence they have remained as before.

As another example I may mention the reversing link of Humphry and Tennant, Fig. 47 (or more rightly of Naismyth),* as compared with the older and more common one of Stephenson, Fig. 48. Here inversions of two pairs have taken place. First, Humphry's bar-link $A B$ is an inversion of the slotted link $A_1 B_1$ used by Stephenson, and hence the element paired with the link in the latter case, the slide $C_1 D_1$, becomes in Fig. 47 a hollow block $C D$, in which $A B$ can slide. Naismyth has also changed the cross pin F_1 into a body $F F$ having a cylindric hole, and the piece E_1, which in Stephenson's has such a hole, into a solid cylinder $E E$, of such a size as is necessary to allow the link to pass through it in the way shown. Kinematically the pieces $C D E$ and $C_1 D_1 E_1$ are completely identical,—both having for their element-forms a curved sector having a prismatic cross-section, and a cylinder normal to it.

These inversions frequently afford great advantages in construction, and on this account they are matters of considerable importance in machine-design. In kinematic science they are examples of the application of a simple general law, which as we have seen affects the simplest element-pairs generally and *a priori*.

§ 17.

The necessary and sufficient Restraint of Elements.

While in the course of our examination of closed pairs we considered the forms screw, revolute, and prism, and examined the relation between the corresponding solid and hollow pieces, we took no notice of the fact that the mutually enveloping geometrical forms were not always equally large or equally extended in the cases we used as illustrations. We found, and find almost always in practice, the nut to be much shorter than its screw,—

* Cf. *Practical Mech. Journ.*, 1862-3, p. 232.

the prismatic sliding block than its guide; in plummer-blocks, spaces are left between the brasses, and in these themselves oil-grooves are frequently made,—all of which cases are equivalent to removing the corresponding parts of the closed figure.

This procedure is so useful and is taken so much for granted in machine construction, where the portions of surface left are always of sufficient extent, that the question how far it can be carried scarcely presents itself to the designer. Where the working of a machine is accompanied by the transmission of forces of considerable magnitude, its designer furnishes it with large bearing-surfaces to prevent wear; but the question here is as to the absolute dimensions of the surfaces, and not as to

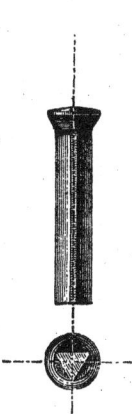 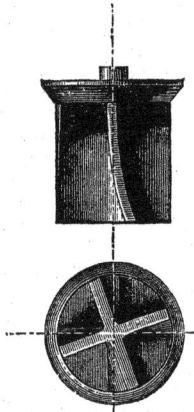 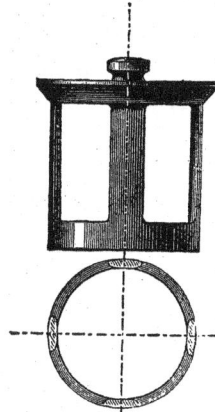

Fig. 49.   Fig. 50.   Fig. 51.

their distribution among the different bodies. If the forces are small, consideration as to wear scarcely enters into the question; the extent to which the surfaces can be diminished is here limited by the consideration that what is left must be sufficient to ensure that the paired elements shall always occupy the intended positions relatively to each other. The common cone-faced valve furnishes one among many illustrations of this. In the above three figures a simple valve, such as might be used to check the passage of water in a pipe, is shown in forms more and more removed from that of a solid cylinder. Such portions of the solid cylinder are always left to form the valve that in neither of the three cases can it be so moved that its axis shall not

coincide with that of the hollow cylinder which forms its seat.   In
the first case (Fig. 49) three small segments of the cylinder are
cut away; in the second (Fig. 50) four screw-formed webs only
are left, and in the third (Fig. 51) four thin strips of the cylinder
parallel to its axis, and connected below by a ring forming part
of its cross section.   There must evidently be some general
principle underlying the arrangement of these small strips
or other portions of the cylindrical surface which have to be
retained in order that the bodies may keep their required mutual
positions, or, as we may say, in order that they may mutually
restrain each other.   A definite number of such points is
necessary, but is at the same time sufficient, in order to ensure
this mutual restraint.   This minimum of points of restraint we
shall now endeavour to find.   It is not an investigation to which
hitherto any special value has been attached, but it is unquestionably one which should be kept in view, not only because no
property of machine-elements can be unimportant in a scientific
examination of the nature of machines, but also because of the
important results which are directly connected with it.

§ 18.

Restraint against Sliding.

We shall first consider the case of a plane figure moving in
a plane,—or, if it be preferred, of a plane section of a cylinder
prevented in any way from leaving the plane in which it lies.
By the expression point of restraint of the figure we shall
mean a point in its circumference towards which the figure is
prevented from sliding along or parallel to a normal to the tangent
at that point.   Sliding of the figure implies here an equal and
similar motion of all points in it.

Single Point of Restraint.—Let the given figure A be prevented from moving freely in its own plane by contact in one
point with a second and con-plane figure B;—we shall examine
to what extent its motion is limited.   The definition of a point
of restraint just given renders it unnecessary that we should

concern ourselves about the shape of the restraining figure *B*
(Fig. 52);—we require only to draw a tangent *T T'* to the restrained figure *A* at *a*, and erect upon this through *a* the normal
N N';—the direction from *A* towards *a* and *N'* is then that in
which the point of restraint renders sliding impossible (Fig. 53).
No sliding therefore which has a component in this direction
can occur. The only motions, however, which have not such
components are those whose directions are included in the angle
T N T', as indicated by the arrows. This straight-angle may be
called the field of sliding for a figure restrained only at *a*, and
having the normal *N'a* to the tangent *T T'* as the direction of
restraint. All the directions in which motion is prevented by the
restraining point fall within the second straight-angle *T N' T'*,—

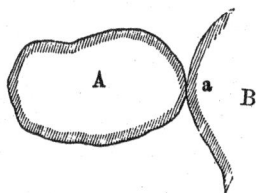

Fig. 52.

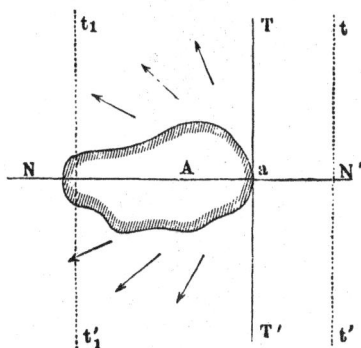

Fig. 53.

this we may therefore call the field of restraint for the
point *a*. The fields of sliding and restraint for any point of
restraint contain together four right angles. They are separated at the point of restraint by the tangent *T T'*; but as the
essential difference between them is a question only of angle or
direction, this line of separation may take different positions, such
as *t t'* or *t₁ t'₁*, so long as it remains parallel to *T T'*. In general
we may therefore say that any normal to the direction of restraint
is a division line between the fields of sliding and restraint.

Two Points of Restraint.—If a figure have two restraining
points, *a* and *b*, Fig. 54, these limit the possible directions of sliding
to the angle enclosed between the two tangents *a T* and *b T*, because all directions falling outside this angle, as those marked

3 or 4, must have a component parallel to one or other of the directions of restraint 1 and 2. If the division line between the fields of sliding and of restraint for each of the points a and b be drawn through the intersection O of their normals, the shaded angle $P O Q$ enclosed between them is the field of sliding, and the exterior angle $Q O P$ the field of restraint, for the case before us. Both points of support would equally prevent sliding in any direction falling within the opposite angle to $P O Q$.

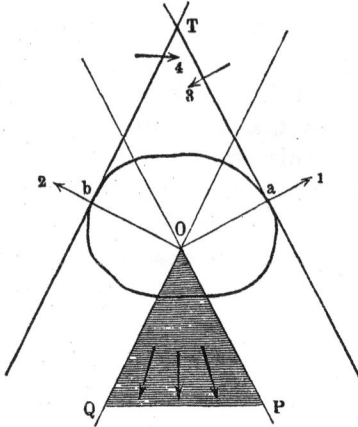

FIG. 54.

By reducing the angle $a T b$ between the tangents the field of sliding is made smaller and smaller. When they become parallel, as in Fig. 55, the angle covered by the field becomes infinitely small. But just as sliding could take place before along the lines of separation $O P$ and $O Q$, it can still occur along those lines now that they have become coincident,—that is, it can take place in a direction parallel to the two tangents. Motion is possible, therefore, not only along $O P$, but also along $O R$, the arms of the now infinitely small opposite angle. In other words, the field of sliding is now reduced simply to a line parallel to the tangents, and along this sliding can take place in both directions.

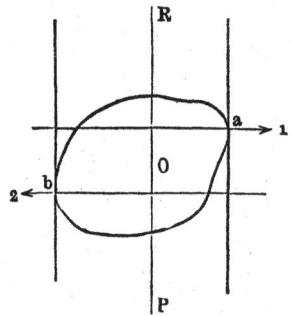

FIG. 55.

If the parallel directions of restraint 1 and 2 had not been opposite, as here shown, but in the same direction, the resulting restraint, so far as sliding is concerned, would not have differed from that given by a single point, so that it is not a case which need be further considered.

Three Points of Restraint.—The result of adding a third

point of restraint, as *c* Fig. 56, to the two already examined, is easily found. We draw the tangent *c U* and the normal through *c*, and also the line *O R* separating the fields of sliding and restraint, placing the latter so as to pass through the intersection *O* of the two first normals. It is then evident from the figure that

sliding is no longer possible through the whole angle *P O Q*, but that the field of sliding is diminished to the angle *P O R*. Here we see at once that we have the means of entirely preventing the sliding of a figure by the use of three points of restraint. For as the field of restraint of each single point covers 180°, nothing more is necessary than to place the third so that its field of restraint covers the sliding field of the other two. Fig. 57 represents this case. The third point *c* is so

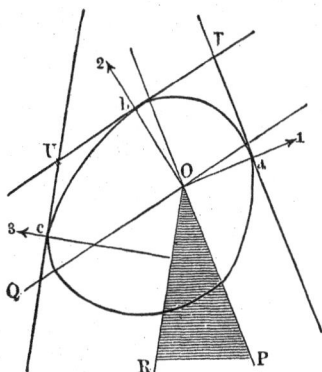

FIG. 56.

placed that its field of restraint,—extending to *R O*,—entirely covers the field of sliding *P O Q* (shown by dotted lines) left by the other points *a* and *b*. The condition for the attainment of this end is that the three points of restraint be so placed that the angle

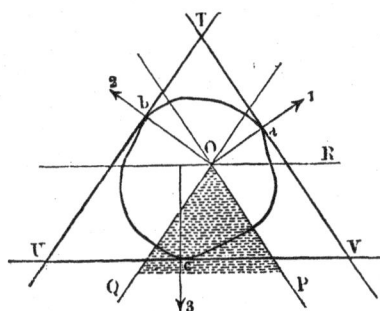

FIG. 57.

between two consecutive normals should always be less than 180°. Figs. 58 and 59 represent separately the relative directions of the normals at the points of restraint in Figs. 56 and 57 respectively, and we see from them that in the first case the angles between the normals 1 & 2 and 2 & 3 are each less than 180°, but that between 3 and 1 is greater; while in the second case each of the three corresponding angles is less than two right angles.

In the case in which the two first directions of restraint are parallel and opposite, Fig. 60, the third point *c* is not sufficient to prevent all sliding;—a fourth point d must still be added. For the directions Oc and OR, parallel to the tangents, in which sliding can take place, are themselves 180° apart. The addition therefore of two points of restraint, one between V and W, and another between T and U, is required in order that the angle between every pair of consecutive normals may be less than 180°. The

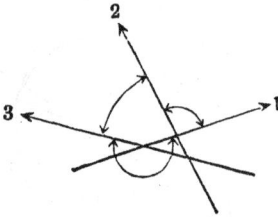

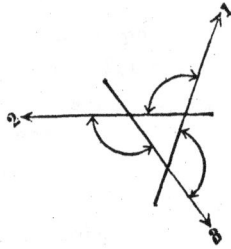

Fig. 58.   Fig. 59.

directions of restraint at c and d may in these circumstances again be 180° apart,—their tangents thus becoming parallel.

The minimum number of points of restraint which can completely prevent the sliding of a plane figure is therefore three, or if two of the directions of restraint enclose between them 180°, four. In the illustrations given in § 17 the sections of the

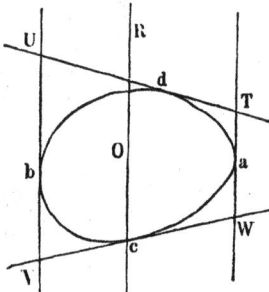

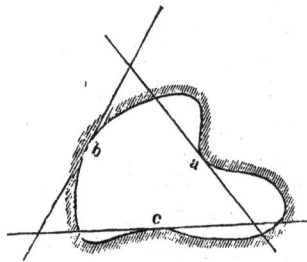

Fig. 60.   Fig. 61.

valve show that an arrangement is used which is in one case equivalent to three and in the others to four points of support. Every plane figure which has an outline returning upon itself can be completely restrained from sliding. It is entirely indifferent for this purpose whether the figure be restrained by external or internal contact, as Fig. 61, in which all the conditions of complete restraint are fulfilled, shows. Indeed, the investigation already made points this out,—although in another way,—for the second figure *B*, Fig. 52, which carries the point of restraint for the first, *A*, must be open or hollow if the first have an external profile (such as we have shown), while the action of the figures as to restraint is reciprocal, just as we have found to be the case with the elements from which pairs are formed.

§ 19.

Restraint against Turning.

Here also we shall consider first the case of plane motions of a plane figure, and shall understand by "turning" such a motion of the figure that some one point connected with it retains continuously or for an instant its position relatively to the plane. Two kinds of turning must be distinguished,—that which takes place in the same direction as the hands of a watch we call right-handed (*R. H.*), and that occurring in the opposite direction left-handed (*L. H.*), turning.

Single Point of Restraint.—If the figure *A*, Fig. 62, have a single point of restraint *a*, at which *T T'* and *N N'* are again tangent and normal, — a right-handed turning may be given to it about any point in the quadrants *N a T* and *T a N'*,

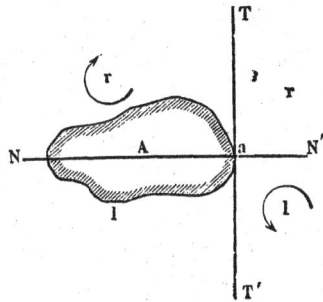

FIG 62.

while left-handed turning cannot occur about any of these points, because the motion of the point *a* in these cases would always have a component in the direction of restraint. *L. H.* turning is possible, and *R. H.* impossible, about every point in the remaining quadrants *N' a T'* and *T' a N*. The turnings possible from points upon both sides of the normal *N N'* are indicated in the figure by the letters *r* and *l*. The whole field *N T N' T'* is thus a field of turning, and is divided by the normal *N N'* into fields of right- and left- handed turning. The normal itself belongs to both fields, so that turning in either direction is possible from points in it. About points in $a\,N$ both turnings are possible to any extent, but about points in $a\,N'$ they can occur only through infinitely small angles. For so soon as turning has commenced about any of the last-mentioned points, the normal passes to one or the other side of the centre of motion, which is thus thrown into the field either of right- or of left-handed turning,— and a glance at the figure shows that it must necessarily pass into that field which does not permit the continuance of the turning commenced.

Two Points of Restraint.—If a figure have two points of restraint, as a and b Fig. 63, and their fields of turning be

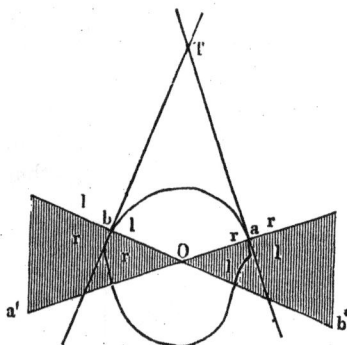

FIG. 63.

separated by drawing the normals $a\,a'$ and $b\,b'$, we find at once that the field of $R.\,H.$ turning of a is covered throughout the angle $a\,O\,b$ between the normals by the field of $L.\,H.$ turning of b, and in this way both turnings are rendered impossible. Similarly, turning cannot occur about points in the angle $a'\,O\,b'$,—where again fields of right- and left-handed turning cover each other. In the angle $b\,O\,a'$ two fields of $R.\,H.$ and in $a\,O\,b'$ two fields of $L.\,H.$ turning coincide. About points in these areas therefore (which are shaded in the figure) right- and left-handed turning are respectively possible. Thus of the two pairs of angles at the intersection of the normals,—one pair (that facing the intersecting point T of the tangents) forms a field of restraint for both turnings, while the remaining pair is the field of turning, one half of it being the field of right-handed and the other half of left-handed turning. The point O being common to both fields, turning in both directions may take place about it.

If the normals to the given points of restraint be parallel and opposite in direction (*i.e.* at an angle of 180° to each other), the angular field becomes a strip between the normals, about points in which either right- (Fig. 64) or left-handed turning (Fig. 65) is possible, according to the relative position of the normals. If the latter coincide (Fig. 66), the strip becomes a straight line, about points in which, as it forms the boundary line of two fields of turning in opposite directions, both right· and left-handed turning can take place.

If the parallel normals have the same direction (Fig. 67), the strip between them becomes a field of restraint, the space beyond

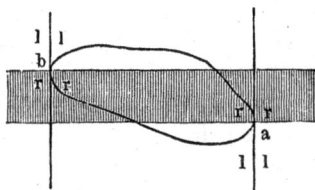

Fig. 64.

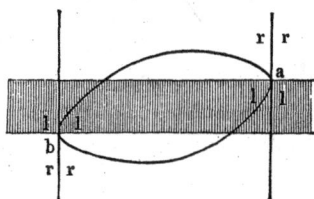

Fig. 65.

it being on the one side a field of *R. H.* and on the other of *L. H.* turning.

Three Points of Restraint.—If a third point of restraint *c* (Fig. 68) be added to any two others *a* and *b*, the tangents at which enclose any angle less than 180°, its influence upon the turning depends entirely upon its relative position. If, for instance, *c* be taken upon that part of the figure lying within

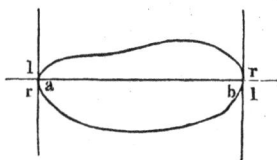

Fig. 66.

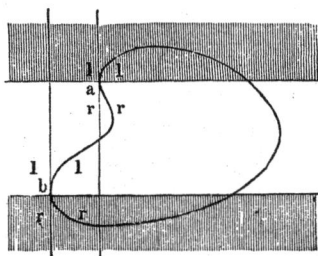

Fig. 67.

the tangent angle *a T b,* its normal cuts the other two in the points *P* and *Q*; its field of *L. H.* turning covers the similar field already existing, *a O b′,* so that this remains a field of *L. H.* turning. The *R. H.* field of *c* covers the part *b P Q a′* of the similar field already existing, which therefore also remains as before. Of the original field of turning only the small triangle *P O Q* is covered by a pair of dissimilar fields,—so that turning about points in it only has been rendered impossible by the addition of the third restraining point.

If c be so placed that its normal passes through both halves of the field of turning (Fig. 69), the triangle $P O Q$ falls in the field of restraint of a and b instead of in their field of turning. The parts $b O P c$ and $c' Q b'$ are each covered by a pair of dissimilar fields, and only the parts $c P a'$ and $c' Q O a$ remain as fields of right- and left-handed turning respectively.

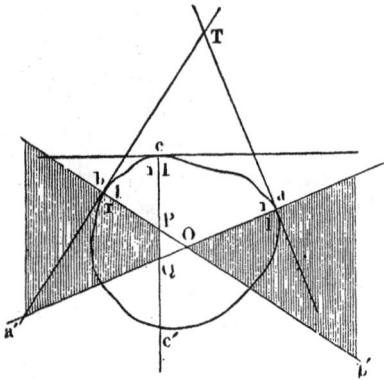

FIG. 68.

If the third point of restraint be so placed that its normal makes a smaller angle than 180° with those next it, as in Fig. 70, the result is widely different. If the normal to c then pass (as shown in the figure), through the original field of right-handed turning, its *R. H.* field entirely covers the *L. H.* one $a O b'$, and its left-handed

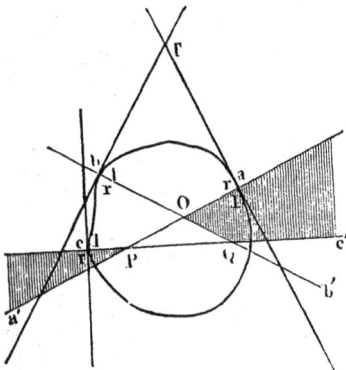

FIG. 69.

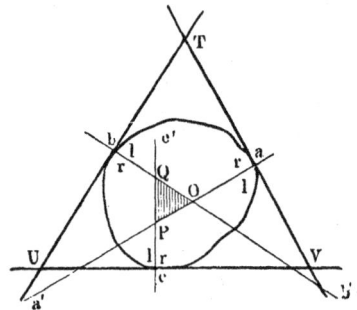

FIG. 70.

one the portion $a' P Q b$ of the right-handed field $a' O b$, so that turning about points in both these areas is prevented. The triangle $P O Q$ only is covered by a pair of similar fields, so that in it alone turning, and right-handed turning only, is possible. If c had been so placed that its normal passed through the left-handed field $a\, O\, b'$,—the case would have been reversed, and a triangle would have been left about points in which *L. H.* turning only could have occurred.

It will now be easily seen how the turning can be still further limited. For this purpose it is only necessary to diminish the size of the triangle $P\, O\, Q$. This becomes a minimum,—that is, a single point,—if c be so placed (Fig. 71) that its normal passes through the intersection O of the two first normals. The turning is then reduced as far as it can be,— but it still remains possible about one point.

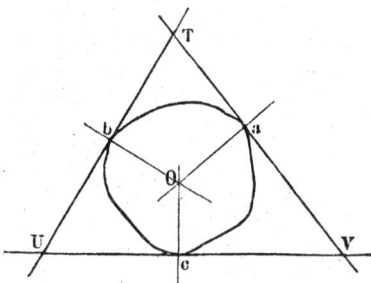

FIG. 71.

If the two first points of restraint have parallel normals we obtain other and entirely different results.

If the parallel normals to a and b have opposite directions (Fig 72), the normal to any point c between them divides the field of turning into two parts, of which one remains a field of turning, being covered by a pair of similar fields, while the other,

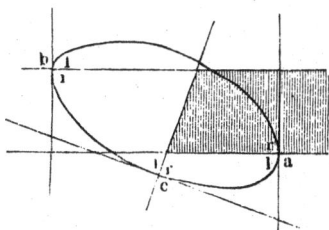

FIG. 72.

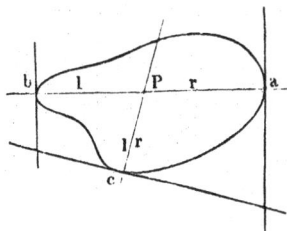

FIG. 73.

which is covered by dissimilar fields, becomes a field of restraint. If the field of turning of the first two points be a line only, (the normals, coinciding, Fig. 73), about points in which, as we saw above by Fig. 66, both right- and left-handed turning may take place, the normal to c divides the line at P into two parts, about points in one of which, $P\, a$, right-handed turning, and in the other, $P\, b$, left-handed turning, is possible.

If the normals to the first two points of restraint be parallel and have similar directions, the addition of a third point may, as Fig. 74 shows, convert a part of each of the fields of turning into a field of restraint;—or it can, if its normal be parallel and opposite to the first two, entirely neutralize one of the fields of turning, and reduce the other to a strip of limited breadth (Fig. 75);—or, lastly, if its normal be placed between the other two, and opposite to them in direction, the whole of the original fields of turning will be covered by fields of dissimilar name, so that turning will be entirely prevented.

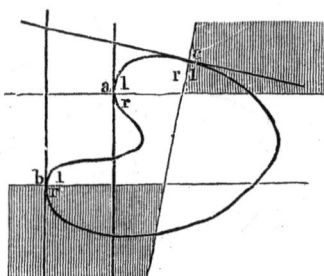

FIG. 74.

Four and Five Points of Restraint.—In cases where turning cannot be prevented by three points of restraint, and we have seen that this is the rule, the object can be attained by the addition of a fourth point, if it be so placed that its field of turning covers those of the other points dissimilarly (that is to say; a right- over a left- and a left- over a right-handed field.) If for instance a fourth point d be added to the three shown in Fig. 70, in such a way that

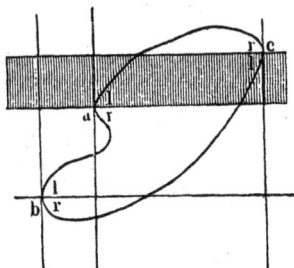

FIG. 75.

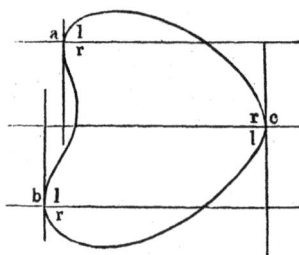

FIG. 76.

its normal passes to one side of the field of turning $O P Q$, so as to cover it with a field of dissimilar name (Fig. 77), no turning can take place. In a case like Fig. 72 a fourth point of restraint d, covering with its $L. H.$ field the remaining field of $R. H.$ turning converts it, as shown in Fig. 78, into a field of restraint. In the cases shown in Figs. 74 and 75 this can also be done.

In the case shown in Fig. 73, however, the end cannot be reached in this way. For if the fourth point d (Fig. 79) pass to one side of P, and has its *R. H.* field to the left, and its *L. H.* field to the right of it, the portions $Q\,b$ and $P\,a$ of the line are covered by unlike fields, so that no turning can take place about their points, but the piece $Q\,P$ is covered with a pair of like fields, and remains a line of centres for left-handed turning. To make $Q\,P$ disappear, the normal to d must pass through P itself, in which case turning can still take place about that point alone,—which is the point of intersection of all the normals, and is therefore common to the boundary lines of all the fields. In order to make turning impossible in such a case, a fifth point

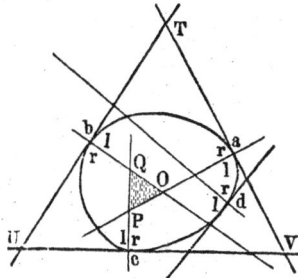

FIG. 77.

must be added to those shown in Fig. 79, so as to cover dissimilarly (as shown in Fig. 80) the still remaining field of turning $P\,Q$. If the fourth point of restraint be so placed that P and Q coincide, two more points are necessary to prevent turning, one for left- and the other for right-handed turning.

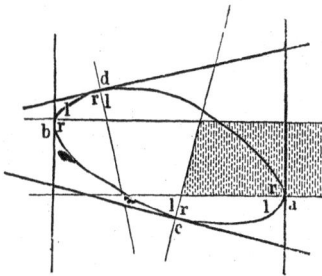

FIG. 78.

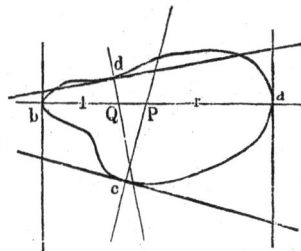

FIG. 79.

The same thing holds good for the case shown in Fig. 71, where the normals to three points of restraint intersect at one point ; for here also one additional point of restraint is required to prevent *R. H.* and another to prevent *L. H.* turning. This and the foregoing case may be stated generally in the proposition : If the normals to three points of restraint of any figure cut one another in one point, at least five points of restraint are required to render turning of the figure impossible.

It will be seen that it is a much more difficult problem so to restrain a figure that it shall not turn than to prevent its sliding. As a rule, it requires at least four points of restraint (three suffice in an exceptional case, Fig. 76, only), while very fre-

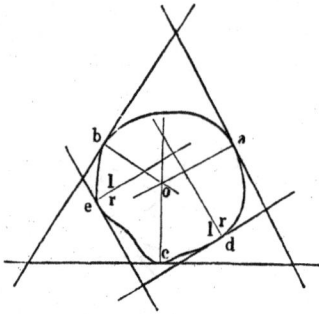

quently five are necessary. The form of the figure, also, cannot vary within such wide limits as when sliding only is to be prevented,—its profile must have a curvature varying in such a way that it may be possible to find normals occupying the required relative positions. Hence the rotation of a circle, as of course can be recognised *a priori*, cannot be prevented by any number of points of restraint. When this form occurs

Fig. 80.

in machine construction,—(as on account of the ease with which it can be made it so often does),—as the cross section of a body which it is desired to restrain from turning, it is necessary in some way to convert its form into one which can be so restrained. We have already (§ 15) looked at this fact from another point of view, and can now examine it in the light of the foregoing investigation.

The fastening of a wheel or pulley upon a cylindrical shaft furnishes us with a very familiar example (Fig. 81). Here, if the original form of the shaft were retained, all the normals would cut at the centre *O*. A rectangular groove is therefore made in it, against the sides of which the key which holds

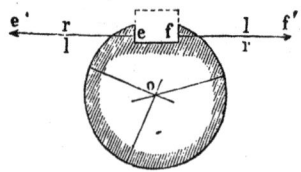

Fig. 81.

shaft and wheel together can press at *e* or *f*. The one normal *e e'* covers *O* with its field of left-, and the other *f f'* with its field of right-handed turning, exactly as we found above (Fig. 80) to be necessary. In similar cases the turning is often restrained simply by flattening a portion of the shaft. The key then exerts pressure at such points as *e* and *f* (Fig. 82), so that the normals *e e′* and *f f′* pass left and right of *O*, covering it with unlike fields as before. The moment of the pressure in the direction of the normal has here, however, a much smaller arm than in the former case, on which account it is employed only where the effort tending to cause turning is small.

Large heavy water-wheels are frequently secured upon the shaft with three, or more often with four keys, a space being left between the solid and hollow cylindric surfaces, which are therefore altogether dispensed with for restraining purposes (Figs. 83 and 84). Such fastenings can offer obviously but small resistance to turning forces, as the normals to the faces of the key pass so very near to the

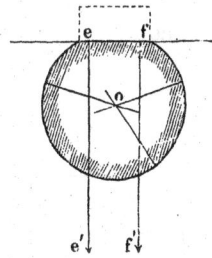

Fig. 82.

centre *O*. On this account such modes of keying the shaft are employed rather as methods of centering, that is, as restraints against the cross sliding of the shaft in the wheel boss in the manner indicated in Figs. 57 and 60, and for this purpose they are correctly designed. Where, however, there is any considerable torsion to be resisted, as in the driving pinions of rolling mills,

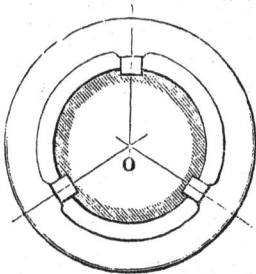

Fig. 83.

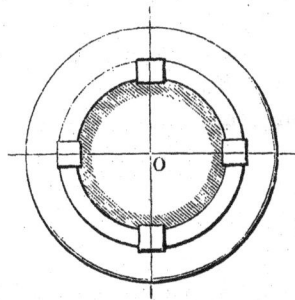

Fig. 84.

such an arrangement as shown in Fig. 85 is used which serves as a restraint both against sliding and turning.

Theoretically, complete restraint might be effected by four of these keys only (as *f*, *g*, *b* and *c*),—or the arrangement of Fig. 86 might be used, where *a*, *b*, and *c* are arranged as in Fig. 76, while *d* and *e* are added as in Fig. 60. But the arrangement shown in Fig. 85 is very much better, for the torsion is resisted at four

points instead of at two, while the arm of the moment of resistance is doubled, so that the resistance of each single key is only
one-fourth as much as in the former case; the placing of the
keys near the corners of each side of the shaft is much better also
than placing them in the centre.

A comparison of Figures 53 and 62 shows at once how it is in

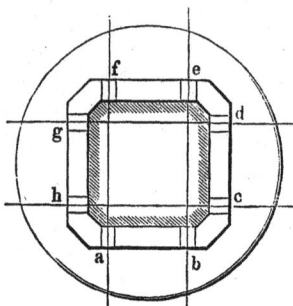   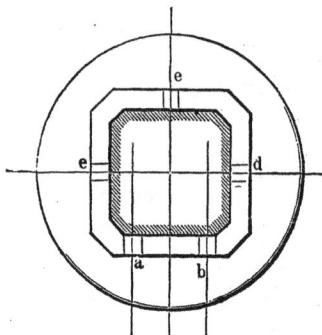

Fig. 85.   Fig. 86.

general so much more difficult to restrain turning than sliding,—
namely, that while one point of restraint can prevent sliding
throughout a field of 180°, it can do nothing more to prevent
turning than to divide the whole field into two parts,—from
points in one of which right-handed and in the other left-handed
turning can take place.

§ 20.

Simultaneous Restraint of Sliding and Turning.

In proceeding now to apply the results found above to cases
where sliding and turning take place simultaneously, we may
first state the following propositions relating to plane figures :—

(1) Neither the sliding nor the turning of a plane figure in
a plane can be prevented by two points of restraint.

(2) By three suitably placed points of restraint (*a*) sliding
can be prevented, but not turning at the same time; and (*b*) turning
can be prevented, but not sliding at the same time.

(3) Only by four suitably placed points of restraint, and with certain profiles only by five, can turning and sliding be prevented simultaneously.

We shall apply these propositions to the closed pairs.

Restraint in the Sliding-pair.—Let it be required with a minimum number of points so to restrain a solid prism that it shall be prevented from having any other relative motion than that which it would have if it formed one element, along with an open prism, of a closed pair. This can be done if, taking any two plane sections of the prism, perpendicular to its axis, we prevent either the sliding or the turning of their profiles. This requires four suitably placed points of restraint on each section (Fig. 87), so that no turning, and sliding parallel to the axis only, can take place,—which is the characteristic of the sliding-pair. This gives eight points of restraint—$a, b, c, d, e, f, g,$

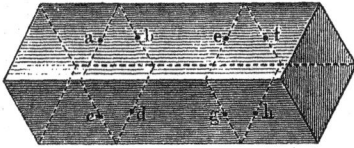
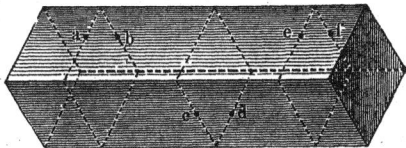

Fɪɢ. 87.   Fɪɢ. 88.

and h,—the four in each section being placed as in Fig. 87. If, however, a third plane section, parallel to and between the two first be taken, two pairs of similarly-situated points, as c and g, d and h, may be placed together upon it. Two out of the eight points are thus dispensed with, and the required restraint is obtained with six—$a, b, c, d, e,$ and f (Fig. 88).

Restraint in the Turning-pair.—To examine what minimum of points is necessary in order that a solid circular cylinder with flat ends (Fig. 89) may be restrained as completely as if it formed one element of a closed turning-pair, we may in the same way take two plane sections of it normal to its axis, and restrain these from cross sliding,—which requires three points in each, $a, b, c,$ and $d, e, f,$—and at the same time restrain a longitudinal section from sliding in the direction of the axis,—which can be done by taking one point of restraint upon each end, as g and h. Here, therefore, eight points of restraint are again required. If, however, the first six points be taken upon the edges of the end surfaces (Fig. 90), and suitable inclinations be given to the restraining surfaces, these six points will also restrain the end-long sliding of the cylinder, and so will suffice for the whole required restraint.

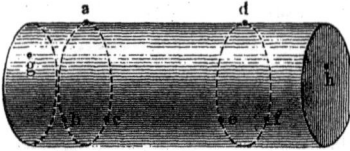

FIG. 89.

Restraint in the Twisting-pair.—In order to restrain a screw spindle (Fig. 91) with the minimum number of points, we must restrain two longitudinal sections of it,— passing through its axis and at right angles to each other, — against both sliding and turning. If this be done the spindle will be as completely

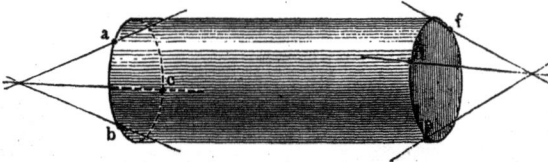

FIG. 90.

restrained as if it were enclosed by a nut. For this purpose each section (see Fig. 87) requires four points of restraint, so that here again eight points of restraint in all are required. Two pairs of these, however, can be reduced to one by a method similar to that shown in Fig. 88, so that the actual minimum number of points of restraint is once more six.

FIG. 91.

We find therefore that for all three closed pairs, eight points of restraint are sufficient,—and that by the use of double points six may suffice,—to hold the moving element in the same position as that which it would have were it restrained by the infinite number of points of its partner element in a closed pair.

§ 21.

The Higher Pairs of Elements.

From the foregoing examination into the restraint of plane figures, we see that pairs of figures may be constructed in which the sliding of the one figure relatively to the other is prevented, while their relative rotation remains possible,—and further, that if the normals of restraint, their number not being less than three, intersect in one point, the rotation which remains possible will be about this point only. Such a rotation is a definite motion, excluding the possibility of all others; and this is just what we have recognised as the distinguishing characteristic of a pair of elements. If, therefore, a pair of figures be so conditioned that after the completion of any indefinitely small turning about a centre *O*, they have again three points of restraint with their normals cutting in a new point,—and that

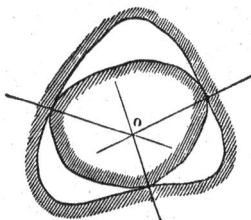

FIG. 92.

this occurs continually for every new mutual position of the figures —such figures may evidently be used as the foundation of a pair of elements. To construct the elements we require, *e.g.*, only to erect cylinders upon them, and provide these with end surfaces so as to prevent axial sliding.

If there be two figures of such form that in all their relative positions the sliding of the one relatively to the other be impossible, then their only relative motion at each instant must be turning. If the normals of restraint have always a common point of intersection, then this turning can take place at each instant about this point only,—but if this be not the case, then turning (if it remain possible) must occur about a point outside at least one of the normals. But such a motion would cause a separation of the figures at the corresponding point of restraint, and is therefore inconsistent with the assumption of continued restraint against sliding. Where, therefore, such continued restraint is required, the normals must always have a common point of intersection, and the single motion possible at each instant will be turning about this point. We therefore have the important proposition :—If it can be shown for any two figures that in all consecutive mutual positions relative sliding is impossible, it follows that the normals to their points of restraint intersect always in one point.

The series of consecutive centres of rotation or points of intersection of the normals for the two figures form the centroids, and the cylinders erected upon these the axoids of the two paired bodies.

Pairs of elements formed in this way are not closed, like the pairs we have before examined, but possess the more general and higher characteristic of mutual envelopment (§ 3). We shall therefore distinguish them as higher pairs of elements from the closed pairs,—which, on account of the smaller variety of their characteristics, we shall term the lower pairs. In order to understand the higher pairs we shall examine in detail one example.

§ 22.

Higher Pairs.—Duangle and Triangle.

If from the ends of any straight line, PQ, with a radius equal to the length of the line, we describe intersecting arcs of a pair of circles, these will enclose a plane figure $PRQS$ (Fig. 93), which we may call a duangle. This will be touched in three points, Q, R, and S, by an equilateral triangle, ABC, of a height equal to $2\ PQ$, if Q be placed in the centre of one side of the triangle. For AB is $\perp$ to QR, ($\angle PRA$ being $= \angle BAQ = 30°$, and $\angle QRP = 60°$), and also Q, R, A and S are all points in a circle described about P with a radius PQ. The normals to the points of restraint Q, R, and S cut each other in Q, the angle between each pair being $120°$. Sliding, therefore, is entirely restrained, and rotation can take place about one point only. The same holds good also for any other position of the duangle,—such, for instance, as that dotted,—as will be seen from what follows.

If we consider in the first instance continuous contact between the duangle and two sides only, AB and BC, of the triangle, it will be seen that if *L. H.* rotation take place the point *P* must move along a straight line *T U* parallel to *B C*, (for *P* being the centre of the curve *R Q S* is equidistant from all points in it), and similarly the point *Q* moves in a straight line *Q T* parallel to *A B*. The two paths intersect at *T* at an angle of 60°—the same angle, namely, as that at which *P S* and *Q S* intersect in *S*. The as yet unknown path of the point *S* relatively to the figure *A B C* is therefore simply that of the vertex of a triangle *P S Q*, of which the ends *P* and *Q* of the base slide upon the arms of an angle equal to the vertex angle of the triangle itself. Let *P Q S*,

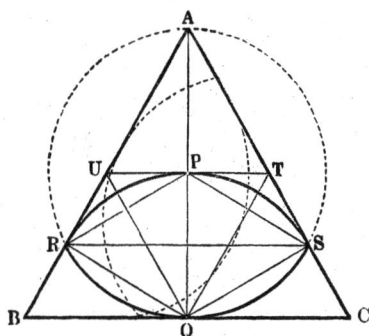

Fig. 93.

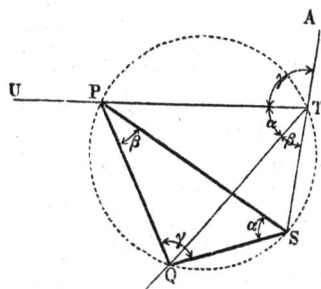

Fig. 94.

Fig. 94, be such a triangle, *a*, *β* and *γ* the angles at its vertex and base respectively. *U T Q* = *a*, is the angle upon the arms of which the points *P* and *Q* slide. The points *S* and *T* are, however, points in a circle passing through *P* and *Q*,—*a* being the angle at the circumference upon the chord *P Q*. If, then, we join *S* and *T*, we have ∠ *Q T S* equal to the angle *β* at *P*, the circumferential angle upon the chord *QS*, and therefore constant for all positions of the triangle. If, further, *S T* be produced through *T* to *A*, the angle *A T P* = 180° − (*a*+*β*), that is, = the base angle *γ* at *Q*. The point *S* therefore moves in a straight line, which makes with the arms of the given angle angles respectively equal to those at the base of the given triangle. This straight line is in Fig. 93 above the third side *A C* of the triangle, which makes at *T* with *U T* and *Q T* the angle 60°—the base angle of the equiangular (and also equilateral) triangle *P S Q;* all three sides of the triangle therefore touch the duangle continuously.

The contact of the side $C\,A$ of the triangle with the vertex S is continuous, along with that of the other two sides with the arcs which form the sides of the duangle; the angle between each pair of consecutive normals is therefore always 120°, because they remain always perpendicular to the sides of the triangle. Thus the conditions necessary to the continuous restraint of sliding are fulfilled by this pair of figures, and consequently (by § 21) the normals must always intersect in one point, so that the figures will serve as the basis of one of the higher pairs of elements. We have now to find the corresponding centroids.

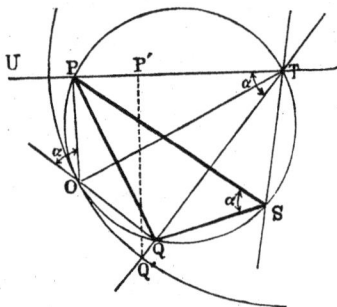
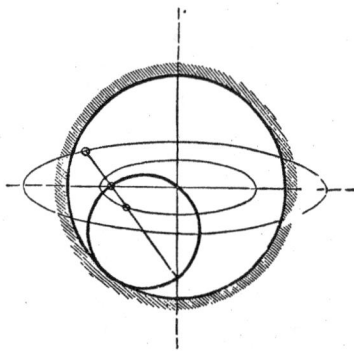

Fig. 95.   Fig. 96.

(*a.*) Centroid of the Triangle.—In order to make our investigation as general as possible we shall take the problem in the form used above,—of angle and triangle,—considering the line $P\,Q$ as a plane figure (see § 5), the motion of which relatively to the figure for which the angle $U\,T Q$ stands is the same as that of the duangle to the triangle. We require to know,—as was shown in § 8,—the paths of at least two points in the moving figure. We do know, however, the rectilinear paths $P\,T$ and $Q\,T$ of the points P and Q. The normals to these paths cut each other in O, Fig. 95, and this point must lie in the circle already found, for the angle $P\,O\,Q$ enclosed by the normals is obviously equal to $180° - a$. Further, both $O\,P\,T$ and $O\,Q\,T$ being right angles, the line joining O and T must be the diameter of the circle $P\,T S Q$. The chord $P\,Q$ and the angle a being constant, the size of this circle is fixed, the distance $T\,O$ of the instantaneous centre O from the point T is constant;—the centroid is therefore a circle described about T

with the radius $T\,O$. To find the magnitude of this radius in terms of quantities already known, suppose PQ to slide until it stands perpendicular to either of the arms $T\,P$ or $T\,Q$, as, for instance, at $P'Q'$. Then one of the normals coincides with PQ itself, and the other has become zero, and it will be seen at once that

$$T\,Q' = T\,O = \frac{P'\,Q'}{\sin a} = \frac{PQ}{\sin a},\ \text{or if we denote}\ O\,T\ \text{by}\ R\ \text{and}\ PQ\ \text{by}$$

$$a,\ R = \frac{a}{\sin a}.$$

(*b.*) Centroid of the Duangle.—If now, in order to find the second centroid, we suppose the line PQ stationary, and set the angle $P\,T\,Q$ in motion, the points passing through P and Q must move always in the direction of the arms $T\,P$ and $T\,Q$ themselves. The normals cut in O as before. The locus of this point is now, however, that of the vertex of a triangle having a base PQ and a vertex angle $180° - a$,—which is evidently the circle $Q\,O\,P\,T\,S$ having a diameter $T\,O$, and circumscribed about the given triangle $P\,Q\,S$. If we denote the radius of this circle by r, we have

$$r = \frac{T\,O}{2} = \frac{a}{2\,\sin a} = \frac{R}{2}.$$

The centroids of our supposed pair of figures, angle and triangle, are therefore, if completely constructed, two circles, having the relative magnitude 1 : 2, of which the smaller rolls in the larger. The relative paths themselves are therefore trochoidal, the hypotrochoids for the rolling of r in R, becoming ellipses (Fig. 96), of which the one described by any point in the circumference of r has a semi-axis major equal to R, and a semi-axis minor equal to zero, and therefore coincides with the diameter of R. For the rolling of R upon r the point-paths are peri-trochoids, of which the common form is the cardioid. The common, curtate and prolate forms of these curves are shown in Figs. 96 and 97.[13] The former of these cycloid problems was first treated—(although by no means completely)—so far as my knowledge goes, by the celebrated mathematician Cardano, in the sixteenth century.[14] As I shall frequently have to refer again to this pair of circles I shall, for the sake of shortness, call them Cardanic circles. In the figures actually

before us, the duangle and triangle, the relative motions correspond
to certain parts only of that motion of the angle and triangle which
would give us these centroids complete.   The actual sequence of
the motions of the pair is as follows:

So long as the point P, Fig. 98, moves towards U, T is the vertex
of the angle on the arms of which PQ moves, the diameter
$R = \dfrac{PQ}{sin\ a} = \dfrac{PQ}{sin\ 60°}$ is the line $T\,Q$, and its half, $V\,Q$, the radius r,—
so that the arc $Q\,U$ belongs to the larger, and $Q\,WP$ to the smaller
Cardanic circle.   Further, as $\angle\ UT\,Q = 60°$ and $\angle\ P\,V\,Q = 120°$,
the arc $U\,Q$ is equal to the arc $Q\,WP$.   From U onwards P moves
upon the chord $U\,Q$ to W, and Q along the half chord $V\,T$,—this
time Q has become the vertex of the angle along the arms of which

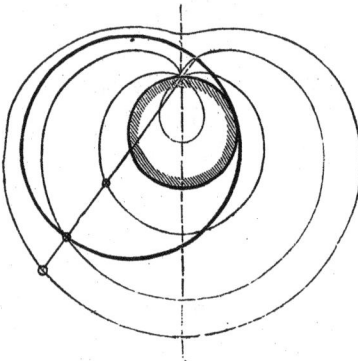
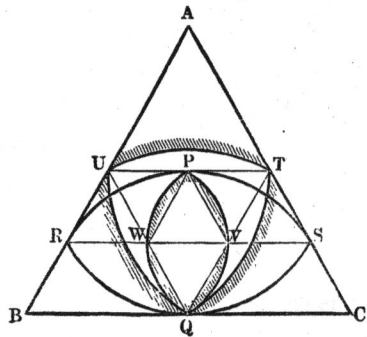

FIG. 97.   FIG. 98.

$P\,Q$ slides;—$Q\,U$ is the radius R and $W\,Q$ the radius r, by which
we obtain the arcs $U\,T$ and $Q\,V\,P$.   Proceeding from W, P moves
next along the half chord $W\,Q$, while Q moves from T to P; U is
now the vertex-angle, from which (with radius R) the arc $Q\,T$ is
described, on which again the curve $Q\,WP$ rolls.   After these
motions P has reached the position Q, and *vice versâ*, and the
duangle has turned through an angle of 180°.   With its further
rotation through two right angles the curve-triangle $Q\,U\,T$ makes
another complete revolution, and the duangle $P\,V\,Q\,W$ one and a
half revolutions.   Thus, when the duangle has returned to its
original position, the instantaneous centre has twice traversed the
three sides of $Q\,U\,T$, and three times the two sides of $P\,V\,Q\,W$,—

continuous rolling having occurred between the two centroids.[15] These centroids we have now found completely. They are (*a*) for the equilateral triangle an equilateral curve-triangle inscribed within it,—(*b*) for the duangle a similar duangle, which has the minor axis of the first for its major axis, and which rolls in the centroid of the triangle.

§ 23.

Point-paths of the Duangle relatively to the Equilateral Triangle.

(Plates I and II.)

The paths described by points in the duangle relatively to the triangle can now be completely determined; for we know the centroids of both figures, and can fix that of the triangle and set that of the duangle in motion upon it. As these paths are formed by the rolling of one centroid upon another, they all belong to the class of curves known as roulettes. We have already determined the paths of two important points of the duangle, the points *P* and *Q*. These points always belong to the smaller Cardanic circle, and so describe always parts of hypocycloids coinciding with portions of the diameter of the larger circle. These portions form, as has been already noticed, two coincident equilateral triangles, *U T Q*, Fig. 1 Plate I. All other points of the duangle describe necessarily arcs of prolate or curtate hypocycloids, which are, as we have mentioned, ellipses. All these prolate and curtate curves are known by the common name of trochoids.[13] We may therefore say that all the remaining points in the duangle have for their paths hypotrochoids, of which the equilateral triangle *U T Q* is the foundation. As this triangle consists of six portions of hypocycloids,—so all the other point-paths must consist of six hypotrochoidal arcs. The figures thus built up take very various forms with different positions of the describing point. Fig. 1 shows three of them external to the triangle. The describing points themselves lie upon the production of the minor axis *Q P* of the duangle, that is, of the major axis of the centroid *Q m_1*, *P m_2*, and are numbered 1, 2, 3, commencing with the outermost point 4, which

coincides with P. The figures are all three-cornered, and approach
more and more nearly the triangular form, which is that actually
described by the point 4. In Fig. 2 the paths of three more points
5, 6, and 7 are shown on a larger scale; the last of these is the
centre M of the duangle. The path of 5 contains three loops; in
the case of point 6, which is so chosen as to coincide with the
centre M_1 of the triangle $A\ B\ C$, the loops have a common point of
intersection. For any describing point between 6 and 7, the curves
which intersect at M_1 in the former case open out, enclosing
between them a triangular space; and lastly, point 7 gives the
three loops fallen together into a continuous curve, which is the
smallest curve which can be described by any point of the
duangle. This curve is two-fold,—in a whole period, that is,
the describing point passes twice through M; this can be seen by
an examination of the curve 6, the tangent to which twice turns
through four right angles.

If the describing point be taken further from P than 7 we
simply obtain repetitions, in reverse order, of the curves already
described.

By choosing describing points upon the major axis of the
duangle we get a further series of curves, of which some examples are shown in Plate II. Point 1 again gives us an
elliptic triangle; point 2, coinciding with the end S of the axis, gives
a three-cornered figure, bounded partly by straight lines and partly
by elliptic arcs; point 3 gives an elliptic triangle with concave
sides, which is shown on a larger scale in Fig. 2. The point 4
coincides with the end m_2 of the short axis of the smaller centroid.
It describes the remarkable figure No. 4 shown in Fig. 2, consisting
of three circular arcs (described by m_2 as centre of the arc $Pm_1\ Q$),
and three (twofold) rectilinear continuations of them (described by
m_2 as a point in the circumference of the arc Pm_2Q). The
point 5 gives a curve with three loops, intersecting in the point
M_1; point 6 gives a three-looped curve with an inner open triangular space, and point M as the centre gives again the curve
shown in Fig. 2 Plate I., and there marked 7. It is to be noted that
the trochoidal triangles which form the paths of the points in the
major axis are turned through an angle of 60° relatively to the
point-paths of the minor axis.

The paths of all points between these two axes lie between

PLATE I.

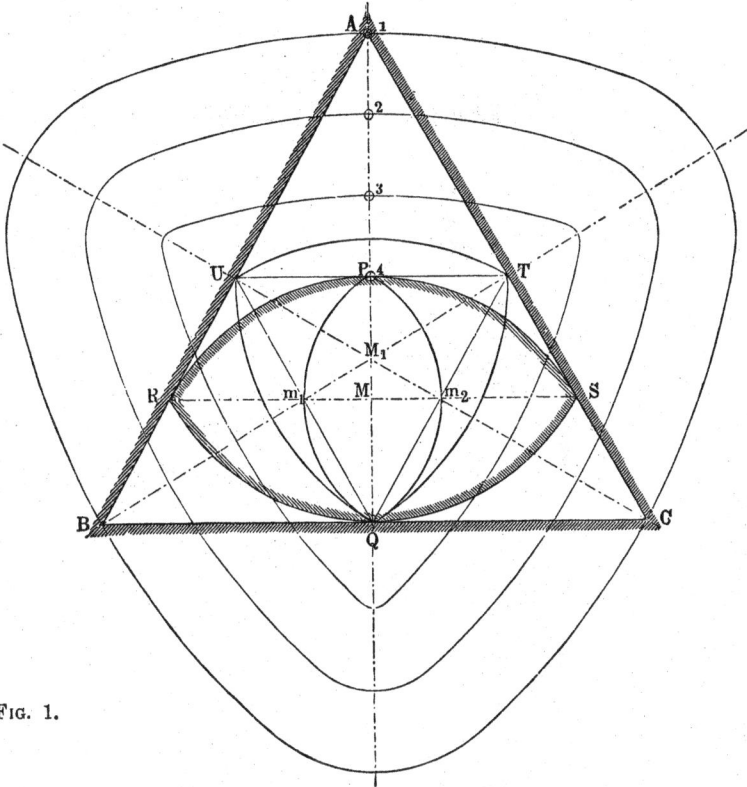

FIG. 1.

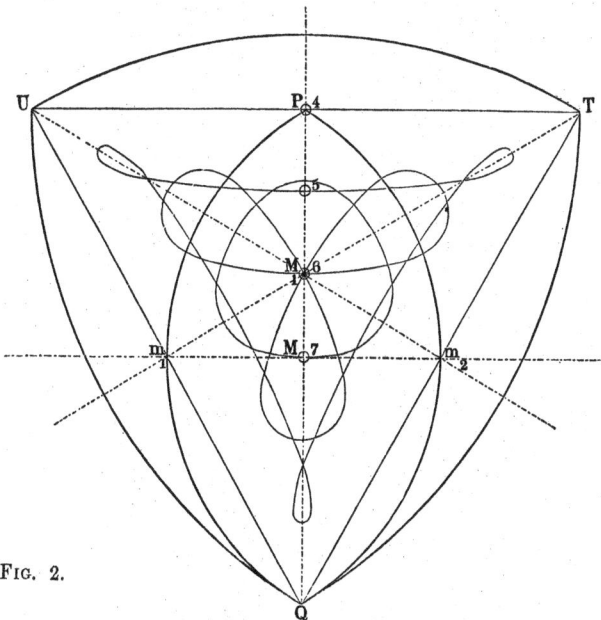

FIG. 2.

PLATE II.

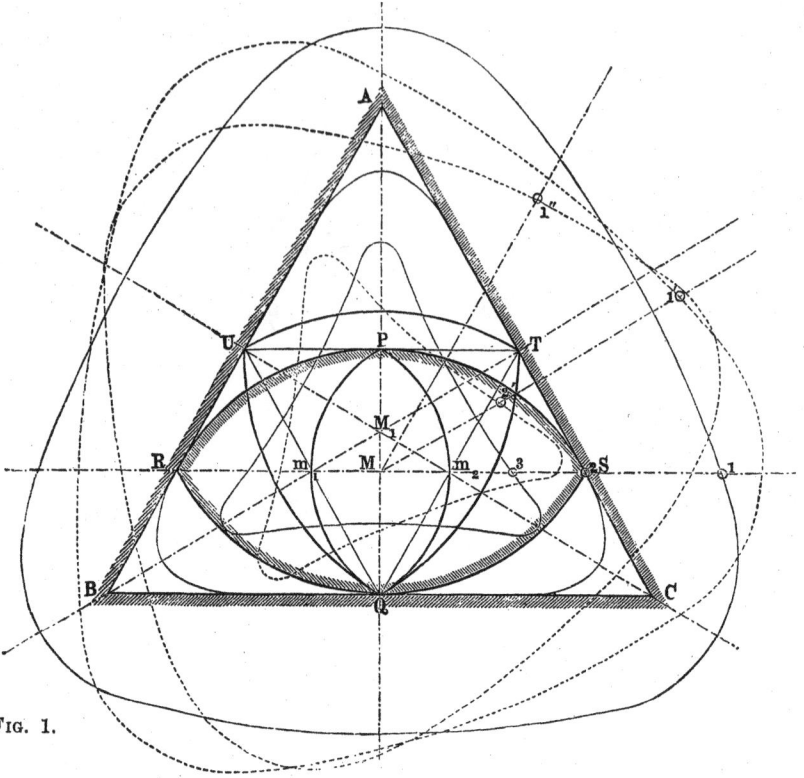

FIG. 1.

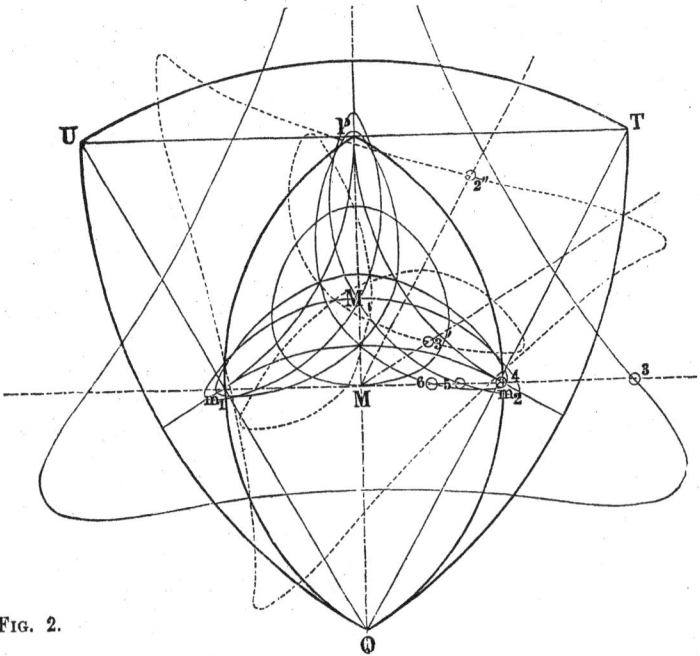

FIG. 2.

these two positions,—they are in general of an elliptical three-cornered
form. Four of them are shown dotted in Plate II. and indicated by the numbers 1′, 2′, 1″, and 2″. These figures are no
longer symmetrical about their three axes, as the former were;—
this can be specially seen from No. 2″.[16]

§ 24.

Point-paths of the Triangle relatively to the Duangle.

(Plates III. and IV.)

To determine the point-paths of the triangle relatively to the
duangle the latter must be fixed and the former set in motion
The centroid, $U T Q$, Plate III. 1, then rolls upon the centroid, Pm_1
Qm_2. The figures described are formed of arcs of peri-trochoids.
All describing points lying upon the rolling centroid here describe
arcs of cardioids, as we have seen in connection with Fig. 97.
It is at once noticeable how greatly these figures differ from the
former ones. This forms an illustration which exactly meets a
mistake made by many former writers on this subject, that the
inversion of such a pair of figures, although it produces the greatest
alteration, in the manner of turning, does not alter the form of the
point-paths.[17] This circumstance has given me occasion to construct these pairs of elements, to which, otherwise, I cannot ascribe
any particular use.
The figures in Plate III., show the paths of points in the axis
$M A$ of the triangle. The point 1 describes a rounded oval, consisting, like all the other figures, of six peri-trochoidal arcs. Point 2,
coinciding with the vertex A of the triangle, gives an oval with
concave sides, as does also point 3; the path of point 4 consists
of two simple cardioids joined in the points m_1 and m_2 of the
stationary centroids. The path of 4 is repeated in Fig. 2 upon a
larger scale. The paths of 5 and 6 each have two loops, which
in 7 fall together into one oval curve. Point 7 itself coincides
with the centre point M_1 of the triangle $A B C$, and it must be noted

that its path is really three-fold; that is, is traversed thrice in each period by the point M_1. This can be recognised from the looped paths 5 and 6, the tangents to which turn three times through four right angles. The path 1 is also remarkable, for the three homologous points 1, 1', and 1" lie continually in it, so that complete restraint occurs, as in Fig. 59.

Plate IV. shows the further point-paths obtained by choosing points in the line $M_1 Q$, or (what is the same thing) in the lines $M_1 T$ or $M_1 U$. It is seen at once that the principal axis of the figures is now turned through 90°, and also that the loops form themselves about an axis perpendicular to the original one. The curve $T S$ (Fig. 1) and its symmetrical repetitions are characteristic, —the former is the circular arc described by the centre T of the rolling curve $U Q$.

If describing points be taken upon radii lying between $A M_1$ and $T M_1$, paths are obtained which are not, as before, symmetrical about two axes. It has not been thought necessary to give examples of these; their nature will be made sufficiently clear by the analogy of the paths 1', 2', &c. in Plate II.

We have found that the point-paths of the pairs of elements which we have considered possess extraordinary variety of form,—they can, however, be somewhat systematised by the use of a method and nomenclature similar to that employed for trochoidal curves. Our curves form themselves into two series, corresponding to the fixing of one or the other element, and each series divides itself into groups according to the position of the line on which its describing points are taken. The paths of points in the centroids themselves, —as, *e. g.,* the triangle $U T Q$, Plate I. 1, the three-cornered paths of the point m_2 in Plate II. 2, &c.—are specially characteristic. These paths may be called the c o m m o n f o r m of the roulettes concerned,— as in the case of the cycloid. By the same analogy we may call all paths of points which lie beyond or within the rolling centroids, c u r t a t e or p r o l a t e point-paths respectively. Among the last, one is specially characteristic, and common to all groups of point-paths,—the path of the centre point of the moving centroid, M in Plates I. and II., M_1 in Plates III. and IV. This roulette is always the smallest of its series; the point-paths concentrate themselves upon it as the path of the point relatively nearest to the centroidal curve itself, just as a circle concentrates itself upon its centre

PLATE III.

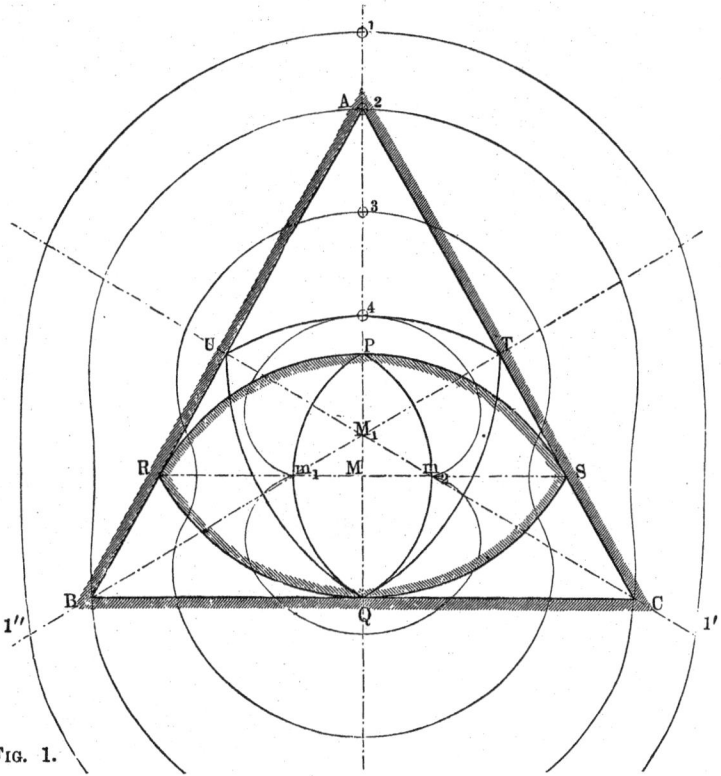

FIG. 1.

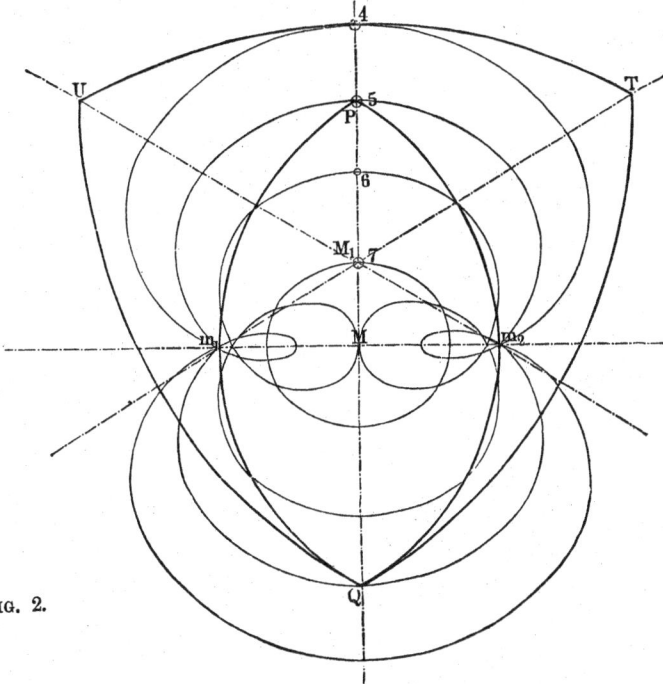

FIG. 2.

PLATE IV.

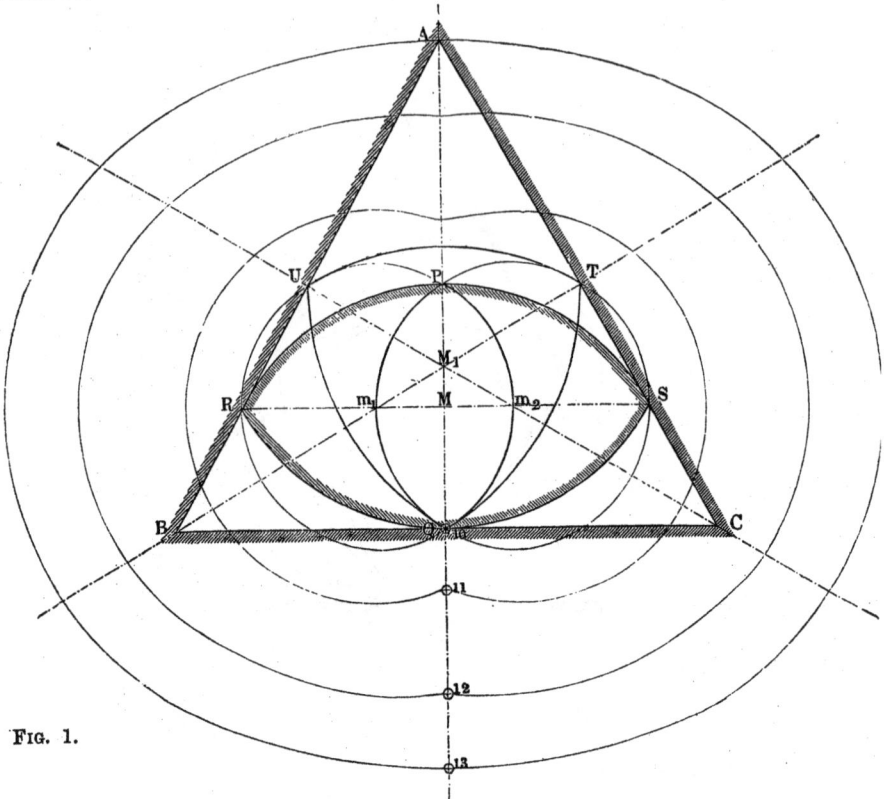

FIG. 1.

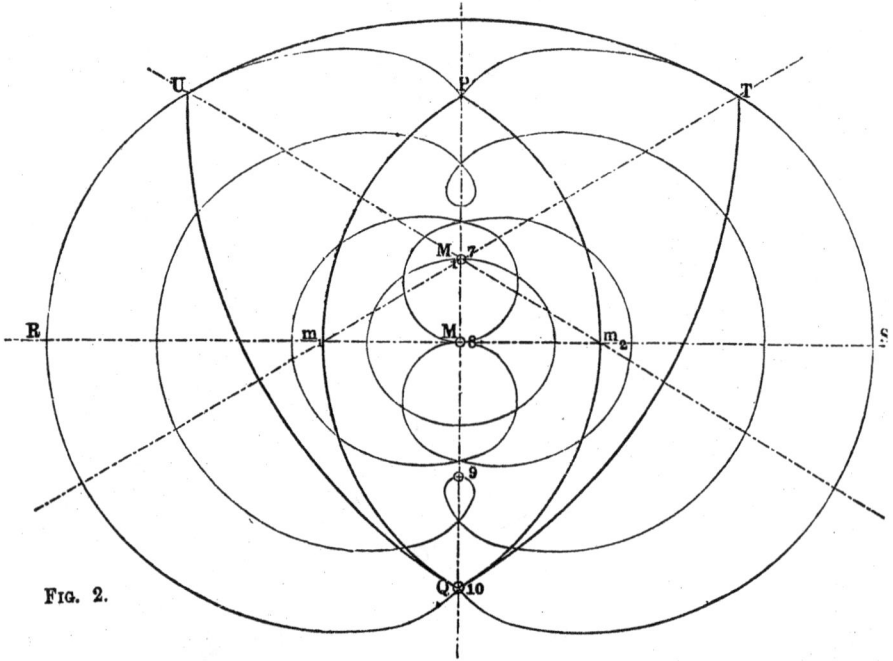

FIG. 2.

point when its radius is diminished to zero;—this form we may therefore call the c on c e n t r a l form of the common point-path.

Further, those roulettes also are noticeable which pass through the middle point of the whole series of curves, as No. 6 in Plate I., 2. These roulettes we shall call h o m o c e n t r a l. In our example they take a series of forms of which some are shown in the second figures of Plates I. to IV. It must be noticed that homocentral point-paths can only be described by those points which, as the moving centroid revolves, pass through the centre of the stationary centroid; or conversely by those points through which the centre of the stationary centroid might pass if the pair were inverted. Such points are, however, only those of the c on c e n t r a l point-paths. In other words —the points of concentral point-paths describe homocentral point-paths if the pair be inverted. Thus the homocentral curves 6, Plate I., 2, and 5, Plate II., 2, are described by points in the concentral curve M_7 of Plates III., 2 and IV., 2. The points of the concentral curves which have been used can easily be seen from the figures.

This way of looking at the curves may also be extended to the examination of trochoids,—which, indeed, we should actually obtain if the centroids of the pair of elements were circles. Here the concentral roulettes would be the circles described by the centres of the rolling circles, and the homocentral paths those star-shaped figures which are described by points in the circumference of a circle concentric with the rolling circle, and having a radius equal to the difference between the radii of the two centroids.[18]

§ 25.

Figures of Constant Breadth.

The conclusions of § 21 lead us synthetically to a series of other pairs of elements, of which we may examine a few here. If upon any plane figure two parallel tangents be laid, as $A B$ and $C D$, Fig. 99, the distance between them, c, measures the extension of the figure in the direction of the normals of restraint. This extension may be called the b r e a d t h of the figure,—in general, it is not constant for the same figure. There are figures, however, in which the breadth *is* constant; in which, that is, all pairs of parallel tangents on opposite sides are at the same distance apart. The circle gives us a familiar example of this. If on any figure having this property we place two pairs of the restraining tangents supposed, they touch it in four points and completely restrain it—as was shown in § 18—from sliding. This restraint, however, does not prevent the turning of the figure, and this turning may be so arranged that it can take place about one point only. That this may be the case the normals to the four points of restraint must intersect in that point —the opposite normals, that is to say, must coincide, as in Fig. 100, where the normal on a passes through c, and that on b through d.

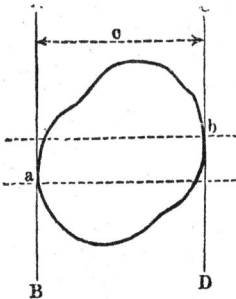

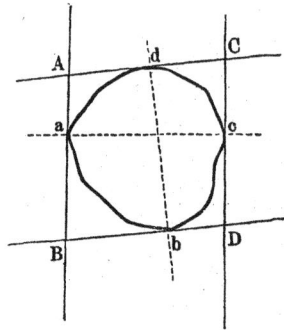

Fig. 99.   Fig. 100.

Then, the breadth of the figure being constant, the restraint is un-interrupted or continuous for all alterations of its position within the four tangents, from which it follows (see § 21) that the normals always intersect in one point. This shows that figures of constant breadth have the property that on any radius there lies not only the centre of curvature of the element of the circumference to which that radius belongs, but also the centre of curvature for the opposite element. The four tangents enclose a square, or more generally a rhombus, as $A\,B\,C\,D$. The foregoing shows also that every figure of constant breadth can be constrained in such a rhombus, so that from it and the rhombus a pair of elements may be formed.

§ 26.

Higher Pairs of Elements.—Equilateral Curve. Triangle and Rhombus.

Figures of constant breadth can easily be constructed of circular arcs. If from the corners of an equilateral triangle $P\,Q\,R$, Fig. 101, arcs be drawn with radii equal to the length of one of the sides, we obtain a figure which we may call an **equilateral curve-triangle.** This has everywhere a breadth equal to the side $P\,Q$, so that it can be constrained in a square or rhombus $A\,B\,C\,D$, the distance between whose opposite sides is $P\,Q$. In the square the normals intersect at right angles, in the rhombus obliquely. If we

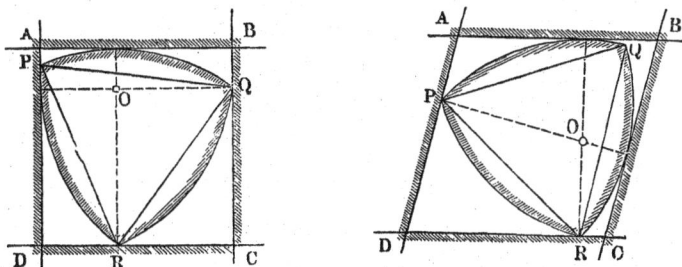

Fig. 101.

take these figures as cross sections for cylinders, and give to the latter such profiles as will prevent end-long sliding, we have constructed a higher pair of elements.

We may examine the centroids of these figures,—taking first that of the square. In Fig. 102 the instantaneous centre of motion is the point O, which in the position shown in the figure falls upon the vertical bisector $R\,O$, of the square, and is also the centre of one side of the triangle. Let the curve-triangle make *L. H.* rotation about this point. Its corner P then slides downwards along the side $A\,D$, while R moves to the right along $D\,C$. The normals from P and R always intersect at right angles, so that the locus of the instantaneous centre is that of the vertex of a right-angled triangle, of which the ends of the hypothenuse slide upon the arms ($D\,A$ and $D\,C$) of a right angle. O is thus always the corner of a rectangle $P\,D\,R\,O$, of which the diagonal $(O\,D)$ is constant and is equal to $P\,R$. The locus of O, or centroid, is therefore a circular arc having D for its centre and $P\,R = P\,Q = A\,B =$ the length of the side of the square, for its radius. The centre continues in this curve until R has arrived at the same distance from C at which P is shown from A in the figure —*i.e.*, up to the point 2. The chord $P\,Q$ then slides in the same way on $A\,B$ and $A\,D$, giving the arc 2, 3, similar to the former one, as the continuation of the centroid, in the same way the arc 3, 4 is obtained, and lastly the arc 4, 1. The centroid for the square is therefore a curve-square having for its sides four circular arcs drawn from the four corners of the square with radii equal to its sides.

To find the centroid of the triangle we invert the pair,—that is, imagine the triangle stationary and the square moving upon it. The

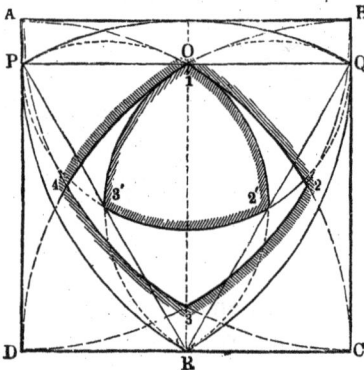

FIG. 102.

centroid is then the locus of the vertex O of a right angled triangle having its hypothenuse $=P\,R$, *i.e.* a circle with the diameter $P\,R$ described from its middle point 3'. The arc of this circle which forms part of the centroid ends at 2', the middle point of the side $Q\,R$. Then follows the similar curve 2' 3', and lastly,—returning to O,—the similar curve 3' 1. Hence the centroid of the curve-triangle is itself a curve-triangle, and is equilateral, its sides being arcs described from the centres of the sides $P\,Q$, $Q\,R$, and $R\,P$, and having radii equal to half their length.

As the one figure rolls relatively to the other, the curve 1.2' rolls on 1.2, then 2'.3' on 2.3,—3'.1 on 3.4, and so on. In order again to reach its initial position the instantaneous centre must traverse equal distances on both centroids,—must therefore traverse the three sides of the curve-triangle four times, and the four sides of the curve square three times. For each completed rolling of the former on its three sides the element to which it belongs turns through an angle of 90° relatively to the square, so that after the

first revolution its corner 1 comes to the point 4, after the second
to 3, after the third to 2, and after the fourth back again to 1.

§ 27.

Paths of Points of the Curve-Triangle relatively to the Square.

(Plates V. and VI.)

The point-paths of this pair have, as follows from the nature of
their centroids, a close relationship to those of the pair shown in
Plates I. to IV. All paths of points of the curve-triangle relatively to
the square consist of arcs of hypocycloids or hypotrochoids, which are
in this case ellipses,—while all point-paths of the square relatively
to the triangle consist of arcs of peri-trochoids (including the
special case of cardioids, as on p. 123). Let us first suppose the
square fixed and the triangle in motion.

Plate V. shows a series of point-paths for which the describing
points lie upon a line drawn from the centre M of the triangle perpendicular to the chord $P Q$. Point 1 gives a four-sided figure
composed of elliptic arcs, its corners being elliptically rounded off
by a pair of similar arcs having a common tangent; at 1 for instance
one of these is given by the rolling of $m_1 m_2$ on $O_4 O_1$, and the other
by the rolling of $m_1 m_3$ on $O_4 O_3$. When m_2 reaches O_1, $m_2 m_3$
begins to roll on $O_1 O_2$. The point 1 is however so chosen that $m_1 1$ is
equal to the radius of the curves of the centroid $m_1 m_2 m_3$, and being
therefore upon the circumference of the smaller Cardanic circle,
it describes a straight line. Thus the portions of the point-paths
passing through A and B,—the centres of the greater Cardanic
circles $O_2 O_3$ and $O_2 O_1$;—are straight lines,—or more strictly, are
elliptic arcs which have become straight lines. The continuation
of the curve can easily be understood. It is completed when each
side of the centroid $m_1 m_2 m_3$ has rolled on each side of the
centroid $O_1 O_2 O_3 O_4$, and consists therefore of twelve (four times
three) separate arcs.

The point 2 describes a four-cornered figure with slightly concave
sides, and the point 3 a similar figure in which the concavity is
more distinct; in both cases all the curves are elliptic. The end

point m_1, fourth in the series of describing points, describes immediately right and left from O_4 straight lines directed towards the
centres C and D of the base circles O_4O_1 and O_4O_3, like the straight
lines O_1m_2 and O_3m_3 from the homologous points m_2 and m_3; joining
each pair of straight lines is a circular arc described by the centre
of the circle m_2m_3 rolling in O_1O_2, m_3m_1 in O_2O_3, etc. This
point-path is the common form for this series of curves.

The fifth point describes a four-cornered figure of elliptic arcs, in
which loops make their appearance. This figure is shown on a
doubled scale in Plate V., 2. In the point-path of 6 the loops have
separated and intersect each other, while with point 7, which is the
centre M itself, the loops run over each other in a dumpy figure
which in each period, or whole revolution of the element, traverses
M three times. (See § 23). The centre M_1 of the square is also the
centre of this point-path, which is the smallest of those which can
be obtained by the motion of the curve triangle, or what we have
called the concentral form of its point-path.

Plate VI. shows curves described by points on the prolongation
through M_1 of the line on which were the points just considered.
Point 1 gives us a four-cornered figure consisting of elliptic arcs—
point 2 a straight-sided quadrilateral, covering part of the square
$ABCD$, but having elliptically rounded corners;—the points 3 and
4 elliptic quadrilaterals with concave sides. The last figure is shown
to double the scale in Fig. 2; within it is the path for the point 5,
which being a point upon the centroid gives us again a common
form for this series of curves;—it consists of four elliptic arcs
with tangential prolongations at each cusp. The path 6 is described
by the point which in Plate VI., 1, coincides with the centre point
M of the square ;—it is therefore the homocentral form of this series
of curves; the point M, lastly, again gives us the concentral
curve 7.

The point-paths 1', 2', and 3', shown in dotted lines, are examples
of those described by points which do not lie in either of the three
principal axes of the centroid.

PLATE V.

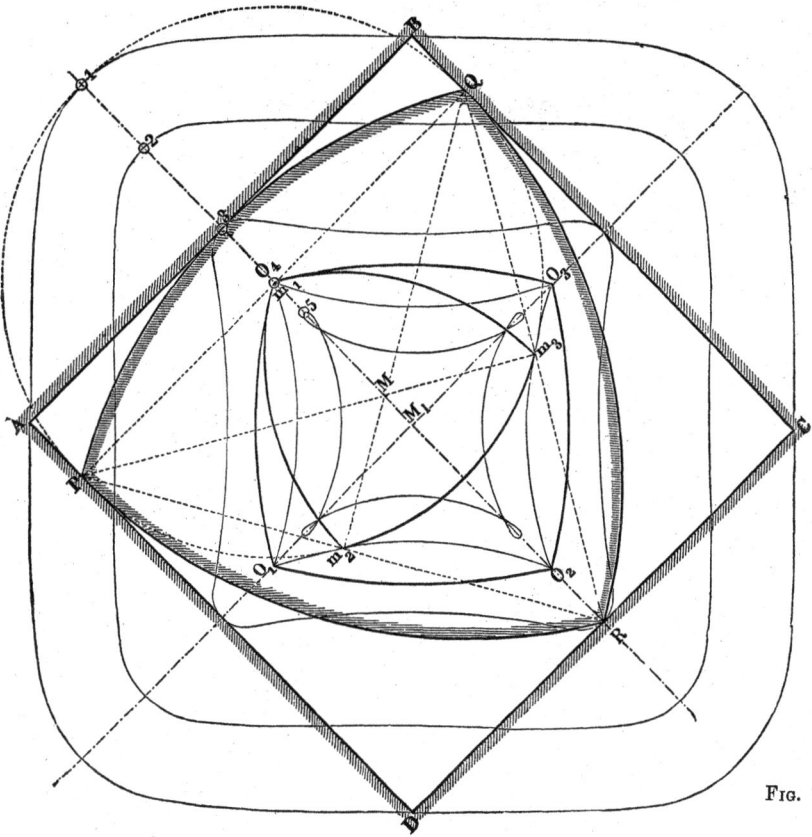

FIG. 1.

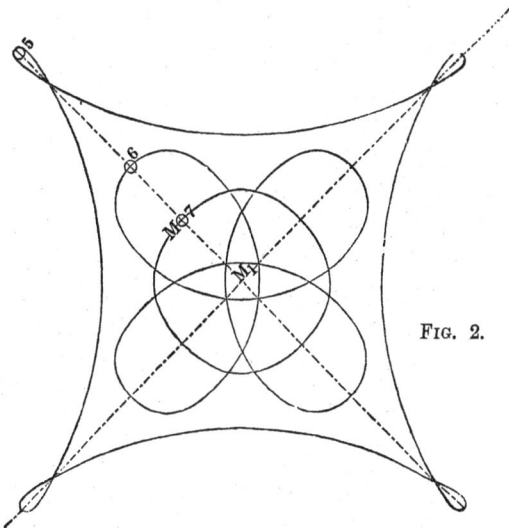

FIG. 2.

PLATE VI.

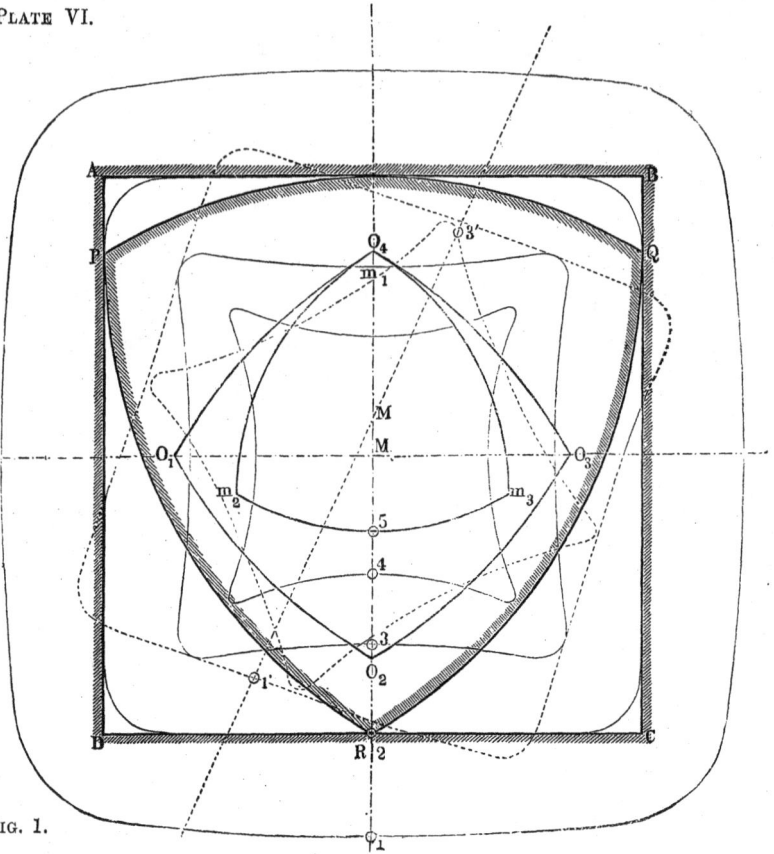

FIG. 1.

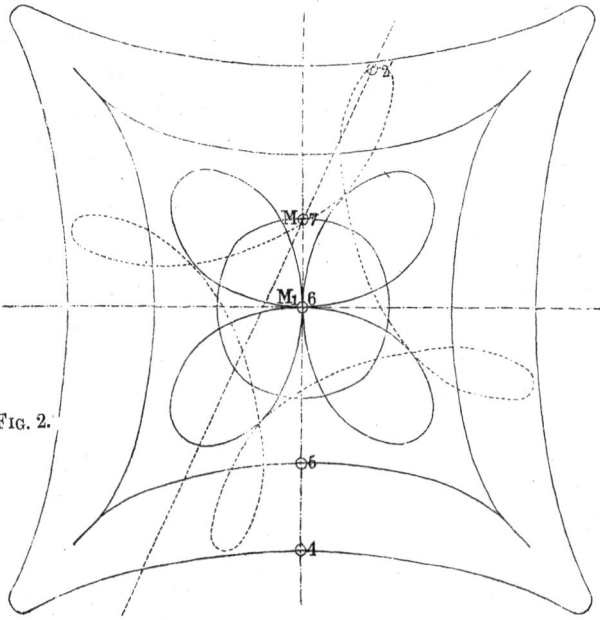

FIG. 2.

PLATE VII.

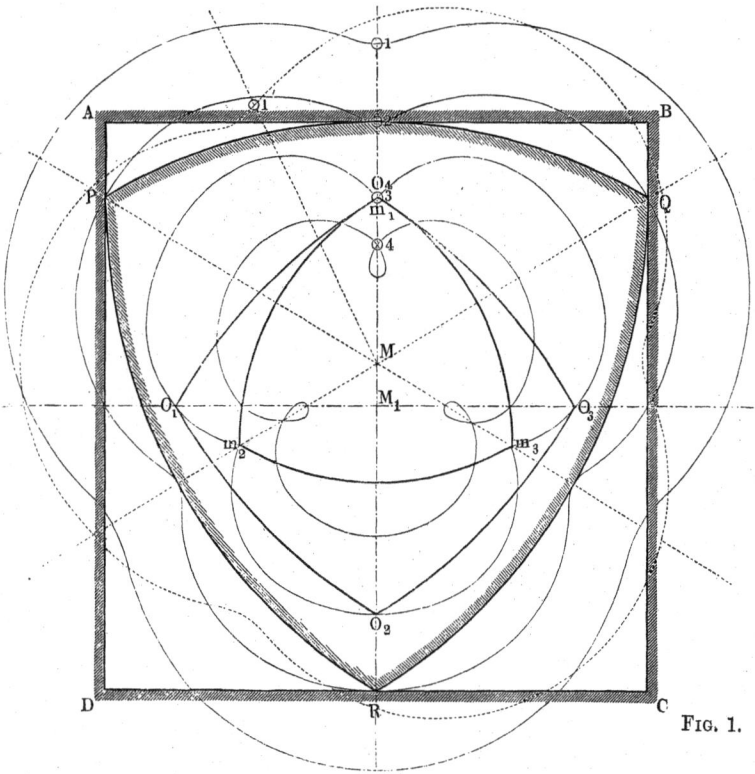

FIG. 1.

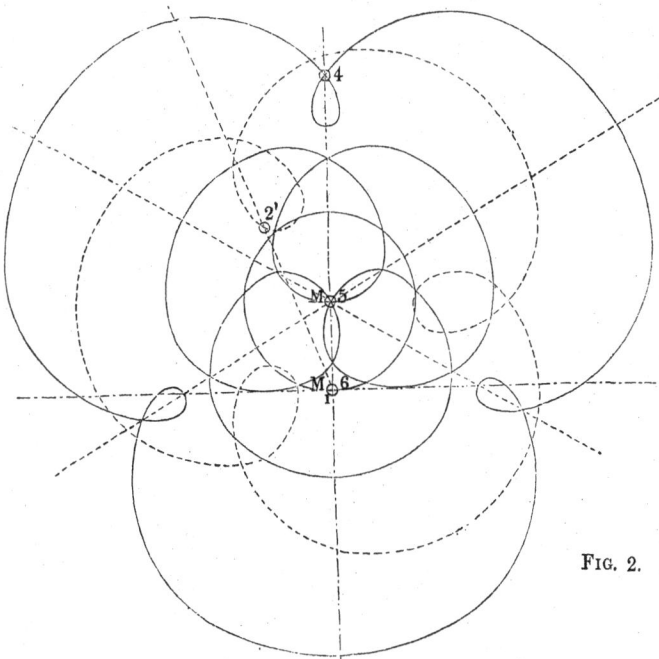

FIG. 2.

PLATE VIII.

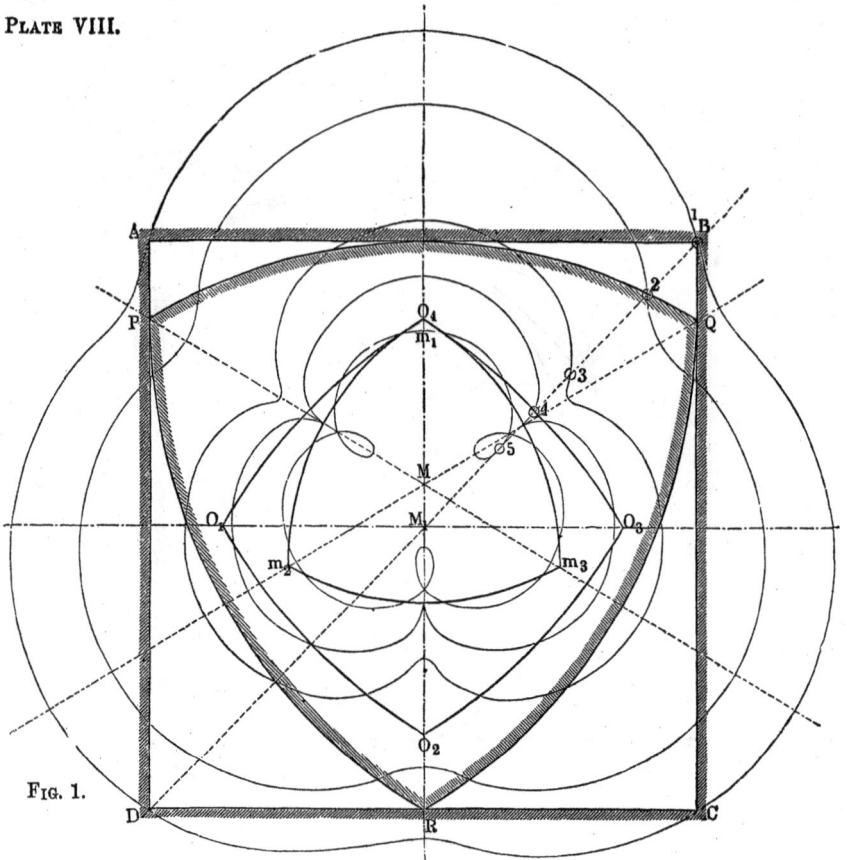

FIG. 1.

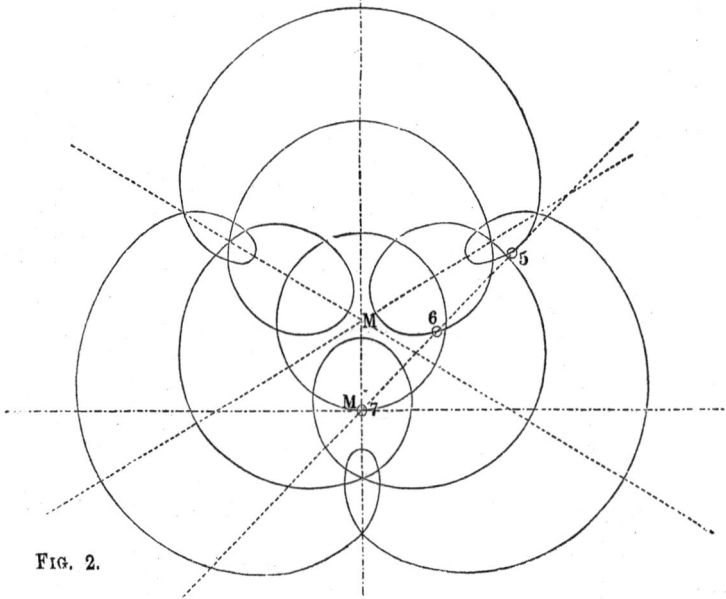

FIG. 2.

§ 28.

Paths described by Points of the Square relatively to the Curve-triangle.

(Plates VII. and VIII.)

Let now the triangle $P\ Q\ R$ be fixed, and the square $A\ B\ C\ D$ moved upon it. In Plate VII., six paths described by points upon the line $M\ O_4$ are shown. Nos. 1 and 2 give curtate roulettes built up of peri-trochoidal arcs,—No. 3 is the common form of the curve, No. 4 a prolate curve. This is repeated on a double scale in Fig. 2. No. 5 is the homocentral, No. 6 the concentral curve. The last resembles a circle very closely, but consists of peri-trochoidal arcs which so cover one another that the curve is fourfold,—being traversed by its describing point four times in every period.

The curves 1′ and 2′ belong to points in a line lying between two principal axes, the first is a curtate, the second a prolate roulette. Plate VIII. shows seven roulettes corresponding to points on the line $M_1 B$. Nos. 1, 2, and 3 are curtate roulettes, No. 4 the common form of the series, Nos. 5 and 6 prolate roulettes, No. 7 the concentral form, the same as No. 6 in Plate VII., 2.

§ 29.

Higher Pairs of Elements:—other Curved Figures of Constant Breadth.

(Plates IX. to XIII.)

We found in § 25 that every figure of constant breadth can be constrained in a circumscribed rhombus, so that a pair of elements may be made from it and the rhombus. Eight examples of such pairs are given in Plates IX. to XIII.,—they are chosen as specially suited for showing the extraordinary variety of constrained motions to which this proposition leads us. On account, however, of the completeness with which we have examined the pairs already considered we shall be able to dismiss these more shortly.

In Plate IX. we have the already known equilateral curve-triangle enclosed in a rhombus with angles of 60° and 120°. The form of

the centroids here differs very greatly from that of the centroids
in Plates V. to VIII. The centroid for the curve-triangle is a three-
rayed figure, built up of three circular arcs of a radius $C\,Q = C\,R =$
half the length of the side of the rhombus;—the centroid of the
rhombus is an equilateral duangle, its sides arcs described with
the radius $B\,A = B\,C =$ the side of the rhombus;—or twice the
magnitude of the radius with which the sides of the first figure
were described. The centroids are therefore again arcs of Cardanic
circles, and the point-paths built up of trochoidal arcs.

Some of the paths described by points of the triangle relatively
to the rhombus are given. Point I, on the centre of the perpendicular $A\,Q$ to one of the sides of the rhombus, gives a figure
symmetrical about two axes, and resembling in profile a double-
headed rail;—the centre, II, of the triangle describes its concentral point-path, which is nothing else than the diameter $E\,F$, of the
duangle—(in the direction $D\,B$)—this line being traversed three
times in each whole period. This concentral point-path coincides
with the homocentral,—and is at the same time (as the path of a
point on the centroid) a common form of the curve. The point-
path I′ is a curtate roulette of the rhombus; the point-path II″ is
a prolate roulette for the same figure. Every point in the diameter
$E\,F$ describes a homocentral curve in the curve-triangle;—one of
these—that corresponding to the points E and F—is given. The
variety of the forms here taken by the roulettes shows that it is
impossible to draw conclusions from analogy alone as to the general
character of the forms of any series of point-paths.

Plate X. 1. Equilateral curve-pentagon in Square.—The
curve pentagon is constructed by describing arcs of circles having
a radius equal to the diagonal about each of the corners of a regular
pentagon, and a figure of constant breadth is thus obtained. The
centroids are:—for the square $A\,B\,C\,D$ the four-cornered figure 1′ 2′
3′ 4′, consisting of arcs of circles having the corners of the square
for their centres and the side length $P\,Q$ of the pentagon for
radius;—for the pentagon another equilateral curve-pentagon,
described with radii equal to the half side-length of the pentagon
from the centres m_1, m_2, m_3, m_4, m_5 of its sides. The centroid of the
square rolls within that of the pentagon, and in every period
each side of the one centroid must roll upon every side of the
other, so that the instantaneous centre traverses the one five times

PLATE IX.

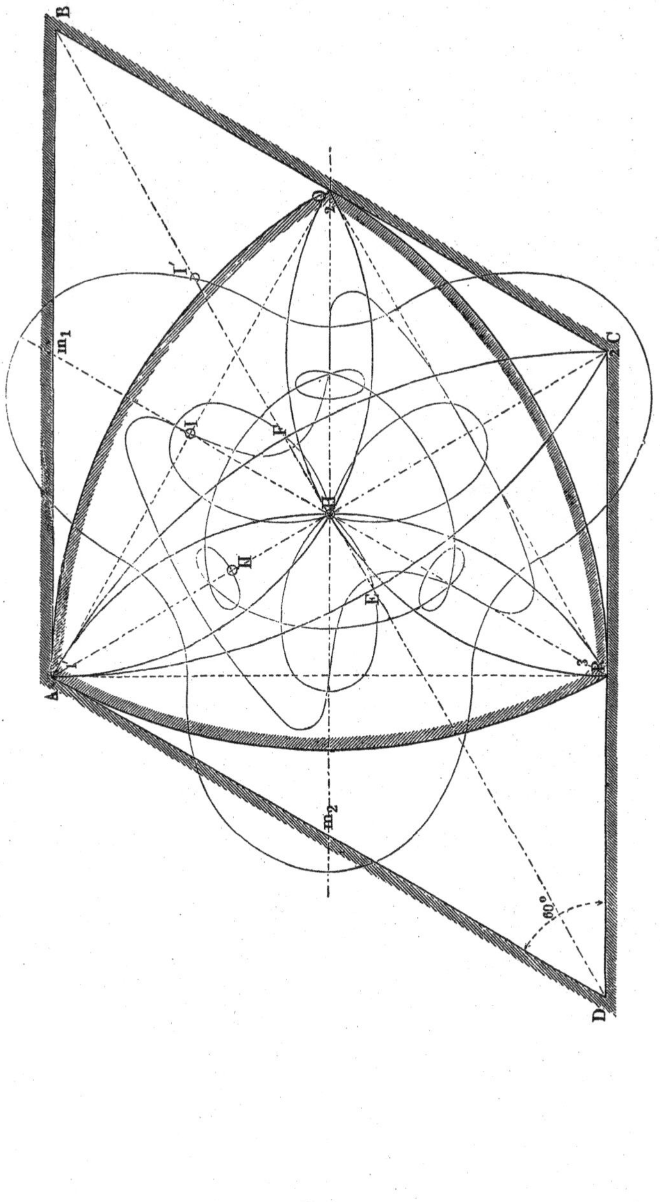

PLATE X.

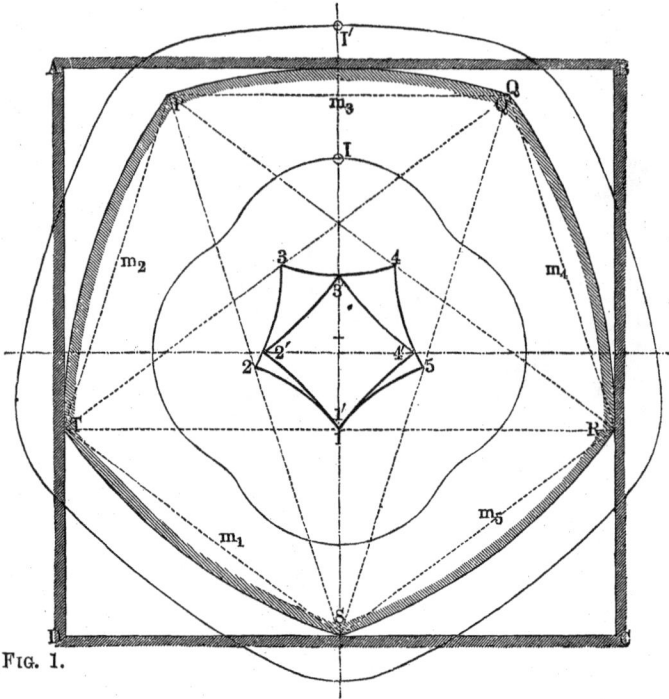

FIG. 1.

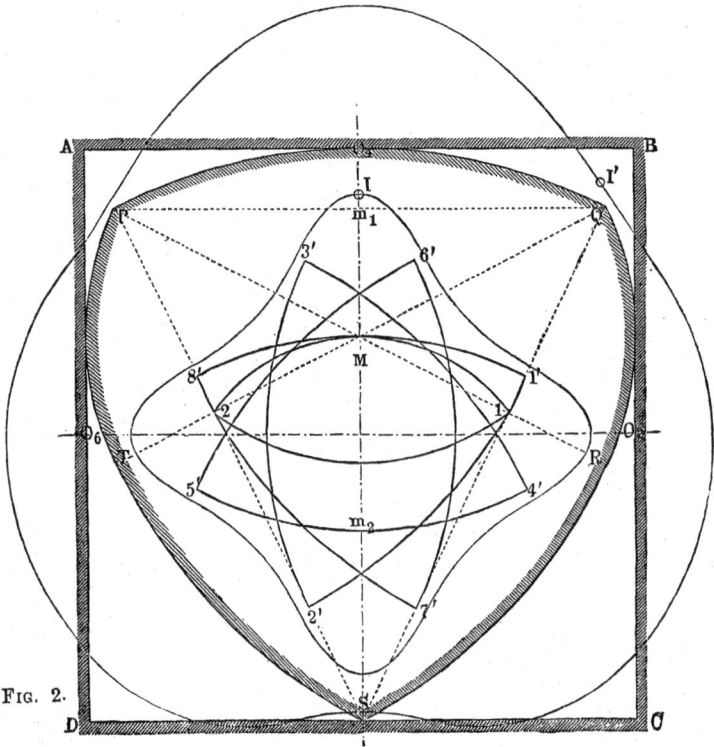

FIG. 2.

PLATE XI.

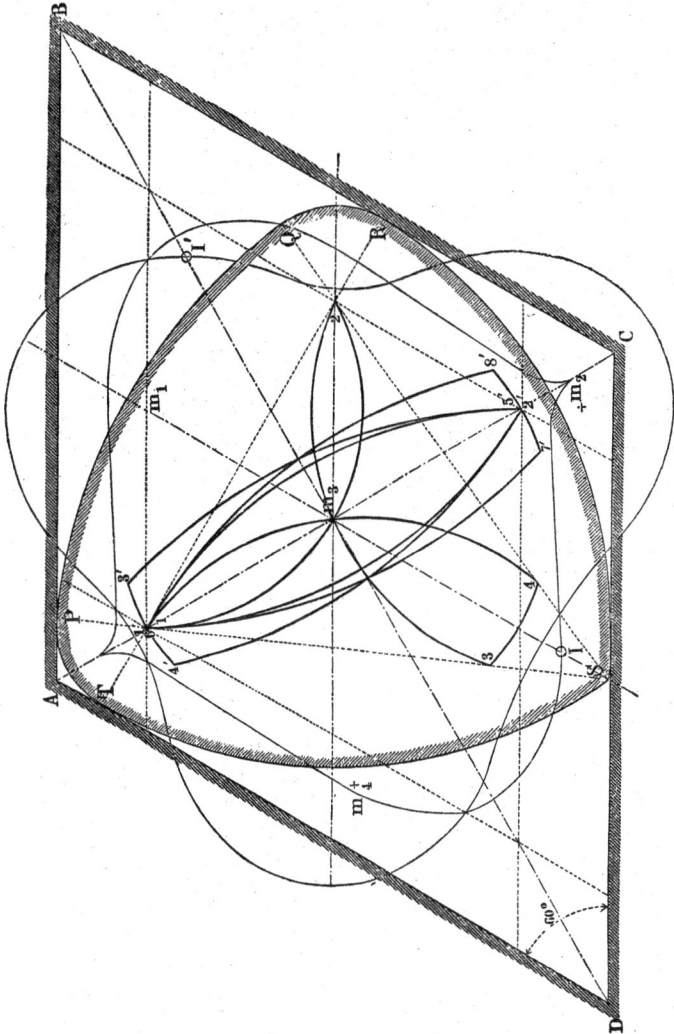

PLATE XII.

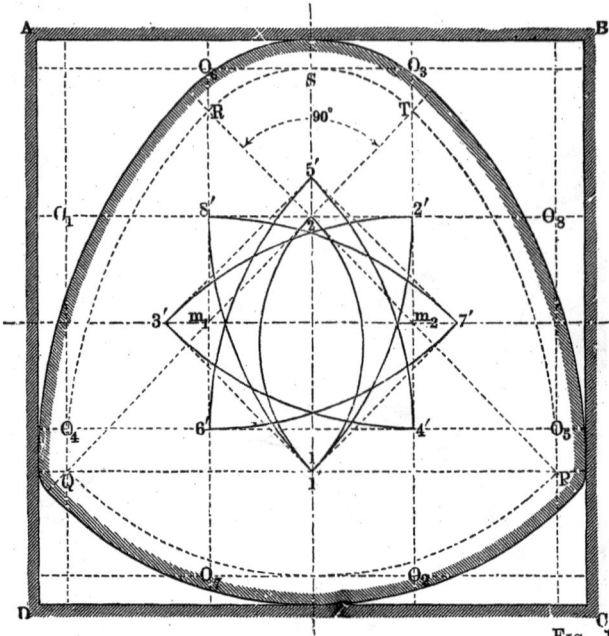

FIG. 1.

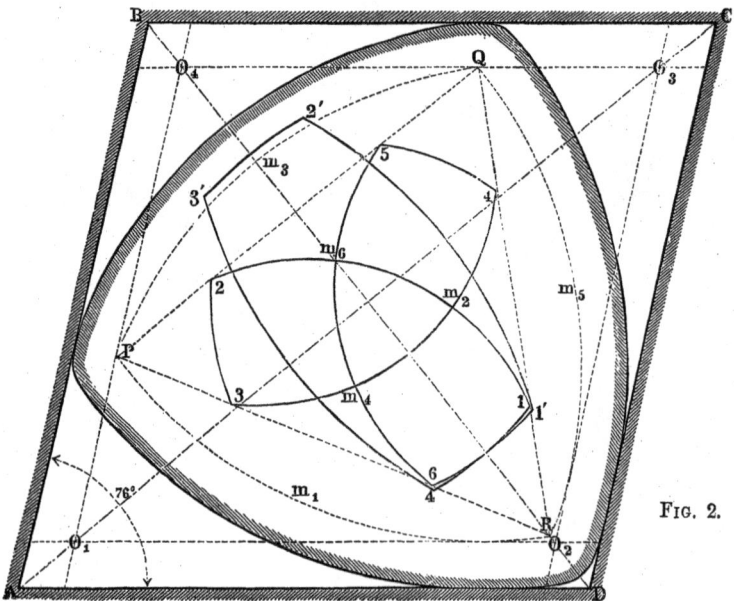

FIG. 2.

and the other four times in each period. The point-path marked I is described by the pentagon moving relatively to the square and is (the describing point lying beyond the centroid) a curtate curve; the I′ is also a curtate roulette, it is however described by a point of the square moving relatively to the pentagon.

Plate X., 2. **Heart-shaped figure, made up of five circular arcs, in square.**—PSQ is an isosceles triangle with vertex angle $PSQ = 53°$. The arcs PQ, ST, and SR are described from S, Q, and P with a radius equal to the side AB of the square,—the arcs TP and QR from the intersection M of the lines PR and QT, with radii equal to half the side of the square. The figure has therefore the constant breadth AB. The centroids for figures of these proportions are: for the curved figure a duangle 1.2 not equilateral; for the square an eight-rayed star of curves of different radii. The arcs of the centroids which roll upon each other belong always to Cardanic circles. Two point-paths I and $I′$ are shown. The roulettes in their common form,—described by the corners 1 and 2 of the duangle (centroid) are specially characteristic. They are squares having for their corners the points 1′ 3′ 5′ 7′ and 2′ 4′ 6′ 8′.[19]

Plate XI. **Isosceles curve-triangle in rhombus.**—Upon an isosceles triangle 1 S 2, having a vertex angle $< 60°$, the circular arcs ST and SR, having radii $S2$ and $S1$, are drawn to their intersections T and R with 1.2 produced;—from the same centres, and with radii 1 T and 2 R, the arcs TP and RQ are drawn until they intersect in P and Q the sides $S1$ and $S2$ of the triangle produced; lastly, P and Q are united by a circular arc drawn from the centre S. The figure thus inclosed has the constant breadth QS. It is here paired with a rhombus having angles of 60° and 120°. The centroids are somewhat complex, but consist as before of arcs of Cardanic circles, the centroid of the triangle having four such arcs, that of the rhombus eight. Two point-paths I and $I′$, belonging respectively to the triangle and the rhombus, are shown.

Plate XII., 2, shows another curve-triangle in a rhombus. The former is equilateral as in Fig. 1, but the radii of its sides (which are as before arcs described from the three corners PQR) are somewhat longer than the sides of the triangle,—and the corners are rounded off with radii equal to this excess of length. The motion which occurs is exactly the same as would be given by a pair consisting of the dotted curve-triangle PQR of the normal form and the enveloping rhombus $O_1O_2O_3O_4$. The rhombus is drawn with angles of 76° and 104° instead of 60° and 120° as before. This difference makes a notable alteration in the centroids, which resemble those of Plate IX. with the corners removed. The nature of the changes of form in the point-paths from those obtained before can thus be readily traced.

Plates XII., 1, and XIII. show three more pairs constructed analogously to those we have already examined. The first of these is remarkable, both on account of the regularity of its centroids and because the end points 1 and 2 of the smaller centroid again describe squares. The curved element in Plate XIII., 1, is formed like that in Plate XI., but with a smaller vertex angle at S; that in Plate XIII., 2, is similar to the one in Plate X., 2. The difference between the centroids in Figs. 3 and 8 is very remarkable. The manner in which the one centroid rolls upon the other is indicated as distinctly as has been possible by the numerals. These examples show clearly the multitude of motions which can be obtained by means of the higher pairs, and show at the same time how wonderfully the use of centroids simplifies the comprehension of these complicated motions.

§ 30.

General Determination of Profiles of Elements for a given Motion.

The forms of the pairs of elements considered in the foregoing investigation were found synthetically ;—starting from the general solution of the problem of restraint we built up constrained pairs in accordance with its conditions, and afterwards ascertained their relative motions by the construction of their centroids. This latter part of our investigation was again analytical. It furnishes us however with the means of solving a further synthetical problem, —the construction, namely, of pairs of elements for a given motion,—i.e. for given centroids. For the forms which we found necessary for the continuous reciprocal restraint of the elements are reciprocally envelopes for one relative motion, and for that one only which is determined by the centroids. We have chosen hitherto forms conditioned only by the property of reciprocal restraint, and have from them determined their centroids. We may now reverse the problem, by assuming the centroids as given

PLATE XIII.

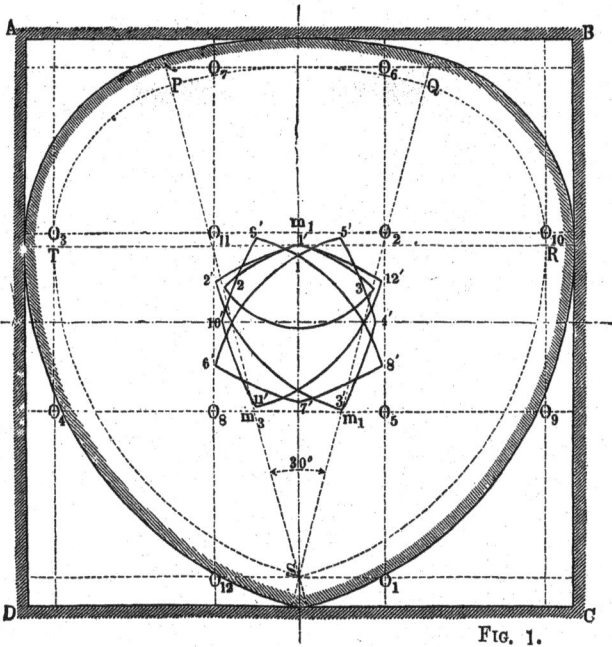

FIG. 1.

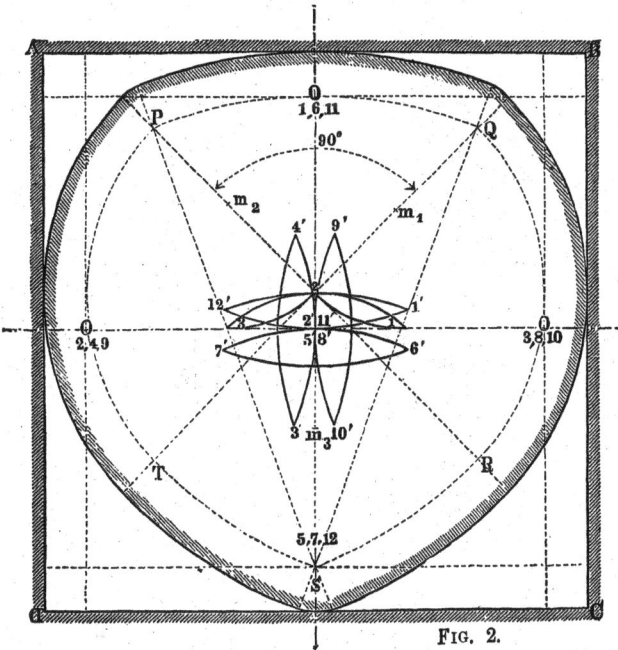

FIG. 2.

and determining by their means the reciprocally restraining figures. This problem is the one which occurs by far the most often in connection with machine design, and has frequently to be solved both for simple and for complex motions.

We shall here simply indicate the general methods of procedure in this case. These are very numerous, but admit of being classed under the seven headings examined in the following paragraphs, in which we shall consider cylindric rolling only in the first place.*

§ 31.

First Method.—Determination of the Profile of one Element, that of the other being arbitrarily assumed.

If the profile of one element of a pair of which the centroids are known be arbitrarily assumed, the centroid of the unknown element may be brought to rest, and that of the assumed one rolled

---

* The following note may make clearer to some readers the nature of the problems treated by Prof. Reuleaux in §§ 31–37. Let *A* and *B* be any two centroids, and *a a'* and *b b'* the profiles of bodies whose relative motions the centroids represent. It is required so to form these profiles that during the rolling of the centroids they shall remain continuously in contact. The necessary condition for this may be thus shown. Let *O* be the point of contact of the centroids, and let the two profiles be touching at *a* ; draw their common tangent *t t'*, join *O a* and draw *T T'* perpendicular to it. Then suppose *B* fixed, and *A* free to roll upon it. The instantaneous motion of the point *a* can only take place about the instantaneous centre *O*, that is in the direction *T T'* perpendicular to *O a*. But if *a* move towards *T'* it leaves the profile *b*, while it cannot move towards *T* because the point *b* restrains motion in that direction. Hence the assumed profiles *a a'* and *b b'*, in no way fulfil the requirements of the problem. They show very clearly, however, the condition necessary for this fulfilment, for it is obvious that the point *a*, moving about *O*, can remain in contact with *b* only if the tangent *t t'* to that profile coincide with the line *T T'*. *O a* is normal to *T T'* by construction,—we may therefore express the condition generally by saying :— in order that two elements may remain in contact during the rolling of their centroids the normal to the common tangent of their profiles must always pass through the instantaneous centre, or point of contact of the centroids. It must be remembered that (except in one special case), the profiles themselves do not roll upon one another, but slip or grind to a greater or less extent.

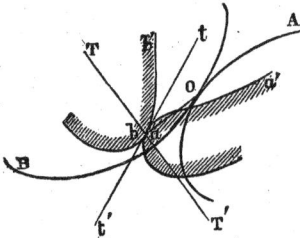

upon it,—any number of consecutive positions of the latter may be drawn, and a curve enveloping these, if rigidly connected to the stationary centroid, gives a figure with which the known element remains in continuous contact during its motion. Such a figure will serve as the profile of the stationary element, if it can be provided with a sufficient number of points of restraint. This may readily happen, indeed the figure may have even more such points than are necessary. In these cases only such portions of the figure need be constructed as suffice to make the restraint continuous. We have seen this in the case of the curve-triangle and square, and also with the duangle and triangle, where the enveloping curves were not drawn in the corners of the square and triangle, their omission not affecting the restraint. The method is therefore available ;—we must examine the manner in which it can be practically carried out.

If A and B (Fig. 103) be the two given centroids, and $a\,b$ the assumed profile of the element corresponding to A, then if any point, as b, of the assumed profile be also a point in the centroid, the corresponding point of contact O of the centroid B gives at once one point in the profile to be found. In order to determine a second point in it,—that for instance which shall correspond to a,—let a normal $a\,c$ be drawn to the given profile at a. It cuts the corresponding centroid A in c. If we roll the centroid B upon A until c becomes the point of contact,—*i.e.* the instantaneous centre,—then a must be the point of contact of the profiles, and $a\,c$, which we may call the central distance of a, must be the common distance of the two touching profile points from the contact points of their respective centroids. The distance at which the point of the centroid B originally at O must then be from a can at once be seen when B is in the position $b_1\,c$ (in which contact occurs at c); it is simply $a\,b_1$, the centroidal arc $b_1\,c$ being $=b\,c$. If now we have $b\,c_1 = b_1\,c = b\,c$, and describe from c_1 with

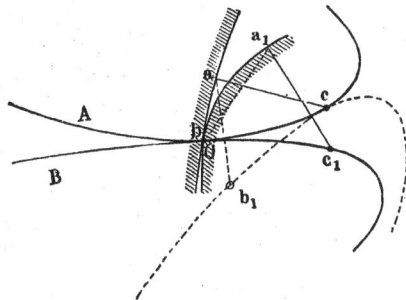

Fig. 103.

radius = the central distance $c\,a$, and from b with radius = $b_1\,a$ arcs of circles, the intersection of these gives us the required profile-point, a_1. In this way the profile $a_1\,b$ . . . can be exactly determined point by point. If great accuracy be not required the following approximate method (by Poncelet), can be used. Erect normals in a sufficient number of points a, c, e . . . of the assumed profile, and continue these to their intersections b, d, f . . . with the corresponding centroid. Find the corresponding points b_1, d_1, f_1 . . on the centroid B, and describe from them circular arcs with radii equal to the central distances $a\,b$, $c\,d$,

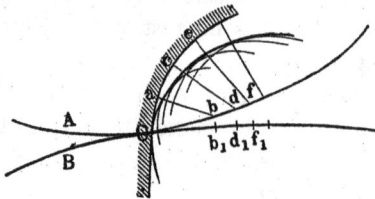

FIG. 104.

$e\,f$, etc.; a curve enveloping these arcs gives a near approximation to the required profile.

The foregoing method in its application to spur-wheels is known in Germany as the "general method" of drawing teeth,

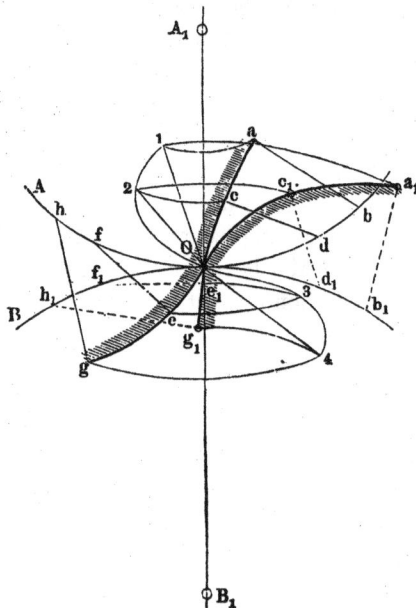

FIG. 105.

(*allgemeine Verzahnung*). In this case the centroids are commonly circles, which renders some simplification possible. The following is the way in which I have used* this method for the construction of wheel-teeth upon circular centroids.

In Fig. 105 $a c O e g$ is the given profile, A the corresponding centroid drawn about the centre A_1 ; B is the centroid, drawn about B_1, of the element whose profile is to be found. After drawing the normal $a b$ to any point a in the profile we must next find the position in which a will be when it itself becomes the point of contact with the as yet unknown profile, for which purpose we may suppose both centroids to be turning about their centres A_1 and B_1, assumed to be fixed points in the plane of the paper. The point of contact 1 lies necessarily at the intersection of a circle described about A_1 with a radius $A_1 a$, with a second circle described about the instantaneous centre O with the central distance $b a$, for at the moment of contact the normal $a b$ must pass through O. B has meanwhile turned through an arc $O b_1 = O b$. The new profile point a_1 which corresponds to a must therefore be at the intersection of a circle drawn from B_1 with radius $B_1 1$ with another circle drawn from b_1 with radius $O 1$. In the same way the points of contact $2, 3, 4 \ldots$ and the corresponding points $c_1, e_1, g_1 \cdot \ldots$ in the profile, can be found. The series of points $1, 2, 3, 4 \ldots$ give us the line of contact, or locus of all the successive points of contact of the two profiles. The line joining the point of contact with the instantaneous centre O is for each instant both the direction of restraint, and the direction of the pressure between the two profiles.

The method of which we have here given three applications furnishes an immense variety of profile forms, among them many which are of little or no practical use. Those curves especially which contain cusps or loops, or form contracted spirals etc.,—(see Fig. 106)—are commonly unsuitable; they are not useful although they are geometrically correct,— fulfilling the required conditions as to continuous restraint in motions determined by the given pair of centroids. If the applica-

FIG. 106.

tion of this method furnish us in any case with such impracticable profile-curves it becomes necessary to choose some more suitable

* Published first in *Der Constructeur*, 2nd Ed. 1865.

form for the assumed profile,—proceeding thus by trial. Absolute freedom in the choice of the form of the first profile is to this extent limited, and a further limitation arises from the fact that such curves as have normals which cut the centroid at the point of contact at too great an angle (as *e.g.* the normal 01 in Fig. 105) are not suitable for the profile, for with these most injurious friction will occur, if not complete "jamming." Those parts of profiles, lastly, whose normals do not pass through the centroid, and therefore cannot be normals of restraint, are entirely unusable. Thus in the employment of this method in Applied Kinematics a number of unsuitable and unusable forms must be withdrawn from those which can be used for the (otherwise) arbitrarily chosen form of the assumed profile.

§ 32.

Second Method.—Auxiliary Centroids.

The method just described gives the profile of a single element only; by that which we have now to examine the two profile forms possessing the required property of continuous restraint are determined at the same time.

Let *A* and *B*—Fig. 107—be again the pair of centroids in contact at *O*. If any third curve *C* touch *A* and *B* at the same point, and roll with them as they roll, the three curves will always have a common point of contact, and that point will always be the instantaneous centre. Then any point *D* fixed to *C* describes a roulette relatively to each of the centroids *A* and *B*. The two curves thus obtained,—*a D* . . . and *b D* . . . .have in any position a common point, as *D*, and also a common normal, as *D O*. They may therefore serve as profiles for elements,—for their common normal always passes through the point of contact of the centroids. Their practical usefulness depends on the same conditions as those mentioned in the last section.

The centroid *C*, by the help of which we have obtained the profiles, we may call an **auxiliary** ce.ntroid. If the describing point *D* be taken on this curve or within it, the roulettes obtained remain always upon one side of the primary centroid to which they belong; there is then always space upon the opposite side of the same curve to construct similarly another pair of roulette profiles. The whole procedure may therefore be there repeated,—a second auxiliary centroid, similar or dissimilar to the first, being employed to give the two new profiles c and d—Fig. 108.

If the describing points are points upon the auxiliary centroids, all the roulettes must extend to the primary centroids,—we can therefore join the profiles a and c, and also the profiles b and d, into one piece. Then by repeating the process for a number of positions on each of the centroids we can obtain profiles which may serve as those of tooth-formed projections upon the element to which they belong. A regular series of such projections with

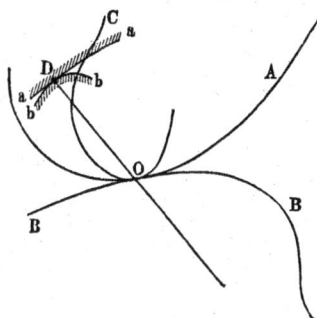
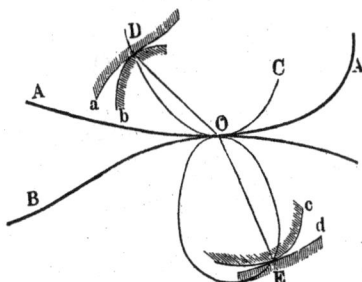

FIG. 107.   FIG. 108.

corresponding hollows between them gives us the familiar spur-wheel. The portion of the centroid lying between homologous points of two consecutive teeth is there called the pitch, and the centroid itself is the pitch-line, or if circular the pitch-circle. Circles (" describing " circles) are used as the auxiliary centroids. The teeth must carry such portions of the roulette profiles that the restraint is never interrupted; such portions must be at least so large that the restraint by each tooth lasts while the centroids roll through a distance equal to the pitch ; the contact between each pair of teeth, in other words, must continue for at least this period.

With spur-wheels or toothed-wheels having cylindric axoids it may be further required that all wheels of the same pitch should gear with each other, that is that their tooth-profiles should communicate a motion to which the corresponding centroids (pitch lines) are circular.* Wheels so arranged we may call Set-

* If wrongly formed profiles be set to work with each other the motion of the

w h e e l s ;* any pair of them, equally pitched, will gear truly together. Willis† appears to have been the first to point out both this problem and its solution. Looking at it from the general point of view which we have here reached, it is evident that the solution of the problem is the use of s i m i l a r auxiliary centroids for those shown on opposite sides of the primary centroids in Fig. 108.

The delineation of wheel-teeth very early led geometricians to the use of roulettes as profiles of elements. Camus laid down its fundamental principles very clearly in 1733 in a little-known treatise, as Willis has shown.²⁰

Camus' predecessor, De la Hire, had apparently used this method before him, and he himself refers to the still earlier Desargues ‡ (1593—1662) as having used epicycloidal teeth, and thus preceded by many years Römer (1664—1710), so often mentioned as their inventor.

In the distinctness and completeness of its results, this method of forming element profiles by roulettes greatly excels the method first described, which indeed, in a certain sense, it includes. For we may conceive of the assumed profile of the first method as having been itself found by means of an auxiliary centroid. The second profile then becomes a roulette drawn by the same curve, which thus may be considered as actually the describing curve of the profiles, although it has not itself been drawn.

De la Hire also enunciated the general proposition as to the describing of roulettes which we are here applying, and which is of so great importance in Kinematics ; and it is to the same geometrician that we owe their name.²¹ The methods and propositions relating to them have hitherto hardly received their due development. The delineation of the auxiliary centroid in the method of § 31 is interesting—but not necessary,—the method itself is practically useful chiefly where a single result is all that is required.

---

bodies to which they belong will have different centroids from those originally assumed. Thus in the special case mentioned above, if the wheel teeth be not of the right shape, the wheels will not have a constant, but a varying angular velocity ratio. Their centroids will therefore no longer be circles, but irregular figures more or less nearly resembling them.

* I cannot find that any name has hitherto been used for them in this country. In Germany they are called S a t z - r ä d e r.

† *Trans. of Inst of Civil Engineers*, 1837, vol. ii. p. 91.

‡ Chasles, *Geschichte der Geometrie*, p. 83. (Sohncke.)

§ 33.

Third Method.—Profiles described by Secondary Centroids.

We have already mentioned (§ 9) the employment of secondary centroids instead of the original primaries, and have found that by their means problems which otherwise contained some difficulties could be easily solved,—that in certain cases we could use the secondary centroids interchangeably with the principal ones.

Among these secondary centroids one class especially is useful in the delineation of element profiles. This class consists of those in which two curves and their tangent are used to represent the motion. Such centroids are obtained if through a sufficient number of points in the two primary centroids we draw a series of secants making a constant angle to the tangents at the points through which they are drawn;—these secants envelope a pair of curves which, touched by a line rolling upon them, form together with it secondary centroids.

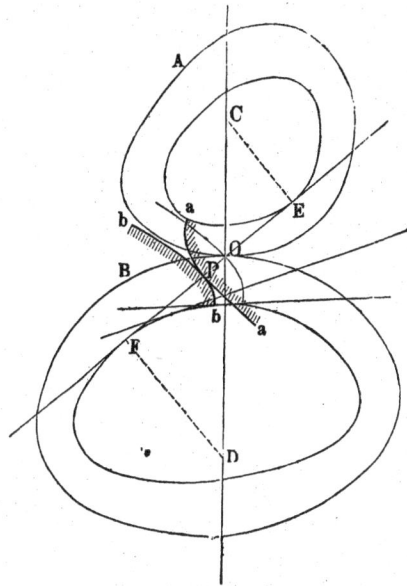

Fig. 109.

If for example C and D be the centres of curvature for the portions of the centroids touching at O, the ratio $\dfrac{C\,E}{D\,F}$ of the perpendiculars is equal to that $\dfrac{C\,O}{O\,D}$ into which the point O divides the line of centres; for any very small motion of the line $E\,F$, therefore, the same small angular motion occurs as if the primary centroids were rolling on each other. If now any point P in the straight line describe a curve relatively to A and another upon

B,—$a\,P$ and $b\,P$ in Fig. 109 —these must by construction have
a common normal passing through the instantaneous centre—the
describing line $E\,F$ itself.   Thus the two curves may serve as
portions of profiles for the pair of elements which it is desired to
construct.   In the special case in which the primary centroids
A and B are circular, the secondaries E and F are circles also,
and the profile-curves $a\,P$ and $b\,P$ are involutes of those circles.
This gives in spur-wheels the involute teeth which have been
sometimes employed.   Set-wheels can be made by making the
angle $F\,O\,D$ constant for the whole series of wheels.

The profile-curves $a\,P$ and $b\,P$ are in this case roulettes obtained
by rolling a straight line upon the two curves E and F.   It must,
however, be possible to describe them, as in the former case, as
roulettes upon the primary centroids A and B.   For circular centroids the auxiliary curve by rolling which upon A and B the
involutes $a\,P$ and $b\,P$ can be respectively obtained is a logarithmic
spiral.*   If the middle curve of the three secondary centroids be
not a straight line, the roulettes described by its points have not a
common normal passing through the point of contact, and therefore
are unsuitable for the profiles of elements.

§ 34.

Fourth Method.—Point-paths of Elements used as Profiles.

The auxiliary centroids employed in the second of the methods
which we have discussed may take the most various forms.   A
special case occurs when the auxiliary centroid coincides with one
of the primaries.   Here it no longer describes a curve in the latter,
but each point in it describes there one other point only ;—
relatively to the other centroid, however, it describes some point-
path.   If the latter be taken as the profile of an element, the profile
of the element with which it works must be a point.   This method
of constructing profiles has also been used for wheel-teeth.   Fig.
110 is an illustration of the contact between teeth profiled in this

---

* This can be seen without difficulty.   A demonstration is given (*e.g.*) by Willis
p. 92, another by Haton (*Mécanismes*), p. 101.

way. The two auxiliary centroids coincide with the two circular primary centroids A and B; ab and bc are point-paths (here epicycloids) described by the points a and c of the circle B; de and ef are epicycloids described by the points d and g of the circle A; ghd is a portion of a curtate epitrochoid described by the point e of the wheel B; aid a similar curve described by the point b of A. Simultaneous contact takes place in a, b and c, the normals to these points of restraint all passing through the point O; after a very small rotation in the one or the other direction the point e comes into contact with gh, or k with al.

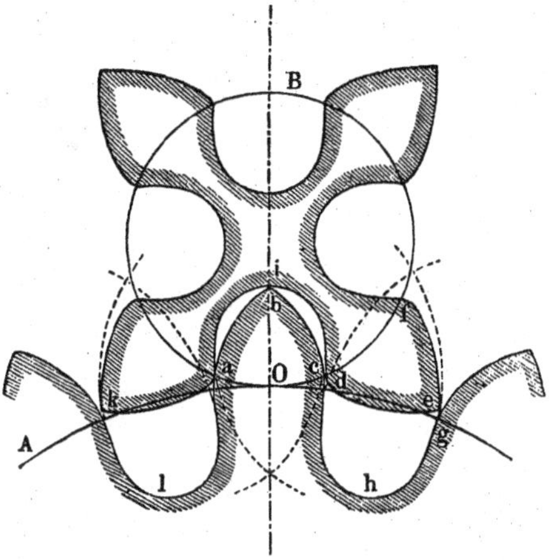

Fig. 110.

It may sometimes be required to combine this method of constructing profiles with one of the others,—such mixed methods are occasionally used in drawing the teeth of wheels.

§ 35.

Fifth Method.—Parallels or Equidistants to the Roulettes as Profiles.

If we have obtained by any of the methods now described the profiles aP and bP corresponding to the centroids A and B, and from the centre of curvature of the element P of aP describe a circle with a radius larger by any amount, PP_1, than the radius of

curvature of that element, and from the centre of curvature of the
corresponding element of bP a circle with a radius smaller by
the same amount, we obtain two circular arcs touching each other
on the normal at P_1, and having their normal passing through the
instantaneous centre O, in common with the elements at P of the
original profiles. This procedure, carried out for every point in
the two profiles, furnishes two new profiles, $a_1 P_1$ and $b_1 P_1$, which
are equidistants or parallels to the first curves, and may equally

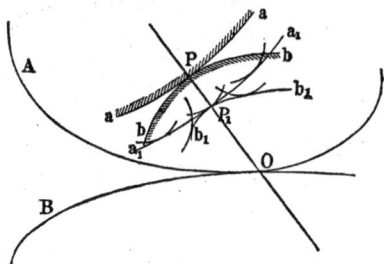

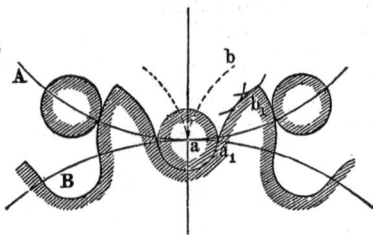

FIG. 111.   FIG. 112.

serve as profiles for elements.* This gives us a further immense
variety of profile forms, which are only limited by the conditions
as to usefulness mentioned in section 31. These equidistant curves
often possess advantages,—as in the case when they are employed
to represent the point-profiles already described. They give us
then a circle or circular arc for a profile instead of a point. We
have this applied practically in the "pin" teeth of "lantern"
pinions, which once were frequently used, and even now are
occasionally seen. Fig. 112 is an example. Instead of the point
a and the epicycloid ab, the circle of radius $a a_1$ and the line $a_1 b_1$
equidistant from the epicycloid are employed.

As a further illustration we may employ the forms already treated
in another way, the curve-triangle and square. In Fig. 113 $O 2' 3'$
is the centroid of the one and $O 2 3 4$ that of the other element
of a higher pair, whose profiles we wish to determine. Using
the method of § 34 we place in $O 2' 3'$ an auxiliary centroid of the
same figure as itself. We choose a describing point at R, a point

* In the figure the equidistants are found by drawing arcs about points in the
profile with radii equal to the difference between the original and the intended radii,
and not by drawing arcs from the centres of curvature of each element with the
actual increased or decreased radii in the way described. The method of the figure
gives the same result as the latter, and is obviously much more simple.

upon the normal bisector $O\,3$ of the arc $2'\,3'$, at the intersection of the curves $O\,2'$ and $O\,3'$. Relatively to the three-cornered centroid R describes a point only; relatively to the other centroid however it describes a straight line right and left of R, being a point in the circumference of a smaller Cardanic circle rolling within a larger. $R\,C$ and $R\,D$ are these lines and their prolongations. If we complete the rolling of the inner centroid upon the other we obtain the four sides of the square as the profile for the outer element. Only the necessary restraint for the inner element is now required. For this purpose we must find

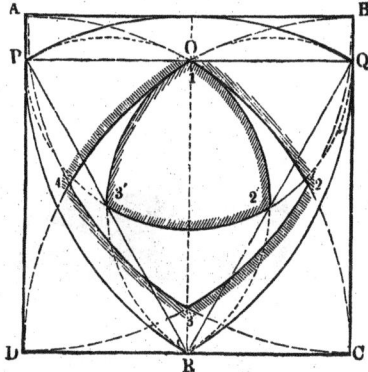

Fig. 113.

a point homologous to R for each of the two remaining sides of the inner centroid; such points we obviously have in P and Q (as shown

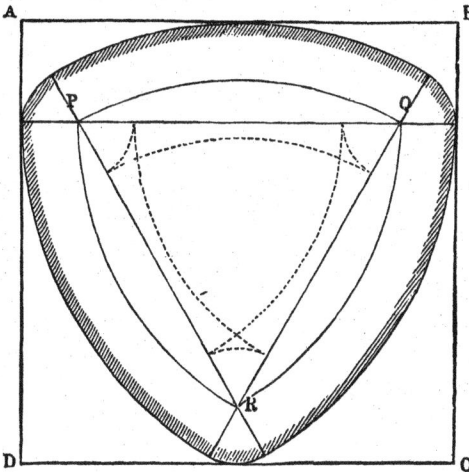

Fig. 114.

by the dotted arcs), and these like R have their paths along the central portions of the sides of the square. To obtain the restraint we have then only to draw from R, P and Q the equi-distant curves $P\,Q$, $Q\,R$, and $R\,P$, and we obtain the curve triangle as a profile.

There is nothing to prevent our choosing larger radii for these equidistants,—in Fig. 114, for instance, they are $1\frac{1}{5}$ times as long as the chord $P\,Q$. The profile form obtained only differs from the

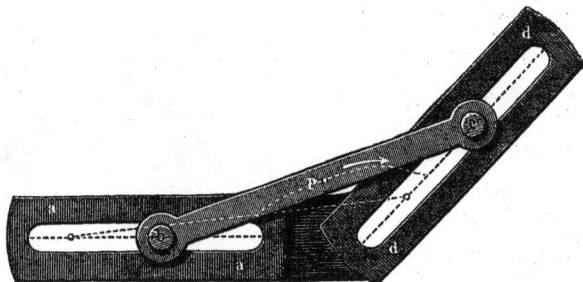

FIG. 115.

former one in having the vertex angles at P, Q and R rounded by arcs having those points as centres. From a practical point of view the resulting profile is a great improvement on the former one, on account of the removal of its sharp edges at P, Q and R. We have before used a similar construction to this in Plates XII. and XIII. If the radii of the equidistants be chosen less than $P\,Q$ we get unusable forms, such as the one shown in dotted lines.

A third illustration of the use of equidistant profiles is furnished to us by the higher pair of elements shown in Fig. 115, which has already been described. Here the nature of the motion is known from two given point-paths,—the straight lines in which the points b and c move. Equidistants to these lines give us the profiles of the prismatic slots in the piece $a\,a\,d\,d$, while the equidistants to the two points are the circular profiles of the pins b and c. Here we do not even require to know the centroids in order to construct the pair of profiles. They have, however, as follows from § 22, the form of Cardanic circles, or their arcs.

§ 36.

Sixth Method. Approximations to Curved Profiles by Circular Arcs. Willis' Method.

If the profiles of elements be curves of varying radius their construction is somewhat troublesome, and it may be very convenient to represent them by circular arcs. This can always be done where an approximation to the true curve will suffice, and when only a small portion of each curve is used, as is in general the case with wheel-teeth. As substitutes for such portions of the curve suitably chosen arcs of circles having the same curvature are employed: for finding these there are several methods in use. For set-wheels with cycloidal teeth I have recommended* the following method:—*A* in Fig. 116 is a circular centroid—the pitch circle of the wheel for which teeth are to be constructed,—*B* its centre,—*C* and *D* the centres of two equal describing circles (auxiliary centroids) by which the cycloidal arcs *a* and *b*, which it is desired to represent by circles, are drawn; their radius is equal to 0·875 of the pitch. *O* is the point of contact of the circles *A*, *C* and *D*. Make the angle *O C a* and *O D b* = 30°, find the peripheral points *a'* and *b'* on the auxiliary centroids opposite *a* and *b*; draw

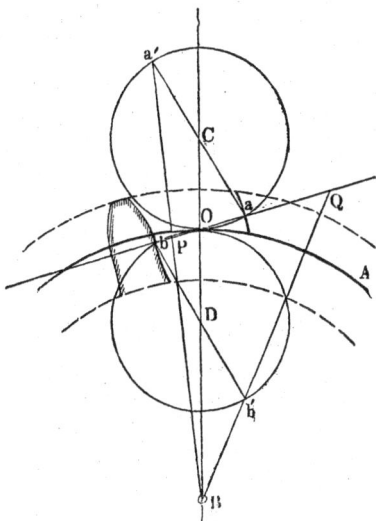

FIG. 116.

through *a* and *b* a straight line, which it is evident must pass through the point of contact and must therefore be a normal to the elements of the curves at *a* and *b*,—and join *a'* and *b'* with the centre *B*,— then the lines *B a'* and *B b'* (produced if necessary) cut the normal in the required centres of curvature *P* and *Q*. The circular arcs for the faces and flanks of the teeth are then drawn beyond and within the pitch circle respectively, joined together on *A* to make a fair profile (as in the figure), and repeated symmetrically round the pitch circle.

In the well-known method given by Willis, he attempts to determine the circles best suited for the teeth profiles directly, that is without the use of auxiliary centroids or roulettes. The nature of his approximation, in which he follows out some suggestions

* *Der Constructeur*, 3rd Ed. pp. 419.

of Euler, is shortly as follows. Let A and B (Fig. 117) be
the centres of rotation of two bodies which can drive each other
by means of the circular profiles touching in R, and drawn from the
centres P and Q; then the intersection O of the two lines of centres
PQ and AB is the point of contact of the centroids corresponding
to the relative motion of the figures A and B (see § 8), which therefore have for their angular velocity ratio $OB : OA$. In order that
this may remain nearly constant for a short interval of time, PQ

Fig. 117.

must in its motion continue
to pass as nearly as possible
through O. The instan-
taneous centre, however, of
PQ relatively to AB is
the point C at the intersection of PA and BQ produced, and if C be a point
upon a perpendicular to
PQ at O, as C' in our
figure, then the instantaneous motion of PQ will be
in fact through the point O.
If therefore one of the
centres, as P, be chosen, the
position of the other must
be the intersection Q' of
PQ and the line BC'. We thus obtain in the distance PQ, or
rather PQ', the sum of the required radii of curvature, but may
take the point of contact R in any position, as R' for example,
as follows from what we have said in § 35.

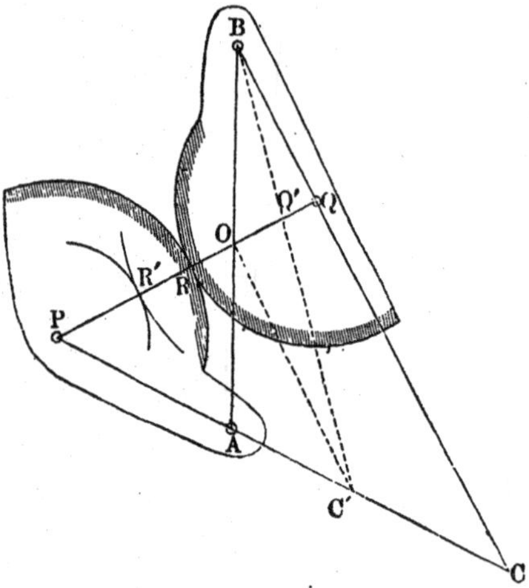

In order to adapt this elegant method to set-wheels, Willis
chose three constant magnitudes, the distances OC' and OR
and the angle POA; the latter he made 75°. If the teeth were to
be profiled by one arc only he took $OC'' = \infty$, $OR = 0$, the
circular arcs becoming approximations to involutes (see § 33). If
two circular arcs were to be used, joining at the pitch line into
an S-shaped figure,—as is usual,—the method was applied twice
over, once for the portion of the tooth on each side of the pitch
circle. Fig. 118 shows this;—OR' and OR'' are each equal to
half the pitch, and OC', OC'' are so taken that for a pinion

of twelve teeth the flanks become radii. This occurs if $O\ C' =$ the pitch $\times \frac{6}{\pi} \sin 75°$. A closer examination shows this construction to be identical with that given in Fig. 116, when equal constants are used. $R'\ C'$ and $R''\ C'''$ are the diameters of our auxiliary centroids, which are here dotted so that the constructions may

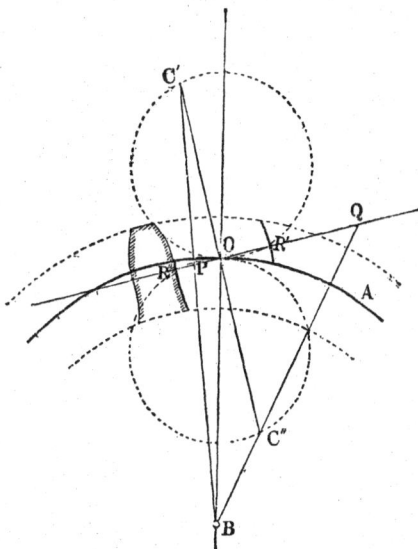

Fig. 118.

be compared; the points C'' and C''' correspond to a' and b', P and Q are found as before. Willis * himself recognised and mentioned this coincidence.†

§ 37.

Seventh Method.—The Centroids themselves as Profiles of Elements.

If we suppose the auxiliary centroids of the method of § 32 to be closed figures, and to be made smaller and smaller until they

---

* *Principles of Mechanism*, 1st Ed. p. 107, 2nd Ed. p. 142.

† The approximation proposed by Professor Unwin, and described by him in *Engineering*, May 29, 1874, is perhaps more exact than either of these. He finds two points in each roulette (epi- or hypo-cycloid), one upon the pitch circle, and one

become points only, the paths they describe in rolling become the centroids themselves. These may be used as profiles if the cylindric axoids bearing them be so pressed together that sliding at the point

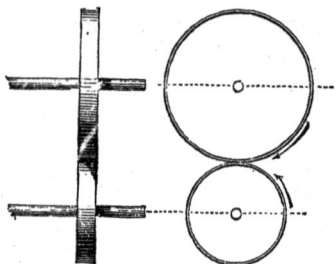

of contact is prevented by friction, so that they are compelled to roll on one another. This is the only case where the profiles of elements have a pure rolling motion. Circular centroids give us cylindric wheels, like those known as friction wheels (Fig. 119). The applications of this method are not few, the most important and most familiar being perhaps the wheels of railway car-

Fig. 119.

riages. We shall in the next chapter consider in detail the subject of axoids pressed together, or restrained, by a force.

§ 38.

Generalisation of the foregoing Methods.

In the foregoing paragraphs we have throughout limited ourselves to axoids for general cylindric rolling, but the methods employed are equally applicable to the case of non-cylindric axoids. For conic axoids this is easily seen, but not so easily with the higher rolling and twisting axoids (see § 13). Considerable difficulties appear here in the theoretic examination even of motions occurring according to simple laws, and still greater difficulties in their practical presentation. It is a part of Applied Kinematics to consider so far as may be necessary the more important cases. It must be admitted that in general the actual forming of profiles for the higher axoids, even those for instance for the teeth of hyperboloidal wheels, presents as yet considerable

upon a circle larger or smaller than that by a distance equal to two-thirds of the intended depth of the face or the flank of the tooth. The first of these points is assumed and the second found from it by a very accurate approximation. A normal to the roulette at the second point is drawn, and the intersection of this line with the normal bisector of a line joining the two points gives the centre for the circular arc to be substituted for the roulette. This arc passes through both the points, and has a common tangent with the roulette at one of them.

difficulties, a circumstance which helps to explain the existing imperfect comprehension of those even which have been hitherto made. The modern sewing-machine manufacture, and in part also that of agricultural machinery, have empirically and unknown to themselves made very satisfactory progress in the employment of the higher axoids;—the former with special skill, for it has already brought to considerable perfection the formation of complex enveloping surfaces.

The illustrations which we have used in the foregoing paragraphs have been in great part, although not entirely, drawn from the methods used for constructing wheel-teeth, and must be therefore more or less known, if not entirely familiar to those readers who have made machine-construction a subject of scientific study. The methods of procedure, however, deserve renewed attention, for they have now been developed in the special light of the general fundamental principles upon which they rest. The question has here been treated as one not of rules for constructing wheel-teeth, but of their general correspondence to a great principle. We found that by a quite small extension of the ideas contained in them, methods are available generally which are commonly stated and understood as very limited rules. I trust therefore that such previous familiarity with particular instances will only have made it more easy to understand the general case and their relation to it.

I hope now to have made completely intelligible the fact that the construction of pairs of elements is possible for any motion, however complex,—that is, that in all cases suitable profiles can be determined for those elements. We have seen also that this problem may be solved in an immense variety of different ways, even in the case of the simpler motions and those more often occurring. While in former centuries the most distinguished geometrician occupied himself with separate solutions of detached problems, and necessarily regarded them as important propositions, he has to-day presented to him a limitless perspective, which appears almost more simple in its universality than the single case appeared before, and which affords rich opportunities to the practical mind in the determination of the best solution among the immense number of possible ones.

Perhaps I must fear that I have wearied my readers by these investigations, in which we have apparently progressed more point by point than step by step, and have rather courted than shunned the difficulties of the problem. By degrees, however, we have completely proved the special laws upon which constrained pairs of elements can be constructed, and this was the end we had in view. These laws are not simple,—not lying on the surface,—but they are fixed—and within the assumed limits universally applicable—general laws. It may be well on this account to pause for a moment and to refer the reader once more to the nature of the ideas on this subject which have hitherto been held and which were sketched in the introduction,—and especially to the somewhat extended system of Laboulaye of which I there promised further mention.

Let us ask ourselves what it is that Laboulaye's three systems levier, tour, and plan, which bear so great an apparent impress of geometric generalization, really represent. We must look at this question in the light of the special acquaintance with the subject which it has been our object in the foregoing chapters to obtain.

In the first system the moving body has one fixed point,—(*le corps a un point fixe*), in the second two fixed points (*le corps a deux points ou une droite fixe*), in the third three fixed, or rather restrained points (*l'obstacle consiste en trois points fixes ou en un plan passant par ces trois points*). We remark in the first place that Laboulaye bases his classification not upon kinematic chains, into which, as we have seen, the machine separates itself, but upon the pairs of elements themselves. For he speaks always of the restraint of a single body, not of one forming a part of a whole sytem of bodies. Let us then limit ourselves to this, although Laboulaye himself considers his system to represent machines generally. Which pair, however, has only one fixed point? We have seen (§ 5. IV.) that the fixing of one point only leaves the motion of the body to which it belongs quite indeterminate; that neither a constrained chain nor a constrained pair of elements can be so formed. Laboulaye, however, chose as his illustration a swinging lever, one, that is, which turns about an axis. His system "levier" therefore coincides with the second of our lower pairs of elements, the turning-pair (§ 15). But such a pair requires not one, but at least six points of restraint. It may of course be said that the fixed point in the système levier represents a geometrical axis, and that Laboulaye's meaning, strictly

rendered, is that two points in this axis,—which might be considered as a kind of idealisation of the body,—should be prevented
from altering their position; the two points coinciding of course
in their normal projection. This cannot, however, be the meaning,
for it is the description Laboulaye himself gives of his second
system, the système tour,—he has therefore not fallen into this
error; he always speaks, too, of bodies, and not of their ideal
representation by axes. But if any body is to be restrained we
know that it must have a definite form, and that, being suitably
formed, it must be restrained at at least six points. If we have one
fixed point only the body must be spherical,—it will require at
least four points of restraint, and the motion which occurs is only
so far constrained that the centre of the sphere cannot alter its
position, and that the other points must move on spheric surfaces;—
with this limitation however they may have any possible motion.

To proceed :—Laboulaye includes under système levier those
pairs of elements which move by conic rolling, § 11. That also is
said distinctly. He does not however leave his chosen illustration
the lever. " Le mouvement d'un point quelconque, appartenant au
levier, sera de nature circulaire, en chaque instant et de plus
en général alternatif dans une machine, se produisant le plus
souvent dans un plan." We see that this definition fails altogether
in clearness and certainty. Apparently it shadows forth in dim
outline a pair of elements with a swinging motion,—its appearance of deep and categoric generalisation has sometimes brought it
into favour with mathematicians; it falls altogether to pieces,
however, on a closer examination.

Nor can the two other systems, tour and plan, fare better.
In no one of the three systems is it made completely clear on
the one hand what in strictness is meant by point fixé or plan
inébranlable, or on the other hand what it is that distinguishes
absolutely one system from another. Let us reverse the question,
and attempt to find in which of Laboulaye's classes one or other of
our higher pairs of elements must be placed. We may choose the
curve-triangle and square for instance. About this pair we know
that if the square be fixed, then on motion taking place no point
of the triangle remains in its original position. Every point in it
moves. According to Laboulaye at least one point must remain
stationary. The pair might perhaps be best placed under the

système plan, the surfaces of the triangle being supposed to be caused to move always in the same planes,—which are therefore such plans inébranlable as are considered to be peculiar to this third system. We might look in this direction for the real essence of the system, but then the turning pair would also on these grounds have to be put in the same category, while it must be claimed on the other hand as distinctly belonging to système tour. So we lose our clue, and fail in discovering what the real difference between things varying so essentially is,—the very thing which it was and must be the object of the " system " to point out.

But I will stop: it is evident that Laboulaye's position is untenable. The question here, too, is not one of criticism of Laboulaye himself; many other writers have followed him without putting his ideas sufficiently to the proof, and they would therefore be open to the same criticism. That I do not undervalue Laboulaye will be seen from the Introduction, where I have paid my tribute of recognition to him as an investigator. I do this the more freely, that he has not drawn any special deductions in the applied part of his work from his first propositions,—deductions which would necessarily have led to error. I have been desirous only to show upon what insecure and feeble supports it has been attempted to build up a science of Kinematics, a science having ostensibly a logical basis of its own. I wished to place again before the reader, in a form that could be readily grasped, a proof that if anything whatever is to be accomplished by means of axiomatic propositions, they must be subjected to, and be able to bear, the most inexorably strict, exact and penetrating examination.

CHAPTER IV.

INCOMPLETE PAIRS OF ELEMENTS.

§ 39.

Closure of Pairs of Elements by Sensible Forces.

IN our examination both of the lower and higher pairs of elements, we have hitherto assumed that the reciprocal restraint of the two elements forming the pair was complete ; *i.e.*, that each of the two bodies by the resistant qualities of its material and the form given to it, both enveloped and constrained the other. We have made this assumption also, expressly or tacitly, in investigating the forms to be given to bodies in order that any sensible forces tending to alter their required relative motion might be balanced by latent forces. Under certain circumstances, however, the strictness of this condition may be somewhat relaxed,—when, namely, precautions are taken to prevent the possibility of sensible forces having certain directions ever affecting the pair. If this can be done, it is obvious that it is no longer absolutely necessary to make the pair entirely self-closed, bodily envelopment being no longer essential for restraint in those directions.

In order thus to prevent the disturbing action of sensible forces acting in any given direction upon an element a which is to be constrained, we allow another sensible force to act upon it continu-

ously, and make the direction of this force opposite to, and its magnitude not less than, those of the former. If, for example, the
greatest anticipated disturbing force be = P, then in order to neutralise it we must cause an opposite force of the magnitude P to act
upon the supposed element a. This force restrains a in the place
of a portion of the enveloping partner-element b; if for the sake
of security we give to it the magnitude $P+Q$, then in the worst
case a is held at the opposite point of restraint by the force
$P-P+Q=Q$, this pressure being balanced by latent forces in the
element b. The general conditions of equilibrium are thus fulfilled.
Any such force as the supposed one $P+Q$ closes, as it were, in
the direction $-(P+Q)$, the pair of elements left unclosed or incomplete in that direction; we shall therefore call it a closing force.
Such pairs of elements as require closing forces are evidently incomplete in themselves; their usefulness depends upon the applica-

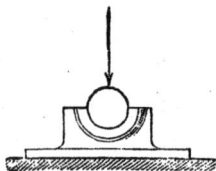

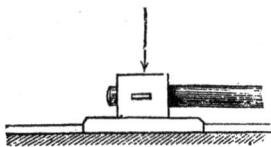

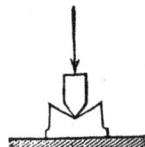

FIG. 120.   FIG. 121.   FIG. 122.

tion of the closing force, or upon what we may call in one word
force-closure.

Force-closed pairs occur frequently in machinery. The shafts
and bearings of most water-wheels are illustrations of them, where
the great weight of the wheel is almost always sufficient in itself
to prevent any vertical motion of the axle without the employment of a plummer-block cover (Fig. 120). The crosshead sometimes used for large horizontal engines gives us another illustration
(Fig. 121); the heavy pistons and rods here prevent any rising of the
crosshead, which is guided only beneath and at the sides. The
knife-edge of a balance (Fig. 122) is also kept in continuous contact with its bearings by the weight of the beam and scales. The
railway turn-table is another example, the whole table being here
held by its own weight and that of the load upon it on a roller
path completely open above; a similar thing often occurs in wharf-
cranes. Railway-wheels, lastly, are common and well-known.

illustrations of force-closed pairs; they are kept in continual contact with their partner-elements the rails, by vertical downward closing forces.

In all these and similar cases force-closure presents itself obviously and naturally, and often greatly simplifies the construction. This, however, is only one of the ways in which it occurs ; we must proceed to consider others.

§ 40.

Force-closure in the Rolling of Axoids.

While in the cases just mentioned the object of the closing force is essentially to prevent the separation of the incompletely formed elements, it may under certain circumstances have a much wider purpose. This occurs when the action of a force derived from it, friction, is used to complete the reciprocal restraint of the elements. An illustration of this is furnished by the friction wheels (Fig. 123) already mentioned in § 37. Here the force-closure has not merely to hold the two cylinders in contact, but also to press them so strongly together as to prevent sliding under a given tangential force,—in other words, to cause the cylindric surfaces, which here are the axoids themselves, to roll upon one another.

By a more strict examination into this case, we can see that the closing force presses the small roughnesses of the surfaces together so as to make the cylinders work like spurwheels, such force-components as tend to separate the wheels being resisted by the closing force. It is this consequence of the pressing together of two bodies which is considered in mechanics under the name friction.[22]

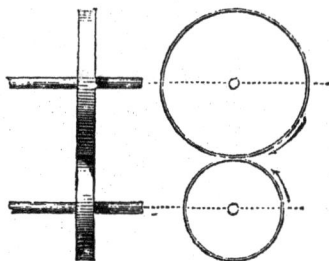

Fig. 123.

This employment of force-closure occurs very frequently. Its application in the case of the driving-wheels of locomotives is important in the highest degree. The whole development of our

railways has been directly dependent on it. Every one is familiar
with the fact that at the first the notion of "adhesion" between
the wheels and the rails was thought so illusory that it could
scarcely obtain a trial,—restraint being obtained by the use of
suitably profiled pairs of elements. Blenkinsop's toothed-rail, and
those of the Liverpool and Manchester Railway, Brunton's revolving legs, and other even less practical constructions all illustrate
this.

The constrained rolling of axoids under force-closure differs
essentially from the mere closure of an incomplete pair of elements.
The two things may, however, occur together as well as singly. In
the driving-wheels of locomotives they are united ; in the wheels
of the carriages there is nothing more than closure of elements by
pressure.

In the latter case it would be possible to bring about the element-
closure by the addition of a second pair of elements, an additional
rail, for instance, which could be so embraced by a suitably formed
piece connected with the carriage as to render any rising of the
latter from the main rails impossible; but this would not in any
way make the wheels more like the driving-wheels. Such an arrangement has indeed been employed on the Rigi railway. With
it the motion of the carriage on the line may be more accurately
described as that of an ordinary closed pair than as occurring with
force-closure.

We see that force-closure finds important and numerous applications. It always retains, however, a certain incompleteness. If the
closing forces be not sufficiently large, or if unforeseen disturbances
occur, the constraint may be destroyed or temporarily broken. Notwithstanding this, force-closure—as the examples given show—is of
most essential service in machinery. It leads us besides to an
entirely different kind of pairs of elements, which are in machinery
of even greater importance than those just considered. We shall
examine these more closely in the following paragraphs.

§ 41.

Flectional Kinematic Elements.

We have hitherto supposed the capacity for resistance, which we recognised as an essential for those bodies from which a machine could be constructed, to be attained by giving to the elements complete rigidity, molecular immovability. We assumed that the material and dimensions of the element had been suitably chosen for this purpose, this being the function of machine design. The admissibility of force-closure, however, shows us that bodies may serve for the formation of elements which cannot be considered as sensibly rigid. If, namely, we choose such bodies as, while not resistant in all directions, can maintain approximate molecular immovability under the action of sensible force in at least one direction, and employ these bodies under such force-closure as

FIG. 124.

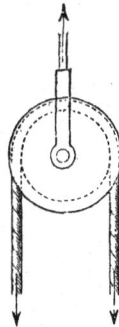

FIG. 125.

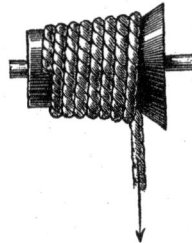

FIG. 126.

corresponds to their special capability of resistance, they will act exactly as if they were completely resistant.

As bodies possessing these peculiarities we have ordinary cords or ropes, woven or leathern bands and belts, bands or ropes of metal or wire, every kind of chain, all those organs in short which, offering no sensible resistance except to tension, can yet be made sufficiently rigid under the action of tensile forces of any magnitude. We may include them all under the name of tension-organs.

On account of their want of rigidity in other directions, the

tension-organs can be very readily united into pairs of elements
with rigid bodies of various forms. Thus we may have a rounded
bar (Fig. 124) over which they slide, force-closed in both directions,—a pulley (Fig. 125), the tension-organ moving upwards on
one side of it, downwards on the other;—a drum (Fig. 126) upon
which it can be coiled, and so on. We find these pairs of elements used and applied in the most various ways, in pulleys
and cranes, in belt-trains, in rope-trains, in submerged rope towing,
etc. The elements always reciprocally envelope each other,—but
although the envelopment passes sometimes into enclosure, they
have not the latter as an essential characteristic; they must be
classed therefore as higher pairs of elements.

Directly opposed to tension-organs there stand others which
possess molecular immovability only in reference to compressive
forces, and which may therefore be called pressure-organs. To
this class belong all fluids, liquid and gaseous,—water, oil, steam,

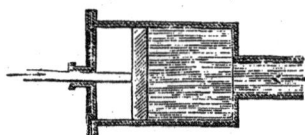

FIG. 127.   FIG. 128.

air, etc. The force-closure applied to them must be such as continually presses their molecules together. In order, however, that
they may not alter their form by extension on either side, all the
surfaces of the fluid body besides those normal to the direction of
motion must be pressed together with the same force. This is
done by the help of latent forces,—that is, by enclosing the fluid
in vessels of suitable form and resistance. This occurs, for example,
in steam- or water-pipes (Fig. 127), and in the cylinders of pumps, or
of steam- or blowing-engines (Fig. 128), and so on. It hardly needs
to be pointed out how extremely important a part pressure-organs
of this kind play in the construction of machines.

Another class, differing but little from the one just described, has
been formed of late years from some of the tension-organs, by
enclosing them in suitably shaped vessels, and having thus rendered
sideway motion impossible, using them as pressure-organs. The

flat-link chain of the Neustadt cranes (Fig. 129) is enclosed in a tube, so that it can be pressed forward; the thin brake band of steel has been made capable of resisting pressure at its free ends by bedding it in a hollow cylinder (Fig 130), its particles then acting like the stones of an arch. Wire-rope has also been used in a somewhat similar way.

The pressure-organs form closed pairs of elements like the lower pairs; on account, however, of their molecular moveability, they must be classed with the higher pairs.

If we compare the two classes of pairs of elements to which the consideration of force-closure has led us, we see that they are closely related. They both have the peculiarity that' they can be used in one way only; with a closing force, that is, of one particular kind or direction, the tension-organ only with tension, the pressure-organ only with thrust. If the rope in Fig. 126 be pushed

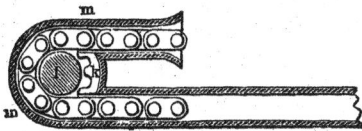

FIG. 129.

FIG. 130.

upwards, it does not set the drum in motion, nor can the piston (Fig. 128) be moved by withdrawing the fluid from behind it. The pair are closed on one side only; they are mono-kinetic, a peculiarity which we shall find further on to belong to other pairs also. They owe this to their molecular yieldingness, or want of fixedness in all except a single or a small number of directions. This quality is what is known in pressure-organs as fluidity, in tension-organs as flexibility or pliability. As a common designation for both, we may use the word flectional, and therefore call both tension- and pressure-organs, when employed as kinematic elements, flectional elements.

The two sets of organs stand opposed to each other as positive and negative, a relation directly indicated also by the nature of their closing forces. The pipe filled with water, Fig. 127, stands opposite to the rope in Fig. 124; the cylinder having a piston moved forward by a fluid pressure on one side, Fig. 128, corresponds to the drum in Fig. 126. The application of a column of water for transmitting pressure, which has lately been made in mining operations, furnishes again the opposite of a rope used in tension. Thus the tension- and pressure-organs are contra-positive. They must therefore be equally reckoned among kinematic elements. Willis' exclusion of all mechanisms of which fluid organs form a part, to which we alluded in the Introduction, was therefore incorrect. If belts, pulleys and so on, are to be considered as forming portions of " pure mechanism," it is logically impossible to omit water- and wind-mills, or steam-engines. We have only to think of the importance of the latter to be astonished that one of the most valuable and most extensively applied of machines, one possessing also the greatest delicacy and accuracy in its motion, should ever have been considered unkinematic,—incapable of scientific kinematic treatment,—"impure." We shall on the other hand be able to see further with what scientific force and with what important consequences kinematic science can be brought to bear upon these machines. Although Willis' view of the matter is not acknowledged as a principle, its incorrectness has not been specially pointed out; practically it has had the result that this class of machinery has scarcely ever been treated kinematically by English writers, and by others only seldom, and even then generally not with the requisite thoroughness.

§ 42.

Springs.

While in the tension- and pressure-organs we have had elements in which the application of force could occur only in a certain very simple manner, there is in machinery another class of elements which can be arranged so as to be used with any possible application of force. These elements are springs. They are familiar in many forms and for many purposes; always, however, under the condition which we found to be necessary in the case of the flectional elements,—with the limitation, that is, that in each special case a single force-application only can be used. The various constructions of springs may be classified according to the nature of this force, so that we can distinguish them as springs for tension, thrust, bending or torsion. Bending- and torsion-springs are most often of metal, but also sometimes of wood ; for the thrust- and (less often) the tension-springs india-rubber and other organic materials are much used.

The ease with which the most various forms can be given to springs permits them often to have a form which, considered as a whole, is adapted for working under a force-application quite different from that for which the cross-section of the material from

FIG. 131.

which they are made would seem to adapt them. The helical spring (Fig. 131) for instance is closed, when used as a whole, by a force in the direction of its axis,—in other words, is employed as a tension-organ; as regards its cross section it is suited for working with torsion.* If the same spring be so formed that its coils are not in contact in its normal position it can be used as a whole as a pressure-organ ; for this purpose, however, it must be enclosed in a suitable case to prevent lateral deflection. A spring which is itself adapted for tension can be, and is

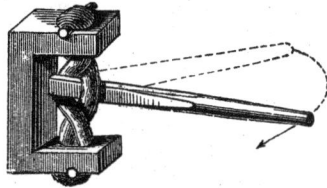

FIG. 132.

used as a whole as a torsion or wrenching spring in the well-known form shown in Fig. 132.

Springs, both simple and compound, working under a bending force are familiar in their many applications to railway work, as are several kinds of torsion-springs, and among the pressure-springs one of a peculiar kind (*Strebe-feder*)—the steel ring used between the tread and tyre of the wheel in the proposed "spring wheels" of Mr. Adams.†

Springs are very well suited for supplying the force-closure of

* See *Der Constructeur*, 3rd-Ed. p. 59.
† See Colburn's *Locomotive Engineering*, pp. 99, 100, etc.

incomplete pairs of elements and are much used for that purpose,
as in the case of spring-packed pistons, spring pawls for ratchet-
wheels, etc.; they play also a most important part in the storing of
energy, to which we shall return further on. Springs of organic material, india-rubber, vegetable fibre, skin, etc. greatly resemble tension-
organs; those of hard material seem more like the rigid elements.
In their mode of action they rather resemble the pressure-organs, as
being, within somewhat wide limits, elastic. They differ essentially
from the rigid elements, however, notwithstanding the apparent
resemblance, for in these the flexibility is supposed to be reduced
to so small an amount that it may be neglected, while in the springs
it is intentionally made very considerable.

§ 43.

Closure of a Pair of Elements by a Kinematic Chain.

An incomplete pair of elements may also be closed by kinematic
linkage. Two bodies a and b (Fig. 133) having for their axoids
circular cylinders, and having their surfaces fluted,—in other words,

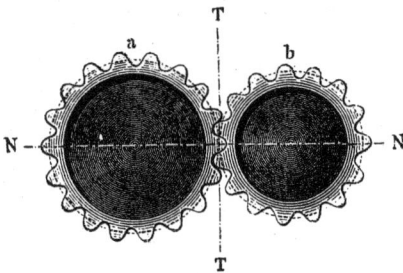

FIG. 133.

two spur-wheels furnished
with accurately fitting teeth,
—have by their profile forms
the necessary restraint in the
direction of the tangent $T\,T$;
the teeth may also be so made
that the wheels cannot be
moved nearer together; there
is required only restraint
against divergent motion in
the direction $N\,N$ of the nor-
mals. This can be supplied by one of the methods considered in a
former article. Let us take the fifth method (§ 35), in which parallels
to roulettes serve as profiles, then we have as a parallel to the
(point-)path of the centre of one of the wheels a circular ring, and
as a parallel to the other centre itself a circle, these giving us the
profile shown in Fig. 134, an annular groove for the wheel a, and a
cylindric pin moving in it for the wheel b,—and by these the

necessary restraint both against convergent and divergent motion is obtained. If we suppose further, that means are provided for preventing lateral motion, we have before us a closed pair.

This method of closure is not practicable in the cases commonly occurring; possibly it seems here somewhat far-fetched; we shall

FIG. 134.

see presently, however, that it is in no way without precedent. In practice we use rather an easily arranged kinematic linkage between a and b. If, for instance, we attach both to a and to b (conaxially with their axoids or pitch surfaces), solid cylinders, enclose these

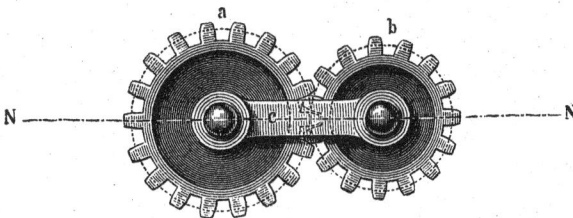

FIG. 135.

in open cylinders, and connect the latter by a rigid bar (Fig. 135), the restraint in the direction NN of the normals is perfect (we may here, as also in the last case, allow the points of the teeth to be quite free, as they are no longer required for restraint). Instead of a closed kinematic pair, we have now a closed kinematic chain

of three links. The two wheels *a* and *b*, with their conaxial cylindric pins, form two of these links ; the third is the bar *c*, with
its two parallel cylindric holes, which form the bearings for the
axles of the wheels.

The chain-closure by which the given incomplete pair has
here been closed, or made into a constrained pair, is used constantly not only for cylindric wheels but also for bevel-wheels,
hyperboloidal-wheels, screw-wheels, etc.

Very frequently it is employed merely in order to simplify a
construction,—the element necessary for the pair-closure being
actually present, although incompletely formed. The screw mechanism shown in two forms in Figs. 136 and 137, for example,
consists in each case of the three following links: *a* a screw with
conaxial cylindric journals, *b* a nut profiled externally as a prism

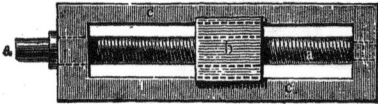

FIG. 136.

FIG. 137.   FIG. 138.

parallel to the axis of the screw, *c* a guiding prism for the nut,
carrying bearings for the journals of the screw. In Fig. 136 each of
the three pairs is completely closed, in Fig. 137 the closure of the
pair *b c* is incomplete. Restraint against the upward motion of the
nut *b* is here given by the link *c;* which is of a suitable form for
that purpose ; a conveniently arranged chain-closure that is,
occurs, besides the incompletely applied pair-closure.

Chain-closure also occurs along with force-closure. The common
ratchet work (Fig. 138) furnishes an example of this. The motion
of the working end of the pawl is here force-closed in moving
backwards over the teeth ; the motion of its jointed end is chain-
closed, taking place in circular arcs about the axis of the wheel.
This mechanism is at the same time single-acting only, or mono-
kinetic, a property already pointed out in connection with Fig. 126.

The hydraulic press gives us an instructive example of a mechanism combining chain- and force-closure. In Fig. 139 the pump valves are omitted for the sake of clearness. The vessel *d* carries the chain-closure between the pistons, *a* and *b*, with which it is prismatically* paired. At the same time it encloses the fluid *c*, which is force-closed by the pressures upon the two pistons in the only direction of motion left possible to it. The hydraulic press forms the contra-positive of a machine apparently most unlike it, the pulley-tackle, the pressure-organ water in the one being replaced by the tension-organ rope in the other. If we substitute rigid rounded bars at the top and bottom of the tackle for the usual pulleys, as in Fig. 140, the logical correspondence

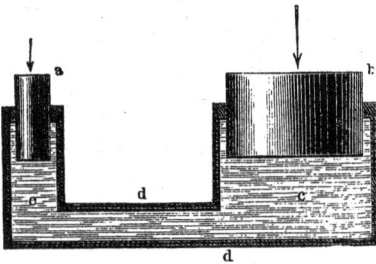

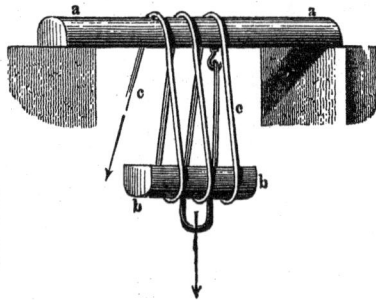

FIG. 139.   FIG. 140.

between these two mechanisms becomes even more obvious.† The chain now, indeed, has three links only, but this is merely because the tension-organ does not need a confining vessel. The motion of the piece *b*, as a whole, is still force-closed by the load. If we used a prism pair to compel *b* to move in a straight line in reference to *a*, the similarity would be still more complete. It is worth noticing how many illustrations occur in modern engineering of what we may call the interchangeability of tension- and pressure-organs, of which we have had here an illustration. The arrangement for ringing bells by air-pressure, now becoming extensively

---

* Each piston, although cylindric, forms a s l i d i n g p a i r with its stuffing-box, turning being prevented by some external means.

† An error is frequently made which stands greatly in the way of understanding this matter, that namely of supposing the action of the tackle to be absolutely connected with the rotating block, or pulley, which in reality has no other object than that of lessening destructive friction. In anatomy the trachlea surface over which a tendon slides is rightly called a *Rolle* or pulley.—*R.*

used, is obviously the contrà-positive of the old bell-rope; the (single-acting) "water-rods" used in mines, of the iron tension-rods,—and partly also the hydraulic crane of the rope or chain crane.

The water-wheel (Fig. 141) gives us a further illustration of a force-closed mechanism. The enclosure of the water in the channel *c* is again two-fold. It is due in the first place to the action of latent forces in the channel walls, and then further to the force of gravity, which prevents the water moving upwards. In the bent portion of the channel the water is paired both with the floats,—virtually a toothed rim of the wheel—and also with the

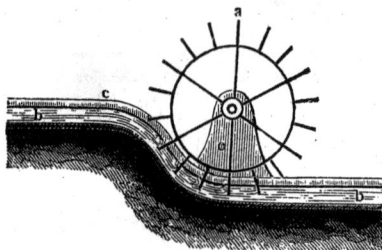

FIG. 141.

channel; the wheel itself, on the other hand, is again pair-closed with the channel through its shaft and bearings. The kinematic chain has three links,—the wheel *a*, with its shaft, the water *b*, and the channel *c* with the shaft bearings. The pairing of the flectional element *b* with *a* is effected by the chain; the force-closure produces at the same time the envelopment (here enclosure) of the floats with the flectional element *b*, and the confinement of the latter within its channel, which we must look upon as a vessel only partially closed.

There are many cases in which the axoid rolling (§§ 37 and 40) takes place, or at least is rendered possible, by chain-closure instead of direct force-closure. The Fell railway is a good example of this. Here the driving-wheels, instead of being held on the rails, as usual, by the weight above them, are made horizontal and held against a central by an arrangement of springs forming a kinematic chain. In this and many similar cases the spring can be used with great advantage, for as an element at once flectional and elastic it admits of advantageous employment under varying forces. In this special arrangement, however, there is often a certain indeterminateness of motion; the possibility occurs of a sort of overbalancing of the force-closure.

§ 44.

Complete Kinematic Closure of the Flectional Elements.

We have seen that the flectional elements may receive, by means of pair-closure and force-closure, important, practical and in the highest degree valuable kinematic applications. We found force-closure to be always necessary, but to a different extent in different cases. In the case of the cord in Fig. 124 we required the absolute prevention of every force not acting as a direct pull upon the cord itself; for the force-closure in the hydraulic press nothing more is necessary than that downward force should be caused to act upon the pistons. If we go one step further we can make even this unnecessary; we can, that is, admit the flectional elements unreservedly into what we have called (§ 1) machinal systems.

The removal of the force-closure is effected by the use of suitably

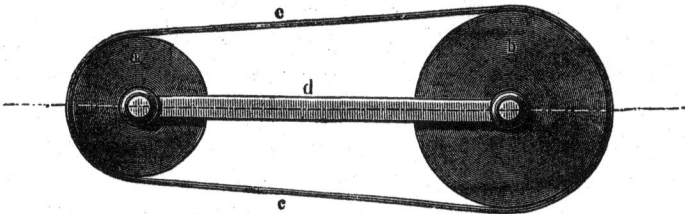

FIG. 142.

combined force-closed chains. The common arrangement of belt pulleys (Fig. 142) gives us a familiar example of this. Here we have two cylindric pulleys, a and c, fitted with conaxial shafts, and connected by a bar or frame d which carries the bearings for the latter; the belt belonging to a is made in one piece with that of c,—the two forming the common endless band b. By this means we obtain a kinematic chain in which the flectional element no longer requires external force-closure. We might regard it as a combination of two chains, each of the form of Fig. 125, their closure acting continuously in opposite directions.

The chain-closure here is twofold. In the first place, as has been mentioned, it makes the force-closure of the flectional element b

unnecessary by substituting for it the action of the latent forces in
the frame *d.* In the second place, it entirely constrains the axoid rolling, the securing of which by force-closure we discussed in § 40. The
pulleys being cylindric, the axoids here are on the one hand the surfaces of the belts continually running off and on, and on the other the
peripheral surfaces of the pulleys. These surfaces are brought so
firmly into contact by the latent forces acting through *d,* that they
cannot slide upon each other. The chain obtained has four links :
the two pulleys *a* and *c* with their shafts, the band *b,* and the frame
d, with the shaft bearings. The motions in the chain occur exactly
as if its elements were completely rigid. Every angular motion of
the one pulley is accompanied by a
corresponding motion of the other ;
the chain can also be inverted, that
is, any one of its links may be fixed,
assuming, of course, that no lateral
disturbing force be allowed to act
upon the belt itself.

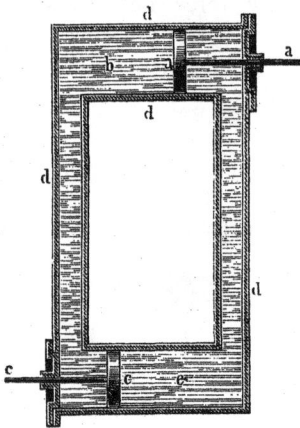

The kinematic chain shown in Fig.
143 may be considered the contra-
positive of this mechanism. The
two pistons *a* and *c,* fitting tightly
into a connecting vessel *d,* are made
to form sliding pairs * with it, and
are kinematically linked by means of
the fluid columns *b* and *e* enclosed

Fig. 143.

in *d.* If the piston *a* be moved to the left, the descending
column of water *b* causes the piston *c* to move to the right; just
as much water is moved up one column as moves down the other.
If care be taken to fill the vessel *d* perfectly, all air being removed from the water, the action of the mechanism is complete,
and the chain is as perfectly closed as if it consisted of rigid
elements only. Herr Anderssohn has shown this convincingly in
his interesting experiments with tubes as long as 3000 metres, used
as water-rods, as in Fig. 139.† Every small sliding of one piston
is accompanied by the corresponding motion of the other. This
mechanism, the tardily acknowledged contra-positive of the familiar

* See note p. 168.
† *Zeitschrift des V. deutscher Ingenieure.* Vol. xiii. (1869) p. 402.

endless band, is now coming into more extended use as a double-acting water-rod.

Springs also, like tension- and pressure-organs, can be completely constrained and used in mechanisms by means of chain-closure. The common clock-spring, Fig. 144, illustrates this. The spring a is connected at one end with the cylinder c, and at the other with the barrel b, which is paired with c, and connected to the clock-work. It thus forms a link in a closed kinematic chain, and this is one of its special uses in machinery, where in some cases a constrained motion of special exactness becomes of great importance.

FIG. 144.

CHAPTER V.

INCOMPLETE KINEMATIC CHAINS.

§ 45.

Dead Points in Mechanism,—their Passage by Means of Sensible Forces.

THE incompleteness possessed by certain pairs of elements, which we have now found means for rendering harmless, occurs also in many kinematic chains; and there also similar means can be employed to neutralise it. The common weighing-machine furnishes to a certain extent an illustration of this (see Fig. 122), for it must, so far at least as concerns the mere connections of its links, be regarded as an incomplete kinematic chain. Incompleteness of this sort, however, is nothing more than the incompleteness of a pair of elements; and we already know how it can be rectified. The real question rather concerns the incompleteness of chains which consist wholly of closed pairs, or in which, by any suitable means, the closure of every single pair has been completed.

The application of a driving-force to one link of a mechanism in such a way that that link may have the motion especially belonging to it, and the whole mechanism be moved, is not sufficient of itself to insure that continuous motion shall be possible. In the mechanism shown in Fig. 145, for example, a force always normal to *a* may be applied at the pin 2, and by its means the slider *c* will be continually driven backwards and forwards in the slot *d*, by the connecting-rod *b*. If, however, the driving-force be applied to *c* instead of to *a*, so as to cause it to reciprocate between 3′ and 3″, there is nothing to insure the continuance of the motion beyond the positions 3′ or 3″ of the block, and 2′ or 2″ of the crank pin, for in these positions the driving effort passes through the axis of the fixed bearing 1, and is therefore received direct by the stationary link *d*. These two positions form what are called the "dead points" of the mechanism. The fundamental idea of this expression is that the chain opposes itself as a rigid body fixed to its stationary link, to the action of the driving force; that its moveability, its life, is lost,—it lies as it were dead. This idea we shall generalise by applying it, not merely to the crank mechanisms, for which Watt first used it, but to all other mechanisms in which a similar condition occurs.

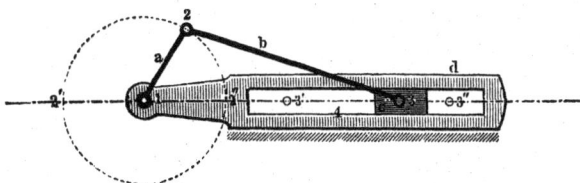

Fɪɢ. 145.

Several means are employed for carrying a mechanism over its dead points. In the case of Fig. 145, and others similar to it, it is usual to employ a revolving mass, connected with some link of the chain which is moving with suitable velocity when the driven link is at its dead points, and to use a portion of the energy stored in such a mass to carry the mechanism over the dead point. In the application of this train to the steam-engine, everyone is familiar with the fly-wheel,—rigidly connected with the crank *a*,—which is so often used for passing the dead points. The fly-wheel furnishes the sensible force required to continue the motion of the machine;—the continued motion of the chain is therefore effected by force-closure. In the locomotive, so soon as the train is once set in motion, the whole mass, moving directly onward, serves the same purpose as a fly-wheel; but in general rotating bodies are used. Now and then we find simple weights

which act statically, and not as reservoirs of actual energy, in
effecting the passage of the dead points by force-closure.

§ 46.

Passage of the Dead Points by Chain-closure.

The use of force-closure in passing the dead points is not suitable in every case. In the steam-engine especially it has frequently
been necessary to resort to another principle. In this case the
dead point has to be crossed repeatedly when a portion of the
mechanism has a very small velocity,—is just passing from rest
to motion,— as is the case in locomotives, marine-engines, winding
engines, etc. This other principle consists in the employment of a
second kinematic chain so connected with the first that it is always

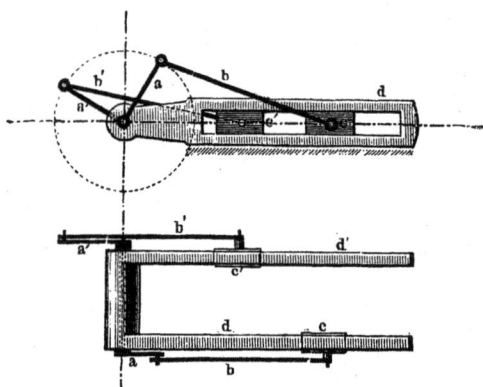

Fig. 146.

in an advantageous position for action when the first is dead. In
most cases the two chains are similar, so that their action on each
other is reciprocal.

In order to close the above crank train, for instance, at its
dead points, it is so combined with a second similar mechanism
(Fig. 146) that the two cranks have a common axis and are 90°
apart,—the directions of the guides being parallel. This form of
mechanism has of course most extensive application to locomotives
and double-cylinder engines generally. If instead of using cranks

at right angles the guides themselves be placed 90° apart (Fig. 147), a single crank may serve for both mechanisms. This is frequently * used for screw-steamers, and has also been applied to land-engines. In this case, as in the former, the driving force (steam-pressure) in the chain is assumed to act upon the bodies c and c'.

Another mechanism which is carried over its dead points by chain-closure is sketched in Fig. 148. Two equal and parallel cranks a and c are connected by a link b having a length equal to that of the fixed link d. The figure 1 2 3 4 is therefore a

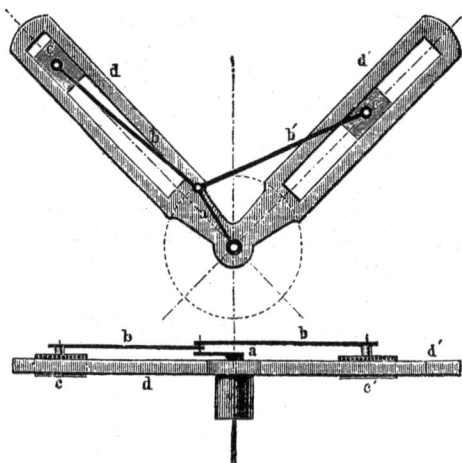

FIG. 147.

parallelogram having dead points in the positions 1 2′ 3′ 4 and 1 2″ 3″ 4, and this whether the driving force be applied to a or c. The dead points may be passed by the addition to the first chain of a second one a' b' c' d' similar to it, in such a way that a has a common axis with a', c with c', and that each pair of cranks encloses an angle of 90°, and further that d is connected with d', *i.e.*, that both are made stationary (Fig. 149). Locomotives with coupled wheels give familiar illustrations of this arrangement.

---

* The arrangement with cylinders equally inclined to the vertical has been very little used for marine engines, at least for many years, in this country : I have seen it applied to oscillating cylinders. One of its most recent applications has been at the Panteg Steel Works, Pontypool, where it forms the driving mechanism of large direct-acting Rolling Mill Engines (see *Engineering*, vol. xix. p. 249). The same mechanism has lately been used for large compound engines in the Guyon

Another method which can be employed with the same mechanism is the placing of a third equal crank, b' (Fig. 150), in the

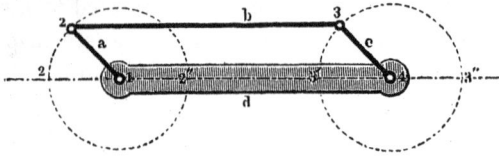

FIG. 148.

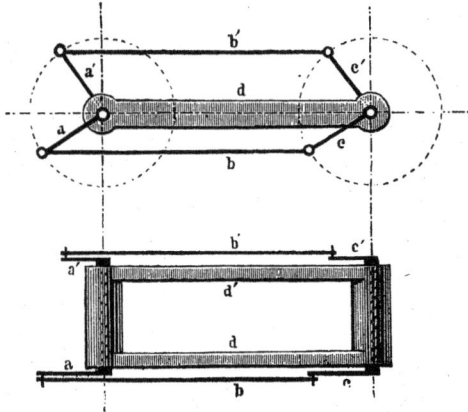

FIG. 149.

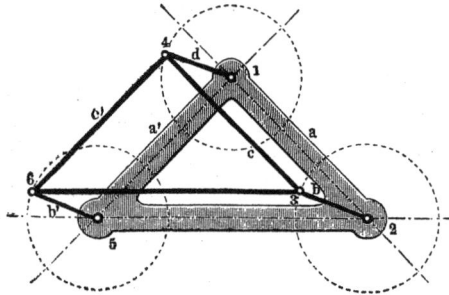

FIG. 150.

plane of the other two (b and d), and connecting it with them by the links 6,3 and 6,4, while the shaft bearings are all connected with

Company's Steamers, with one cylinder vertical and the other horizontal. With the addition of a third chain similar to the first two and making equal angles with them, Fig. 147 would represent the driving mechanism of Brotherhood's "three-cylinder" engine, and the same treble chain has also been used in much larger machines, as in the three-cylinder engine of H.M. Gunboat "Waterwitch."

the fixed link a in such a way that 1465 and 3256 are parallelograms. Here three similar chains are combined, of which—if the points 4,3 and 6 form a triangle—only one can at any instant be crossing its dead points. Applications of this mechanism occur not unfrequently.

§ 47.

Closure of Kinematic Chains by Pairs of Elements.

We have already found (§ 42) that in certain cases pairs of elements are closed by kinematic chains. We must now examine the converse of this—the closure of kinematic chains by pairs of elements. This does not supersede the closure proper to the chain, but simply acts along with whatever closure already exists in positions where it is insufficient.

Even in the mechanism which is once more represented in Fig. 151, we may note one thing about the sequence of the separate phases of the motion which must not be entirely disregarded. Let us suppose the dead point 2′ to be reached by the left-handed

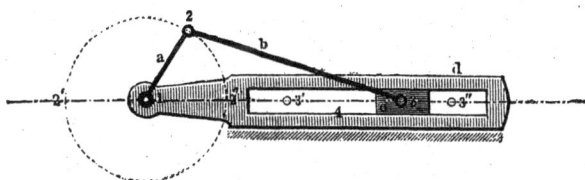

Fig. 151.

rotation of the crank from the position 2, driven by the application of some force to the block c. If now this force be reversed, without the intervention oɪ either force- or chain-closure to carry a over the dead point, it remains indeterminate whether the crank shall move forwards in the lower half circle or backwards in the upper. Motion in the chain is therefore not absolutely constrained; and the chain is not in the strict sense of the word closed, if by that expression we understand that for every motion of any link only a single motion of any other link be possible. At the same time the supposed backward motion of the crank a is not in itself different from the motion before made; we have only to consider the motion of *c* from left to right as itself a reversed motion to leave the whole conditions just what they were before.

Under similar circumstances, however, the case is quite different in the mechanism with two parallel cranks, which was described above. Let us suppose here (Fig. 152) that by turning in the direction of the arrow the link 2,3 takes the position 2′3′ ; then in the absence of any provision for carrying it across the dead point, any forward motion of the crank *a* may cause *c* to move either backwards or forwards ; thus, for example, when *a* has reached 2‴, *c* may have returned to 3, and so when *a* has arrived at 2″ it may have again resumed its dead position 4 3″.

Thence the backward motion may be continued through the other semicircle, *a* again passing through 2 and so on. In short the L. H. rotation of *a* may give to *c* either a right- or a left-

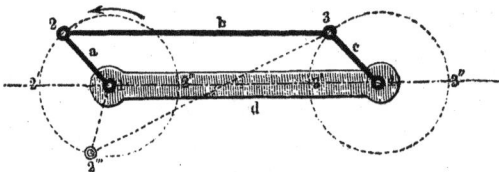

F**ig. 152.**

handed rotation, and the latter would take place in a manner entirely different from the former. The change from the one motion to the other may occur at either of the dead points. Obviously here again the chain is not closed at these points, and the methods for passing them represented in Figs. 149 and 150 furnish also the necessary closure at them. If such closure be not provided, a complete change in the motion may take place at each dead point ; the nature of the mechanism, that is, may be entirely altered. Any position of a link of a mechanism at which this is possible we shall therefore call a c h a n g e - p o s i t i o n or change-point.

The chain-closure of which we have shown two forms in Figs. 149 and 150 is arranged for constraining the two cranks to rotate similarly. Another question now presents itself ;—how to make the closure complete and at the same time to constrain *a* and *b* to revolve in o p p o s i t e instead of in s i m i l a r directions.

This closure can here be effected, as in all chains which have change-points, by c o n n e c t i n g t h o s e l i n k s w h i c h a r e u n c l o s e d in the change-positions by pairs of elements formed so as to correspond to the nature of the motion which is required.

In order to do this we must, as we know from Chapter III., determine the axoids or centroids of the bodies to be paired. These bodies are here, for example, the two cranks a and c, or otherwise the two links b and d;—two opposite links, in other words, of the four-linked chain containing four parallel turning pairs shown in Fig. 153, the centroids of which we have already examined in §§ 8

FIG. 153.

and 9. In that case the centroids were very complicated figures, here they are made extremely simple by the equality of the opposite links. Remembering always that the cranks are to revolve in opposite directions, the centroids will be found to have the forms shown in Fig. 154.

For the links a and c, the shorter pair, they are ellipses, having their foci at the ends 1,2 and 3,4 of the cranks, and their major axes $A B$ and $C D$ equal in length to the links b and d. The instantaneous centre moves backwards and forwards along these links (being always at their intersection). For b and d the centroids are hyperbolæ, their transverse axes $E F$ and $G H$ lying on the links themselves, and being equal to a (= c); their foci are the points 2,3 and 1,4. The instantaneous centre traverses each branch of the curve to infinity, turning from − ∞ along the other branches.

If it be required to pair two opposite links at their points of change, a higher pair must be employed in each case; such pairs need not, however, go further than corresponds to the elements of the rolling conics in contact at the change positions. If the links chosen be the two shorter ones, a and c, these are the elements of the ellipses at the extremities of their major axes, viz. A, B, C and D. By putting a pin and a gab * at these points, as shown in

---

* I use the common technical word for a fork or open eye of this kind.

Fig. 155, our mechanism becomes really a closed chain.* To this
I shall have to return again, looking at the question from another
point of view.

If the two longer links b and d are to be paired instead of the
shorter ones, we have only to notice that it is the vertices F and G,

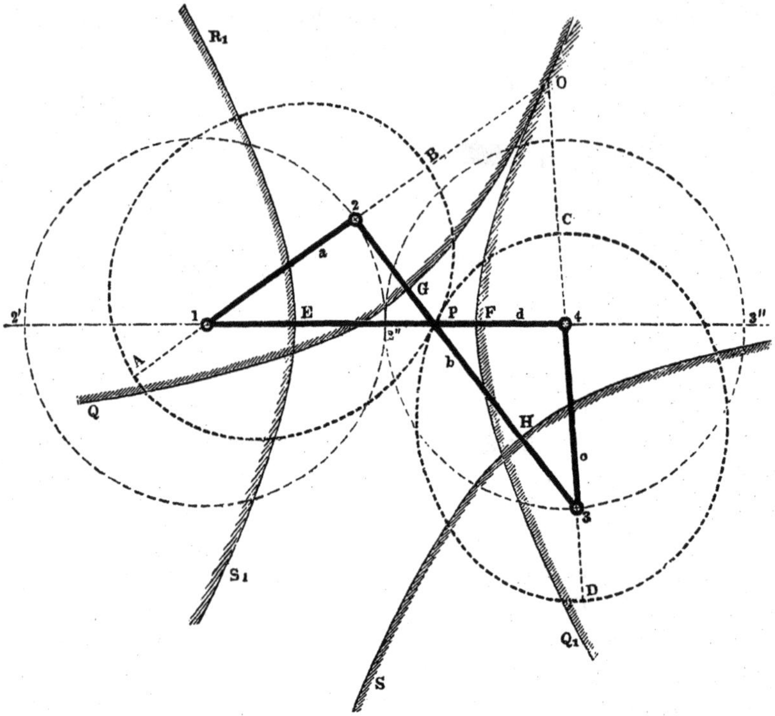

FIG. 154.

and E and H, which touch each other in the change-positions. By
placing in these points again a pin and corresponding gab, Fig. 156,
we have again a pairing which effectually closes the chain. We
have here then two solutions of the problem before us. If it were
wished it would be possible to close the chain at one dead point
by one method, at the other by another.

The following is another interesting example of the pair-closure
of a chain having a point of change. If in the mechanism shown

* This mechanism was first described by Prof. Reuleaux, who called it "Gegen-
drehungs-Kurbel" (reverse-cranks), in the *Civil Ingenieur*, 1859, p. 99.

in Fig. 145, the length of the link b be made equal to that of the crank a, then when the crank makes a quarter revolution the centre of the block c has been brought up to the axis 1 of the crank, and it would be moved forward symmetrically as the rotation continued, were it not that a change-point here occurred. The centre of the

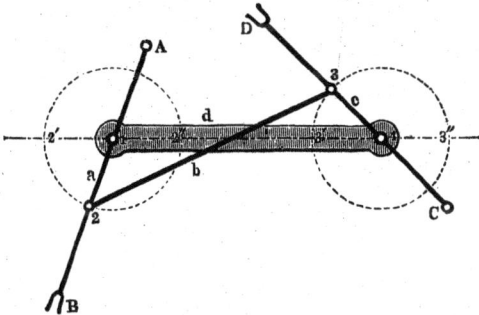

FIG. 155.

pin 3 coincides with the centre 1, a and b cover each other, and can together turn about their common axis 1. The chain thus becomes at its point of change simply a pair of elements. This can be prevented by pairing the link b with the opposite one d, for which purpose we must find the centroids of both links.

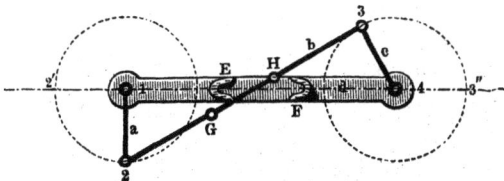

FIG. 156.

An examination of the relative motions of b and d shows at once that these centroids are Cardanic circles (Fig. 157), the smaller having a radius $a = b$, and described about the point 2 (which is a point always of the link b), the larger having a radius twice as great, described about the point 1 of the link d. It is easy to see that the pairing at the point of change can be effected by supplying b and d with such small portions of toothed profiles as can come into action at that point.

The diametral ratio 1 : 2 of the centroids, and their simple figures, make this extremely easy. If, for instance, we connect with b at the point 5,—opposite to 3,—a pin, this will be at 5′ at the first change point and at 5″ at the second, and if suitable gabs be fixed in those positions, it will gear with them, and the mechanism will be completely closed, as can be readily seen from the figure. It is worth noting that, for every half revolution of the crank, the point 3 moves through a distance equal to four times the crank radius, while in the mechanism in Fig. 145 its path has only half this extent.

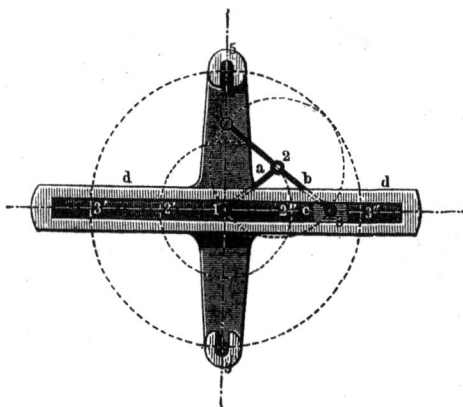

FIG. 157.

This mechanism has been long known. It was applied by Dawes with little alteration in 1816 to a steam-engine, which I shall have to mention again. It must be remembered that this closure in no way helps the crossing of the dead points 3′ and 3″ which occur if the driving effort act upon the block c. Neither here, nor in the case of Fig. 145, is this possible, for at these points the instantaneous centre lies upon the line of motion of the block, so that the driving force passes directly through it.

Generally, the passage of the dead points by higher pairs is only possible if the centroids have suitable forms. These must be such that, at the moment when the dead point is reached, the driving effort must not pass through the instantaneous centre, for in that case the arm of its moment about that point, and therefore the moment itself, have become = 0.

As a further example we may take a remarkable mechanism, obtained from the chain of Fig. 153 by altering the lengths of its links. If we make two pairs (a and d, c and b) of adjacent links equal, the opposite links being unequal, as in Fig. 158, we obtain, by fixing one of the links, a mechanism which has two change-points. The first occurs if a be brought by a left-handed rotation into the position 1, 4; 3 will then have come to 3', and if no closure be arranged, the chain becomes simply a turning-pair, with 4 as its centre. The second change-point occurs if a makes a further complete rotation about 1, so as again to come into the position 1, 4, but with 3 at 3'' instead of 3'. The result is the same; the links b and c, coinciding, can turn about 4.

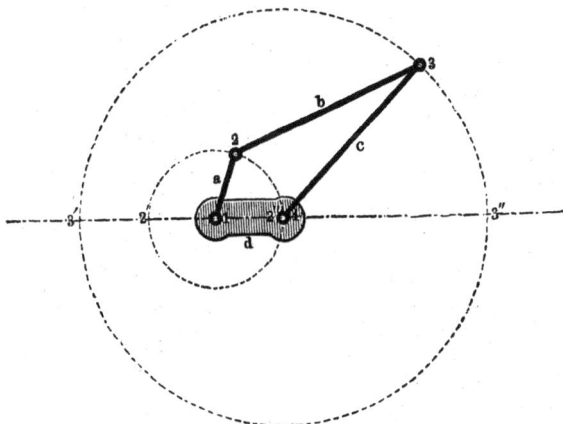

FIG. 158.

Thus for one whole revolution of the crank a, suitable closure being supposed to exist, c makes half a revolution, or conversely one whole revolution of c causes a to make two revolutions. This mechanism was first described by Galloway, who used force-closure derived from a fly-wheel for the passage of the dead points. He assumed the ratio of the lengths of the cranks a and c to be 1 : 2, a limitation which we see to be unnecessary.*

* This particular form of the four-link chain with four parallel turning-pairs forms the mechanism called by Prof. Sylvester the "kite." The centroids are the curves known as Limaçons, a special (nodal) case of Cartesians. It is worth noticing that if the centroids be placed with their axes coinciding, the node 2 falling upon the point 4 and A upon C, then any line drawn through 2 will cut off equal arcs from both centroids, *i.e.* will cut them in a pair of points

To find the pair-closure for the chain we must know its centroids. As the opposite links 2, 3 and 3, 4,—2, 1 and 1, 4,—are

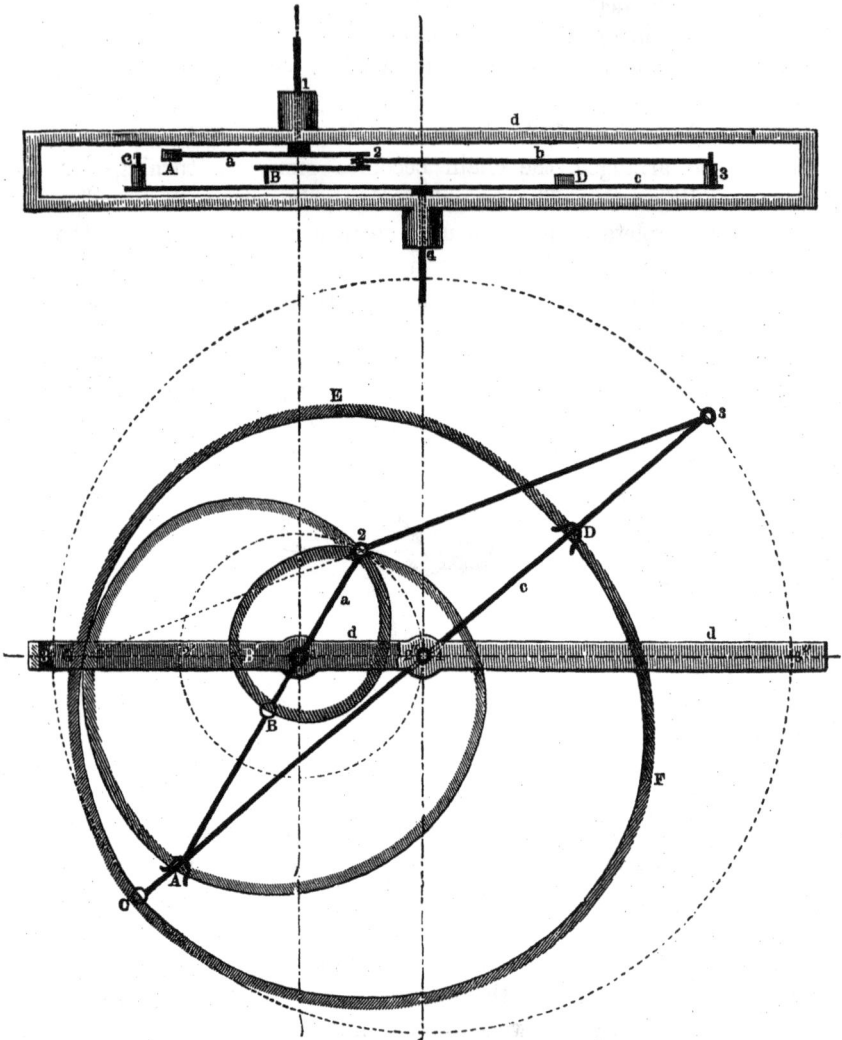

FIG. 159.

equal, the two pairs of centroids are similar,—we need therefore

which fall upon each other during the rolling. The tangents to the curve at 2 will
of course pass through E and F.

examine only one of them. These curves have several points in common with the centroids in the last case, the internal rolling and the ratio of their peripheral lengths 1 : 2.* Their forms, however, correspond to the unequal velocity-ratio of the two rotating arms, *a* and *c*. The lightly shaded and larger branch 2 *A* 2 of the centroids of *a* rolls in the arc *E C F* of that of *c*, the loop 2 *B* 2 in the arc *E D F*. In the two change-points *A* and *C* and *B* and *D* respectively come into contact. By once more using pins and gabs, placed at these points, we can here also obtain complete closure. The plan above our figure shows in outline how this may be carried out constructively, in a way which might replace with advantage the force-closure of Galloway. Otherwise the mechanism is not described as one which has any practical application,— hitherto at any rate its interest, although great, is purely theoretical. We cannot tell, however, if it may not find in the future some useful application. It need only be further remarked that here, as in Fig. 154, the second pair of centroids might equally well have been used for the pair-closure.

Among those arrangements which have for their primary object the carrying of a mechanism over its dead points must be included the air vessels of fire-engines, steam-pumps, and hydraulic machinery,—those, too, of various kinds of bellows, simple hand-bellows and those of the smithy or organ, and the wind-chests of blowing-engines. The air vessel at the same time—corresponding exactly to the fly wheel of a steam-engine—acts as a means of making the motion of the out-flowing pressure-organ uniform,—it acts, that is, not only at the dead points, but before and after them. Its action depends—like that of the fly-wheel—upon force-closure ; in which relation however it goes even further, for the organ by means of which the action takes place, the air, is itself force-closed.

It is interesting to observe how, in machine construction, chain-closure is more and more superseding force-closure as a means of passing the dead points. One whole class of engines, those which have to be reversed, are almost always made with two cylinders as coupled engines, *i.e.* with chain-closed dead points. This has even extended itself to the clumsiest of all our steam-engines, those for

---

* It will be seen further on, § 70, that the two mechanisms are in reality the same.

rolling mills,* where formerly the fly-wheel was considered a *sine qua non.* There was a necessity for reversal in the case of the plate mills, but the application of chain-closure there seems to have led to its use in other cases where no such necessity existed. Coupled engines are used for non-reversing factory engines also,— locomotives and marine engines may have suggested the arrangement, while its own intrinsic value has confirmed and extended its adoption. In blowing-machines, too, no less than in steam-engines, we see the same thing ; by increasing the number of cylinders, by adding separate regulating cylinders having pistons driven by cams, as well as by other arrangements, frequent endeavours have been made to substitute chain-closure for force-closure.† We have unquestionably in all these things a distinctly recognisable tendency shown towards certain alterations of older machine-forms into others which furnish more direct solutions of the actual problem,—the production of certain determinate motions.

* See for instance the reversing rolling mill engine on Mr. Ramsbottom's system in *The Engineer*, July 17, 1874.

† A special, but very interesting, case of the use of a second cylinder instead of a heavy moving mass, has been treated in detail (but from a somewhat different point of view) by the Translator, under the heading "Steam Economy in Pumping Engines" in *Engineering*, Nov. 13 and 27, 1875. Mr. Arthur Rigg's proposed "turning gear" for single marine engines is also a case in point. Instead of using two identical chains, as in Fig. 149, he makes the second chain auxiliary only. See *Trans. of Inst. Naval Architects*, 1870.

CHAPTER VI.

SKETCH OF THE
HISTORY OF MACHINE DEVELOPMENT.

" How many creations of Art, how many wonders of Industry, what light in every field of Knowledge, since man ceased to waste his powers unprofitably in a weary struggle for existence, since he became master of his own destiny, making terms with that necessity which he could never entirely escape ; since he attained to the dearly-bought privilege of freedom to rule his own faculties—to follow the call of his genius. What eager activity everywhere, since multiplied wants gave new wings to the spirit of Invention, and opened up new fields for Industry !"—SCHILLER.

" In possession of the idea of World - development History no longer lies within a bounded horizon, it no longer repeats the self-same story in wearisome iteration century after century. In its unmeasured depth one form of existence follows another, Nature reveals to us its infinite succession of wonders, and the soul rises up in heavenly majesty, and with mighty wing-strokes speeds through the Æons !"—GEIGER.

§ 48.

The Origin and Early Growth of Machines.

THE investigations which have occupied us in the last two chapters have led us involuntarily to a subject which certainly does not appear to be necessarily connected with a deductive theory of machinery; which, however, concerning as it does the manner in which the machine idea originated in the minds of men, claims our

highest interest, and which, besides this, is indeed essentially of
importance to the right comprehension of our work. This subject
is the History of Machine Development. I can only enter
into the matter here very briefly; partly because, as I have said,
it does not lie directly in our way, and partly because the materials
existing for such a history are as yet too few to enable me to enter
into details with any degree of certainty.

This history of a development must be distinguished from a
history of the ordinary kind. The latter gives us in chronological
order a series of individual phenomena, which may be retro-
gressive as well as progressive. The former seeks only to find the
steps by which some known position has been reached,—it repeats
itself anew with each nation's growing civilisation, it even reflects
itself in the development of each single individual. The history
of machinery, at least of certain special machines—mills, apparatus
for transport, and the steam-engine for instance—has been already
somewhat completely written, and more and more, fortunately,
is being done in this direction. The history of machine development, on the other hand, was not possible without a previous
distinct acquaintance with the real nature of the machine itself,—
the subject, that is, which we have been investigating,—and this
must form the foundation of as yet unmade investigations. Reflection from things known and existing may always, however,
throw some light upon the past,—it is by this means principally
that I must attempt to work from the stand-point which we have
now reached.

At the commencement of a study of machine development it is
first of all necessary to know distinctly what it is that makes a
machine complete or incomplete. It is only possible to judge of
the completeness of a machine from the excellence of the work
produced by it, if we are able to estimate separately what part of
the result is due to the skill of the workman. Certain Indian
fabrics, for instance, are of extraordinary excellence and delicacy,
although they have been made in most defective looms; throughout
the whole manufacture of these it is the weaver's dexterity that plays
the most important part. In no machine, however, can we absolutely
do away with human action, if it be for no further purpose than to
start and stop the process. It appears, therefore, that the most
complete machine is the one fulfilling best its own share of the

work, and having for this share the greatest proportion of the whole task. It is in both these directions, in fact, the intensive and the extensive, that modern machine industry tends to progress. The two interact, and so for every improvement in the method of accomplishing any given purpose, there follows, as if instinctively, an extension of its share in the whole work.

Our aim now is not so much to describe the gradual growth of the machine in capacity, and the extension of its applications, as rather to find out by what means it has become what we now see ; —what, that is, has been the real nature of the means by which its gradual improvement has come to pass. The clearer we can make this to ourselves, the more completely we can divest it of all connected ideas, and place it by itself objectively before us, the sooner shall we succeed in the future in making conscious progress in perfecting the machine.

For our purpose we must, so far as it is possible, trace the machine from its very origin.

If we search history for the beginnings of the machine, we find ourselves carried further and further back into the past. Every people that appears in history shows itself more or less familiar with machines, of however imperfect a kind. We do not find the actual beginnings with them, their traditions only give us information as to progress and improvement. We must, therefore, forsake the historic for the prehistoric period. We must enter the domain of ethnology, the study of primitive peoples, of nations still in those first stages of development through which the now civilised portions of humanity must at one time have passed. For inquiry points more and more distinctly to the conclusion that the human race as a whole has everywhere grown through similar stages, progressing according to great natural laws.[23] The further we examine this matter, collating what we have found with the vestiges of long-perished prehistoric cultures and semi-cultures which have lately become accessible, the more distinctly shall we see that we cannot trace the machine backwards alone, but that it interweaves itself continually with the whole development of nations, indeed of the human race. We see, in other words, that we must carry our inquiries into the dim distances of the history of the development of mankind itself in order to find the first germ, the earliest rootlet, of the ideas which have slowly grown through unnumbered centuries into our complex civilisation, through high cultures and between declining ones, until at last in the western lands during the last two hundred years they have received the impulse which still urges them in their upward flight. It is, therefore, partly from Archæology, partly from Philology, partly from Ethnology and Anthropology, that we must obtain materials for the investigation of our subject.

In addition to material found existing in fact or in tradition, we may assist our research by the examination of those delicate indications of the course of the development of men's capacities which are revealed in their language, by means of which notable results have been obtained in the so-called linguistic or glottistic Archæology. We are, indeed, indebted to a philologist for a remarkable attempt to trace out the earliest appearance of the machine. Geiger, whose death has been so great a loss to students of the history of language, has laid down certain base-lines in two published chapters on the Origin of Tools and the Production of Fire,* which the mechanician, who wishes to trace his subject through prehistoric and historic times up to to-day, cannot leave unnoticed. Geiger, in his excellent little work, comes to the conclusion, after very thoroughly considering the matter from all sides, that rotary motion was the first that men produced by any arrangement which could be called machinal. The pieces of wood rubbed together for the production of fire, which as "double wood" play no unimportant part in the religious ceremonies of the ancient Indo-germanic peoples, and are still so much used by uncivilised races, form according to him one of the first, if not the very first, arrangement deserving the name of machine.† This was at a time so early that fire was apparently not yet used in houses, but appeared only in religious observances. For important reasons seem to point more and more distinctly to the conclusion that the human race has passed through a fireless time, a time when it had not yet learned to employ the "friendly element" in its dwellings, although it reverenced or worshipped it in sacred places.

A piece of wood, roughly pointed at one end, is placed perpendicularly upon another in which a small hole or recess has been made, and is caused to twirl rapidly backwards and forwards by

---

* Geiger, *Entwicklungsgeschichte der Menscheit.* Stuttgart, 1871.

† Klemm also mentions this in passing. *Kultur-Wissenschaft;* III. § 392.

the hands, until the shreds rubbed off the wood, or some fibres of cotton or fragments of pith placed upon it, catch fire (Fig. 160).* The hands have not merely to twirl the wood (made somewhat too European-looking in our illustration), but also to press it continually downwards upon the other piece—held fast horizontally with the toes or knees—so that they gradually move downwards upon it. For woods which are comparatively uninflammable two men therefore must work together, the one beginning to twirl the upper end of the stick when the hands of the other have reached its lower end.[24] From the descriptions which we have the apparatus appears to have been very similar to the " fire-drill " even now used by the Brahmins, although distinct and not altogether unimportant differences may really exist.

In later times, long after the first use of the " double-wood," a cord was wound once or twice round the twirling-stick, its ends held in the hands and pulled backwards and forwards, and by this means the stick received its motion (Fig. 161).

The upper end of the twirling-stick was now also sharpened, and the required downward pressure was given to it through a third piece of wood similar to the under piece, held and pressed down by a second worker.† It will be quite evident to all who have studied primitive history with any degree of thoroughness that any such improved application

Fɪɢ. 160.

of such an important apparatus as this presupposes for it the most widely-extended use. Some notion of this may be obtained from the fact that at the great sacrificial feast of the Hindoos, the people from whom the Europeans have sprung, it was prescribed in the

* See Tylor, *Early History of Mankind.* London, 1870, § 241 ; and Klemm, vol. II. § 66.

† I take figures 160 and 161 from the work of Tylor's quoted above. The latter represents two Esquimaux making fire, and is copied from a drawing of the last century. Tylor describes also another series of these " fire-drills " belonging to a later period. R.

ritual that the sacred fire should be lit 360 times, with nine different kinds of wood, upon each day of a whole month.

The discoveries made in recent excavations allow us to conclude with great certainty that primitive nations used some such arrangement as we have described for working those borers with which to our astonishment they were able to drill holes in wood, bone, horn, and even in very hard stone. The inhabitants of some portions of the world still use this twirling apparatus for boring. Indeed it has lately appeared that the deeply-bored stones, emeralds, rock crystal and nephrite, carved with figures of animals, which were discovered in South America by Humboldt and Bonpland, and looked upon by

Fig. 161.

them as belonging to some higher culture long destroyed,—were fashioned by means of the simple twirling-stick of Fig. 160, or perhaps by that of Fig. 161. The secret was that sand and water were used along with the boring-piece. A single piece of carving required years, perhaps more than one generation, for its completion.[25] This whole question is put in an entirely different light by the fact that the words now in use for boring were in no way originally connected with the making of a round hollow in, or hole through a piece of material with which we now think them inseparably connected. They all rather imply, more or less, the idea of rubbing, stirring, or grinding which stands in close connection to polishing, smoothing, or forming by friction.*

How long a time elapsed before the to-and-fro turning of the twirling-stick became a continuous rotation is mere matter of conjecture. In any case its length must have been such as to leave far behind it our historic period. It is certain that undershot water-wheels, which may well be the first representatives of machines containing continuous rotation, appear in very early times: they imply, however, the existence of some degree of culture. Where they were used for irrigation at least, they presuppose a settled tribe building on the ground. We have traditions of their use in Mesopotamia in a form corresponding to that still to be found there, a wooden wheel with clay buckets.† The scoop-wheels of the oldest kind still used in China, of which one is represented in Fig. 162, have a wooden axle, but all their other parts are of bamboo and basketwork, no metal being used; the shaft rests in V-shaped wooden bearings. They have a diameter of from 6 to 12 metres, and discharge the water

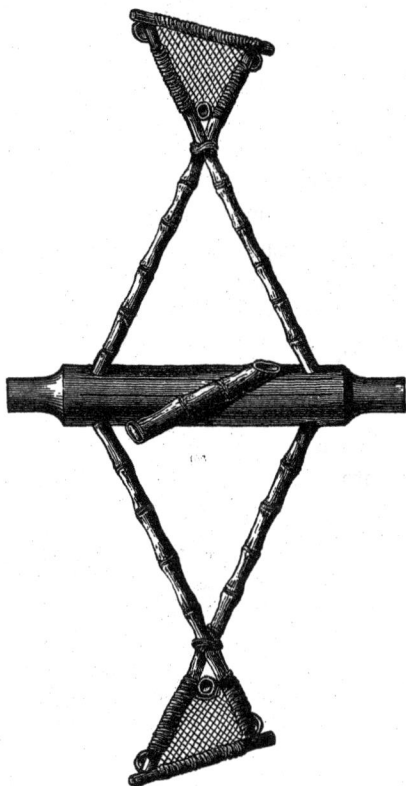

FIG. 162.

raised by the bamboo scoops or corfs into a wide trough from which a channel conducts it to the land. Geiger, who does not mention these wheels, gives an earlier date to the praying wheels still used

* See Geiger, *Ursprung u. Entwicklung der menschlichen Sprache u. Vernunft.* (1872) ii. p. 54,

† Pliny, xix. 22.

in the temples of Thibetan and Japanese Buddhists. These are in part windmills, in part undershot water-wheels; their use in worship Geiger traces with great skill to the awe with which the continuous rotary motion was regarded, but does not go further into the consideration of the interesting question before us.

The potter's wheel, still unknown to the lake-villagers, is not apparently of an earlier date than these water-wheels; it represents, however, an earlier employment of more or less continuous rotation; it is very possible that before the employment of a rotating mass to carry on the motion once it was started, the potter may have had an assistant to keep the wheel continually in motion by whirling its spindle with his hands.

The question as to the origin of the carriage and carriage wheels is very interesting to us, for the latter as kinematic contrivances permit important conclusions to be drawn as to the previous existence of other machinal arrangements. Among the Greeks, Egyptians and inhabitants of Western Asia we find two-wheeled vehicles very early in use.[26] Their introduction travelled apparently from East to West; for a long time they served both Greeks and Egyptians as the chief, indeed almost the only, means of horse transport, whether in war, commerce, or public processions. Riding came late into use among these nations, brought from the countries East and North of them. The Homeric heroes do not ride, but drive in battle; indeed riding men were in those days thought of as wild uncivilised barbarians, as the myth of the Centaur shows us. The bas-reliefs of the Assyrians, on the other hand, indicate that among them both riding and driving were known. The chariot was in that time an invaluable instrument of warfare, the possession of which in preponderating numbers—as with us of cannon—gave an army an enormous advantage over its antagonists. We read for instance in the Bible (Judges i. 19) that the Israelites on their entry into Palestine felt greatly their want of chariots. Although Judah took possession of the high lands he was not able to drive out the dwellers in the valleys, "for they had chariots of iron"; see also Deborah's Song (Judges iv.). The want does not appear to have been permanently supplied until long afterwards, first perhaps in the battles of David, in one of which he is said to have taken 700 chariots from the Syrians (2 Sam. x. 18).[27]

Egyptian sculptures and paintings, and those also of the ancient Assyrians and Greeks, frequently show us some details of the construction of these primitive wheeled vehicles. The Assyrians and Egyptians used for the most part wheels with six spokes,— those of larger diameter and more clumsy construction having sometimes eight, and even twelve. The Greeks preferred wheels with four spokes only. The wheels constructed with the smaller number of spokes are the more perfectly made ones ; originally of wood, they appear in time to have been made of wood and metal together, and at last entirely of metal (bronze).[28] Simple disc-shaped and rudely-constructed wheels are indicated by these pictorial records as used by the less civilised nations of Asia Minor. In India to-day also we find rough disc-wheels, besides many with a large number of spokes ; these are used with axles both of iron and of wood. In the Plaustrum, the older form of the ancient Roman freight waggon, disc-wheels were also used with the notable peculiarity that the two wheels had square holes through which the wooden axle was placed, the latter having cylindrical journals with bearings in the waggon frame.* Waggons of just such a form are used even now in Portugal,[29] and in Formosa also the aborigines use the same construction.[30]

The miniature bronze vehicles which have been found in burial mounds on the North German plain, in Schonen, &c. (and of which the Mainz Germano-romance Museum, among others, possesses some excellent copies) can scarcely belong to a higher antiquity than the ancient vehicles just described, if indeed they be not younger. They are generally supposed to have had a religious signification,† and are considered as parallel to the "laver"-carriages of Solomon's Temple ; but there are not wanting important authorities who consider them to have had a quite different nature. It is noteworthy that their wheels have mostly four spokes, like those of the majority of the Grecian vehicles.

Wheeled vehicles were known, however, before the times in which those we have described take their origin. The oldest Indian literature mentions them repeatedly. The following are some of the references to them in the *Riksanhita :*—

---

* The wheels of "Puffing Billy" (1813), which is now in the Patent Museum, are secured in this way.

† L i s c h,—*Ueber die ehernen Wagenbecken der Bronzezeit.* Schwerin 1860.

(X. 89, 4) The mighty Indra supports the heaven and earth
Separating them like two wheels upon their axle . . .

and in another passage :—

(VIII. 6, 38) As the wheel behind the horse, so behind thee roll both worlds . , ,

and again in describing day and night :—

(I. 185, 2) The Day and Night turn like two wheels,
Their own power supports the Universe ˙. . .
These two, themselves un-moving, foot-less,
Have moving, footed spokes innumerable . . .

and so on, many passages showing that already great attention had
been given to the decoration of vehicles and to the breeding of
horses. From this free use in poetry of the image of a wheeled
carriage we must conclude that such vehicles were already of great
age at the time at which the Vedas were written, more than 1700
years B.C.

Thus if we trace all its intermediate steps we find the use of the
carriage going even further and further back into the dim times
before the historic period, to change the darkness of which even
into twilight we have only the single torch of philological research ;
and this research points to the conclusion that the original part of
the carriage was not the framework or body, and that the whole
has been built up from the rolling piece, the wheel itself.[31] The
gradual development may have taken place from the circular tree-
trunk placed as a roller under a load, to which the disc-shaped
wheel, and especially the Formosan pair of wheels, stand very near.[32]
However this may be, the use of the carriage extends back to the
very beginnings of civilisation, to the time when men first lived in
villages of their own building, and it ranks therefore among the
most primitive inventions of the human race.

The kind of boring apparatus which we have described extended
itself into historic times. Homer gives us (Od. IX, 384 *et seq.*) a
clear description of it :—

—" I standing above them,
Bored it into the hole : as a shipwright boreth a timber,
Guiding the drill that his men below drive backward and forward,
Pulling the ends of the thong while the point runs round without ceasing. "

This method of working, used by the ancient carpenters and
obviously very common in the Homeric times, required three

workmen, as was probably the case also in the boring of large stone axes by the prehistoric men.

It requires here to be pointed out that the making of a cylindric boring, *i.e.*, a hollow cylinder, must have been immensely older than the reverse operation of making a solid circular cylinder. The hole could be bored with a very incomplete form of cutting tool, for it is easy to make the hollow body tolerably regular in shape whether its generator, the cutting edge of the tool, be regular or not. Almost any piece of flint could be used for boring

FIG. 163.

n wood, bone, or horn, if it were only sharp and of such a shape that it could be gripped.[83] The turning of a body, on the other hand, requires that it shall be placed in some kind of machinal bearings, which shall render possible its rotation about a fixed geometric axis, before a chisel can be successfully applied to it. It seems to me probable that it was the potter's wheel which led the way to the lathe. It can be shown certainly that the operation of boring was older by ages than the use of the potter's wheel, and that this again long preceded the lathe.

Perhaps one of the oldest forms of the latter machine is the one still in use among the Kalmucks. It has, as the figure * above shows, a horizontal spindle of wood resting in bearings, and caused to rotate by an assistant of the turner, by means of the cord between the bearings, exactly like the fire-drill in Fig. 161. The object to be turned is fixed to the free end of the spindle. The turner simply spikes his machine to the ground, places the little

* From Klemm's *Kulturwissenschaft*, i. p. 387.

knee-shaped tool-rest in front of it, and begins his work. Considering the simple nature of his machine he makes exceedingly good work.* It may be noted that he turns by preference bowl-shaped pieces of wood, horn, metal, &c. This arises from the construction of his lathe, in which the want of a second support for the piece which is being turned,—or what we call a movable headstock,—compelled him to choose either flat or hollow pieces for his work. Thus both the construction of the machine and the kind of work produced by it point out its descent from the potter's wheel. We are led to a similar conclusion by the circumstance that among the Romans[34] the turner was not only called tornator but also vascularius,† that is, dish- or bowl-maker. Whether or not the two expressions were used interchangeably, we know at least that the vascularius used the lathe often, and with great skill, as various vessels and fragments show. Thus this bowl-lathe appears as a connecting link between the potter's wheel and the lathe for turning long bars. In later times the Kalmuck turner has made himself independent of assistance, at least for light work, by taking one end of the driving cord with his left hand, and the other between the toes of the left foot, keeping his right hand still free to handle the chisel.

The skill of the Roman turners displayed itself in many ways; they could, for example, turn stone bowls with sides of the extremest thinness, they even knew how to turn glass vessels, as appears from fragments of them in the Minutoli collection. Nor were they skilled only in the more delicate kind of turnery; for we find that they could and did also turn immense stone columns or drums in the lathe.‡

Although we know that the old Egyptians employed the art of turning, we know little as to their modes of working in it.[35] Perhaps we may venture to think that the lathe still used in Egypt, which is shown in the following engraving, is descended directly from that used in Pharaoh's kingdom. The simplicity of this instrument is primitive. *a* is the piece (of wood) to be turned ; it rests between two iron points at *b* and *c*, and can be turned by

---

* See Bergmann, *Nomad. Streifereien u.d. Kalmücken* (Riga, 1804) ii. 171.

† Cic. Verrin. iv. 24.

‡ See Ottfr. Müller's *Archäologie der Kunst*, after Klenze, in Böttiger's *Amalthœa*, ii. p. 72.

the bow d. The wooden cross-bar at b is fastened to the table e, the second cross-bar c is quite separate, and movable upon e. It is held in any required position by means of the iron bar f, which is weighted with a stone at g. The rod f is itself linked to b by means of a pin at h. The iron "centres" at b and c are simply bent nails driven into the cross-bars. In using this machine the turner squats on the ground behind it, moving the bar with his left hand, while he holds the chisel at i with his right. He presses the tool against the rest at k with the great toe of his right foot. His skill is most remarkable.* In this instrument, in which we notice a very extended use of force-closure, we cannot trace any distinct development from the potter's wheel. It is, however, improbable that Southern Europe should have worked in any way

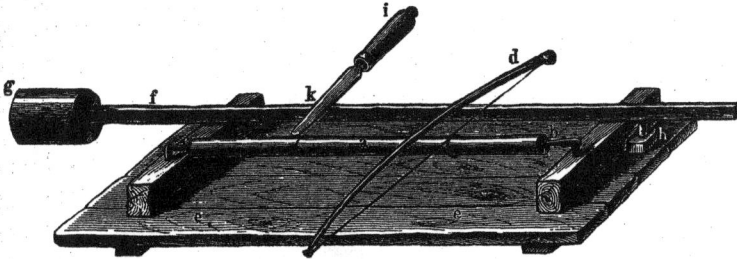

FIG. 164.

upon the type of lathe used in Egypt, for its construction is directly connected with the squatting position of the worker, a position which is foreign to the habits of the Western nations.

In the middle ages we find a lathe used which has beyond question developed itself from the ancient form; its motion is received just as in the fire-drill, but with notable improvements. It still remains in use in Italy, and indeed, in various parts of Southern Europe. One man only is required to work it. A cord or band is turned round the lathe spindle, or often round the piece to be turned (this method is certainly the older), and is attached above to a spring beam of wood and below to a tread-plate, Fig. 165. The latter is pressed downwards with the foot, and the cord is then raised again by the elasticity of the spring,—the piece to be turned receives rotary motion alternately in two directions,

* See *Descript. de l'Egypte* (2nd Ed. 1823) vol. xii. p. 452. Plate XV.

and during the forward turning the workman uses his chisel upon it.* An Italian turnery, its ceiling covered with a labyrinth of cords and wooden beams, impresses a modern engineer as something very strange ; it may be conjectured that it is a tolerably true copy of a turner's workshop in Ancient Rome.

Our continued use of the alternate rotation of the fire-drill for small turning and boring work is familiar to all. We secure the two ends of the driving cord to a bow, having given it in the middle a turn round a small drum. The simple drawing to and fro

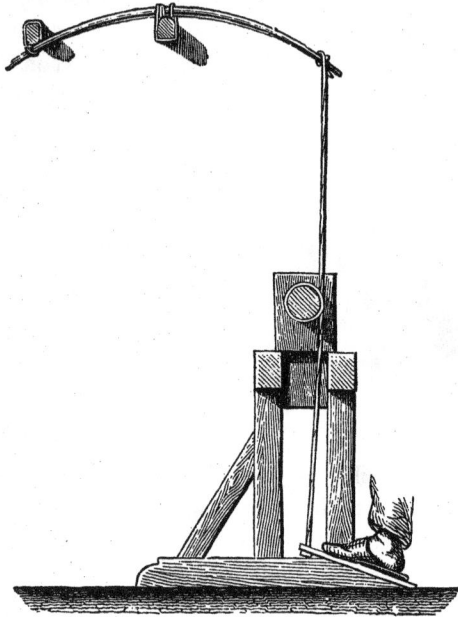

Fig. 165.

of the bow, which, unlike the spring-beam just described, is always equally tense, causes the rotation of the drum. This bow-lathe is still used by watchmakers for turning, both with centres and bearings ;—for mere boring the more primitive method of hand or breast-pressure is used. The bow-drill must be of very high antiquity, for it is very widely used in a homely form by the Chinese,† —by the Kalmucks it is employed in two different forms,‡—and it

---

* Laboulaye, *Cinemat.* p. 463. calls this lathe "*tour en l'air.*"
† Klemm (as above) i. 385.   ‡ Bergmann (as above) ii. p. 93.

can be proved to have been used by the Egyptians at least some 1500 years B.C.* Late discoveries, indeed, have made it appear extremely probable that the nations which peopled America before the epoch of its present Indian inhabitants, were acquainted with it and used it.† The bow disappears very slowly from the lathes of our own clock and instrument makers, making way for an endless cord mechanism of the kind shown in Fig. 142. From all this

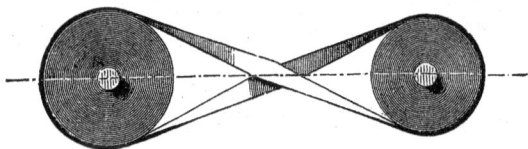

FIG. 166.

we can form some idea of the difficulties which have surrounded the transition from the alternate to the continuous rotary motion produced by a cord or band.

How soon this latter came into use we cannot say with certainty Its extended use among almost all Asiatic nations, as a means of

FIG. 167.

driving a spindle, bespeaks a great antiquity for it. The crossed belt, Fig. 166, appears to be the older, and this again was preceded by an arrangement in which the belt or cord, crossed or open, was wound more than once round both pulleys, Fig. 167, in order to prevent slipping.[36] With this last arrangement a very imperfect and incompletely constructed frame sufficed to make transmission of motion possible, for of course the friction of the cord

* Weiss, *Kostümkunde* i. (1860) p. 96. Wilkinson, *Ancient Egyptians* (1871) i. p. 56.

† Rau. Drilling stone without metal, *Smithsonian Report* 1868, p. 398 ; where the existence in the stone age of stone and bone drums for drill-spindles is pointed out.

upon the rollers was very great. The change from the twirling
cord, wound several times round the drums, to our own endless band
may well have occurred through this mechanism. The two pulleys
of Fig. 167 are two whirling rods with their cords connected to each
other. Gradually the number of turns on each pulley has been
diminished to one, or indeed, to a half turn, as in Fig. 166. At the
same time the broad flat band is substituted for the small round
cord. At last by some chance the cord has been used uncrossed.
We cannot forget how firmly the tightening pulley has been held
to, even to our own day; to this hour it is a favourite example in
many mechanical text-books. Or lastly, that my readers may
realise how extremely gradual the change of form has been, let them
remember what difficulty many people in our own day have in understanding that in driving by means of a wire rope, the rope has
merely to be laid upon the surface of the pulley, the application of
the cord of the whirling-drill being here carried to its extreme limit.

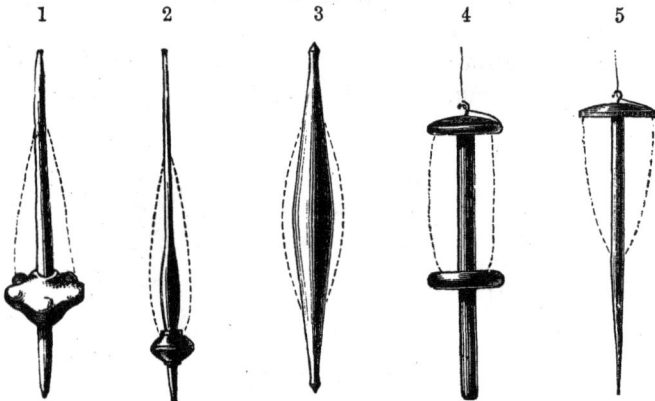

FIG. 168.

A very important employment of rotary motion, one which we
may well regard as an early step in the direction of continuous
turning, is that for twisting the thread in spinning. In primitive
times this operation may have required two persons; later on it was
found possible for one man to accomplish the task by twining the
fibres into thread by rolling them with the palm of the hand on
the thigh, as is even now done by certain Indian nations. From
this it would appear that the introduction of the spindle, and especially of the hand-spindle, belongs to a much later time, coming

perhaps within the historic period. The discoveries in the lake-villages, however, undeceive us in this matter. Complete hand-spindles of the stone age have been there found ; one of these is shown in Fig. 168 (1). The bob is made of burnt clay, and occurs in various ornamentally-moulded forms. This may point to the earlier part of the stone-age, a time when the possession of such a thing was a special matter ; or, on the other hand, may indicate the later period of the stone age.

With the hand-spindle comes the notable dynamical discovery that if the rotation be in any way once started it can be kept up for a time by means of a fly. The spindle still used here and there in Bohemia and Silesia, Fig. 168 (2), has obviously grown up from pre-historic times. It has a fly of wood, pewter, or clay, and is pointed at both ends. The spinner takes the upper end between two fingers, in order to start the turning motion, and by repeated twists to continue it until the spindle reaches the ground, a method of working which we can recognise in the story of the Princess *Dornröschen.* This hand-spindle of German life and story is by no means the only one still used in Europe ; in Lower Italy and in Greece different forms, also of great age, are in continual use. One of these is the Tuscan spindle, Fig. 168 (3), in which the bob is re-placed by a considerable thickening in the middle, but which is used like the German one. The spindle of the Neapolitan and Sicilian peasant, to whom our spinning-wheel is entirely unknown, is made of wood. Apart from small local variations, it consists of a cylindric spindle carrying two discs,—one at the top and one in the middle,—between which the thread already spun is held, Fig. 168 (4). The spinner generally sits at her work,—she places the spindle upon her knee, over which she gives it a quick rolling motion with the palm of her right hand. She makes use of this motion to draw out from the distaff a new thread, which is twisted by the spindle by means of a little hook attached to its upper disc. The rotation ceases gradually, as the spindle sinks lower and lower ; the thread spun is then wound on, fastened in the hook and the process repeated anew. In Egypt, also, the distaff is still employed. Fig. 168 (5) shows the form now used there, which almost exactly coincides with one of the forms used in ancient Egypt.[37] It is set in motion by the fingers of the right hand at its ·lower end,—this being made necessary by the squatting position of the worker (in

Egypt men spin as well as women), who in his left hand holds up
the flax.

The method here used, by which a rotation in a certain degree
continuous, and always in the same direction, is produced, stands
in both instances very near to the primitive rotation by whirling.

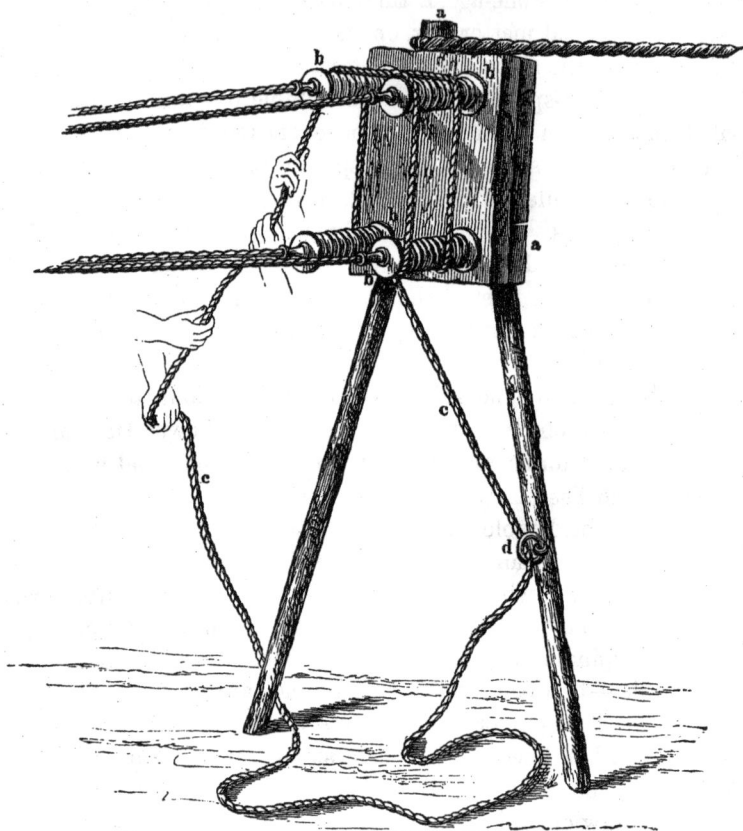

FIG. 169

Ropemaking belongs to spinning in the wider sense of the word.
The Egyptian ropemaking apparatus shown in the above figure has
a special interest for us, as forming a remarkable connecting link
between our own endless band motion (Fig. 142) and the original
whirling cord. It is still used in Egypt for twisting the already
prepared strands into a rope (here one of four strands) but is

apparently primitive. The strands run from the two-footed frame *a,* (which is kept from moving forwards by a strong stay rope) to the place where the twisting is to be done. The European ropemaker uses for that purpose a wooden cone fitted with guiding grooves so formed as to give the strands the required twist, the part of the rope which has been finished being caused to rotate continuously about its own axis. The Egyptian ropemaker, however, uses his hand instead of the cone; he guides the strands most dexterously with his fingers, always walking slowly towards the frame *a.* The four spindles to which the strands are attached, must, at the same time, in order to maintain the required twist, be continuously turned, like the rope, about their axes. This is done by the help of the endless cord *c* carried twice round them, and pulled continually by two men. At *d* this rope passes through a ring, which is so made as to offer a certain resistance to its motion, so as to keep its tension sufficiently constant.

The hand spinning-wheel is in such widely extended use among the Indians, Chinese, Japanese and the peoples of the Malayan Archipelago that a considerable antiquity must be ascribed to it It was not unknown, either, to the Romans. Among us, on the other hand, it appears that spinning was not done by the help of the wheel before the middle ages—by the foot-worked wheel not before the sixteenth century; the latter somewhat altered the method used, but the greater portion of the work was still left for the hand of the worker.

The working of the spun threads into a web by some kind of loom was done as early as the time of the lake-villages. The apparatus used was not, however, a machine in our sense. According to the remarkable reconstruction of it by Herr Paur in Zürich it is more like a pillow lace apparatus, possessing no more than the germs of specially machinal characteristics.

The Picota or Kuppilai of India, shown in Fig. 170, is a very old machine, although scarcely one that reaches back to pre-historic times. Here we find a lever used. A rod carrying a water-bucket is fixed to one end of a rocking-pole or beam, and a sort of counter-balance to the other. The beam rests in forked bearings, where it is further secured by cords. Men stand upon the beam, and by stepping backwards and forwards upon it cause it to swing up and down, so that the bucket can be dipped by a third man into the water and emptied again into the channel which carries it over the land. The Picota is still used not only in India, but also in Northern Africa, in Spain, by Belgian brickmakers, and here and there even in Germany. It is by a very similar arrangement that the Chinese worked the rope borers for driving their wonderful artesian wells. The Egyptians have for ages used an arrangement

Fig. 170.

something like the European Picota, called a Shadoof[38]; it is, how-ever, worked by one man, and the lifting is done essentially by the counter-weight. A man standing by the bucket manages the machine. The figure on the opposite page shows a large Shadoof arrangement with three lifts.

We see a very early use of rectilinear motion in the bow and arrow, with the help in this case of an accumulator of energy, the

FIG. 171.—Egyptian Shadoof.

a, *b*, *c*, pillars of Nile mud; both at *a* and at *b* there are three columns, at *c* two. Both at *a* and at *b* there are four swing beams, at *c* two. The latter work twice as quickly as the former, having only about half their lift; *d* the buckets, *e* counterweights of dried mud. The levers are fastened to wooden axles. Ten men are required to work it, working to a rhythmic chant.

elastic bow itself. In this clever use of the elasticity of the bow,—the principle of which may have been long anticipated by the primitive slingers,—we certainly have an indication of a somewhat advanced development. The blow-tube, which is in use among the South American tribes in the form of the twelve feet long Sabarkan, is younger than the bow, although certainly very old. With it both leaden bullets and feathered arrows are shot with great precision ; as regards the kind of motion received by the projectile, the blow-tube is a precursor of our modern weapons.

In both these cases of the use of rectilinear motion a projectile is discharged, the motions of which come under the action of kosmic forces so soon as it leaves the machine, so that the purely machinal part of the motion is by far the smaller. Indeed the rectilinear motion, which appears so primitive to our geometrical ideas, occurs but seldom in the first growth of civilisation. The nearer a people were to their aboriginal condition so much the less do they appear to have been acquainted with motion in straight lines, so that again we have to learn that we must separate our judgment of what is really near and what really distant, from our own experiences of to-day.

In some of the instruments of war of the Greeks and Romans (which, it may be noted, came from the East), the machinal side has already been distinctly developed, the storing of energy for throwing the projectile, especially, has been very effectually carried out. The cross-bow, which superseded the older form of the

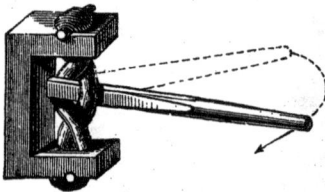

weapon, formed in general the foundation of the ballistas and catapults ; instead of the elastic bow, rigid arms were employed, along with wrenching springs (Fig. 172) of the kind described in § 42, made from skin or hair.* The remaining

Fig. 172.

parts,—guides, winches and gear were arranged in a way that showed considerable skill both in design and construction.

It remains yet to be decided when the pair of elements screw and nut first made their appearance ; they were certainly known

---

* See W. Rüstows and H. Köchly, *Geschichte des griech. Kriegswesens.*

to the Greeks and Romans, as *e.g.* in carriage building, even if they were not very frequently used by them. In our antiquities the screw—understanding this always as including the nut—is exceedingly rare. The way in which what we call right-handed screws have been chosen so universally in preference to left-handed ones is extremely remarkable, and is worth closer examination. I cannot attempt an explanation of it here, for too little material exists. It seems to me improbable that it can always have been as now, where people in general scarcely know of the existence of a left-handed screw. There are ancient representations of left-handed screws :— besides several from the middle ages, there are those of fulling-presses in the Pompeian Fullonica which show both a right and a left-handed screw.

It is, meantime, very difficult to determine the way in which this notable pair of elements was arrived at. I cannot think the notion tenable that it was immediately suggested by some natural form, as for instance a spiral shell. There are indeed many circumstances which seem to support this notion. First, that such shells, with few exceptions, have a right-handed twist. Then that in Greek the words for screw and for such a shell (κοχλίας, κοχλίον, κόχλος) are almost or entirely the same. Neither of these, however, can decide the case. For once the screw had been found—in whatever way—it might very well on account of its form have afterwards received the name of the shell;—the Greek word for spoon also, (κοχλιάριον) has been derived from the name of the shell, obviously from its hollowness and not its spiral shape. The idea of a form immediately suggested by nature would presuppose such a leap forward in the course of machinal development as would entirely contradict all that we have observed, in every other case, of the almost sedimentary formation of ideas. Besides this, the shell gives a model of a conic screw and not a cylindric one,— it requires translation into the latter form, the only one used. Before all, however, the shell gives no illustration whatever of the paired elements, with their definite relative motions and their striking capacity for exerting pressure.

The hollow screw,—the nut, must have formed part of the pattern from which the idea of the screw motion was derived. I venture the conjecture—I can hardly call it more—that it was the fire-drill of Fig. 161 that led up to the screw. With long-continued use, the cord may have worn or pressed spiral grooves in the spindle, and these formed in a manner screw-threads, the enveloping cord itself being the nut. The frequency with which this accidentally formed screw action was observed may have led gradually to its useful employment. The forms of the word screw in the Germanic languages greatly strengthens my suggestion. We cannot take into account the fact that in English and the Romance languages, the characteristic portion of the screw is still called " thread" (filo, filet), for this name may have been subsequently given to it. It is at present difficult to see,—although it may in time become clear,—whether the screw was first used for causing forward motion, or for a fastening, or for exerting pressure; it is difficult also to say in what manner the nut was originally formed. It is for the philologist, as the explorer of primitive times, to solve this question as to the original form of the screw and nut.

A variety of force-actions and causes slowly developed themselves in the machine, besides the ever-increasing variety in its motions. In taking the fire-drill, in which the expenditure of force is comparatively trifling, as the first machine, we directly contradict the very popular notion that the lever occupied this position. Apart from the fact that it is by no means clear what precisely is to be understood by the term lever,—this notion shows in my opinion a mistaken idea of the way in which human capacities have generally grown, and must have grown from the first. In taking the lever as the first machine, we think of men's attempts to deal with or overcome great resistances. It is not this, however, which first attracts the opening consciousness, but is much rather the accompanying phenomenon, motion. The child shows the most lively interest in the sails of a windmill, in the mill-wheel and such other portions of machinery as have distinctly regular motions: at first, however, he thinks nothing of the forces applied or brought into action by them. The separation of the idea of force from that of motion is a very difficult mental operation, and we find it occurring late and gradually. We find accordingly that the machines coming first from the unaccustomed hands of their-makers are those in which forces play a comparatively subordinate part,—for they do not exceed the exertions which the worker himself makes, imperfectly conscious of what he is doing.

This really lies at the root of the continual recurrence of the problem of perpetual motion ;—certain minds are always irresistibly attracted by the motion itself, by the first impression gained solely from external appearances, from the overpowering influence of which even the most accomplished cannot boast themselves to be entirely free. From attempts to cause motion the direct production of the corresponding force-actions slowly and step by step developed themselves. The popular idea which reverses the process makes the error of assuming the primitive inventor to have been a kind of Robinson Crusoe, endowing him with a full acquaintance with modern ideas, while in reality he has both to find out the need of improvement and to recognise its possibility, before he even attempts to carry it out.

Men certainly required an enormous period before they began to develop what might be called the motoral side of the machine,—before they attempted to use for working it other forces than their own muscular efforts. For this purpose they naturally turned first to the animals beside them, and made use of their muscles to save their own, but even this could not occur before the end of the long period during which the domestication of these animals was gradually taking place. Meantime men's energies were directed towards such improvements of their machinal arrangements as should enable the necessary number of workers to be diminished, thus increasing the capacity and efficiency of each single workman.[39] The primitive man looked only with fear upon the incomprehensible forces which he saw acting in the lifeless universe around him ;— only very gradually did he lose his timidity sufficiently to attempt their utilisation. He used boats propelled by oars,—as Curtius has shown philologically with great acuteness,—long before he ventured to employ the wind-force beside him by using sails.

The rushing waterfall may well have appeared to him the most living thing in nature ;—first, however, he noticed only its restless motion, the apparent unendingness of which led him to employ it, *e.g.*, in the Thibetan sacred wheels already mentioned. By degrees the idea of using the energy of this easily obtained motion came to him, and he carried it out in the scoop-wheel, as we have seen.

Meanwhile he gained some experience in such applications as the bow, of the great principle of storing energy in order that it may be used suddenly at the instant when it is required.

The bow of the archer is a machinal organ in which energy is stored, the sensible force of the muscles is made latent in it, and it is this latent energy stored in the elastic bow which actually propels the arrow. In the ballista and the catapult this principle receives still more extended application, for in them kinematic means are employed to store the muscular energy of many men, so as to employ it concentrated with correspondingly increased effect. Later on the same principle extends itself to primary forces, and it is to-day more used than ever, from the tiny watch-work or the spring of a gun-lock through innumerable mechanisms up to the Armstrong accumulator, or the air-vessels of the Mont Cénis borers.

The discovery of the motive-force of steam occurred late, long after that of some explosive materials was known; in each we use simply the latent forces which nature has distributed upon the earth in such enormous profusion in her decomposable materials. This discovery gave to men a source of energy of which the importance was at first not seen, but which has raised the machine into a power in nature sufficient to have made an entire revolution in the life of the human race.

§ 49.

The development of the Machine from a Kinematic point of view.

The question now arises :—what is the special kinematic meaning or nature of the changes by which the machine has advanced to its present degree of completeness? What has been near to, and what far from the spirit of invention,—if we indicate by that name the recognition,—becoming clearer and more distinct gradually from the remotest times,—of the mechanical in the machine? I believe that the answer to this question is :— the line of progress is indicated in the manner of using force-closure, or more particularly, in the substitution of pair-closure, and the closure of the kinematic chains obtained by it, for force-closure.

The notion which the gradually expanding mind of the primitive man first connected more or less dimly with his machine was the constrainment of certain motions in lifeless bodies for his own purposes. The forces necessary for these motions he sought in himself and his fellows. The idea of making the forces of nature do his work was far beyond him. He was contented, happy, if by any exertions he could even in scanty measure carry out his intentions. Force-closure was his most ready auxiliary in effecting the desired action among the bodies which he had chosen for his purpose.

The twirling-stick of the fire-drill,—the most early form of what we know as a " turning-pair,"—was not only force-closed by pressure in the direction of its length but also in all transverse directions, the hands set it in motion by force-closure, the piece of wood beneath it was kept stationary by force-closure. The introduction of the cord and the upper bearing-piece for the twirling-stick (Fig. 161) marked a great machinal advance,—for by the addition of two new kinematic elements it enabled the whole muscular force of the worker to be applied to the pulling backwards and forwards of the cord, while formerly his hands had to be pressed together as well as moved to and fro. The cord itself, however, is again a force-closed element, and is kinematically paired with the stick by force-closure.

If we trace this twirling mechanism onwards from the boring-tool of the lake-villagers, and of Homer's ship-carpenters, who used it almost unchanged, to the ancient lathe, we see that it has made notable progress. The double force-closure of the bearings of the revolving-piece has been superseded by pair-closure, by the addition of the second head-stock. The motion of the driving-cord has been greatly improved. Not only is the upper end of the cord made to move in a definite manner by the spring-beam, and the lower end by the treadle, (for which purpose two additional links have been added to the kinematic chain)—but the backward motion of both is effected by energy stored in the spring. The workman has thus to produce directly only the one motion, for-wards. The machine being arranged so that this can be done by the foot, the hands are at liberty to hold and direct the turning-tool. But again the new element, the spring, is force-closed, and the action of the foot upon the treadle, whether driving it downwards or driven by it upwards, is force-closed also. The Kalmuck lathe already mentioned may be regarded as one of the too little known intermediate steps between the Homeric drill and the Italian lathe;—although it may here be difficult to estimate separately the influences of the different epochs in which the machines have existed. We may note, however, that it has here been rendered possible by kinematic means for one man to work the machine instead of two or three.

We see something similar to this in the hand-spindle. Although force-closed in almost every direction we must yet consider it as a machinal arrangement, consisting of the elements spindle and thread (tension-organ);—by it, however, the troublesome twisting of the thread, which before must have required, like rope-twisting, at least two persons, has been made possible for one worker,—or at least the incomplete spinning of the Indians has been replaced by a quicker and better method.

In the Indian Picota, although this machine is of much later origin than the fire-drill, or even than the hand-spindle, we can trace similar processes. We have here forces of much greater amount to deal with,—there were Picotas with six or eight workers and correspondingly heavy buckets, but force-closure still governs the whole machine. There is force-closure in the main bearings, vertically by the weight of the beam itself, in other directions partly by cords but principally by the skilfully directed motion of the worker's feet;—force-closure in the joints of the bucket-rod both top and bottom ; force-closure lastly in the body to be moved, the water itself. The motions of the beam itself are regulated by a double force-closure, for of the two men one always presses downwards the end of it, while the other, maintaining his position by holding fast to his end of the bamboo frame, simply guides with his feet the end which is rising. In the Egyptian Shadoof the joint of the beam has become a wooden axis,—pair-closure, that is to say, has come into use ;—and by the use of a counter-weight it has been rendered possible for a single man to work the machine—with a not too heavy bucket—by a suitable hanging rope.

In the long-bow, in which rectilinear motion is obtained by means which we have considered to be machinal, all the movements of the archer himself are force-closed ; we see also the bent finger guiding the arrow essentially by force-closure, while bow and string are alike force-closed elements. In the cross-bow and the ballista, a great portion of this force-closure is superseded, for the arrow is guided by a straight groove; the winding apparatus also is a kinematic chain which takes the place of the hand stretching the cord by force-closure. In the blow-tube the prism or sliding-pair is used for guiding the projectile in a form already very complete,—in the musket it is smoothly bored out,—in modern ordnance the bullet and barrel are formed accurately into a twisting-pair,—the force-closure of the former being at last completely done away with.

The Chinese scoop-wheel, which we above described, carries the stamp of antiquity on it in its use everywhere of force-closure. The driving element itself is force-closed both in the bed of the stream into which the wheel dips, and in the basket-work paddles which it drives before it;—force-closed in the bamboo scoops which carry it upwards, and in the channels which direct its

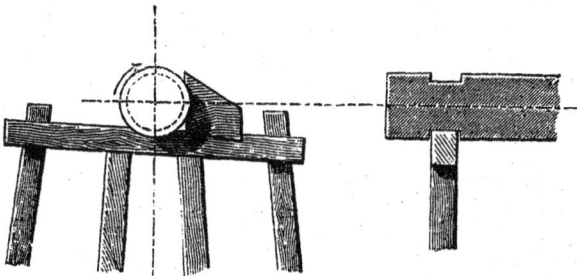

Fig. 173.

course over the land ;—the shaft of the wheel, too, is force-closed in its angular bearings. To what an extent the mere struggle against or counteracting of disturbing motions has lain at the base of the invention of kinematic closure, we may see for instance in the very ancient Noria, a Spanish water-wheel.[40] Its shaft lies upon somewhat inclined frames without any notching whatever being made to receive it (Fig. 173). The tendency of each journal, as it turns forward, is to roll further down the frame, and this is prevented by small projecting pieces. We know that this and similar bearing-arrangements have by degrees become transformed into carefully fitted turning-pairs;—that the water has been enclosed in a channel, then in guides surrounding the wheel and so on, and has thus more and more become used with pair-closure. Still there can be no doubt that in water-wheels generally there are still left distinct traces of the former use of force-closure in all their parts.

In the pumps, mills, and other machines used in the Middle Ages, we find crank or lever mechanisms very frequently employed,—mechanisms, that is, in which turning-pairs are used for joints. If we examine these closely,—and we have drawings which enable us to do this,—we find in them a continual use of force-closure. The joints of cranks and connecting-rods are round bars enclosed in wide round eyes like the links of a common chain. Clumsily made collars, placed far apart, suffice to prevent excessive lateral motions. There was play enough left to allow turning to take place, when it was wanted, about axes oblique to the axis of the shaft itself;—thus in many cases where we would use a universal joint it was then unnecessary,—the older form had fewer parts than the newer one.

In isolated survivals from former centuries, such as the ancient wine-presses which are even now to be found here and there in the Rhine and Moselle vallies and in Switzerland, we can still see force-closure in extraordinary completeness. A horizontal lever, generally made of oak,* working everywhere under force-closure, is used for transmitting the pressure ; in the oldest form its free end is loaded with a millstone. A screw (of wood) is then also applied, but this is not used for applying pressure, but for raising or lowering the loaded end of the lever.[41] In the Rhenish press, which must be considered the younger, the screw is used for the application of pressure through the lever, the whole machine being an arrangement something like a screw-vice in which the pressure acts between the screw and the joint, and very near the latter, instead of upon one side of both.

The primitive iron hammers, which the pedestrian may still meet with in the busy little valleys of the mining districts of Westphalia, are also very remarkable. A little wooden water-wheel roughly made drives the tilt-hammer, and another the blowing-apparatus. Both machines are driven by force-closure, by means of projecting tappets or wipers, which act by downward pressure, the return motion in the latter being effected by a wooden spring-beam. There is scarcely a pair of elements in these machines,—an inheritance preserved through so many centuries,—which is not force-closed. By slow degrees our modern blowing-engine has developed itself from this primitive contrivance.

* The "Kelter-baum" of so many songs.

From all these examples of ancient machinery we see how force-closure has gradually made way for pair- or chain-closure. This process has first converted the complete force-closure into the force-closed pair of elements, by degrees these have been more and more closed until at last constrained pairs have been reached, and chains have been unnoticed been built up from them. One thing, moreover, has helped another, each machine devised to aid hand-work furnishes the means of completing a part of some new and more extended machine. Thus,—as we have already seen,—simplicity or fewness of parts does not itself constitute excellence in a machine, but increased exactness in the motions obtained, with diminished demands on the intelligence of any source of energy,—and this even at the cost of a considerable multiplication of parts, or in the language we have employed, of links in the kinematic chaining.

Looking at the kinematic principle as a part of the higher unity of human development, we can recognise from all this that the first machinal arrangements were of a kind which we may designate as make-shifts. Certain constrainments of motion were required. Men obtained these as best they could, and by the necessity of the case,—for our investigation has shown that no other equally simple solutions were possible,—they used pairs of elements in their first incomplete form. Very gradually each invention came to be used for more purposes than those for which it was originally intended, and the standard by which its excellence and usefulness were judged was gradually raised. An external necessity thus demanded its improvement, and from this cause machinal ideas slowly crystallised themselves out, and gradually assumed forms so distinct that men could use them designedly in the solution of new problems. These attempts resulted in further improvements, and these in their turn led once more to new applications and more extended use.

We recognise here that wonderful tendency towards extension of the limits within which men can work which appears in such different degrees among different races, and which has therefore led to such unequal development among them. Some races possess this tendency in small measure;—their development makes but a few small steps in thousands of years;—they have remained more true to their original nature and submitted more readily to its conditions than others ;—these, on the other hand, as if driven by ever new inner forces, have disputed with nature one department after another, and have found their reward in the growth of their capacities and in the magnitude of the work they have been able to perform.

§ 50.

The growth of Modern Machinery.

Modern machinery came into existence with the invention of the Steam Engine, and with it and by it has developed itself with

Fig. 174.

a rapidity not even approached in former times. This has not been, in my opinion, by any sudden leap, by any discontinuousness in the sequence of ideas ; it is due rather to an acceleration in the rate at which one has followed the other. The curve has risen suddenly without any change occurring in the law according to which it is formed. We must here not forget how difficult it is in all cases to form an opinion about matters occurring in our own time, for we ourselves are subject to the influence of the time, and must judge it while we form part of it. The immense number of cases existing, on the other hand, and the exactness of our knowledge of them, here help us very greatly. An examination of the way in which the gradual perfecting of machines is to-day going on teaches us, however, one thing,—as we shall presently see,—namely, that the process of the replacement of force-closure by pair- and chain-closure goes on quietly extending itself further and further to this hour. We may therefore consider this process as showing the

Fig. 175.

essential general tendency of the whole machine-development up to our time;—we may even go further, and say that we must consider it as an essential characteristic of future machine-development.

In Newcomen's steam-engine, Fig. 174, force-closure still predominated, and it remained thus through the whole eighteenth century. The machine was force-closed in its pit-work, in its beam-chains, in its steam-piston and in its valve-gear,—although in the latter Potter's invention had substituted a machinal arrangement for the hand-gear. Watt introduced pair- and chain-closure by degrees into the machine. Thus, for instance, the force-closed beam-chains became the imperfect but still kinematically far more complete "parallel motion." Even to our own time the venerable pumping machinery used in our mines remains partly in the fetters of force-closure; it is only very lately that direct-acting steam pumping-engines have begun to dispute its position.

FIG. 176.

The well-known "Sun and Planet" wheels of Watt give us an interesting illustration of the course of the change from the one kind of closure to the other. The form in which Watt originally put the mechanism was not the familiar one of Fig. 175, but the entirely different one shown in Fig. 176.* In order to maintain

* See Muirhead's *Inventions of James Watt*, Vol. III. p. 50 ; also Bourne's *Treatise on the Steam Engine*, pp. 20-21.

continuous contact between the wheels *c* and *d* he employed exactly the kinematic pairing which we described in § 43. The idea present to his mind seems evidently to have been that a mechanism was "simple" just in proportion to the fewness of its parts. It was later on that he adopted the arrangement of Fig. 175. In this the chain does indeed contain one link more, the bar *e*, but the required motions are obtained with greater precision, and the·fly-wheel with its force-closure and the accompanying destructive wear have disappeared.[42]

In our various means of transport the change from force- to pair-closure has continued to the present time. After all had been done in improving the construction of the vehicle itself, furnishing it with a suitable fore-carriage, making better roads for it to move upon, etc., force-closure still remained, if nowhere else at least in the preservation of the direction of motion, which still demanded accustomed animals and an intelligent driver. Men naturally attempted to replace this force-closure by pair-closure. In the Railway the rails are paired with the wheels,—force-closure is used only to neutralise vertical disturbing forces. The step thus made in the direction of machinal completeness,—which it required half a century to make,*—was a most important one ;—it was in reality no other than the uniting of the carriage and the road into a machine. The rail forms a part of this machine, it is the fixed element of the kinematic chain of which the mechanism really consists. The further improvement of the pair-closure, the removal of any remaining disturbing force-closure whether in the rails, in the axle-boxes, in the arrangement of the springs of carriages and of locomotives and so on, still engages most careful attention. In opposition to this we have the problem of steam locomotion on common roads, which has been so feverishly taken up again within the last few years, but the solutions of which seem doomed to eternal incompleteness, for they are self-contradictory. It is desired to make something which shall be a machine, but in which at the same time the special characteristic of the machine,—the pairing of elements,—may be disregarded. On the other hand, attempts have been made,—as in Boydell's Traction Engine,—to carry with the machine at least a portion of a transportable element which

---

* Wooden rails were in use at pits near Newcastle as early as 1676,— the first iron rails were laid down in 1738.

could be paired with the wheel, all indicating the general tendency towards the limitation of force-closure. Thomson's India-rubber tyres have essentially the same object;—the inner side of the ring of vulcanised India-rubber, externally flattened upon the road, serves as a smooth uniform surface for the rigid tread to run upon— thus corresponding generally to the rail of the railway.*

The development of the Turbine has followed the same course ; —it has grown out of the primitive wheel of the Tyrolese and Swiss mountaineers in the hands of the mechanicians of our century. In the latter the water dashed and eddied against its irregular blades in vehement force-closure ;—in the Turbine it is already combined into a pair of elements with the accurately shaped wheel with very considerable completeness.

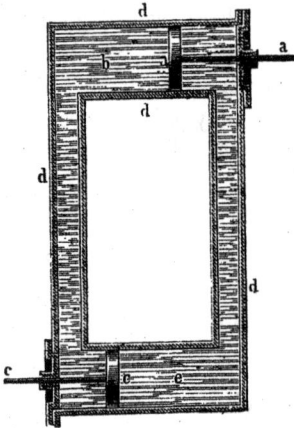

FIG. 177.

The progress of instruments for breaking stone into small fragments,—from the ancient stamp to the crushing rollers with which we were so long content, and then almost abruptly to the stone-breaking machine, without which to-day a blast furnace would scarcely be thought complete, — has been from closure by sensible forces in the mass alternately lifted and allowed to fall, to closure by the latent forces of a system of cranks and levers. There are many old iron-works which have seen the transition from tilt- and lift-hammers to their present crushers, rolls, and forging-presses.

The still young agricultural machine industry is attempting the very difficult problem of superseding, or at least limiting, a very complex force-closure by pair- and chain-closure.

How much we are still engaged in replacing force-closure by kinematic closure, and how strongly we feel in the matter that we are striving after something new, trying to reach a better position, the hydraulic press, invented at the end of the last century, shows.

---

* It was this action unfortunately, the motion of the tread inside the tyre, which caused the failure of many of these engines. The excessive wear which took place in the India-rubber made the cost of repairs enormous.

I have already pointed out its direct parallelism with the primitive pulley-tackle[43] (§ 40); but men have not yet ceased from admiring it and wondering at it, while its older counterpart is no longer thought worthy of special notice. The appearance of the "water-rod" (Fig. 177) which has been before described (§ 44) and its warm acceptance, show the same thing. I have already shown that this mechanism is simply the contrapositive of the endless band of Fig. 178, although it has followed it so late.

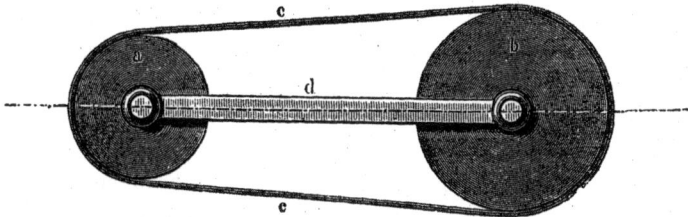

Fig. 178.

The ever-extending endeavour to do away with force-closure shows itself in interesting forms in some of the more refined outcomes of our modern machine construction, as for instance in mechanical lubricating-apparatus, and in various "safety" appliances, lock-nuts, nut-guards, split-keys, etc.

The mechanical lubricator replaces the force-closed supply of the fluid by the use of kinematic chains—sometimes of very complex form—which often solve the problem with great delicacy. The other arrangements mentioned substitute carefully made pair- or chain-closure for closure by friction or gravity.

In number of parts modern steam engines greatly exceed the older ones. In the engine of a common man-of-war, for instance, the parts required simply for securing the nuts increase the number of pieces in the machine by from 200 to 400.

Toothed wheels furnish us with another example.[44] Although they have been known for thousands of years, their improvement to-day is still essentially in the direction of excluding force-closure, that especially which has remained with the "clearance" or "freedom" allowed between the surfaces of the teeth, and which has often enough made itself disagreeably felt. In the Chinese winding mill (gin) and in the similar machine used by the Egyptians, and worked by water (the Sakkiah),[45] there is a large amount of play left between the teeth, which were merely such rough blocks as rendered it possible for one wheel to drive the other.* But we see that during the Middle Ages, and in the last few centuries, the freedom has been more and more reduced, as greater care has been taken to find the kinematic condition to be fulfilled by the form of the teeth-profiles, until we have now succeeded in reducing it to a very small fraction of the pitch. During the last century, the wheel and its teeth gradually came to be understood as forming together one whole, and the teeth-profiles were then looked at in a new light. I believe that in a few decades it will be the rule to employ spur-wheels working without any clearance between the teeth.†

We have already noticed that the contest between pair- and force-closure continues in full activity in the department of prime-movers, where the question is one as to the mode of crossing the dead points of mechanisms. We saw that the double cylinder-engine was more and more taking the place of the single one. Even twenty years ago, capable although indiscriminating "practical" men said distinctly that the employment of the coupled engine in connection with mines was a mistake,—that the single machines, if only on account of their "simplicity," were greatly to be preferred; that men would soon change their minds, tire of these novelties, which were only fashions, and return to the old machine. To-day, however, the double engine, in spite of the far greater number of its parts, is almost invariably used. In rolling-mills force-closure was, and is still, obtained by the use of colossal fly-wheels, which have too often, as is well known, been the cause of fatal accidents. As we have already mentioned in § 47, the latest forward step here, too, has been the substitution of the double for the single engine. Indeed it is very probable that another decade may see the coupled engine universally employed in spinning- and weaving-mills, machine-works, and manufactories of all kinds in which it

---

* See for instance Eyth's *Agricultur Machinen-wesen in Ægypten.* Stuttgardt, 1867.

Six months after these remarks were published they were corroborated,—unexpectedly soon,—by the appearance at the Vienna Exhibition of Sellers' Wheel-cutting Machine, which makes teeth of which the clearance is only $\frac{1}{100}$ of the pitch.

has now just made its first entry, Here also, as everywhere else, force-closure is given up to make way for the more complex but kinematically more complete closure by pairs and chains.

In addition to this internal difference between the modern and the ancient machine, there is another very important but external difference. This lies in the improved construction of the single parts of the machine,—the links, that is, of the kinematic chains of which it consists. The introduction in the last century of cast iron in place of wood led gradually to the making of beams, wheels, levers, frames, etc., each in as few pieces as possible, when practicable at last each out of a single piece. In our own time cast steel has commenced to act in the same way upon construction in wrought iron. In designing machinery, increasing care is taken that the intensities of the latent and the sensible forces correspond, and the dimensions given to each part are calculated with precision. These dimensions are, in consequence, much less than they once were; and this has caused what is essentially a decrease in the number of parts, or more correctly of pieces, which has extraordinarily simplified the external appearance of the machine. For this reason the modern machine often appears more simple than the ancient one, although in reality it is generally far more complicated. The old wharf cranes, for example, by which but very moderate loads could be lifted, were much more striking in appearance than our slighter-looking machines, which are really so much stronger, as well as so much more complex in their construction. The same thing would be noticed in comparing the old and the new pumping-machinery, mill-work, steam-engines, etc. This important external simplification of the machine—a process which is still continually going on—has been the means of making its actual construction more and more easy. It must not be forgotten that this increased facility of execution depends upon the enormous capabilities for work which we have stored up in the resources of our workshops. It is the interest of this capital that we find in the external simplification of the machine. It must not be allowed to conceal from us the simultaneous increase of internal complexity. That there are limits to this increase, pointed out by kinematic conditions and already in some instances nearly approached, we shall find as we proceed. At present I cannot enter into this very important question.

Putting in a few words the results of our examination in their relation to the fundamental idea laid down at the beginning of the chapter, we may say that the limitation of force-closure has essentially been the means by which machines have been made capable of better carrying out their own share of work. This limitation led gradually from the make-shift first attempts at machines to the accurately working pairs of elements and the simpler mechanisms. This at the same time creates the possibility, and becomes the cause, of further extension of the limits within which the machine acts,—of obtaining larger results by human intellect,—or as we expressed it before, of making the share of the machine a larger fraction of the whole problem.

The endeavours after this lead to the invention of new mechanisms, and in these again force-closure—which seems always to be nearest to our hands—is at first employed. This shows itself every day, especially in machines invented by workmen or others whose knowledge of their subject is merely empirical. Of such machines we have many ; not unfrequently they have been pioneers to open up a new region. They contain such a combination of weights, springs, tappets, catches, stamps, fly-wheels and so on, clattering and jerking in their force-closed working, that they might be a little representation of all the steps in the development of the machine seen through a reversed telescope. The experienced and scientific designer sets them aside with a smile, and replaces them with accurately working elements. But in spite of his experience and knowledge, if the same man have to design an entirely new machine, he too will at first employ force closure in many places where he might better have used pair-closure, and where in time he will use it. The Corliss valve-gear is a capital example of this ; in its earliest form it was everywhere force-closed, and all the subsequent improvements have been unconsciously in the direction of the replacement of this by something better. In the intensive growth of the machine we thus see that the removal of force-closure is also continually going on, by restricting its employment within narrow limits, so distinctly that we cannot wish, nor indeed dare, to attempt to return again to its use.

We must not overlook the fact that to a certain extent the general development of the machine has hitherto gone on unconsciously, and that this unconsciousness which has characterised the older method of production has left its special mark, it prevents that method indeed from being distinctly understood. The way in which the modern machine is designed is different, lying as it does from the beginning in the hands of experienced and more or less scientific men. Here some things at least, if not a large number, are clearly and deliberately grasped. Here we do not so much see the improvement of old and defective arrangements as the bringing into existence of new ones, enabling the machine to perform operations which had previously been considered quite beyond its province. The mechanism, although new, is presented to us complete,—a faultlessly constrained and closed system of bodies,— ready to be put to practical proof; as we see, for instance, in sewing-machines, in the new guns and projectiles, and so on. There can be no doubt that in some of these there are tokens of a new tendency a very striking one, very distinctly differing from that which gave us the older machines. The difference somewhat resembles that between the processes of integration and differentiation. Formerly the fundamental idea of alteration or extension was improvement, a word which says much in itself of the nature of the process. Now, on the other hand, we have a direct production of new things, a sudden bringing into being of so far complete machines. We see the beginnings of a perception which will some day apparently be universal among those who have to do with all classes of machinery. Upon this growing sense I believe that our polytechnic machine-instruction should act with increasing certainty. The nature of men's talents meanwhile remains as a whole unaltered. The idea must be developed in each individual afresh microkosmically from its beginning onwards. For this reason, and also because incomplete solutions may still be real solutions, the existing antagonism between pair- and force-closure will never become quite extinct.

The whole inner nature of the machine is, as our investigations have gradually made clear, the result of a systematic restriction ; its completeness indicates the increasingly skilful constrainment of motion until all indefiniteness is entirely removed. Mankind has worked for ages in developing this limitation. If we look for a parallel to it elsewhere we may find it in the great problem of human civilization. In this the development of machinery forms indeed but one factor, but its outline is sufficiently distinct to

stand out separately before us. Just as the poet contrasts the
gentle and lovable Odyssean wanderers with the untamable
Cyclops, the "lawless-thougthed monsters," so appears to us the
unrestrained power of natural forces, acting and reacting in limitless
freedom, bringing forth from the struggle of all against all their
inevitable but unknown results, compared with the action of forces
in the machine, carefully constrained and guided so as to produce
the single result aimed at. Wise restriction creates the State, by
it alone can its capacities receive their full development; by
restriction in the machine we have gradually become masters of
the most tremendous forces, and brought them completely under
our control.

§ 51.

The Present Tendency of Machine Development.

In the foregoing paragraphs I have had to oppose the customary
and very widely diffused notion that the first requirement in the
primitive machine was the execution of certain work, and I have
shown that this view is essentially erroneous. We have seen rather
that this first requirement, the one out of which machinal ideas
gradually formed themselves, was the production of motion. It
in no way follows from this that the requirement of work did not
influence the matter. We found, on the contrary, that questions
of force entered very distinctly into the history of machine-
development, and that they left their impress upon its inner and
more characteristic kinematic manner of growth.

The external impulses affecting the growth of the machine move
therefore in two lines : the first and earliest in its action was the
want of various kinds of motion, the other, that of the execution
of work. These impulses run side by side, uniting here and there
and then separating again, both continually helping forward the
perfecting of the machine. Apparatus for war and for construction,
especially for the moving of heavy loads, demands always an
increase of its capacity to deal with forces ; manufacturing instruments, on the other hand, those for time-measuring, and many
others, require the extension of the number of motions which they
can execute. The two directions can to-day, in spite of our more
advanced scientific position, which has shown us the right relation-

ship between force and motion, be distinctly separated, for we have always one class of machines in which the forces have chiefly to be considered, while in another the motions are more important. The ways in which men's ideas have developed in regard to the two questions have also been quite different.

Man has always had before him in nature moving forces, but so far as these were beyond himself they were in the beginning unrecognized and unknown ; he had first to learn how to distinguish and separate them from the multitude of accompanying phenomena, —to discover them. Thus the development of the machine on its dynamic side has been closely connected with men's knowledge of nature, with what grew later on into natural science, with which it became more and more closely connected. In inventing the steam engine Papin was as much a physicist as a mechanician, and the same may be said of Watt when his searching genius grasped the subject. And so to-day the most exact resources of mathematical and experimental physics are employed in the discovery and accurate scientific investigation of the various sources of energy.

In the same way, but in even greater variety, man has always had motions before him in nature; but these have always been either kosmically free motions or such as were directed by some animated intelligence,—never, or extremely seldom, those closely limited and regularly interdependent motions which we find in the machine. This constrainment is obtained only as the result of thought, man has had to create it through an intellectual act, in other words to invent it. Discovery on the one side, invention on the other ; in this antithesis we have the difference between the dynamic and the kinematic development of the machine. The discovery of each new source of energy leaves it to invention to supply means for utilizing it. The discovery of the dynamic properties of steam, for instance, may rather be said to have rendered progress possible than to have been itself a step forward. It called forth the most energetic activity of thought, the most careful reflection and study, in order to create, by invention, means by which the new sources of power might be utilized. Invention remained unceasingly active in its endeavours to extend the applications of these, its consciousness of its own object becoming gradually more and more distinct, until to-day it is in part recognized and directly aimed at. As we have before stated, it is by the science of Kinematics that the laws governing the means for attaining this object are determined.

We have recognized and examined in certain pairs of kinematic elements the property of force-closure, by which a certain amount of kosmic freedom is left in the machinal system, and seen that it has been for thousands of years the aim of invention to limit or destroy this freedom, and that many new tasks have been made possible only by its complete restriction. We can now see in this force-closure also the borderland between the kosmic and the ideal machinal systems. In force-closure we have that distinct although not sharply-defined boundary-line,—discussed already in speaking of the limits of the machine-problem (§ 1),—which divides the two systems. In this line of contest between pair-closure and force-closure I believe that the future historian of machine development will find the thread to guide him through the complex but not altogether planless course of his subject. It will, besides, be of the greatest value in the further designing of machinery if the problem be entered upon from the beginning with the distinct knowledge that in the substitution of kinematic- for force-closure there lies the very central idea of progress, and that the more rigidly this idea is carried out the sooner will the desired end be reached. To impart a clear and distinct understanding of this process should be, in my opinion, the function of polytechnic school instruction. We have here undoubtedly an idea before us, put in a separate and distinct form, which is of the greatest and most urgent importance to the inventor, although often not understood by him. It is therefore in every way right and necessary that the study of it should be closely entered into.

By this means another important end may also be gained. It will partly strengthen and partly create a sense of the fundamental connection between the special work of the machine-designer and the whole region of practical mechanics, and hence with the whole domain of human activity. Hitherto the tendency has been to weaken this sense, until now in some places it has almost disappeared. The popular cry of "division of labour" has, entirely in opposition to its own principles, contributed to this.

This principle has been applied,—wrongly,—beyond the limits

within which alone its action is really confined. An attempt has been made to base a formal division of knowledge upon it. We have already gone so far in this direction that there are whole departments of machine construction scarcely intelligible to each other. For a practical mechanician to know something of those regions of intellectual life which lie beyond the industrial circle has become rare. Yet nothing can be more certain than that the endless isolation of efforts must be detrimental to the whole. This division of knowledge cannot be carried further without harm; it is rather our duty to join together once more the sundered departments, resting their connection upon a higher unity, and so bringing to view the real scope and purpose of the whole. The sense of the community of human efforts should find expression not merely in the scientific consciousness of individuals, but also in the form in which the perceptions are cultivated and extended.

The idea upon which the foregoing sketch of the growth of machinery has been founded, the very notion of development, does in itself act powerfully to strengthen this sense. All our later investigations have made this idea more or less their own, in the region of historical research as well as in that of natural history, into which it has infused such life. It alone both demands and renders possible the looking at a whole department really as a whole. It compels a far-reaching view, a looking beyond the present time and place—it at once deepens and heightens the comprehension of single phenomena. It has given to the science of to-day a power which could scarcely have been imagined two generations ago. To the inquirer at that time a series of phenomena was a series of isolated facts; the order in which they arranged themselves was only that of a string of pearls, the casual connection between them was nothing more than the thread which tied them togther. To-day, on the other hand, we look upon this same causal connection of thoughts, with their growth and unbroken flow, as that which is most essential; we see in it not so much the thing which links phenomena together as that to which they owe their very life and being.

I have attempted to place this antithesis before the reader in the two mottoes which head this chapter. Between the sentences of Schiller and of Geiger lies the deep contrast between the former spirit and the present. The passage from Schiller,—interesting indeed, but cool and contemplative,—moves us but little ; Geiger's telling words compel us to attend, and seem to clothe our thoughts with form and colour. They are in the deepest and best sense of the word modern, and for this very reason doubly effective. They are at the same time so true and universal, that although written with reference to the investigation of entirely different subjects, they are none the less wholly applicable to the abstract region of inquiry which lies before us.

CHAPTER VII.

KINEMATIC NOTATION.

§ 52.

Necessity for a Kinematic Notation.

THE investigations concluded in the last article have conducted us again through the lower and higher pairs of elements to the kinematic chain, the form which, as we have already seen (§ 3), represents the general solution of the machine problem. What we found before to follow directly from simple fundamental propositions, we have now been led to a second time, indirectly, in the course of natural development, and we have seen further the employment of the kinematic chain extending rapidly in all directions. The glance which we have taken at the history of machine development has shown us the course of the mental processes which have produced the chain, and by a continuation of which—we may suppose—it may be still further improved and applied. We must now turn to the direct consideration of the thing itself.

Such an immense variety of cases—existing and possible—here present themselves, that it becomes increasingly difficult to comprehend them all. This difficulty presents itself specially in the indication by names of their separate characteristics, and in distinguishing between cases which ought to be separated; and it,

appears likely to become still greater in the future with the increase
in the variety of chain-forms employed. It has become, too, equally
necessary to be able to survey the inner relationships of mechanisms
as well as their differences. We are here led involuntarily to look
for some means of facilitating the expression of both.

In similar circumstances Mathematics, and afterwards Chemistry,
have taken to their aid special symbolic notations, which have now
become so essential to both sciences that neither could proceed
without them. Both adopted them so soon as the real nature of
their fundamental operations had been determined. The ideas
connected with our subject are now so distinctly and individually
before us, their mutual relations can be so definitely determined,
that their concise expression by means of simple signs becomes not
only justifiable but practicable. We shall therefore use these important aids to the furthest extent possible in our work.

It is very easy to see what an immense advantage there is in the
possibility of so expressing a complex idea that when it is employed
along with another of the same kind they may both be expressed
by a single sign. The continual returning upon already defined
conditions becomes unnecessary, while the conciseness of the expression allows conclusions to be arrived at as to the mutual relations of the parts combined, which with the common method of
expression can only be formed with great difficulty, and can scarcely
be communicated at all. The reader need not fear that any continual alteration of his accustomed ideas will be demanded from
him in making himself familiar with the system of contractions
which we are about to describe. For a scientific symbolic notation
is in essence nothing else than a systematised method of contraction, it is not a hieroglyphic system, mysterious to the uninitiated.
Our examination of it here will not be simply parenthetical, but
will give us opportunity for examining more closely the real
nature of several important kinematic chains.

§ 53.

Former Attempts.

Attempts have not been wanting to express machine combinations in some concise form. Clockmakers, among others, and

writers upon subjects connected with horology, have employed a kind of notation for showing the sequence of the wheels and arbors in clockwork. Willis has entered somewhat closely into this method of symbolization. The following, for example, shows the arrangement of the wheel-work of a common clock in a form which he himself adopted :—

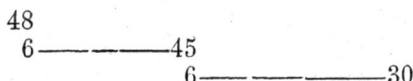

```
        48
        6————45
                  6———————30
```

Here the numerals stand for numbers of teeth, the lines indicate the connection of two wheels by an arbor, the placing of one figure over another shows that the corresponding wheels gear together. Putting the names of the wheels beside the numbers of their teeth we should have :—

```
Great Wheel 48
        Pinion  6————45 Second wheel
               Pinion  6—————————30 Balance-wheel.
```

Other writers have used methods somewhat differing from this.[*] It is evident, however, that the object here is the representation of a portion only of an isolated case,—and even that portion is not intended to include the general kinematic nature of the spur-gearing, but to cover merely the indication of its velocity-ratios, a very important matter, of course, in itself.

The method proposed by Mr. Babbage " for expressing by signs the action of machinery " was more important than these, and was indeed intended to be quite general in its application. Babbage, to whom no doubt the subject was suggested during the extremely difficult construction of his calculating machine, described his system in a small book, not much known,[†] in which he has illustrated it by two large examples,—a clock with working and striking trains and a hydraulic ram. His method is as follows :—The names of the whole of the moving parts are first put down in order, and then signs are placed in tabular form beside the name of

<hr/>

[*] Willis instances the following for the case supposed :—

Oughtred (1677)	Derham (1696)	Allexandre (1735).
30	48)6—45)6—30	48
6)45		45--6
6)48		30—6

[†] *A Method of Expressing by Signs the Action of Machinery,* by C. Babbage. London, 1826.

each part, to indicate its motion. The symbols employed are arrows of various kinds, full lines, dotted lines, brackets, crosses, hooks, and so on. It is certainly possible to interpret the action of the machine by the aid of the signs when the meanings of these have once been completely mastered. Notwithstanding this the method has never been used. No notice was taken of it by those practically interested in machinery, and by this want of attention they added unconsciously to the great irritation which displayed itself in the work which Babbage published shortly before his death· In this he struck about him most vehemently, like Timon of Athens with his spade, accusing his contemporaries of their want of comprehension and appreciation of his work. Without in the least depreciating, however, his most important labours in other directions, it must be said that the cause of the non-acceptance of his system of notation was due to its own defects, and not to those of the public.

What the symbolic memoranda of Babbage express, and were intended to express, is not the essential constitution of the machine, its different parts scientifically defined and recognizably indicated by the stenographic symbols, but merely the general nature of the motion of those pieces which were themselves described at length or by their names in the usual manner. We learn whether such and such a piece turns backwards or forwards, moves continuously or discontinuously, uniformly or with varying velocity, and in cases where there is turning about axes we have the velocity-ratio and so on, given. It is at once evident, however, that under this system mechanisms of completely different constructions might be represented by one and the same set of symbols. These extend merely to the external conditions accompanying certain characteristics of the single organs, not to their full meaning ; they form simply a concise description of the action of the machine, not in any way showing its dependence upon general fundamental principles. If the symbols proposed by Mr. Babbage were placed upon the necessary drawing of the machine itself, they would express their meaning much more clearly than when used in the more abstract form of a table.*

* In a small pamphlet of half-a-dozen pages published in 1857 Mr. Babbage again proposed a very complicated " Mechanical Notation," no doubt the offspring of his own requirement in connection with his machines ; but here he appears to have.

For our purpose—the representation of kinematic chains by symbols,—Babbage's method is of no service; I therefore pass over Willis' attempts to make it more useful by certain alterations.*

§ 54.

Nature of the Symbols required.

The object of the kinematic notation which we wish to form is, like that of mathematical symbols, to express certain operations performed with, or supposed to be performed with, the bodies indicated by signs or otherwise ; partly also its province is similar to that of chemical notation, for it must afford information, and indeed somewhat full information, as to the q u a l i t y of the thing named. The symbols for kinematic bodies must not, therefore, be in themselves meaningless like those of Mathematics, where different letters indicate only the variations in m a g n i t u d e of known,—and so far as their measurability is concerned s i m i l a r,—things ; but each letter must stand, as in Chemistry, for a particular class of bodies, the differences between the classes being here their g e o m e t r i c a l pro-perties. The letter must therefore stand for the n a m e of the body— that is, of the kinematic element, the definite characteristics of which are sufficiently indicated by that name. The letter used in this way we shall call the c l a s s- or n a m e-s y m b o l of the element. The sign for the general name of a kinematic element, as *e.g.* the sign for "screw," "revolute,"† "prism," and so on, is seldom sufficient by itself. Most frequently some further indication is required as to the form of the body, as for example whether the screw be external or internal, that is whether the screw-spindle or the nut be meant. The geometrical basis figure is the same in both cases, but there is a great difference between the forms in which it is used. Signs serving to indicate this more exactly, which will be used in connection with the name-symbols, we shall call f o r m-s y m b o l s.

In addition to these two classes of symbols, a third kind is

intended that the letters and symbols should be put on the drawings themselves, as Prof. Reuleaux suggests.

* Willis : *Principles of Mechanism*, p. 343. (2nd. Ed. 292.)
† See page 91.

required, its object being to show the mutual relation between two or more elements of a mechanism; whether two neighbouring elements be paired or linked together: if the latter, what their relative geometric position is; whether a link be fixed or movable —its relation, that is, to surrounding space—and so on. Signs for these purposes we shall call s y m b o l s o f r e l a t i o n.

The more exactly and explicitly the signs explain the kinematic elements and their chaining the better will they serve the purpose for which they are intended. We shall, however, be content with a certain degree of completeness in order to avoid diffuseness; ·in all cases, however, the signs will express the real and general nature of the thing symbolized.

§ 55.

Class or Name-Symbols.

In choosing symbols to indicate the class to which, considered kinematically, a particular body belongs, we shall follow the example set us in chemical notation, and use Roman capital letters, making these where possible the initial letter of the name of the class. The following twelve signs will be used to stand for the bodies whose names are placed after them :—

S Screw,	G Sphere (Globe),
R Solid of Revolution (Revolute),	A Sector or sweep (Arc),
P Prism,	Z Tooth or projection,
C Cylinder,	V Vessel or chamber,
K Cone (Konus),	T Tension-organ,
H Hyperboloid,	Q Pressure-organ.

It may appear remarkable that the number of classes of elements is so small. In fact, however, forms which can be called kinematic all lie within such a limited circle that a greater number of signs is not required, and it is advisable in all cases to be content with as few signs as possible. The letters have been chosen with care so as—so far as possible—to suggest the form for which they stand, and also to be available in other European languages than our own. I can also say from experience that the recollection of the signs is no great tax upon the memory.

§ 56.

Form-Symbols.

In choosing signs suitable for kinematic form-symbols we are struck by the existence of a certain insufficiency, for our purposes that is, in some very usual geometrical ideas or methods. In geometry the name given to a body of a certain form is that of a portion of space limited by the same figure. In general it is the portion of space inclosed within this figure which is considered to be the form of the body having the same name. Evidently there is here a certain indefiniteness ; because the two portions of space, the one outside and the other inside the figure, cannot both be meant at the same time.

For our purposes, however, it is necessary to distinguish between these two, the portion of space inclosed by the figure, and the portion inclosing it. If between two parallel planes, for instance, there be a circular cylinder, its axis at right angles to them, then the space inclosed between the planes and inside the cylinder must be distinguished from that which is between the planes but outside the cylinder ;—in other words we must know upon which side of the cylindrical surface the material forming our element is placed. We call the body inclosed by the cylinder a full cylinder, and that in which it is inclosed an open cylinder. We shall use for the form-symbols of full and open* bodies respectively the ordinary signs for plus and minus.

The plane limitation of a solid of revolution requires also a sign. It lies equally between the limits + and −, and therefore may suitably be indicated by zero.

For curved profiles, that is profiles which are neither rectilinear nor circular, we may use the circumflex ;—we therefore have

<table>
<tr><td>+for full bodies</td><td>⁰ for plane bodies</td></tr>
<tr><td>⁻for open bodies</td><td>˘ for bodies having curved profiles.</td></tr>
</table>

These form-symbols will be placed above and to the right of the class-symbols (excepting the circumflex, which will be placed over the letter to which it refers), and in a smaller type. Thus

---

* I use these words as being at the same time shorter and more expressive than the commoner ones solid and hollow.

for instance we may use the following symbols for the forms named :—

C^+ Full cylinder, C^- Open cylinder,

S^+ Screw spindle, S^- Nut,

K^+ Full cone, K^- open cone,

K^0 Plane cone (Cone having a vertex angle of 180°.)

$\tilde{C}$ Cylinder upon a general curvilinear base,

$\tilde{C}^+$ the same cylinder full,

$\tilde{C}^-$ the same open,

$\tilde{P}$ Prism upon a general curvilinear base.

We have chosen the symbol V for a vessel of any kind. By a suitable form-symbol we can make its meaning more definite in certain special cases,—we can use V^+ for the reciprocal of the vessel,—that is, a body touching it all round upon its inner surface, so that V^- will stand for the vessel itself. V^+, for instance, might represent a piston, V^- the cylinder in which it works.

Small letters similar to those of the class-symbols will also be used as form-symbols. They will be placed below and to the right of the former, and permit a distinct separation to be made between the forms of various elements, so as to shut out all meanings but the one intended. The following may serve as a few examples of this :—

C_z Cylindrical spur-wheel, and from it

C_z^+ Spur-wheel with external teeth,

C_z^- Do. with internal teeth, or annular wheel,

K_z^+ Bevel wheel with external teeth, K_z^0 face wheel,[*]

H_z^+ Hyperboloidal toothed wheel.

H_z^0 Hyperboloidal face wheel,[*]

$\tilde{C}_z^+$ Non-circular spur wheel with external teeth.

P_z Rack.

C_z^+ Cylindrical screw wheel.

T_p Prismatic tension-organ, such as a flat belt.

T_p^+, T_p^- the same moving respectively towards or from its pulley.

T_s Rope, T_c wire, T_z common chain, T_r jointed chain.

---

* See *Der Constructeur*, 3rd. ed. p. 435 and 451.

For pressure-organs we require to distinguish between gases and liquids. We may use the Greek letters λ and γ for this purpose, and thus have:

Q_λ liquid pressure-organ, water, etc.

Q_γ gaseous pressure-organ, gas, air, steam, etc.

In certain cases the pressure-organ consists of more or less round grains, which may with sufficient accuracy be taken as spherical, and therefore we have

Q_g or more exactly $Q_{\tilde{g}}$ for a pressure-organ consisting of more or less globular portions.

We shall form and use further compound symbols as we have occasion.

§ 57.

Symbols of Relation.

Of the relations which one element in a chain can have to another the most important are those of pairing and linkage. The first we may indicate by a comma. C, C will thus stand for two cylinders rolling together, C^+, C^+ would be used if both were full, C^-, C^+ if one were full and one open. We shall always presuppose that the comma indicates both the possibility and the existence of correct pairing. Thus we shall not require any sign beyond C^+, C^+ to show that the axes of the cylinders are parallel, —while C^-, C^- is incorrect, for it is impossible to form a kinematic pair from two open cylinders.

Linkage will be denoted by a dot or dotted line. $C^+ \ldots C$ for instance is a link having two full cylinders for the elements which it connects, $C^- \ldots C^-$ a link connecting two open cylinders or eyes.

The fixing of a link may be indicated by underlining the dotted lines. $P^+ \ldots C^+$ for instance stands for a fixed link connecting a full prism and a full cylinder.

It may occasionally be necessary to indicate that a link is elastic,—namely that it is a spring,—in which case a wavy line may be placed over the dotted one.

A number of other signs are partly the same as the common arithmetical signs, and partly based upon them. They are as follow :—

= equal, > greater than, < less than, ∞ infinite;

| conaxial, ‖ parallel, ∠ oblique, ⊥ normal;

⊢ crossed obliquely; + crossed at right angles;

⧧ equal and con-axial, ⧢ equal and parallel;

≌ coincident;

□ conplane,— lying in the same plane,

Z anti-parallel (in a quadrilateral),

⊴ isosceles, or having adjacent arms equal (in a quadri-
     lateral).

The relations expressed by these signs may exist either between the elements of a pair or the links of a chain. In the former case the comma between the two elements is omitted, and the symbol itself is printed in a smaller type than in the latter.

If a hollow cylinder be paired with a geometrically equal solid cylinder they are equal and con-axial, so that the comma would have to be replaced by the sign ⧧. If however such bodies, being equal, are to be paired, they must also by necessity be con-axial, so special indication of that relation may be omitted without any loss of distinctness, and we may write the pair $C^+_{\underline{}}C^-$. If this be a closed turning-pair, the conditions as to the prevention of cross-motions by a proper sectional profile (§ 15) must be fulfilled. We shall here always presuppose that two elements, the symbols for which are connected by the sign for pairing, form a closed pair, unless the contrary be expressly stated. We shall see further on that in cases where they are not closed the notation of the chain itself always makes it possible to do this. The three lower pairs, then, twisting pair, turning pair, and sliding pair, have for their symbols :—

$$S^+_{\underline{}}S^- \qquad\qquad R^+_{\underline{}}R^- \qquad\qquad P^+_{\underline{}}P^-$$

The curved discs in the triangular, quadrilateral, etc., hollow prisms, (Chap. III), can be indicated generally by the formula $\check{C}^+,\check{P}^-$; they fall therefore in one and the same class of pairs. With respect to the simple turning pair, $R^+_{\underline{}}R^-$, in which the most various profile forms may be used so long as the pair-closure remains, it will be noticed that as far as the relative motions of its elements are concerned it does not differ from the closed cylinder pair $C^+_{\underline{}}C^-$. In most cases it is therefore allowable to write $C^+_{\underline{}}C^-$, instead of $R^+_{\underline{}}R^-$. The idea is somewhat simpler, the cylinder instead of the revolute, and in machines actually the special case C of the body R is almost always used. It is only in very special cases that we shall find it necessary to adhere to the strictly general notation.

Symbols of relation between the elements of a link will be placed in the dotted line. Thus $C^+...\|...C^+$ stands for the linkage of two parallel full cylinders; $C^-...\|...C^-$ for a linkage of two parallel open cylinders, that is, a connecting rod; $C_z^+...|...C^+$ a spur wheel attached to its con-axial shaft; $C_z^+...|...C^-$ a spur wheel with a con-axial open (bored-out) boss.

A special indication is sometimes required for incomplete pairs. The first necessity is here a symbol for incompleteness, and for this we use the ordinary sign of division, so as to allow the method of closing to be indicated by a divisor.

To indicate merely the incompleteness of a pair we may use the divisor 2, considering the piece or element as halved. If it be completed by force-closure, the divisor f (force) may be employed. For closure by a kinematic chain we choose the divisor k; if the chain-closure occur by means of a spring we substitute for this l; and, lastly, if closure be effected by a pair (§ 47), we shall use the divisor p. We therefore have the following:—

$\dfrac{C^-}{2}$ a portion of an open cylinder,

$\dfrac{C^+}{f}$ a full cylinder paired by force-closure,

$\dfrac{C^+}{k}$ do.   paired by chain-closure,

$\dfrac{C^+}{l}$ do.   closed by a spring,

$\dfrac{C^+}{p}$ do.   closed by a pair of elements.

If any link used for chain-closure have a special indicating letter, as a, b, c, etc. (as we shall see to be sometimes the case), this also can be placed in the divisor so as to indicate distinctly the method of closure. We shall find further on frequent applications of these methods of symbolization.

§ 58.

Formulæ for simple Kinematic Chains and Mechanisms.

In describing a complete kinematic chain by symbols written, as they must be, in lines, we cannot represent the returning back upon itself, or closure, of the chain, and must be content with merely indicating this. This disadvantage, if it be one, our notation shares with that of Chemistry, but the matter if fairly examined is seen

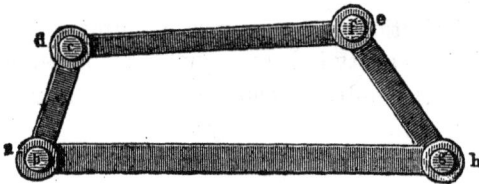

FIG. 179.

to be quite unimportant. In writing down a chain we begin with a link, and thus a single element of some pair must stand first in the formula, at the end of which, consequently, must be its partner element, and the sign of pairing annexed to the symbol of this latter sufficiently indicates the closure of the chain.

This will perhaps be made clearer by an illustration. Let it be required to write down the familiar chain shown in Fig. 179. The cylindric elements are indicated in the figure by the letters *b c, d e,* etc., and for distinctness' sake we may in the first case add these to the symbols. Beginning then with the link *b c* we have to write :

$$\underbrace{C^+...\ \|\ ...\ C^+_=}_{b\,c}\ \underbrace{C^-...\ \|\ ...\ C^-_=}_{d\,e}\ \underbrace{C^+...\ \|\ ...\ C^+_=}_{f\,g}\ \underbrace{C^-\ ...\ \|\ ...\ C^-_=}_{h\,a}$$

The links, so far as the above form-symbols are concerned, appear identical. If the chain be fixed in the way shown in Fig. 180, the formula becomes (omitting the letterings of the special links) :—

$$C^+\ ...\ \|\ ...\ C^+_=\ C^-\ ..\ \|\ ...\ C^-_=\ C^+\ ...\ \|\ ...\ C^+_=\ C^-\ ...\ \|\ ...\ C^-_=$$

At first sight it might appear to be strange, and to show a defect in our notation, that it does not show any difference in form between the links. This I deny at once, for we know from § 16 that the

form-symbols − and + are in the lower pairs absolutely interchange-
able without alteration of the pairs, so that in the formula before
us all the links might be indicated by the same symbols. Closer
examination changes the defect into an advantage. The links
appear alike in the formula because in their actual nature
they are really alike. A clear indication of this is necessary

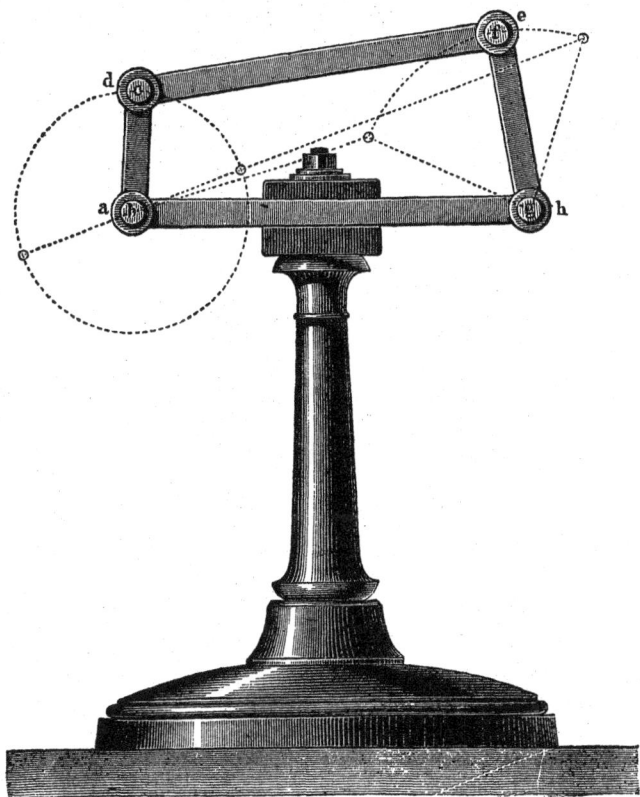

Fig. 180.

for the realization of the abstract form of the mechanism, for the
perception of its essential nature under its material disguise. In
the mechanism of Fig. 180 the ordinary mechanic sees a " beam "
driving a " crank,"—fg is the arm of the beam, de the connecting
rod, bc the crank, and ah the frame of the machine, formed in the
most various ways of columns or castings or timber, and supported

on built foundations. In the actual machine this framework, the fixed link of the chain, is of so varied a figure, so mixed up with portions of buildings, and complex in so many ways, as to afford no indication of its really simple nature; indeed, to very many people the mechanism presents itself as one of three links only,— beam, connecting-rod and crank. This has been carried so far that even in text-books, when a purely schematic or abstract represent- ation of the mechanism has been used, the fourth link *a h* is entirely omitted from the figure. The inconsistency is no doubt explained by the tacit assumption made that the paper forms as it were the frame for the three moving parts whose existence is formally recognised.

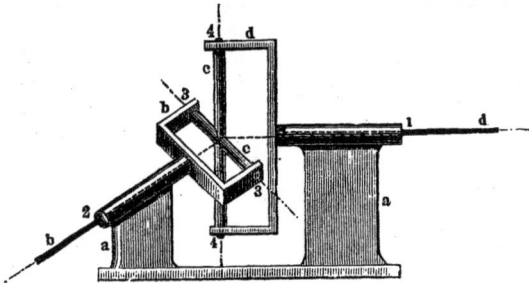

Fig. 181.

The description of the mechanism given by our formula is speci- ally suited to remove this prejudice, or rather this confusion of ideas. It might be asked, however, whether the formula should not show also the relations between the lengths of the different links. This would certainly be possible; it would however be difficult, for of course the four lengths may have between them an enormous variety of numerical relations, and it would at the same time be of little use, for even if the lengths were added to the formula it would require a special study of the mechanism in each case to make any practical use of them. We shall, however, find it possible further on to employ a method of indicating these length-relations which is simple and very easily used.

We may take for a second illustration a universal joint, or Hooke's joint,* of which Fig. 181 gives a schematic representation.

---

* Prof. Reuleaux points out in his *Constructeur* that if this joint is to be named after its inventor it should be called Cardano's joint; for he was the first to point out its possibility.

The chain which constitutes this joint has four links, which are marked in the figure with the letters $a, b, c,$ and d. The link a is paired with b by the turning-pair 2. Normal to this turning-pair is another, 3, which has its open cylinder in the fork of b, and its full cylinder in the sloping arm of the piece c; the link b must therefore be written $C^+ \ldots \perp \ldots C^-$. It must be noted that the lower and upper arms of the fork form together one piece only, and must be reckoned as such; the same is true of the two ends of the arm of c, which kinematically form a single element only. The piece c consists of two solid cylinders, 3 and 4, having their axes crossing at right angles, and it must therefore be written $C^+ \ldots \perp \ldots C^+$. The third link, the fork and spindle d, is similar to b, and will be written in the same way. The fourth link a, lastly, consists of two open cylinders, 1 and 2, oblique to each other, and so must be written $C^- \ldots \angle \ldots C^-$; it is a fixed link, as its form in the figure shows. The complete formula, therefore (to which we have added the letters and numbers used above to distinguish the links and pairs), runs thus:—

$$\underbrace{C^+ \ldots\underset{2}{\perp}\ldots C^=_- \; \underset{3}{C^+} \ldots\underset{}{\perp}\ldots C^\pm_=C^- . \,\underset{4}{\cdot}\perp\ldots C^\pm_= \; \underset{1}{C^-} \ldots\angle\ldots \underset{2}{C^=_-}}$$
$$\underbrace{}_{b} \quad \underbrace{}_{c} \quad \underbrace{}_{d} \quad \underbrace{}_{a}$$

There is one geometrical property of the chain which is not shown by our formula, namely that the axes of the pairs 1, 2, 3, and 4 have a common point of intersection. But unless the chain possessed this property it would not be possible, on our supposition that all its pairs are closed. No special indication of this property is therefore commonly necessary. Our formula shows, however, that the three links b, c, and d are again identical. This circumstance is very notable, and we shall later on have to deal with it in another form; the common construction of the joint so entirely conceals it as to make it almost unrecognizable.

The belt train, the kinematic nature of which we have already examined, will be written as follows:—

$$T^\pm_{\text{p}} \ldots \simeq \angle \ldots T^\mp_{\text{p}}, R^+ \ldots \mid \ldots C^+_= C^- \ldots \parallel \ldots C^-_= C^+ \ldots \mid \ldots R^+,$$

The tension organ used here is a flat band, and is therefore marked with the suffix p (prismatic); it rolls both on to and off each

pulley, and therefore receives the signs ± and ∓, these being reversed because the portion of the belt running on to one pulley

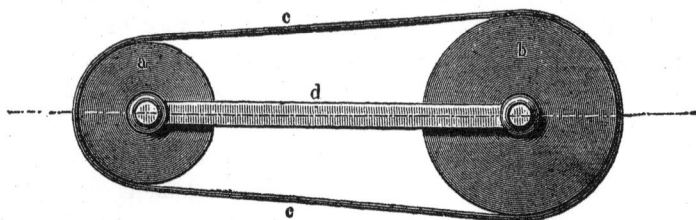

Fig. 182.

is the same as that running off the other. The band for the
pulley *a* is identical—coincident—with that for *b*, the corre-

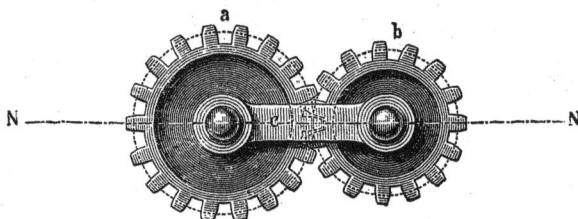

Fig. 183.

sponding sign ≅ must therefore be placed in the dotted line; the
belt is—lastly—open, and its two sides are inclined to each other

Fig. 184.

on account of the inequality of the pulleys, so that we must add
the symbol ∠, oblique, to the sign ≅. If the belt were crossed

we should have had to substitute for $\angle$ the symbol $\vdash$. The rest of the formula is clear from what has gone before. Put into words, the whole stands thus : " Kinematic chain consisting of two unequal revolutes connected by an endless band touching them externally, each being provided with a con-axial cylindric spindle, and the two spindles working in parallel bearings in a stationary supporting piece."

The simple spur-gearing of Fig. 183 is written,

$$C^+ \; \dots \; | \; \dots \; C_z^+,\, C_z^+ \; \dots | \; \dots \; \underline{C_{\pm}\, C^-} \; \dots \, \| \; \dots \; C_{\equiv}^-$$

The links coming first in the formula are the two spur-wheels with their shafts, the last is the (fixed) bar carrying the bearings. The annular gear of Fig. 184 is written :—

$$C^+ \; \dots | \; \dots \; C_z^+,\, C_z^- \; \dots \; | \; \dots \; \underline{C_{\pm}\, C^-} \; \dots \, \| \; \dots \; C_{\equiv}^-$$

The first link is here the wheel a having external teeth, the second is the annular wheel b, and the third is again the bar c, the latter being supposed fixed.

§ 59.

Contracted Formulæ.

If once the separation of a chain into links and the special examination of the latter has been completed—so that they may be assumed to be already known—the formula or symbolic description of the chain may be in many cases greatly shortened. There are several possible forms of contraction, which we shall examine in order.

Firstly, in the case of the lower pairs, and of some others in which the partner elements have the same name-symbol, one letter may frequently be made to suffice for a pair of elements if it be used along with some distinguishing mark. For this purpose a parenthesis can be used, so that we may employ as contractions :

(S)	for the twisting pair		$S_{\equiv}^+ S^-$
(C)	„ „ turning	„	$C_{\equiv}^+ C^-$
(P)	„ „ sliding	„	$P_{\pm} P^-$
(C_z)	for a pair of spur wheels		C_z, C_z
(K_z)	„ „ „ bevel	„	K_z, K_z

and so on. This allows us, for example, to write the chain represented in Fig. 179 as:

$$C^+ \dots \| \dots (C) \dots \| \dots (C) \dots \| \dots (C) \dots \| \dots C_=^-$$

Here only the elements of the first (and last) link require to be written separately with their form signs—the one letter in the parenthesis standing for a pair of elements. The same method can be carried further; it allows us to write certain simple kinematic chains in a still shorter form, for where we have the parenthesis the relations of the linked elements are sufficiently defined without the use of the dotted line, which may therefore be omitted. This is certainly the case in the present instance, where the same relation —parallelism—exists between all the elements forming links. Indeed we may in these circumstances extend the parenthesis so as to include several, or the whole of the links. Thus for certain cases we may compress the above formula, without making its meaning uncertain, into the symbol (C_4'''), in words " C parallel four," or " C four parallel," and meaning "a chain formed upon four turning pairs, consisting, that is, of four links, each connecting two parallel cylindric elements." Such a contraction presupposes in all cases a familiarity with the way in which the chain can be formed out of its elements ; its form, however, is so concise as to leave nothing to be desired in this direction. The chain forming the universal joint, Fig. 181,—to take another example,—allows itself to be written $(C_3^\perp C^L)$, in words, " C normal three C oblique"; the spurgearing of Fig. 183 may be written $(C_z^+ C_2'')$, in words, " C plus z C parallel two," and so on.

These concentrated symbolic forms seem at first suitable only for the kinematic chain, not for the mechanism formed by fixing one of its links ; further on, however, we shall find means for making use of them in these cases also, within certain limits.

§ 60.

Formulæ for Compound Chains.

In the simple kinematic chains the choice of the link with which to begin the formula was to a certain extent arbitrary. This strikes us still more in compound chains, and makes it appear at first sight to some extent difficult to attain the required distinctness. A little experience, however, enables this to be obtained, as a few examples will show.

We have before examined (§ 3) the chain represented in Fig. 185, containing seven cylindric pairs. It is obtained from the familiar chain (C_4'') by the addition of two more links of the form $C...\|...C$, and possesses a certain symmetry of arrangement in having two opposite three-cylindered links twice connected by a pair of two-cylindered links,* altogether, that is, by four such links. This is made more distinct by the schematic representation in Fig. 186, in which also the dimensions are so chosen as to make the chain symmetrical. The turning-pairs are here numbered

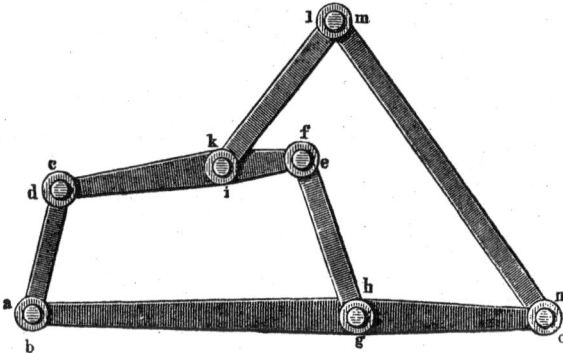

Fig. 185.

from 1 to 7. We may look at the whole chain as consisting of two five-linked cylinder-chains 1, 2, 3, 4, 5 and 1, 2, 6, 7, 5—in which the links 1, 2 and 1, 5 are common, the cylinders 2, 3, 6 united into one link containing three elements, and the cylinders 5, 4, 7 into another. To distinguish between links containing two, and links containing three elements, we may call them **binary** and **ternary** links respectively.

We may now proceed by first writing down these two five-linked cylinder-chains—neither of which is by itself constrainedly closed —singly, and then as it were adding them together—that is putting a single sign only where pieces are common to both chains, and bracketing the elements brought together in the ternary links.

---

* Firstly in the original connection by $a\ d$ and $e\ h$, and secondly in the additional connection by $k\ l$ and $m\ n$.

We obtain the following result (using the contracted symbols for all the inner pairs) :—

$$
\begin{array}{cccccc}
1 & 2 & 3 & 4 & 5 & 1 \\
C^+ \dots \| \dots (C) \dots & \| \dots (C) \dots & \| \dots (C) \dots & \| \dots (C) \dots & \| \dots C^-_= \\
C^+ \dots \| \dots (C) \dots & \| \dots (C) \dots & \| \dots (C) \dots & \| \dots (C) \dots & \| \dots C^-_= \\
1 & 2 & 6 & 7 & 5 & 1
\end{array}
$$

$$
\begin{array}{cccccc}
1 & 2 & 3 & 4 & 5 & 1 \\
C^+ \dots \| \dots (C) \dots & \| \left\{ \begin{array}{c} \dots (C) \dots \| \dots (C) \dots \\ \dots (C) \dots \| \dots (C) \dots \end{array} \right\} \| \dots (C) \dots & \| \dots C^-_= \\
& & 6 & 7 &
\end{array}
$$

The compound formula resulting from the addition—and to which, for explanation's sake, we have added the numbers of the

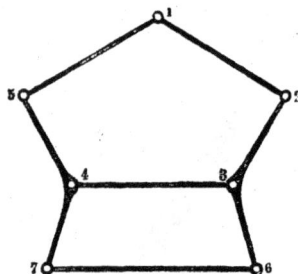

turning-pairs—may be considered as one which really allows the nature of the chain to be seen, for it distinctly reproduces its symmetrical arrangement. It is possible, however, to bring the formula into a still clearer shape, which may be useful in some cases. Noticing, namely, that the four turning-pairs within the brackets form by themselves a simple closed cylinder-chain (C''_4),

FIG. 186.

and at the same time that this whole chain has taken the place before occupied by a pair of elements, we see that the whole formula may be written,—

$$
\begin{array}{ccccc}
1 & 2 & 3,4,7,6 & 5 & 1 \\
C^+ \dots \| \dots (C) \dots \| \dots (C''_4) \dots \| \dots (C) \dots \| \dots C^-_=
\end{array}
$$

so as to take up much less space than before. The formula could be used in this shape for the mechanism, also, if the fixed link were 1.2, 2.3, 1.5 or 5.7—but if the fixed link be one of the inner group, 3, 4, 7, 6, the more extended formula must be employed.

We may choose a train of spur wheels as another illustration. Fig. 187 represents a compound mechanism of this kind with two pairs of wheels a, b and c d. The wheel c is fixed to b, the three spindles con-axial with the wheels have their bearings in the ternary link 1. 2. 3, which is here the fixed link. The formula may be arranged in several different ways.

Starting from the turning pair 2 we have, on each side, simply a pair of spur-wheels with their connecting link. We write them singly and add them together as follows :—

$$
\begin{array}{cccc}
2 & b,\,a & 1 & 2 \\
C^{+}\;\ldots\;|\;\ldots\;C_z^{+},C_z^{+}\;\ldots\;|\;\ldots\;(C)\;\ldots\;\|\;\ldots\;C_{=}^{-} \\
C^{+}\;\ldots\;|\;\ldots\;C_z^{+},C_z^{+}\;\ldots\;|\;\ldots\;(C)\;\ldots\;\|\;\ldots\;C_{=}^{-} \\
2 & c,\,d, & 3 & 2
\end{array}
$$

$$
\begin{array}{cccc}
2 & b,\,a, & 1 & 2 \\
C^{+}\;\ldots\;|\;\ldots\;\begin{Bmatrix} C_z^{+},C_z^{+}\;\ldots\;|\;\ldots\;\underline{(C)\;\ldots} \\ C_z^{-},C_z^{+}\;\ldots\;|\;\ldots\;\underline{(C)\;\ldots} \end{Bmatrix}\;\ldots\;\|\;\ldots\;C_{=}^{-} \\
& c,\,d & 3 &
\end{array}
$$

This shows two binary links a, 1 and d, 3, and two ternary links 2, b, c and 2, 1, 3. The latter are the coupled wheels b and c with

FIG. 187.

their common spindle, and the bar with the bearings, which—as the underlining shows,—is fixed.

We may obtain the formula in another shape as follows. Beginning with the wheel a and its spindle, we then write the wheel b and its connections at once as a ternary link ; continue with d, and write its linkage with 3 ; and then unite the second cylinders of 2 and of 3 with 1. For distinctness' sake we use the uncontracted form.

$$
\begin{array}{cccc}
1 & a,\,b & 2 & 1 \\
C^{+}\;\ldots\;|\;\ldots\;C_z^{+},C_z^{+}\;\ldots\;|\;\begin{Bmatrix} \ldots\ldots\ldots\;\underline{C_{=}^{\pm}\;C^{-}}\ldots\ldots\ldots \\ \ldots\;C_z^{+},C_z^{+}\;\ldots\;|\;\ldots\;\underline{C_{=}^{+}\,C^{-}\ldots} \end{Bmatrix}\|\;\ldots\;C_{=}^{-} \\
& c,\,d & 3 &
\end{array}
$$

The meaning of this formula is exactly the same as that of the last. The part of it within the brackets, however, shows itself at once to be a simple kinematic chain, consisting of the wheels c and d and their connecting link 2, 3. If we use for this the contracted symbols of § 59, and contract also the symbol for the pair a, b, we have:

$$1 \qquad a,b \qquad c, d, 3, 2 \qquad 1$$
$$C^+ \ \dots \ | \ \dots (C_z^+) \ \dots \ | \ \dots \ (C_z^+ \, C_2'') \ \dots \ \| \ \dots \ C_{\underline{=}}^-$$

In some cases it will be quite sufficient to write this in a still shorter form—extending the use of our former method of contraction,—as $(C_{z2}^+ \, C_3'')$; or generally for an n-fold train of spur-gearing $(C_{zn}^+ C_{n+1}'')$, where for n pairs of wheels there are in general $n + 1$ axes, or rather turning-pairs (C), required.

<h2 style="text-align:center">§ 61.</h2>

<h3 style="text-align:center">Formulæ for Chains containing Pressure-organs.</h3>

In order to find the formula for a chain which contains a pressure-organ it is frequently advisable to imagine the substitution of a rigid element for the latter—the pairing being obviously somewhat altered in consequence—and then to transform the formula thus obtained by the re-insertion of the pressure-organ.

In order, for example, to express by a formula the water-wheel (Fig. 188) which we have already looked at, we may first replace the water by a rack with a prismatic guide (Fig. 189) so arranged as to drive the spur-wheel a by its own weight, its action being thus similar to that of the fluid for which it is substituted. The formula will run:

$$C^+ \ \dots \ | \ \dots \ C_z^+, \, P_z \ \dots \ \| \ \dots \ P_{\underline{=}}^\pm P^- \ \dots \ + \ \dots \ C_{\underline{=}}^-$$

If we now change the link $P_z \ \dots \ \| \ \dots \ P^+$ into $Q_\lambda \dots \dots Q_\lambda$,—replacing the water for its temporary substitute,—we must put V^-, the symbol for a vessel of any kind, for P^-, and so obtain as a formula for the water-wheel,

$$C^+ \ \dots \ | \ \dots C_z^+, \, Q_\lambda \ \dots \dots \ Q_\lambda, \, V^- \ \dots \ + \ \dots \ C_{\underline{=}}^-.$$

If it require further to be indicated that the channel is uncovered, the water being paired with it, that is, by force-closure,

we must substitute $\dfrac{V^-}{f}$ for V^-. The constitution of this mechanism is the same as that of the lift- or flash-wheel. If we suppose
the link $Q_\lambda \dots Q_\lambda$ fixed instead of the link $V^- \dots + \dots C^-$, we
obtain a very different but very familiar mechanism. The link
$V^- \dots + \dots C^-$ moves in the (relatively to it) stationary water:
the mechanism is that of the paddle-steamer. It will be seen that
in this case we must presuppose that the link c possesses the
requisite buoyancy.

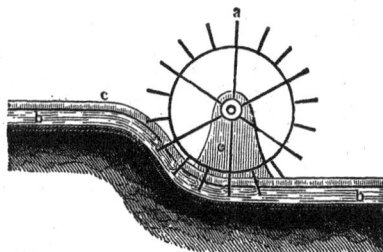 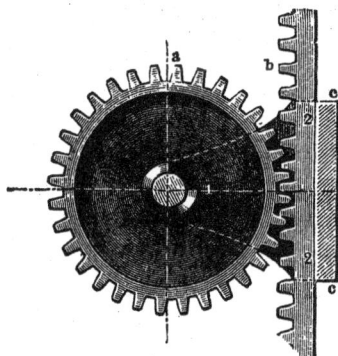

FIG. 188. FIG. 189.

We may here also use the abridged notation. In the first
formula the pair $C_z P_z$ presents a certain difficulty, for it is our
object if possible not to use two capital letters in the contraction
for one pair of elements, in order that there may never be any
doubt as to whether each letter stands for a pair or not. So far as
it is possible we wish that the number of capital letters in any contracted formula—with the addition of course of the repetitions, if
any, indicated by the suffixes—shall show at once the number of
pairs in the chain for which the formula stands. We must for
this purpose have recourse to a convention. Without leading to
any misunderstanding we may denote the pair $C_z P_z$ by the symbol (C_{zp}), and by doing so we obtain an expression for the whole
chain: $(C' \, C_{zp} \, P^+)$. We add the sign $+$ (" crossed ") to the symbol
(P) of the sliding pair, so as to make the position of that pair
quite determinate.

Similar difficulties as to double letters occur twice in the second formula, but treating them similarly to the one just discussed, we may write :

$$(C_{z\lambda}) \text{ for } C_z Q_\lambda$$
$$(V_\lambda) \text{ for } V^- Q_\lambda .$$

The index λ stands as contraction for q_λ, and is sufficient to make the pairing of C_z and V respectively with a liquid quite distinct. For the kinematic chain of Fig. 188 we therefore obtain the concentrated formula $(C' \; C_{z\lambda} \; V_\lambda)$, which, as we have seen, serves as well for the water-wheel as for the lift-wheel and the paddle-steamer.

§ 62.

Contracted Formulæ for Single Mechanisms.

The abridged notation which we have described for kinematic chains cannot be applied to the same chains in the form of mechanisms without some additions,—for it shows only **pairs** and **not links**, and therefore does not in itself furnish any means for indicating the fixing of a link. It is, however, most important that we should have the means of extending these concise,—and yet for so many cases quite sufficient,—symbols to mechanisms.

Although this cannot be done by such logical generalizations as those by which the contractions were arrived at in the first instance, still in the chains which are most important to us the end can be obtained by various special means. These means are the giving of definite name-symbols, settled by agreement in each particular case, to the separate links of the chain. If this be done, and the name-symbol of the fixed link,—as the one about which something special requires to be indicated,—be assigned some particular and conspicuous position in the formula, we have obtained an abridged notation for the mechanism.

We choose the letters of the small Roman alphabet for the link symbols, beginning with a in each case, and going on as far as may be necessary; the letters indicate in themselves therefore no quality or form. To prevent any confusion arising between these letters and the form-signs, we give the former a specially distinctive position in the formula, namely, that of an **exponent** outside the brackets which inclose the symbols of the pairs. Only one letter,

as a rule, will occupy this position, there being only one fixed link
to be indicated. An illustration will make the method quite distinct.

Let it be required to write contracted formulæ for the mechanisms in the form of which the four-linked chain, Fig. 190, can be
used. We first give to the four links the signs a, b, c, d in the way
schematically indicated in Fig. 191 : these signs are arbitrarily

chosen in the first instance, but once chosen they must of course
be adhered to. The lengths of the links are so proportioned that
if d be fixed the link a (the crank) can revolve while c swings in
circular arcs. So long as the chain is unfixed its contracted formula we have already found to be (C_4''). If now d be fixed—as its
form in Fig. 191 indicates—the formula will become $(C_4'')^d$—in
words "C parallel four on d." The particle on indicates that the

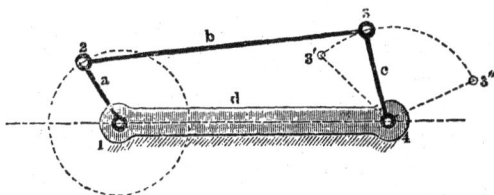

chain is, as it were, placed on the link d, that this link becomes
its base. If d were released and a fixed instead, the new mechanism
would be $(C_4'')^a$; in the same way the water-wheel of § 61 would
be $(C'C_{z\lambda} V_\lambda)^c$, the paddle-steamer $(C'C_{z\lambda} V_\lambda)^b$, and so on. It
will be seen that this very short method of symbolization enables
us easily to distinguish by distinct symbols the different mechanisms which can be made from one and the same chain. It possesses always the limitation, however, that the letters $a, b, c,$ etc., are

here without any general qualitative meaning—there is no special connection between the symbols and the links for which they stand. The only direction in which they have a partly general character is in their alphabetic sequence—their order, that is, may be not without significance. Where possible we may begin with a at some specially distinctive link,—such as the crank in the case supposed,—and so greatly facilitate the recollection of the meaning agreed on for the symbols. The applications of this contracted notation, as we shall find in the sequel, prove it to be of the greatest value.

Our method of symbolization, lastly, allows a further and most useful piece of information to be brought into the formula. It is frequently important to indicate that link of the chain to which the driving effort is applied, or through which the mechanism is moved. For it is evident that there is an immense difference between two mechanisms—otherwise the same—if one be driven by an effort applied to the link a and the other by an effort applied to b. We had a striking example of this in the mechanism of the water-wheel and the lift-wheel. Both would be indicated by the symbol $(C'C_{z\lambda}V_\lambda)^c$, while the transmission of motion in them would be essentially different.

It becomes evident on looking into this matter that this formula is of the nature of a general or indeterminate formula for both mechanisms,—which it must be our object to turn into a special or determinate formula for each of them. We may do this, and supply the information that is wanted, by putting the symbol for the driving link as a denominator in the exponent. The latter will then show the fixed link only in the general formula, but in the special formula it will be fractional, its numerator indicating the fixed, and its denominator the driving link. The choice of the fractional form is justified by the analogy with the symbols for force- and chain-closures which were fixed in § 57.

Thus for example the mechanism $(C''_4)^d$, if the crank be the driving link, will be written $(C''_4)^{\frac{d}{a}}$,—in words " C parallel four on d by a," the latter part being a contraction for "placed on d, driven by a." The same mechanism, if driven by the lever, has for its special formula $(C''_4)^{\frac{d}{c}}$; the general formula $(C''_4)^d$ being of course common to both mechanisms. The water-wheel will now be written $(C'\,C_{z\lambda}\,V_\lambda)_{b}^{c}$, the lift-wheel $(C'\,C_{z\lambda}\,V_\lambda)_{a}^{c}$; the steamer $(C'\,C_{z\lambda}\,V_\lambda)_{a}^{b}$. These last examples already show in the most distinct way the usefulness of our formulæ. For the mere transcription of them is sufficient on the one hand to show the intimate connection between machines which constructively seem to stand so far apart, and on the other hand, to indicate definitely and simply the true differences between them.

The special formulæ of mechanisms are chiefly useful in the analysis of complete machines,—that is, in reference to the applications of mechanisms,—while the general formulæ commonly suffice for their abstract representation. Here, too, however, the special formulæ are often very valuable, as showing which of the link motions is to be considered as the independent variable. We may now proceed to the systematic application of the kinematic notation; in the following chapters we shall have to make extended use of both kinds of formulæ.

CHAPTER VIII.

KINEMATIC ANALYSIS.

§ 63.

The Problems of Kinematic Analysis.

THE analysis of a kinematic arrangement as such consists in separating it into those parts which may be regarded kinematically as elements, and in determining the manner in which these are combined into pairs and kinematic chains. All constructive details are left out of the question. The notation which we have formed gives us the means of representing the results of the analysis in a form which can be easily surveyed, and which distinctly expresses the law of their connection. We shall now undertake a series of such investigations; partly in order to show how the method of analysis is applied, but principally in order to determine clearly the nature of certain important subdivisions of Machine-science. Our work will show us that hitherto there has been an entire want of definiteness about many fundamental ideas, with which nevertheless it has been thought easy to operate. We shall have to rectify many common notions; indeed we shall find necessary the destruction, or at least total transformation, of some propositions apparently universal. As compensation for this, however, we shall be able to place on a really scientific basis other conceptions of even greater meaning and weight.

§ 64.

The "Mechanical Powers" or "Simple Machines."

The mechanical arrangements which go by the name of "mechanical powers" or "simple machines" are familiar to all. Since the time of Galileo, or before it, they have been described in the majority of text-books as those arrangements to which, to a greater or less extent, all machines can be traced back,—of which, in other words, they may all be regarded as compounded. As to the how and the whether, however, there has not been complete agreement; and it is specially noticeable, and at first sight astonishing, that the higher Mechanics has more and more separated itself from any connection with these arrangements. For if they have really the meaning put upon them,—and the contrary, in spite of the sceptics, is nowhere shown,—they should here only acquire a higher value. The highest science could not then venture to overlook them,—however homely or trifling they might appear to be,—while in point of fact the notion seems to be gaining ground that while the "simple machines" are good enough for elementary mechanics, they are worthless for the higher part of the science.

If we look more closely into the question, and compare one text-book with another, we discover everywhere a doubtfulness as to the real significance of the ideas of which they yet retain the outward form.[46] Even as to the number of "mechanical powers" there is no unanimity. Some speak of six—Lever, Inclined Plane, Wedge, Pulley, Wheel and Axle, Screw;—while others would include unconditionally the "funicular machine"* as a seventh. The definition of the "simple machine" fares even worse—no two books can agree upon one. The most various places also are given to them in the treatment of the subject. Sometimes they stand at the beginning, sometimes in the middle, sometimes at the end, sometimes taken in different chapters; sometimes they are treated of without being called by their traditional names, as if with the suspicion that if they were acknowledged nothing

---

* A cord suspended from both ends, and having weights attached to it at different points. I have not noticed this among the mechanical powers in English works, but here generally the "toothed wheel" takes its place— not to mention the "compound wheel and axle," &c.,.occasionally met with.

could be done with them. In short, such a comparison shows
that there is no common idea really underlying the matter, for the
differences are more than superficial ; it rather leads to doubt
as to whether the "simple machines" have any right whatever
to their name.

And yet there is something specially characteristic in these
arrangements,—at least in some of them, as, *e.g.*, the lever and
the inclined plane,—which have so entirely passed from a special
department into common language and ideas. There is something
homely and familiar about them, they excite, I might almost
say, a sentimental interest. Does this merely result from recol-
lections of youthful mechanical study, or is it a breath from the
childhood of science itself playing upon us ? Or has this
sympathy, to which even the most abstract theorist would probably have to acknowledge in his quiet moments, really no deeper

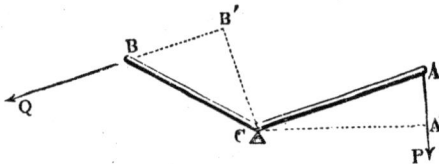

Fig. 192.

ground ? Kinematic analysis must give us a distinct answer to
these questions ; it must show us whether we have really to give
up these old heirlooms of mechanics, and if so it must enable
us to remove them altogether, or whether there is not something really indestructible in them. Let us proceed with the
examination.

The Lever.—A straight bar or knee-shaped body supported
upon a fixed angular bearing, about which it can turn, (Fig. 192);
two forces act on the bar on the one or the other side of the
support; their equilibrium is to be studied. The problem has
been stated thus since the time of Archimedes. In most cases
the description is not exact. It is assumed, but not distinctly
stated, that the support is so arranged that only plane motions
can occur; it remains unsaid that in cases where the direction
of the forces is such as to move the lever from the support, this
does not occur, in other words that it is prevented by suitable

restraint. We have here certainly an incompleteness in the statement of conditions which is very extraordinary in the case of an important fundamental proposition. If we supply these defects we have the bodies, lever, and support so arranged that their relative motions are constrained, and that each is free only to rotate relatively to the other. This, however, is nothing else than the arrangement of the turning pair $R_{=}^{\pm}R^-$, or (see § 57) $C_{=}^{\pm}C^-$, and will be called, according as one or the other elements be fixed

$$\underline{R_{=}^{-}R^+} \text{ or } \underline{R_{=}^{\pm}R^-}$$

or otherwise

$$\underline{C_{=}^{-}C^+} \text{ or } \underline{C_{=}^{\pm}C^-}$$

and the " principle of the lever " is simply the conditions of equilibrium of the forces in a turning pair. The pair is usually represented, however, as incomplete and force-closed, in principle as in Fig. 193, for which the formula stands : $\dfrac{C_{=}^{-} C^+}{f}$.

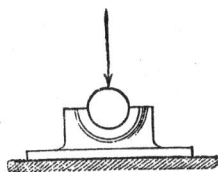

FIG. 193.

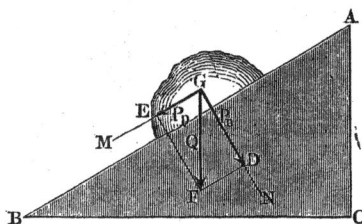

FIG. 194.

The Inclined Plane.—A surface oblique to the plane of the horizon, having a body resting upon it, touching it throughout a plane section, and tending by its weight to slide downwards (Fig. 194) ; the magnitude of the force necessary to prevent this sliding is studied. Here again the description leaves much to be wished. It is, as a rule, left unexpressed that the body can only slide parallel to the greatest slope of the plane,—that is, the necessary bodily restraint in other directions is imagined,—and means are also imagined to exist by which it is prevented from leaving the plane. In other words, it is tacitly assumed that the sliding body with the one below it are paired for rectilinear motion, and the pair under the supposed conditions is simply a

sliding-pair, written, according as one or the other element be
fixed.

$$\underline{P^{\pm}_{\underline{=}} P^-} \text{ or } \underline{P^-_{\underline{=}} P^+}.$$

The complete "principle of the inclined plane" gives the conditions of equilibrium of the forces in a sliding pair. The common
representations show, as in the last case, an incomplete, force-
closed pair, which would be written $\underline{P^-_{\underline{=}}P^+}{\Large /}{f}$,

The Wedge.—This arrangement is commonly represented in a
very primitive form, and one almost entirely wanting in the strictness of machinal motion, namely, as a means for splitting a piece
of wood, Fig. 195. In this very rural-looking apparatus the ratio
of the driving effort to the lateral resistances against the sides of
the wedge is investigated. If we complete the description—which

 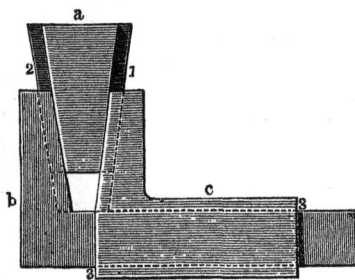 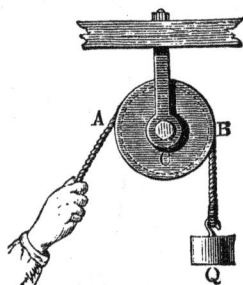

FIG. 195.   FIG. 196.   FIG. 197.

as a rule is so entirely wanting in definiteness—sufficiently to make
it applicable to a machinal system, we may say shortly, that the
two sides of the wedge are imagined to be prismatically paired
with surfaces against which they work,—and further, that the latter
(the halves of the tree stem), having, as they are separated, relatively a rectilinear motion, are also imagined to be paired in the
same way. The whole represents a mechanism formed of a three-
linked prism chain (Fig. 196), of which the formula, taking the
links in the order a, b, c, would be :

$$P^+ \ldots \angle \ldots \underline{P^{\pm}_{\underline{=}} P^-} \ldots \angle \ldots \underline{P^{\pm}_{\underline{=}} P^-} \ldots \angle \ldots \underline{P^-_{\underline{=}}}.*$$

* More strictly, as the chain is force-closed throughout, this should be,

$$\frac{P^+}{f} \ldots \angle \ldots \frac{P^{\pm}_{\underline{=}} P^-}{f \quad f} \ldots \angle \ldots \frac{P^{\pm}_{\underline{=}}}{f} \underline{P^- \ldots \angle \ldots P^-_{\underline{=}}} \text{ or } \left(\frac{P^L_{\underline{3}}}{f}\right)^e_a.$$

The "principle of the wedge," if it be expressed in a sufficiently general form, gives the conditions of equilibrium of the forces in this chain. The traditional representation stands for a combination of bodies, force-closed throughout, which only roughly approximates to the combination really intended.

The Pulley.—A disc turning about a fixed pin, and having a grooved periphery over which rests a rope stretched at both ends (Fig. 197); the equilibrium of the forces acting at the ends of the rope and upon the pin is studied. The pulley takes a remarkable position among the simple machines. In the first place, we have here not two but three bodies used in combination. As a rule no mention is made of the assumption that the bearings of the pulley are supposed to be such as to prevent cross motions. Then again it is remarkable that while here a force-closed element, the rope, is employed, there is very insufficient recognition of its characteristic property of one-sided resistance. If the bracket for the pulley-spindle be considered as fixed, the kinematic formula for the chain is as follows :—

$$C^\pm_= C^- \dots \mid \dots R^+, T \dots \left\{ \begin{array}{c} \dots \dfrac{T^-}{f} \\ \dots \dfrac{T^+}{f} \end{array} \right.$$

a mechanism of three links covering very indefinite motions, which approximate to machinal strictness only in consequence of force-closure.

The mechanism is commonly known as the fixed pulley, but under the head "pulley" another arrangement, the loose pulley (Fig. 198), is usually treated. Here the pulley frame is movable and loaded, and one end of the rope fixed, as in the formula.

$$\dfrac{C^\pm_=}{f} C^- \dots \mid \dots R^+, T \dots \left\{ \begin{array}{c} \dots \dfrac{T^-}{f} \\ \dots \underline{T^+} \end{array} \right.$$

This expression differs from the former only in the link which is fixed. The old mechanicians have busied themselves with the inversion of a kinematic chain! In the loose pulley also force-closure is applied to the fullest extent.

The Wheel and Axle.—Two drums of different diameters
fixed together and having a common shaft, each having one end of
a rope which is loaded at the other fixed to it; the shaft works in
fixed bearings, or at least is imagined to do so, for the bearings are
often enough omitted in the drawing (Fig. 199); the equilibrium
of the forces is studied.   Again the problem is wanting in
clearness, and is only solved by the employment of a number of

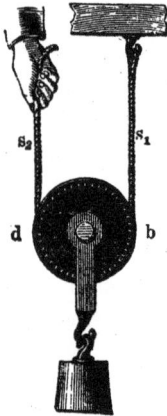

FIG. 198.

FIG. 199.

abstract assumptions, for the most part not expressed.   Supposing
the axle bearings fixed the chain runs

$$C_{=}^{-} \, C^{+} \, \ldots \, | \, \ldots \, \begin{cases} \ldots R^{+}, \, \dfrac{T^{-}}{f} \\ \\ \ldots R^{+}, \, \dfrac{T^{+}}{f} \end{cases} .$$

All the indefiniteness which we saw existing in the assumptions
in the former cases exists also here.   Indeed they are increased by
the helical winding off and on of the cords, which occurs, too, so
that their axes must describe higher screw-lines if artificial means
of preventing it are not supposed to exist, or if the difficulty be not
got over by the supposition of infinitely thin cords.   This last is
very common.   The sense of the necessity for eliminating these
complicated motions of the cord has led many to omit it altogether,
replacing it merely by tangential forces acting upon the peripheries

of the drums. This, however, makes the problem simply a re-petition of that of the lever, which was not its original meaning.

The Screw.—A screw placed vertically working in a fixed nut and loaded by a weight (Fig. 200); the force which has to be applied, normal to a radius and at some point not in the axis of the screw, in order to balance the load, is determined. We recognise at once the twisting-pair, written either

$$S_{\underline{=}}^{-} S^{+} \text{ or } S_{\underline{=}}^{\pm} S^{-}.$$

The "principle of the screw" is a very limited, indeed incomplete, case of the equilibrium of forces in a twisting-pair.

The Funicular Machine.—This, lastly, is a problem which,—apart from its value in pure Mechanics when put into an abstract form, is so far removed from the machinal idea by its extended force-closure and the indefiniteness of its motions, that it obviously has no right to a place among "simple machines," and we need not therefore consider it here.*

As a whole, the result at which we have arrived is very remarkable. We find in the simple machines, which of all others ought to appear harmoniously related, a crude mixture of kinematic problems—closed and unclosed pairs, and chains mistaken for pairs, arrangements mostly force-closed—among them

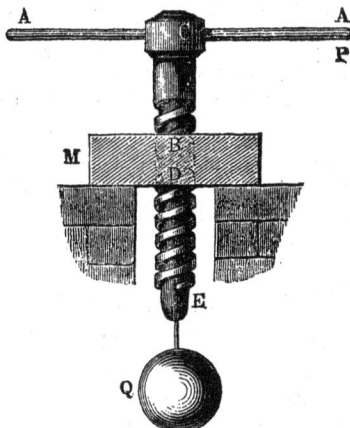

FIG. 200.

the tension-organ with all its difficulties of treatment,—and in addition an experiment in the inversion of a mechanism. We have been compelled to recognise, too, that in their usual treatment there is an extraordinary inexactness in stating the problems, which can hardly tend to give the beginner clear ideas. The explanation of

---

* The "toothed wheel" in the form in which it appears among the "mechanical powers" is really the mechanism $(C_z^+ C_2'')^c$ which is shown in Fig. 183. Precisely the same chain placed upon another link, viz. $(C_z^+ C_2'')^a$, forms an epicyclic train, which is treated not as a simple machine but as a more or less difficult case of "aggregate motion."

all this may be found in the general mode of development of machinal ideas which we have already studied, and under which we have seen the early machines to have grown up gradually from force-closed combinations of fixed and moving bodies. In the history of machine-development the simple machines formed the first experiment at a scientific arrangement of existing material; the same train of ideas which governed its phenomena as a whole repeated itself upon a smaller scale in the early attempts at the scientific explanation of what had been empirically determined.

Beyond this, we may ask further whether, when the necessary strictness of conception and definition has been obtained, the "mechanical powers" do really constitute the elementary parts of all machines? The answer must be most distinctly negative.

Three of the simple machines indeed, stripped of their conventional disguise, are no other than the three lower pairs (R), (P) and (S),—and another the higher pair R, T; but all the other higher pairs are wanting, while there is no representative of the pressure organs, not to speak of the springs. With steam-engines and pumps—the triumphs of pressure organs—before us, how is it possible to assert that the traditional simple machines have formed the foundation for all others? It seems scarcely conceivable that this should ever have been said. It has been so far modified as to be replaced by the statement that all the static problems of machinery were contained in the simple machines, and that it was this that gave them their importance and formed the real connection between them. This also, however, is incorrect. The "principle of the lever" does not teach the relations among forces in the higher cylinder-pairs—for that purpose we have to go back to the infinitely small instantaneous motions—nor in the hyperboloidic pair. There are many dynamic problems in machinery of which the simple machines teach us nothing. In themselves they teach nothing of couples, and they leave entirely without notice the application of fluid-organs as elements in machinery, although they recognise their contra-positives the tension-organs. In short, the assertion that all machines can be traced back to those which have received the name of "simple" is justified from no point of view whatever.

We can now well understand the increasing fear of recognising the simple machines, in spite of their historical position, which appears in modern text-books; and we see also the reason of the neglectful treatment they have received from the higher mechanics, but our investigations have shown us something which helps to explain the attachment to these old and well preserved problems. This no doubt rests chiefly upon the fact that three of them, the lever, inclined-plane and screw, represent pairs of elements,— perhaps also upon the existence in another, the pulley, of a timid step towards a free and exhaustive treatment of a kinematic chain.

It was therefore in the first place an indistinct feeling that the motions of a machine were founded upon those of pairs of bodies, which led to the " simple machines." In point of fact they have, as it were, felt the way in this direction. It is this that has allowed the lever, inclined-plane and screw—to which we arrived by a priori reasoning as the three lower pairs (§15)—to take such deep root. The faint trace of the law of the kinematic chain which appears in the two forms of pulley is both interesting and striking—only to this extent do the venerable problems seem justified. I think, however, that our examination of them has shown that this whole department of elementary Mechanics, whether treated by itself or as a part of Physics,—in text-books or orally,—absolutely requires a very searching revision.

§ 65.

The Quadric (Cylindric) Crank Chain (C_4'').

The kinematic chain which consists of four links connected by parallel cylinder pairs, and which has already repeatedly engaged our attention, is one of the most important chains occurring in practical machine-construction, and we shall now proceed to its analysis. Its complete treatment belongs to applied and not to theoretic Kinematics; our purpose here is not its exhaustive treatment, but simply the examination of the various forms in which it is applied as a mechanism. We shall find that they have very great variety.

We may look first at the train already described in § 62 and shown in Fig. 201,—where the four links are so proportioned that, *d* being fixed, *a* can revolve while *c* swings about its axis. For this we must always have the conditions

$$a + b + c \geqq d \qquad a + d + c \geqq b$$
$$a + b - c \leqq b \qquad a + b - c \leqq b$$

and *a* the smallest of the four links; the letters here standing for the lengths of the links between the centres of the pins.

The parallelism of the cylinder pairs makes all the centroids plane figures, and all the axoids cylinders. In its applied forms the link *a* is always known as a **crank**, and from this we may call the chain a cylindric crank-quadrilateral, or, more concisely, a

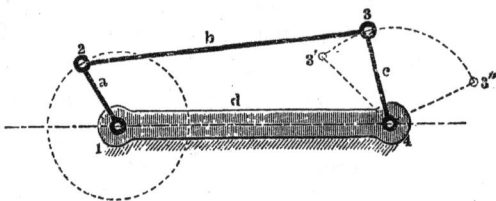

FIG. 201.

quadric (cylindric) crank-chain. The mechanisms obtained by fixing one or other of the links will then be called quadric (cylindric) crank mechanisms or trains. The designation cylindric requires to be retained, as we shall presently become acquainted with crank mechanisms of another kind. The mechanisms occurring are four in number, their contracted formulæ being $(C_4'')^d$, $(C_4'')^b$, $(C_4'')^a$, $(C_4'')^c$. We may take them briefly in order.

The mechanism $(C_4'')^d$. We have met with this mechanism often enough to be now tolerably familiar with it. Its links possess such totally distinct functions that we may venture to use for them distinct names, this will enable us make our descriptions shorter and more exact. We shall call

a the crank	*c* the lever
b the coupler	*d* the frame.*

The characteristic of the mechanism (Fig. 201) is that it has both a crank and a lever among its moving links, the one turning while

---

* I propose to distinguish between links that can turn completely round their centres and those that can only swing to and fro by calling them cranks and leve:s respectively. I do not think this will lead to any confusion, and it often greatly simplifies the nomenclature of the trains, as will be seen further on. For brevity's sake I have used coupler instead of the more common but much longer name c o n - n e c t i n g r o d.

the other swings; from this peculiarity we may call it a lever-and-crank train, or simply a lever-crank.

The mechanism $(C_4'')^b$. If we now place the chain on b, that is, release the frame d and fix instead of it the coupler b (Fig. 202) we obtain a mechanism in which a and c again turn and swing respec-

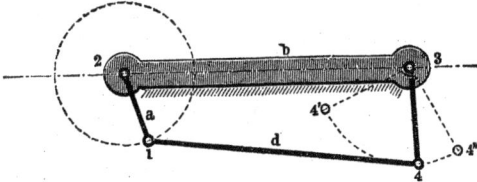

Fig. 202.

tively, but now about the centres 2 and 3 instead of 1 and 4. The frame d has become the coupler, and the coupler b the frame. The whole is still a lever-crank, and differs from the former only in the relative lengths of the coupler and frame. There is therefore no difference in kind between the two mechanisms, and we have $(C'')^d = (C_4'')^b$.

The mechanism $(C_4'')^a$. If the link a be made the frame, Fig. 203, we obtain the entirely different mechanism, one which we have previously examined in § 9. The links b and d rotate about the axes 2 and 1,—that is, they become cranks,—c, on the other hand, becomes the coupler. The mechanism is known in practice as a drag-link coupling, we shall call it the double-crank. The cranks move with varying angular velocity ratio in a way which we were able to represent conveniently by the aid of reduced centroids in Fig. 25.

The mechanism $(C_4'')^c$. In this last arrangement the links b and d swing to right and left about their axes 3 and 4; c has become the frame, and a the coupler. In the position 4 1′ 2 3 shown in dotted lines, b has completed its swing to the right; as it returns, however, d can move somewhat further to the right and then will swing in the same direction until at 1″ it reaches the left limit of its travel. As it returns b in its turn moves further to the left and then returns as d did before:—4 1‴ 2‴ 3 shows an intermediate position with the links crossed. We may call this mechanism,—which is frequently used in the parallel motions of machinery, but then not to the limits of its motion,—the double

lever.   This will indicate its relation to the mechanism $(C_4'')^a$, in
which the arms b and d turn instead of swinging.[*]

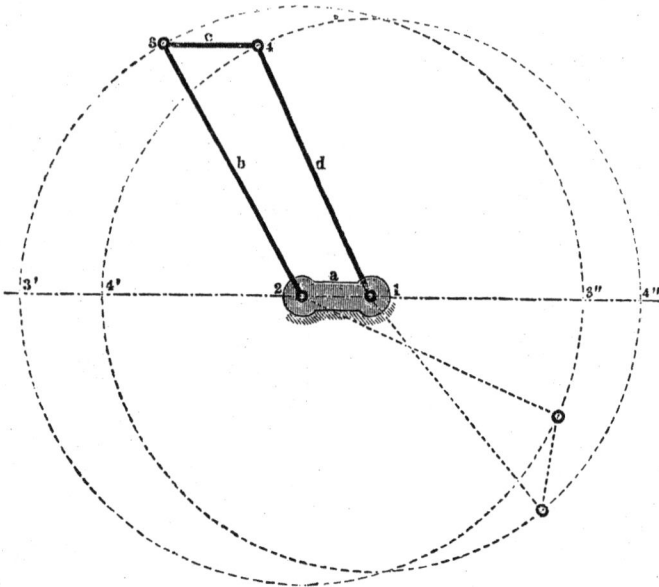

FIG. 203.

Here we have exhausted the methods of placing the chain
(C_4''), and have found that three out of the four mechanisms

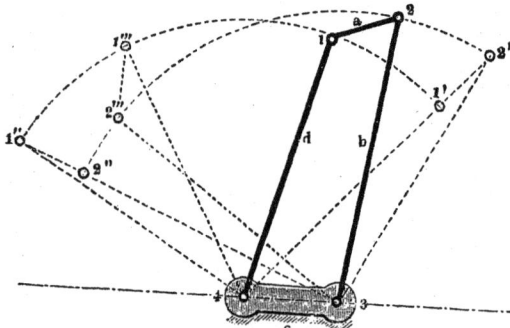

FIG. 204.

[*] Prof. Reuleaux uses " Revolving double crank " and " Oscillating double crank "
for $(C_4'')^a$ and (C_4'') ᶜ respectively.   By using the words crank and lever, as I have
proposed, we can thus greatly shorten the names without, I think, making them indefinite.

belong to different classes. The three different kinds of motion obtained are, as we know, simply those relative motions in the chain which we have made absolute, or more strictly speaking absolute " for us," by fixing one or other of the links (see § 3.) The most frequently used of the four mechanisms is $(C_4)^d = (C_4'')^b$; or putting the two formulæ together, $(C_4'')^{d=b}$.

§ 66.

Parallel Cranks.

It is obvious that by altering the relative lengths of the links in the chain (C_4'') we alter the mechanisms to be obtained from it, and therefore the resulting motions,—for by extending the angle of oscillation we can convert relative swinging into rotation and *vice versâ*. We shall consider the most important special cases which arise here. In the original mechanism we had $a < c$, if the difference between them be reduced until $a = c$, and if at the same time b be made $= d$ the crank chain becomes a parallelogram, as Fig. 205. The lever c becomes a crank equal to a, and (d being fixed) it moves always through the same angle.

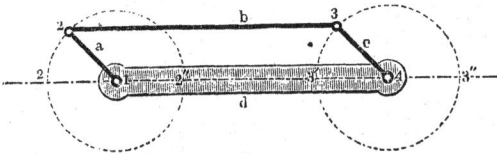

FIG. 205.

The contracted symbol for the chain, the opposite links being always parallel, is $(C_2'' \| C_2'')$. It is unnecessary to use the sign $\#$, for the $\|$ is by itself sufficient to exclude the crossing which, as far as the construction of the chain itself is concerned, is possible (§ 47). The sign of equality, on the other hand, would not be sufficient by itself, for the equality of pairs might be $a = b$ and $c = d$, which would allow $a <$ or $> c$, and would therefore be inconsistent with our conditions. The sign $\#$ may be reserved for the case where the parallelogram is a rhombus.

If the mechanism be placed on d, as in Fig. 205, its formula runs $(C_2'' \parallel C_2'')^d$. It falls into the same class whether it be placed on b or c or a; so that all the four mechanisms with which the chain furnishes us are similar. We shall call them **Parallel Cranks.**

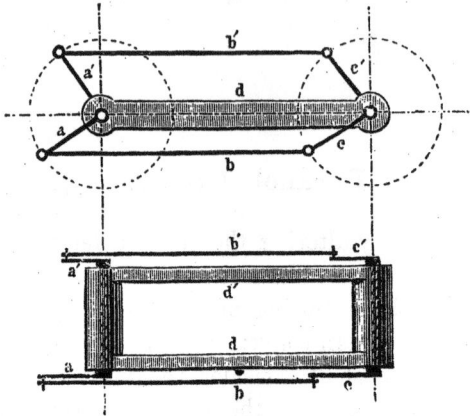

Fig. 206.

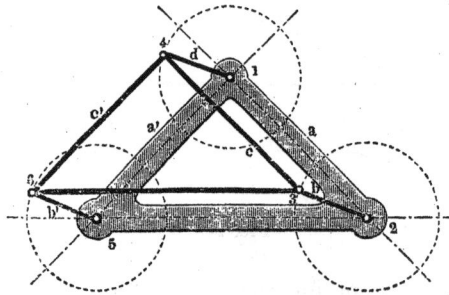

Fig 207

We have already seen that in the dead positions $2'$ 1 $3'$ 4 and 1 $2''$ 4 $3''$ the chain is not constrainedly closed. If then it is to be used so that the points $2'$ and $2''$ can be passed some special closure must be arranged. We have found (§ 46) that this could be done by the addition of another similar chain in the two ways, among others, shown in Figs. 206 and 207. We have now to find means for indicating these in our kinematic notation.

We have here chain-closure. It may therefore be indicated, as mentioned in § 57, by placing the sign k as a divisor below the original formula, so that both chains could be written $\dfrac{(C_2'' \parallel C_2'')}{k}$. But the addition to the k of the sign of equality, and the inclosure of both in brackets will allow us to make distinct that the closing chain is equal to the one closed. The formula would then be $\dfrac{(C_2'' \parallel C_2'')}{(k =)}$, or in words: a pair of parallel cranks closed by another pair of parallel cranks. We may, however, choose a still more convenient way of indicating the combination of two chains which are both equal and reciprocally closing, namely, by adding the factor 2 to the formula for the single chain: $2\,(C_2'' \parallel C_2'')$.

There is, lastly, another doubtful point to make clear,—the difference between the two arrangements of Figs. 206 and 207. In the first case the cranks of the closing chain are rigidly connected into links with those of the other;—in the second, one of the cranks of the closing chain appears to be identical with one of those of the primary chain, the other being separately constructed but connected by a coupler also with the second primary crank. If however we compare the two chains more carefully, and in their most abstract forms,—so as to see distinctly what is actually before us,—we find that the two chains (not the mechanisms) are identical. The ternary links $a\,a'$ and $c\,c'$ of the chain Fig. 206 correspond to the ternary links $a\,a'$ and $c\,c'$ of Fig. 207,—and the binary links d, b and b' of the first to those similarly lettered in the second. If then, as the figures indicate, the chains be made into mechanisms by placing them upon $d\,d'$ and $a\,a'$ respectively, the second is nothing more than an inversion of the first, so that the difference between the two will be indicated in the general formulæ, $2\,(C_2'' \parallel C_2'')^d$ for Fig. 206, and $2\,(C_2'' \parallel C_2'')^a$ for Fig. 207. They are both formed from the same five-linked chain, and they are examples of the only two classes of mechanisms into which this chain can be formed.

§ 67.

Anti-parallel Cranks.

By means of pair-closure we can, as we have already seen in § 47, convert the crank parallelogram into an anti-parallelogram,[47] and this can be so constrained as to retain its special property in every position. Figs. 208 and 209 represent the two forms of this chain, pair-closed, which we have already considered. We may call the mechanisms to be formed from it anti-parallel

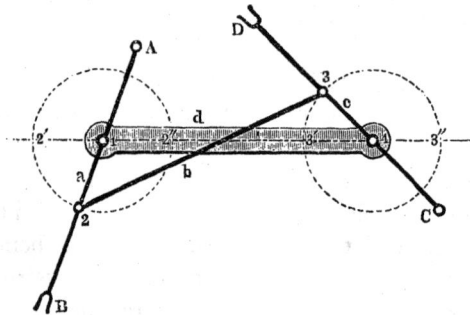

FIG. 208.

crank trains: Two different results can be obtained by the different modes of placing the chain, one if it be placed on d or b, the other if a or c be the fixed link. If d be fixed, as is supposed in both the figures, the two cranks turn in opposite directions, or reversely, for which reason I have already given the mechanism the name of reverse cranks (§ 47, Fig. 155). If, however, the chain be placed on a (Fig. 210), so that c becomes the coupler and the former coupler and frame both become cranks, then b and d both revolve continuously, but in the same direction, or we may say conversely. In the first case the anti-parallelogram gives us reverse anti-parallel cranks, in the second converse anti-parallel cranks. It should be noticed that the nature of the relative rotations is the same in both cases. This arises from the equality, — through the anti-parallelism, of the angles 123 and 143. We may therefore use a pair of congruent ellipses as reduced centroids (§ 9) for the

hyperbolæ (§ 47), the form of which necessarily makes it somewhat difficult to realise the motions they represent.

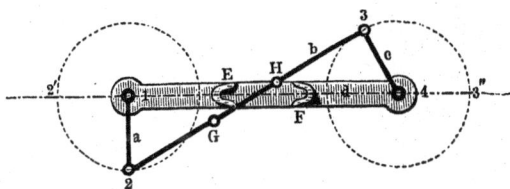

FIG. 209.

The contracted formulæ for these mechanisms must, in the first
place, make their characteristic property of anti-parallelism clear,
—we therefore put its symbol between those of the cylinder

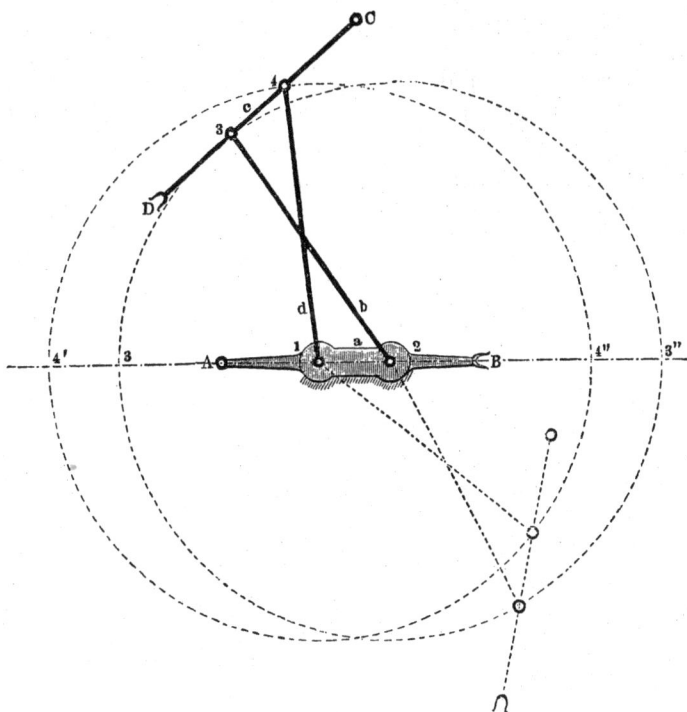

FIG. 210.

pairs. The chain, unfixed, will then be written $(C_2'' \mathrel{Z} C_2''')$. The
reverse anti-parallel cranks will be $(C_2'' \mathrel{Z} C_2'')^d$ or $(C_2'' \mathrel{Z} C_2'')^b$,
of which formulæ we need use only one,—let it be the former,—

unless we wish to combine the two expressions in $(C_2'' \gtrless C_2'')^{\mathrm{d}=\mathrm{b}}$. The converse anti-parallel cranks will be $(C_2'' \gtrless C_2'')^{\mathrm{a}}$, if in the same way we omit the exponent c as superfluous, or $(C_2'' \gtrless C_2'')^{\mathrm{a}=\mathrm{c}}$ if we wish to express the fact that the chain placed either on a or on c gives the same mechanism.

The pair-closure has still to be indicated. This will only be necessary if the action of the mechanism extends over the dead points. If the closure exists, and if it be arranged as in Fig. 208, the formula will run $\dfrac{(C_2'' \gtrless C_2'')^{\mathrm{d}}}{(p)\,a.c}$; if as in Fig. 209, $\dfrac{(C_2'' \gtrless C_2'')^{\mathrm{d}}}{(p)\,b.d}$; where the existence of the pair-closure is denoted by p (see end of § 57), while the brackets and the addition of the symbols for the paired links sufficiently indicate the rest. It will frequently, however, be unnecessary specially to indicate the pair-closure, for the maintenance of the anti-parallelism,—the assumption, that is, of the continued validity of the sign $\gtrless$,—presupposes it. The anti-parallel cranks have here and there been used, but without being recognised; Dübs's locomotive coupling is an instance, and here the ellipses actually serve as profiles for the buffers.[*]

§ 68.

The Isosceles Crank-train.

We obtain a special case of the chain (C_4'') which has very great theoretical interest if we make $a = d$, $b = c$, and, as before, $a < c$. We have already described (§ 47) the pair-closure in a mechanism formed from this chain. Figs. 211 and 212 represent the mechanism first without, and then with, the pair-closure. A diagonal joining the points 2 and 4 of the quadrilateral divides it always into two isosceles triangles, for which reason we shall call the train Isosceles. The writing of the chain is easy after the foregoing; the formula must be,—using the symbol for isosceles given in § 47 ($C_2'' \lesseqgtr C_2''$). If the higher pairing of Fig. 212 have to be expressed this becomes $\dfrac{(C_2'' \lesseqgtr C_2'')}{(p)\,a.c}$. As with the anti-parallel cranks, the higher pairing may here be arranged between

---

[*] Dübs and Copestake's patent coupling was illustrated and described in *Engineering*, vol. xi. p. 318.

d and b instead of between a and c;—or the pair-closure may be partly between a and c and partly between d and b, but this gives us no new results.

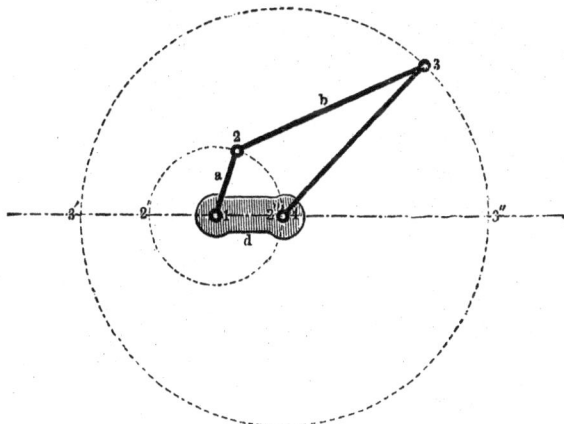

The chain gives us two kinds of mechanisms, one by placing it on d or a, the other by placing it on c or b.

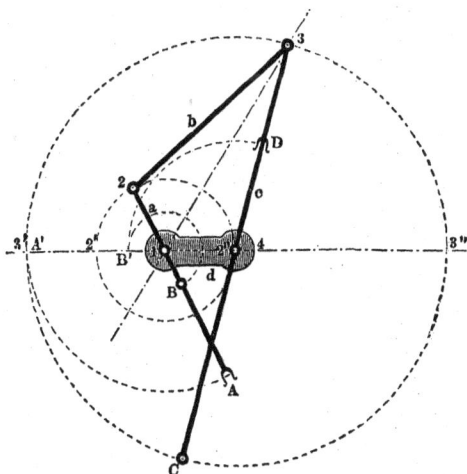

The mechanism of Fig. 211, which is placed on d, has the formula $(C''_2 \lessgtr C''_2)^d$. The motion of c is remarkable, for it now not merely swings but completely revolves,—and it has a mean angular velocity equal to half that of a, as we have already seen in § 47.   By fixing c (or b) we obtain the mechanism shown in Fig. 213, for which the formula (including the expression for the pair-closure) is $\dfrac{(C_2'' \lessgtr C_2'')^\circ}{(p)\,a.c}$. Its motion is no less characteristic than that of the first mechanism.   It is in some respects similar to that of the lever-crank $(C_4'')^\mathrm{d}$.   The link d has become the crank, and a the coupler; b however swings about its axis 3

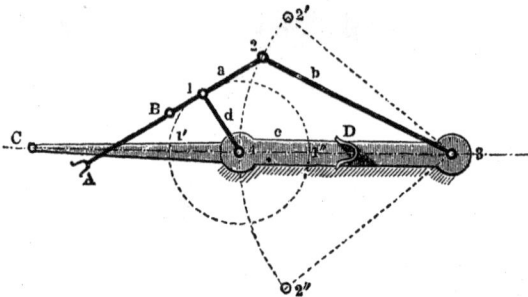

Fig. 213

symmetrically to c through such an angle that the greatest distance of 2 from 4, when the former is in either of the positions 2′ or 2″, is equal to $2a$.   The points 2′ and 2″ therefore are nearly four crank lengths apart, while in the mechanism $(C_4'')^\mathrm{d}$ the end 3 of the lever oscillates through a distance which approximates only to two crank lengths.   We shall further on have occasion to return to this interesting case.

§ 69.

The Cylindric Slider-crank Chain $(C_3'' P^\perp)$.

Continuing our examination of the chain (C_4'') let us now somewhat alter its form.   We can substitute for the lever c a small sector of an annular cylinder, and inclose this in a circular slot, (Fig. 214) rigidly connected with the eye at 1.   If the centre of the slot and of the sector c be placed at a distance from 1 equal to the distance 1.4 in the former case, the sector has exactly the same relative motions as it would have had had it been connected to the lever c. We may therefore allow it to take the place of the latter :—we shall find further on that kinematically the two are identical. The new arrangement may be written, beginning with a,

$$C^+ \dots \parallel \dots C_{\underline{=}}^{\pm}\, C^- \dots \parallel \dots C_{\underline{=}}^-\, C^+ \dots \parallel \dots A_{\underline{=}}^{\pm}\, A^- \dots \parallel \dots C_{\underline{=}}$$

for we have already chosen (§ 57) the symbol A for a circular sector. The contracted formula is $(C_3'' A'')$. This shows even more distinctly than in the former case that the links must be so proportioned that c slides backwards and forwards in its curved path ; for otherwise the pair $A_{\underline{=}}^{\pm} A^-$ would be insufficient.

We can now, without introducing any constructive difficulties,

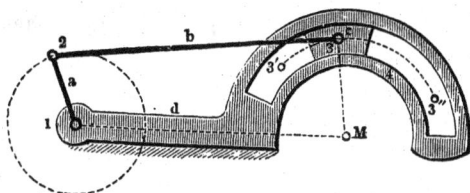

make the radius of A of any required magnitude ; the only alteration will be that the slot and the slider become flatter than before. Let us therefore make this radius infinite. With this the distance of the centre 4 from the point 1, that is the length of the link d, must also become infinite. In other words the links c and d, or the distances 3.4 and 1.4, are made infinite simultaneously ; so that

$$c = d = \infty .$$

Our last formula will then require alteration, for the arc A becomes a prism P, and the pair $A_{\underline{=}}^{\pm} A^-$ is replaced by the prism-pair $P_{\underline{=}}^{\pm} P^-$. It follows from the equality of c and d that the line in which 3 moves relatively to d passes through the point 1, and is perpendicular to both the axes 3 and 1. The new chain, therefore, which is shown in the following figure, and which is already known to us, must be written

$$C^+ \dots \parallel \dots C_{\underline{=}}^{\pm}\, C^- \dots \parallel \dots C_{\underline{=}}^-\, C^+ \dots \perp \dots P_{\underline{=}}^{\pm}\, P^- \dots \perp \dots C_{\underline{=}}^-$$

or more shortly,

$$C^+ \ \dots \ \| \ \dots \ (C) \ \dots \ \| \ \dots \ (C) \ \dots \ \perp \ \dots \ (P) \ \dots \ \perp \ \dots \ C_=^-$$

or in its contracted form $(C_3'' P^\perp)$, In this most important chain
the link c slides along a straight line instead of swinging in an

arc as in (C_4''). We may call it shortly the cylindric
slider-crank chain,—or simply the slider chain.
We have now to examine the four mechanisms corresponding to its four positions.

The mechanism $(C_3'' P^\perp)^d$. If we place the chain on
d, as in Fig. 216, the link c slides backwards and
forwards as the crank rotates, and we have before us
one of the most familiar of mechanisms, one which
appears constantly in direct acting steam-engines, in
pumps, and in slotting and so many other machines.
The link c we shall call the block, and the link d
the slide, when we have occasion to name them. The
whole mechanism we may call a turning slider-
crank, on account of the characteristic rotation of a.
In its applications to the steam-engine the block c

Fig. 215.   becomes the driving link, so that the general formula

$(C_3'' P^\perp)^d$ gives us the special formula $(C_3'' P^\perp)_c^{\frac{d}{c}}$. In
the other applications of it which we mentioned the crank a is the
driver,—their special formula is therefore $(C_3'' P^\perp)_a^{\frac{d}{a}}$. The complex
motion of the coupler b can be exactly determined by its centroids,

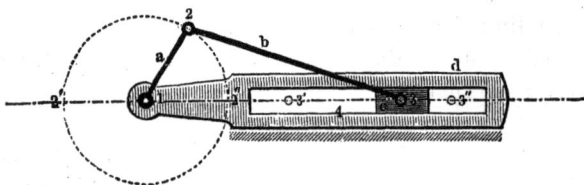

Fig. 216.

but these we must here leave unexamined, merely noticing that
they are symmetrical about the axis 3·1.

The mechanism $(C_3'' P^\perp)^b$. Following the order formerly adopted
let us now place the chain on b, Fig. 217.* The crank a now

* In Prof. Reuleaux's models for the $(C_3'' P^\perp)$ mechanisms he uses the stand with
screw adjustment which is shown in Figs. 11 and 180.

turns about 2, which was formerly the crank-pin; the block *c* oscillates about the point 3, and causes the slide to turn about the same point in addition to following the motion of the crank. We shall call the mechanism a swinging-block (slider-crank). It will be remembered that we can reverse any of the lower pairs (§ 16) without altering their relative motions; by choosing the arrangement of slide and block shown in Fig. 218, therefore, we do not alter the mechanism. In this form it is exceedingly well known, although not so constantly employed as $(C_3'' P^\perp)^d$. A

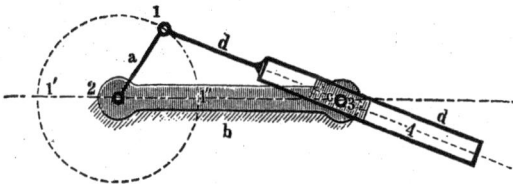

FIG. 217.

familiar illustration is the oscillating engine, of which Fig. 218 at once reminds us. Here the slide *d*, in the shape of the piston, is the driving link, the special formula being therefore $(C_3'' P^\perp)^{\frac{b}{d}}$. There have been various attempts to elucidate the connection between the mechanisms of the oscillating and the common direct-acting steam-engine, the explanations being generally founded on some process of altering the relative dimensions of their parts.

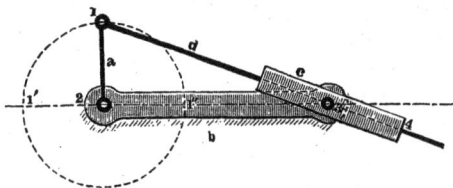

FIG. 218.

It has been said, for instance, that the former is simply the direct-acting engine with the length of its connecting rod reduced to zero, and at the same time, in order that motion may be possible, with its cylinder made so that it can oscillate about an axis. There is here obviously something more than a mere alteration of dimensions, and the whole process remains indistinct. We see now how entirely different and at the same time how completely clear the connection between the two is ;—that the whole matter

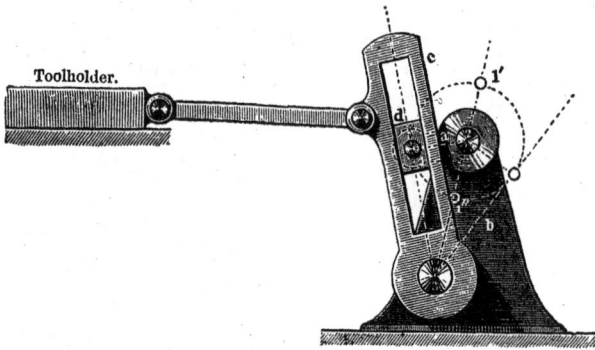

FIG. 219.

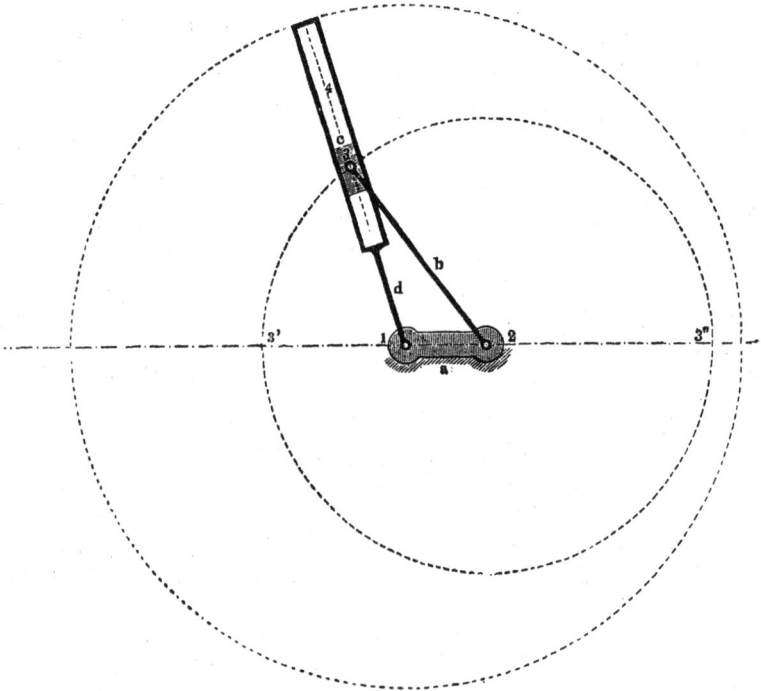

FIG. 220.

lies simply in the inversion of the kinematic chain which forms
the basis of both equally. On the difference between the two

steam-engines themselves we shall have more to say further
on (§ 80).

The mechanism $(C_3'' P{\perp})^b$ in the form $(C_3'' P{\perp})^{\frac{b}{a}}$ has found
another application in shaping and slotting machines. In this
the peculiar motion of the link c,—swinging un-uniformly while
the crank rotates uniformly,—is made use of. The crank a is
the driver,—during the semi-revolution through 1′ it imparts to
the block a much smaller mean angular velocity than during the
semi-revolution through 1″. By connecting the link c with the
holder of a cutting tool, as Fig. 219, we can therefore obtain a
slow (mean) forward motion of the tool while cutting, and a
quick (mean) return. Mechanisms of this kind are known as
" quick return " motions. The mechanism $(C_3'' P{\perp})^b$ is known
and familiar to the engineer in this and other ways.

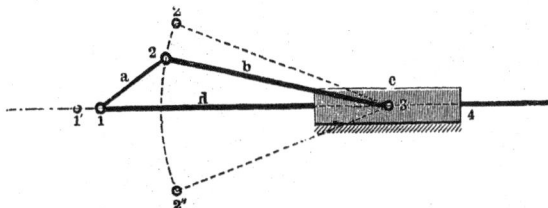

FIG. 221.

The mechanism $(C_3'' P{\perp})^a$. By placing the chain on a we obtain
a third mechanism, Fig. 220. The link b which was the coupler
now revolves about the axis 2,—it has, that is, become a crank ;
the crank a becomes the frame. The slide d, driven by the block
c, turns about the axis 1. Its rotation, if the crank b turn
uniformly, is un-uniform, the latter imparting to it at 3‴ a
minimum and at 3′ a maximum velocity. On account of this
property Whitworth and others have used it in the form $(C_3'' P{\perp})^{\frac{a}{b}}$
as a quick return. According to Goodeve* the mechanism is an
old one, and was long ago used to represent the motion of the moon
relatively to the earth. We shall call it the turning-block
(slider-crank).

The mechanism $(C_3'' P{\perp})^c$. The fourth and last mechanism is
obtained by fixing the block c instead of the link a, Fig. 221. The
coupler b now swings about the fixed axis 3,—the slide d moves
rectilinearly to and fro in the block c, now become the frame,—

* Goodeve, *Elements of Mechanism*, p. 68.

and the crank a becomes a coupler and makes complex oscillations. We shall call the mechanism, on account of the swinging of the link b, a swinging slider-crank. This mechanism is little known, but does here and there find applications. Among others there is the apparatus sketched in Fig. 222, which is used in disc-polishing machines in order to give the polishing wheel a to-and-fro motion axially along with its rotation. The train is set in motion by the link a (by means of the worm), and its special formula is therefore $(C_3'' P^\perp)_{\dot{a}}^{\frac{c}{\bullet}}$. I have elsewhere described* another application of this mechanism in the same form, and I shall come further on to another very notable one.

Our analysis has shown that four mechanisms can be obtained from the chain $(C_3'' P^\perp)$, of which the first is extremely familiar, the last very little known;—the connection between them, however,

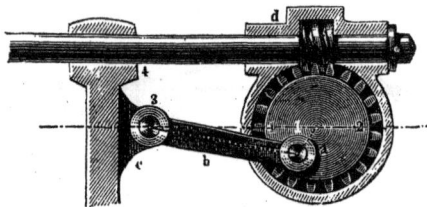

Fɪɢ. 222.

has remained until now completely unseen. At the same time we see that we have exhausted the chain which is before us;—we know that no more mechanisms than these four can be formed out of it. If we now put these together and consider them once more, we shall be able to recognise, by the help of their formulæ, still closer relationships between them. With this object let us write the formulæ at length, one above the other, so that they may be easily compared. We then have:

$$
\begin{array}{cccc}
& a & b & c & d \\
(C_3'' P^\perp)^d = & \overbrace{C^+} \ldots \| \ldots (C) \ldots \| \ldots (C) \ldots \perp \ldots \overbrace{(P) \ldots \perp \ldots C_{=}^{-}} \\
(C_3'' P^\perp)^b = & C^+ \ldots \| \ldots \underline{(C)} \ldots \| \ldots (C) \ldots \perp \ldots (P) \ldots \perp \ldots C_{=}^{-} \\
(C_3'' P^\perp)^a = & \underline{C^+} \ldots \| \ldots (C) \ldots \| \ldots (C) \ldots \perp \ldots (P) \ldots \perp \ldots C_{=}^{-} \\
(C_3'' P^\perp)^c = & C^+ \ldots \| \ldots (C) \ldots \| \ldots \underline{(C)} \ldots \perp \ldots (P) \ldots \perp \ldots C_{=}^{-}
\end{array}
$$

If we recollect that these formulæ are expressions which return upon themselves, that they can also be read, or written, from either end, we see that the second and third mechanisms have absolutely the same formula. Both are placed upon a link of the form *C*...∥...*C* having for its adjacent links one like itself and one of the form *C*...⊥...*P*, these two last being again connected by a link *C*...⊥...*P*. The difference between the mechanisms, which, as we have seen, is very great, lies solely in the ratio between the lengths of the links *a* and *b*. The names we have chosen for the mechanisms,—swinging and turning block

  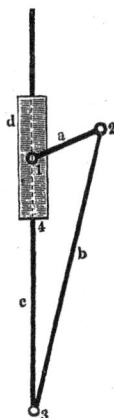 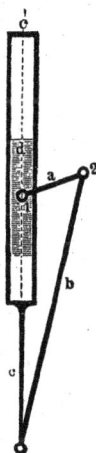

FIG. 223.   FIG. 224.   FIG. 225.   FIG. 226.

slider-cranks respectively,—give expression both to the relationship and the difference between them.

An exactly analogous connection exists between the first and the fourth mechanisms. The fixed link is in each case *C*...⊥...*P*, having on the one side a link *C*...∥ ..*C* and on the other a link *C*...⊥...*P*, these in their turn being connected by one of the form *C*...∥...*C*. Here also the difference between the mechanisms depends upon the relative lengths of *a* and *b*, and we have again employed names which indicate that relationship in calling them turning and swinging slider-cranks.

We notice lastly, what is very striking, that in all four mechanisms the two adjacent links of the form *C*...⊥...*1*, the block and the slide, are represented by exactly the same symbols,—that between them, therefore, there is absolutely no kinematic difference.

However extraordinary this may seem at first, it is perfectly true, and requires moreover to be well remembered by anyone who wishes readily to understand existing mechanisms ;—it is sufficient to cite Fig. 219 as an illustration of this. The chains which are represented in the four figures 223 to 226 are kinematically absolutely identical throughout. The external differences which appear in each case are merely due to that reversal of lower pairs which we emphasised so strongly,—it can now be seen with how good reason,—in an earlier chapter (§ 16).

§ 70.

The Isosceles Slider-crank Chain.

We have seen that the difference between the two mechanisms $(C_3'' P^{\perp})^b$ and $(C_3'' P^{\perp})^a$ is simply due to our having taken $b > a$; the difference between $(C_3'' P^{\perp})^d$ and $(C_3'' P^{\perp})^c$ is due entirely to the same cause. We must therefore obtain an intermediate form for each pair of cases if we make $a = b$;—the chain thus obtained is the one already described in § 47 and shown in Fig. 227. The links a and b are here made equal ; the links c and d are also equal, for they are the two infinite links which always form part of the chain $(C_3'' P^{\perp})$. The equal links are adjacent in each case, so that the general conditions of the chain are the same as in the isosceles crank train of Fig. 211 ; the chain before us is simply a special case of the former, and we shall therefore give it a similar name, calling it an isosceles slider-crank chain.

We have already considered its centroids in § 47. They are two pairs of Cardanic circles, the smaller being the centroids for the links a and b, the larger those for c and d. The peripheral ratio which appears here is a general property of the isosceles quadric crank chain,—we found it before where the centroids had unlike and complex forms, and we find it here also in the limiting case, which is one, as we see, of peculiar simplicity.

The four mechanisms of the slider-crank here become two only, of which the first is shown in Fig. 227. Omitting the symbols for the higher pairing, its formula will be $(C_2'' \angle C'' P^{\perp})^{d=c}$. The same mechanism is obtained whether the chain be placed on c or d, which is indicated by their equality in the exponent. We shall call it,—carrying out the system of nomenclature already adopted,—an isosceles turning slider-crank.

If the chain be placed on a or b we obtain the second of its possible mechanisms, for which the formula runs $(C_2'' \leqq C'' P^{\perp})^{a=b}$. It is represented in Fig. 228. The crank a has become the frame, the coupler b the crank. The block c transmits the rotation of the latter to the slide d, or *vice versâ*. We shall call the mechanism an isosceles turning block.*

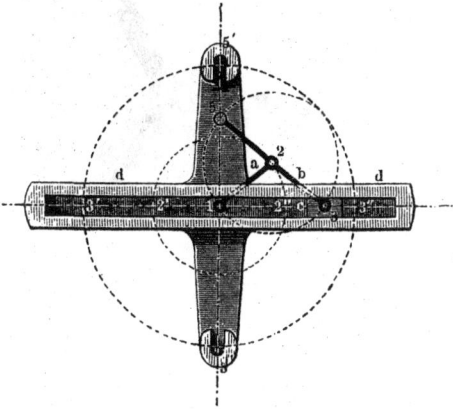 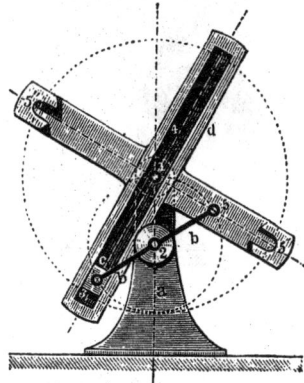

Fig. 227.   Fig. 228.

The links b and d both rotate, they revolve in the same direction, and have the constant angular velocity ratio $2:1$; the motion is exactly what it would be if b and d were two spur wheels having internal contact and having the ratio $1:2$ between the numbers of their teeth. In fact the toothed gearing shown in Fig. 229,—in which the smaller wheel a has two teeth with cylindrical profiles (pin-teeth),—is very similar to the mechanism before us, although it has one link less. The four-toothed wheel b corresponds to the turning slide d. The similarity becomes less apparent if we make the numbers of teeth 3 and 6, as in Fig. 230, and disappears almost entirely if other forms of teeth be used. The real relation between the mechanisms is however very obvious; they have identical centroids. The whole matter gives us an interesting illustration

* Here, as in former cases, the words "slider-crank" can be added to the designation given, should it be necessary to do so. I think that it will very seldom be required.

of the solution of one and the same kinematic problem by quite different mechanisms.

The motion of the block c (Fig. 228) is also remarkable. Its centroid relatively to a is a great Cardanic circle, described about the centre 3, and the smaller centroid with which this rolls must be

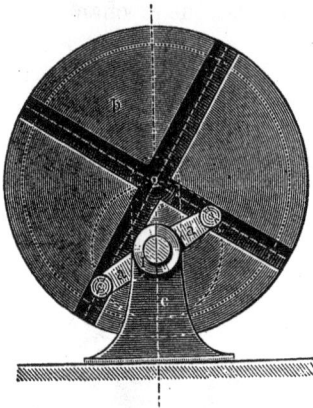

FIG. 229.

FIG. 230.

imagined to be fixed to a, and to have 2 for its centre. It therefore coincides with the circular centroid of the link b. The motion of c can thus be realised by remembering that its centroid rolls about the fixed smaller centroid of a. The point-paths of the block are therefore all peri-trochoids.

§ 71.

Expansion of Elements in the Slider-crank Chain.

We have not hitherto concerned ourselves at all with the diameter of the cylinder pairs in the crank mechanisms. We know that alterations in the dimensions of the elements do not affect their motions, so long as the centroids remain unaltered. It will be well, however, to give them some special consideration here, for these external alterations sometimes so conceal the real nature of the mechanism as to cause much indistinctness in its ordinary kinematic treatment. In considering this subject we shall confine ourselves in the first place to the changes of the relative

dimensions of the three cylinder pairs in the chain $(C_3''P^\perp)$.
The extension of our results to other cases will then be quite easy.

Each of the four links of the slider-crank chain $(C_3''P^\perp)$ Fig.
231 is more or less closely connected with its three cylinder pairs
1, 2, and 3, and their forms are therefore dependent upon the

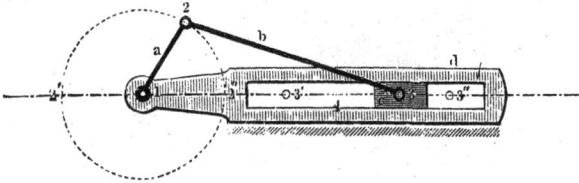

FIG. 231.

relative sizes of the latter, although, as we have said, the nature of
their motion is not affected by the same cause. Evidently, for
instance, we do not alter the chain kinematically if we give to the
full cylinder, or pin 1, on which the crank a revolves, a diameter so
large that the profile of the pin 2 falls within it. Such an enlargement, we shall call it an expansion,* of the pin is shown in Fig.

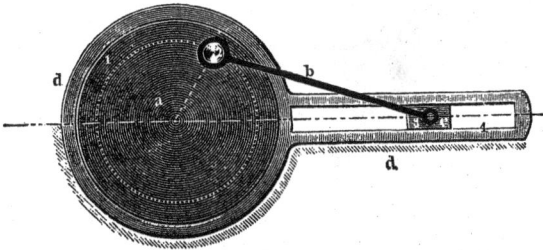

FIG. 232.

232. The open cylinder of d must now obviously be enlarged to
exactly the same extent, so that the pair may still be closed. This
arrangement, which may be shortly described as "2 within 1,"
occurs in practice in some slotting and shearing machines, and in
other cases when a short crank forms one piece with its own
shaft.†

---

* Compare the idea of expansion with that of equidistant profiles in § 35.

† It is a very common arrangement too for working a pump,—on board ship or
elsewhere, from the end of the crank-shaft.

If we expand the pin 2 instead of 1, and make it large enough to include 1 within its profile, we obtain the form of chain shown in Fig. 233. If this be placed on d and driven by a, we have the mechanism $(C_3'' P^\perp)^{\frac{d}{a}}$ in the form which is so familiar to us as an

FIG. 233.

eccentric and rod. It can be seen at once that it differs only in its constructive form from the common slider-crank. This expansion is also used in practice placed on a, so as to give us the turning block slider-crank, $(C_3'' P^\perp)^a$. Fig. 234 shows a form in which

FIG. 234.

Mr. Whitworth [*] has thus applied it, where—b being the driving link—its special formula becomes $(C_3'' P^\perp)^{\frac{a}{b}}$; it is here used as a quick-return motion, and has already been described and named by Redtenbacher.[†] The driving rod and parts connected with it do

---

[*] See Prof. Shelley's *Workshop Appliances*, p. 253.

[†] *Die Bewegungs-Mechanismen* (Basserman).

not concern us here, nor does the spur-wheel. In the body of the latter, however, we can recognise the coupler, the two elements of which are represented by the open cylinder 2 and the pin 3. The latter fits into and carries the block *c*, which in its turn moves in the open prism of the slide *d*.

If the pin 2 be further expanded until it includes also 3, we obtain the arrangement shown in Fig. 235. In this case we have made, as we are always at liberty to do, the element of the cylinder pair 2 which belongs to the crank *a* as an open figure. The

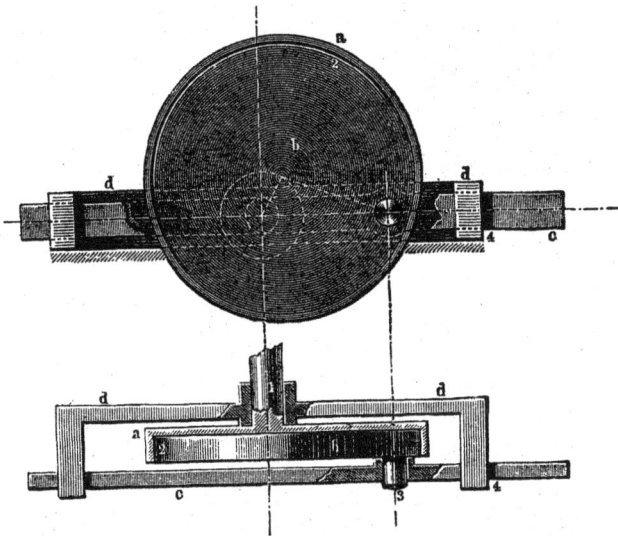

Fig. 235.

coupler *b* becomes an eccentric disc which swings about the full cylinder 3 of the block *c*, while it remains always in contact with the open disc 2 of the crank *a*.

Instead of placing 3 within 2 we may allow 2 to fall within 3, as in Fig. 236. The coupler *b* is again an eccentric disc; but it now oscillates in a ring forming part of the block *c*, while the crank pin drives it by internal contact. The reader whose eye is not yet accustomed to detect the abstract form of such mechanisms as these behind their constructive outline, and to whom therefore they may be somewhat difficult to understand, will find the dotted centre lines which we have shown to be of considerable assistance.

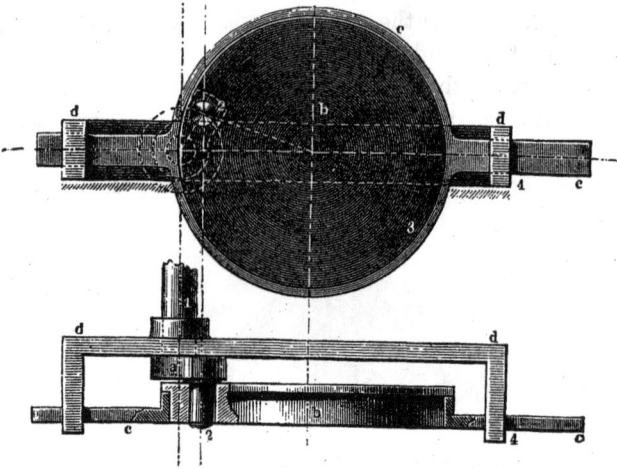

FIG. 236.

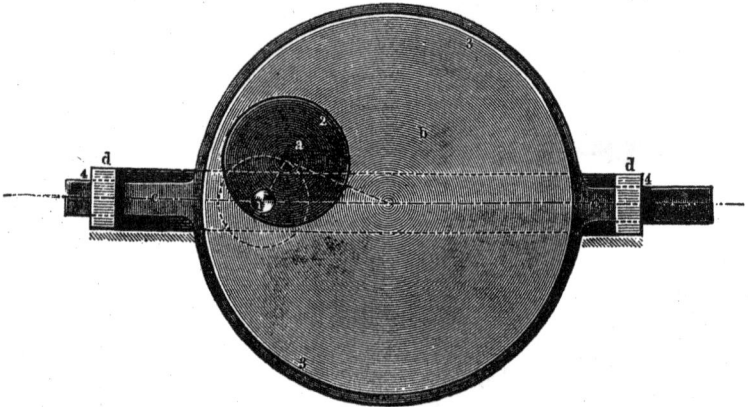

FIG. 237.

We have thus considered four methods of pin-expansion in the slider-crank chain, obtained by placing

2 within 1	(Fig. 232)	3 within 2	(Fig. 235.)
1 within 2	(Fig. 232)	2 within 3	(Fig. 236.)

We have therefore exhausted the praticable combinations of the pins 1, 2, and 3 in pairs. We may, however, go further, and make one of the pins include two others. 1 can be placed in 2, for instance, at the same time that 2 is lying in 3, so that we can place

1 within 2 within 3

and also

3 within 2 within 1.

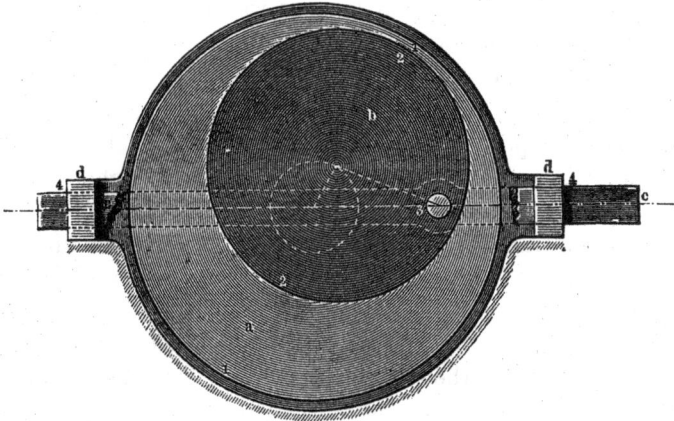

FIG. 238.

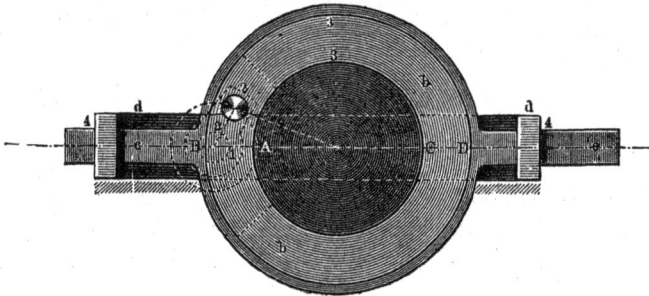

FIG. 239.

These two arrangements are shown in Figs. 237 and 238; both being placed on the frame d; both are turning slider-cranks, $(C_3'' P^\perp)^d$.

The reader may perhaps think that this idea of pin-expansion, carried so far beyond practical limits, can have but little importance in Applied Kinematics. This, however, is not the case, as we shall now proceed to show.

Turning again to the fourth method of expansion, 2 within 3, a closer examination of it shows us that the link c may be made with a concentric cylindrical projection, which can be fitted into a corresponding opening or eye in the link b, Fig. 239. Suppose the mechanism placed on d as before. The coupler b has now become a ring of rectangular cross-section which makes oscillatory motions in the annular groove of the block c. We have in no way altered the mechanism by this, for so long as we keep the pair 3 as a closed turning pair we can alter its profile at will. This condition allows us to go still further. Let us suppose the crank a to be the driver, the mechanism having for its special formula $(C''_3 P^{\perp})^{\frac{a}{a}}$, we then have simply the rectilinear reciprocation of the block c to consider. The coupler b, as it is moved to the right, drives c both at A and at D, and as it is moved to the left both at B and at C, we may certainly replace this double contact in each direction by a single one; and this can be done in several ways. We shall attain the object very conveniently if we substitute a sector of the ring b for the whole of it, choosing the sector so as to include the pin 2, as is shown by the dotted lines. This can then drive the block to the right through A, and to the left through B, the motion of the coupler itself being always an oscillation in the annular ring of c. Of the latter we require to use no more than a piece large enough to afford room on each side of the centre line for the swing of the sector b.

Fig. 240 represents the arrangement altered in this way. It must not be forgotten that b is still the coupler as it was before, and that its motion as a link in the chain remains quite unaltered and completely constrained. Kinematically it consists of just the same parts as before, as does also the link c. The form of the link b is still, $C \ldots \parallel \ldots C$, one of the cylinders being the eye enclosing the crank pin 2, while the other works, with sufficient restraint, in the portion of an open cylinder belonging to c, relatively to which it has exactly its former motion. If we wish to write the links b and c in a manner corresponding to their constructive form we must use the symbol for sector, A, instead of that for the complete cylinders in the pair 2, and we thus have :

$$\overbrace{C^- \ldots \parallel \ldots A^{\pm}_{\pm}}^{b} \; \overbrace{A^- \ldots \perp \ldots P^+}^{c}$$

This shows us that a pair of the form $C_{\pm}^{\pm}C^-$ or $C_{\pm}^-C^+$ moving only in oscillations of small angle, may be replaced by a pair of the form $A_{\pm}^{\pm}A^-$ or $A_{\pm}^-A^+$, so that in such cases

$$(C) = (A).$$

We have already (§ 69) had occasion to employ this substitution of the pair sector and curved slot, (A), for a cylinder pair. It can now be seen that we were fully justified in doing so, that the change did not in any way alter the nature of the chain.

This circumstance is very frequently taken advantage of in practice; pin expansion, that is to say, occurs there very frequently. The mechanism of Fig. 240 is both known and used, although it has not hitherto been considered identical with the turning slider-crank. It has been shown * that the block when driven by the crank moves

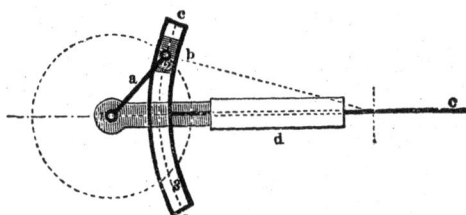

FIG. 240.

exactly in the same way that it would move were it connected with the latter by means of a connecting rod having a length equal to the radius of the slot. It has not been noticed, however, that the little sector b really is itself this connecting rod or coupler. The small space which it occupies makes this form of mechanism very convenient in some cases.† We are very familiar with its employment in reversing gear, both in links of the common form, and in Gooch's link, and others. These mechanisms are compound and not simple crank chains, so that we have not to consider them here, but the replacement of (C) by (A) occurs in them in various forms, forming an essential part of each (compare § 16).

---

* See for instance *Giulio, Cinematica*, p. 109.

† I have added in a note, p. 320, a somewhat interesting example of the use of this form of expansion in such a way as to render it impossible to alter a mechanism from $(C''_3 P\perp)$ to $(C''_3 P+)$ a thing which it is sometimes very convenient to have the means of doing.

The mechanism shown in the following figure, which occurs sometimes in slotting machines, furnishes us with another illustration of pin-expansion. The whole forms a turning slider crank having the formula $(C_3'' P \perp)^{\frac{d}{a}}$. The link b, the coupler, is formed essentially as it was in the last case, but here the profiles against which it works are concave on both sides of the pin 2, the upper profile being of large, and the lower of very small radius, but both forming part of the block c. The block c is in this case so closed by the binder d that the profiles representing the pin 3 lie entirely within the prism-pair 4. We have here, therefore, the elements of the slider crank chain so proportioned that 2 lies within 3 and 3 within 4, an illustration of the way in which the method of expansion can be applied also to the fourth pair. To show how, conversely, we may place 4 within 3 we may take the chain shown in Fig. 242. Here the pin 3 of the link c is made so large that the open prism of the pair 4 can be formed within it. Upon 3 there now oscillates the open cylinder of the coupler b, which, however, must be cut away at the sides so as to allow it to clear the frame d. The extent to which this

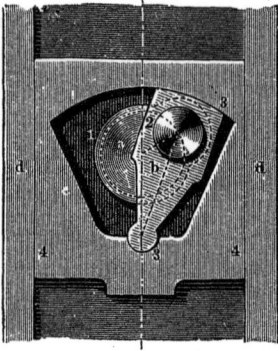

Fig. 241.

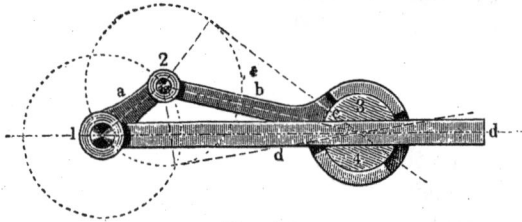

Fig. 242.

cutting away must take place can be found, as the figure shows, by drawing d in the two positions in which it encloses the greatest angles with b. These occur when a and d are at right angles. We shall have to consider some applications of this form of expansion further on.

In general the expansion of elements occasions, as we have seen, extraordinary alterations in the form of a mechanism, alterations which on the one hand tend very much to conceal its original and real nature, and on the other hand frequently offer great constructive advantages. This is true also for other mechanisms besides those we have been considering. Many familiar arrangements appear in a new and unexpected light if we replace slots and sectors by the complete cylindric forms, $C \dots \| \dots C$, which they represent; others again, by the reversed process, can be put into a form which allows of their use in practice where otherwise this would be impossible.[*]

$$\S\ 72.$$

The Normal Double Slider-Crank Chain. $(C''_2 P^{\perp}_{\frac{1}{2}})$.

We have already, in § 68, considered the limiting case of the substitution of the pair (A) for a swinging (C) pair, in taking the lever c of infinite length. If we apply the method there used to the coupler b of the slider-crank chain, which appears already in Fig. 240 as a sector working in a slot, we can make it also infinite. The slot of the block c will then be straight and at right angles to the line 1, 3, the coupler becomes a prismatic slide with a cylinder normal to it, as Fig. 243 shows. If we write the new chain in full, beginning with the crank a, we have:

$$\overset{a}{\overbrace{C^+ \dots \| \dots C^+_=}}\ \overset{b}{\overbrace{C^- \dots \perp \dots P^{\pm}_=}}\ \overset{c}{\overbrace{P^- \dots \perp \dots P^{\pm}_=}}\ \overset{d}{\overbrace{P^- \dots \perp \dots C^-_=.}}$$

The block c has become a pair of prisms at right angles to each other, one of them (as in Fig. 243,) or both (as in Fig. 248) being open, or in the form of slots. We shall call it a cross-block, or in particular a **normal cross-block**, and the whole chain (as it now contains two sliding-pairs) a **normal double slider-chain**. The crank a remains as before, the coupler, however, has assumed the form $C \dots \perp \dots P$.

---

[*] In the *Constructeur* I have for a long time made use of the method of expansion of elements, but I have not there been able to analyse it causally, for this, as we have seen, is a matter which requires a somewhat lengthy investigation. I do not wonder therefore that it has remained greatly misunderstood, and has been sometimes pronounced unimportant, and even superfluous.—*R.*

To put the formula into the contracted shape we have to notice that the chain consists of two parallel cylinder-pairs and two pairs of prisms normal to each other; we must therefore write it $(C_2'' P_{\frac{1}{2}}^{\perp})$.

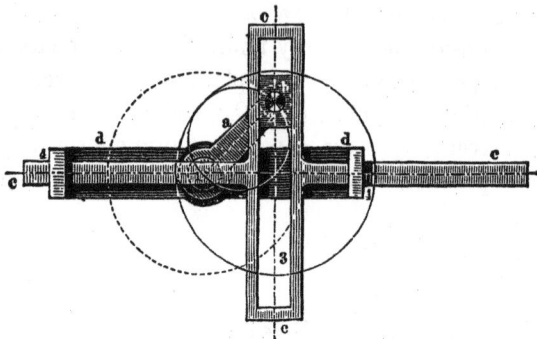

Fɪɢ. 243.

The mechanism $(C_2'' P_{\frac{1}{2}}^{\perp})^{d=b}$. In considering the mechanisms which can be formed from the chain before us, we may begin, as before, by placing it upon d. We notice at once that the coupler b is kinematically exactly equal to the frame d, and lies exactly similarly in the chain, having namely a link $C \ldots \| \ldots C$ on one side of it,

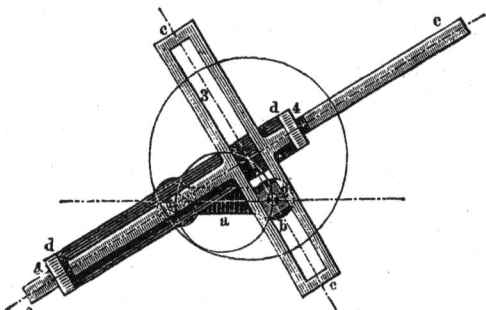

Fɪɢ. 244.

and a link $P \ldots \perp \ldots P$ on the other; the mechanisms $(C_2'' P_{\frac{1}{2}}^{\perp})^d$ and $(C_2'' P_{\frac{1}{2}}^{\perp})^b$ are therefore identical. In Fig. 243, d is made the fixed link. Following the analogy of our former nomenclature we may call the train a turning **double slider-crank**, or shortly turning **double slider**, for we obtain it by the addition of a second sliding pair to the turning slider-crank. The train $(C_2'' P_{\frac{1}{2}}^{\perp})^b$, which is formed from the swinging block, we may call a swinging cross-block. The motion produced is very simple. The centroids of a and c are Cardanic circles, the smaller (for a) having a diameter $= 1.2$, and the larger described from the centre of the cross-block with a radius $= a$, as in Fig. 243. The primary centroids for b and d are infinite, and must be replaced by secondaries, which are here omitted; they would show that every point in b describes circles relating to d, the whole piece moving always parallel to itself.

The turning double slider-crank is not unfrequently used, its most common application being in the driving gear of steam pumps in the form $(C''_2 P\frac{\perp}{2})^{\frac{d}{a}}$. It is often of value also from the fact that if the crank revolve uniformly, it imparts to the cross-block c a simple harmonic motion.

The mechanism $(C''_2 P\frac{\perp}{2})^a$. If the chain be placed on a, the links b and d move about fixed axes 2 and 1. The cross-block revolves, its centroid rolling always upon that of a, as is shown in Fig. 244. We shall call it a turning cross-block.* The links b and d are kinematically identical, although constructively different; their angular motions are always the same.

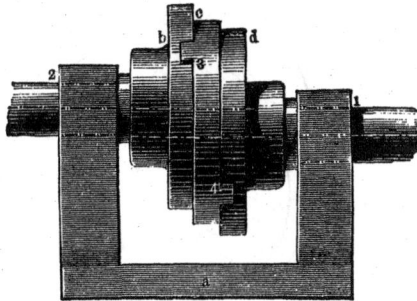

Fig. 245.

There have been many practical applications of this train. The well known Oldham's coupling (Fig 245) gives us one interesting illustration. The object of this mechanism is the communication of a uniform rotation between two parallel shafts, its special formula is therefore $(C''_2 P\frac{\perp}{2})^{\frac{a}{d}}$ or $(C''_2 P\frac{\perp}{2})^{\frac{a}{b}}$. The bed-plate of the two shafts is the frame a (the crank originally), the middle disc is the link c, the cross-block $P...\perp...P$; the two shafts and their connected discs are the links b and d. To make the construction of the coupling clearer, the three last-named links are shown separately in Fig. 246. The special property of the train $(C''_2 P\frac{\perp}{2})^a$ which is utilized in Oldham's coupling is the uniformity of the rotation

---

* Compare " turning block " for $(C''_3 P\perp)$ p. 299.

of the links b and d. It was applied in an original manner in Mr. Winan's " Cigar-boat." *

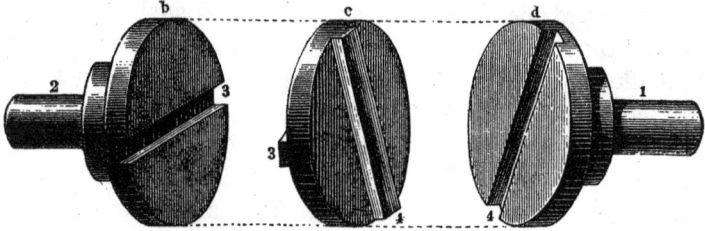

FIG. 246.

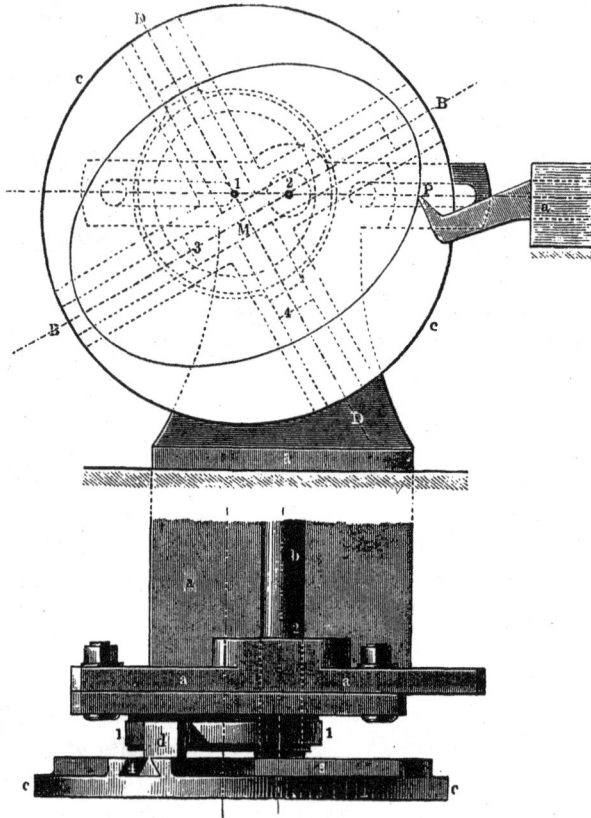

FIG. 247.

The "elliptic chuck" shown in Fig. 247—which so far as we know was invented by Leonardo da Vinci, and was certainly

* *Practical Mechanic's Journal*, vol. xix., (1866-7), p. 271.

investigated by him—is a very remarkable application of the mechanism before us. Use is here made of the fact that all points connected with the smaller centroid, that is in this case all points connected with the fixed link a, describe ellipses * relatively to the piece to which the larger Cardanic circle belongs. In the apparatus itself the cross is formed upon the back of the disc c. In one of its two slots it encloses the full prism 3, which is attached to the lathe-spindle. The headstock a forms the cylinder pair 2 along with the spindle b. The cylinder of the pair 1 which belongs to a is attached to the headstock by screws ; it is made annular, so that the spindle b passes through it, in other words, it is expanded sufficiently to allow 2 to lie within 1. The piece d is made as a ring :—its inner surface forms the hollow cylinder paired with a, while it carries outside the full prism of the pair 4 (divided into two) which works in the second slot of the cross c. The describing point or tool P forms a part of the fixed link or frame a. The ellipses which are described relatively to the disc by the point of P, have— if P lie beyond a—a difference between their semi-axes equal to the length a ; if P lie between 1 and 2, a is equal to the sum of the semi-axes. The enlargement of the pin 1 allows the magnitude of a (that is the distance 1, 2) to be varied within certain limits, and this, together with alterations in the position of P, allows very great variety in the ellipses produced by this apparatus. The link b being the driving link, the complete formula is $(C''_2 P \frac{\perp}{2})^{\frac{a}{b}}$. The mechanism might also be so arranged that d, which is kinematically equal to b, became the driving link. · (Compare § 76). It must be remembered that the point-paths of the disc c are those determined by the larger Cardanic circle, and are therefore peri-trochoids, in-cluding the particular case of cardioids. The path of the centre M of the disc c is the smaller centroid, through which it passes twice for each revolution of b or d.

The mechanism $(C''_2 P \frac{\perp}{2})^c$. We have now left only the train ob-tained by placing the chain on c. This may be called the swinging double slider-crank, or shortly s w i n g i n g d o u b l e s l i d e r ; it is the mechanism familiar to us as the "trammel" used by draughts-

---

* Laboulaye (*Cinématique*, 1861, p. 863) attempts to show that the curves de-scribed in this apparatus are not ellipses, but he is mistaken. I shall afterwards (§ 76) come to the form of the mechanism given by him, which differs somewhat from the one represented above.—*R.*

men, or in the clumsier form of Fig. 248 employed for drawing on plaster ;* and so on. Its special formula is $(C_2'' P_2^{\perp})_n^0$. The connection between this mechanism and the last—which we call the inversion of a chain—was discovered by Chasles, as I have already mentioned in a note to § 3 ; he missed, however, the principle really underlying it. The chain $(C_2'' P_2^{\perp})$ is, as we have seen, frequently used in machinery. Its real nature, however, is even more hidden than that of most chains by its constructive form ; and on that account its real connection with other chains has hitherto remained unrecognised.

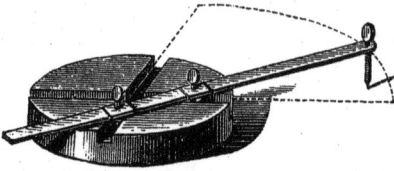

FIG. 248.

§ 73.

The Crossed Slider-crank Chain, $(C_3'' P^+)$.

The very considerable number of forms in which we have now seen the quadric crank chain by no means exhausts it, for in the slider-crank chain $(C_3'' P^{\perp})$ and those derived from it it is always possible to make a difference in length between the infinitely long links. By using different points in our mechanism as starting points, as it were, for the infinite lengths, we can make between these a finite difference of any desired magnitude. We shall very briefly consider the alterations which can thus be made in the chain.

If in the crank chain (C_4''), Fig. 249, we make the link c infinitely long, and therefore at the same time make d infinite, but arrange them so that c be longer than d,† we obtain the chain shown in Figs. 250 and 251. The direction of motion (relatively to d) of the pin 3 no longer passes through 1, but at a distance from it equal to the finite difference between c and d. We shall call these

---

* Called in Germany *Stuckateur zirkel*, and in France also *Compas de menuisier*.

† Perhaps it would be better to say rather that the links are so arranged that if their point of intersection were at any imaginable finite distance, c would be longer than d.

mechanisms crossed slider-cranks, and their uncontracted formula will be

$$C^+ \dots \| \dots C^{\pm}_{\doteq} C^- \dots \| \dots C^-_{\doteq} C^+ \dots \perp \dots P^{\pm}_{\doteq} P^- \dots + \dots C^-_{\doteq} \quad \text{(Fig. 250)}$$

or

$$C^+ \dots \| \dots C^{\pm}_{\doteq} C^- \dots \| \dots C^-_{\doteq} C^+ \dots + \dots P^{\pm}_{\doteq} P^- \dots \perp \dots C^-_{\doteq} \quad \text{(Fig. 251)}$$

In place of one of the former symbols for normal we have here, either in the link c or in d, the symbol for normally crossed, or crossed at right angles.

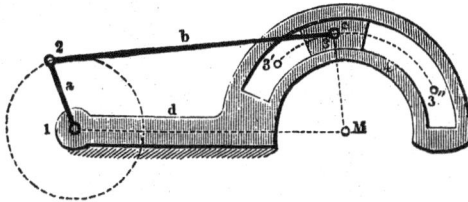

FIG. 249.

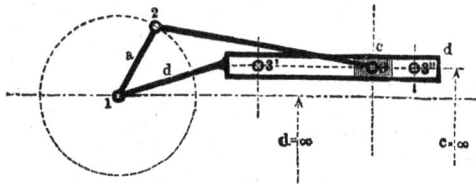

FIG. 250.

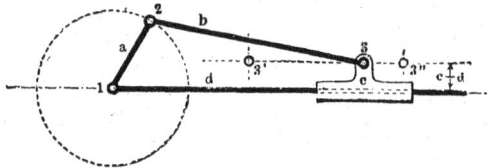

FIG. 251.

In its contracted shape the formula will run $(C''_3 P^+)$. The chain, like the more simple one $(C''_3 P^{\perp})$, gives us four mechanisms, corresponding to its four positions on d, b, a, and c. We may use for them the same names as before, prefixing the word crossed in each case. We have thus the crossed turning slider-crank, $(C''_3 P^+)^d$ the crossed swinging block $(C''_3 P^+)^b$, and so on. The motions occurring in these chains are more complex than in the former cases, the links being no longer symmetrical; necessarily, however, they are very closely related to them. Their applications are far less common. Schwartzkopf's adjusting spanner, Fig. 252, is a very original example of one of them. It is a self-acting universal spanner. To produce the required pressure on the movable cheek c, the mechanism of the crossed turning slider-crank is used in the form $(C_3'' P^+)^d$. The link a, a portion of which serves as a handle, is the crank, turning about the pin 1 carried by the frame d. At 2 it is connected by a pin with the coupler b, which joins the block c at the cylinder-pair 3, and moves it in the direction CD. 4 is the prism-pair between the links c and d. If the handle be pressed in the direction of the arrow, the mechanism grips the nut between the cheeks of c and d, and holds it the more

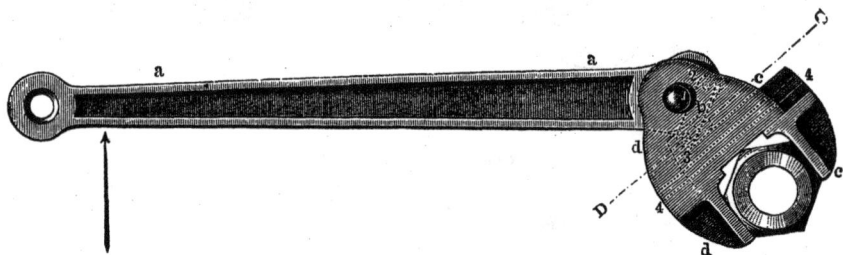

Fig. 252.

firmly the harder a be pressed. The nut and the spanner thus become virtually one piece, and the action of the handle a is simply that of a lever attached to the nut, by which it can be tightened, or by reversing the spanner, loosened, at will. Each time the spanner is placed upon a nut, we have the mechanism $(C_3'' P^+)\frac{d}{a}$ in action for a short period.*

---

* Another illustration of the mechanism $(C_3'' P^+)^d$ is shown in the accompanying figure, which represents an arrangement proposed by Deprez, (*Engineering*, June 18, 1875), and before him by Mr. Henry Davey, and very possibly by others also, as a reversing gear requiring one eccentric only. Here the link c is made with a curved slot having a radius equal to the length of the coupler, and in this there is placed a sector b' which carries the pin or full cylinder of the pair 3. From what has been already said (p.310) it will be recognised that this sector is kinematically identical with the coupler b. When the mechanism is working it retains a fixed position relatively to the slot in c. If it be fixed in the position shown by the full lines the mechanism is simply $(C_3'' P\perp)^d$, but by giving it any other position, as for instance, 3', the train becomes $(C_3'' P^+)^d$ and the motion received by c from a (and therefore, in the machine itself, by the valve from the eccentric) undergoes

The crossed-chains formed from the isosceles slider-crank chains are of less importance than the foregoing. They are formed by a method analogous to that described in § 69, and will be called isosceles crossed turning slider-crank $(C_2'' \leqslant C''P+)^{d=c}$ and isosceles crossed swinging block $(C_2'' \leqslant C''P+)^{b=a}$.

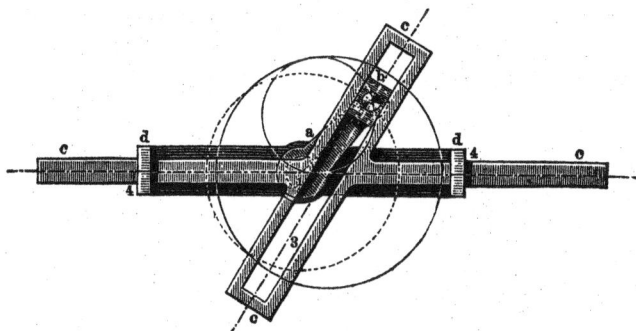

Fig. 253.

If we make b also $= \infty$ however, we obtain some remarkable special cases. The normal cross-block then becomes oblique, or to just that alteration of phase which is required to change the cut-off or the direction of the engine's motion. The excessive friction in the pair 4, when the amount of crossing is large, has prevented any great use being made of $(C_3'' P+)$ for this purpose.

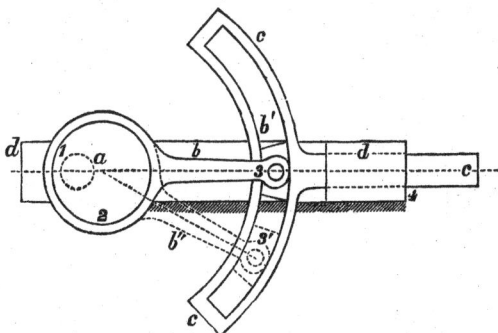

The mechanism $(C_2'' P\frac{L}{2})$, Fig. 253, might be used in the same way if it were constructed so that the angle of skew of the cross could be altered. Such an arrangement would be in some respects better than the one just described, but it presents some constructive difficulties in cases where an eccentric takes the place of the crank.

use a shorter word well known to engineers, skew, as in Fig. 253. The chain will be written:

$$\overbrace{C^+\ldots\|\ldots C^{\pm}_{=}}^{a}\;\overbrace{C^-\ldots\perp\ldots P^{\pm}_{=}}^{b}\;\overbrace{P^-\ldots\angle\ldots P^{\pm}_{=}}^{c}\;\overbrace{P^-\ldots\perp\ldots C^{-}_{=}}^{d}.$$

The symbol for crossed disappears, and makes room for that of oblique. If, as in former cases, we make prominent in the contracted formula the characteristic symbol of relation of the links, it will in this case be $(C''_2 P^L_2)$. The links b and d are again equal and similarly placed, so that the chain gives us, like $(C''_2 P^L_2)$, three mechanisms, namely—

The turning skew (double) slider or swinging $\Big\}\;(C''_2 P^L_2)^{d=b}$
   skew cross-block . . . .

The turning skew cross-block . . . . . . $(C''_2 P^L_2)^a$

The swinging skew (double) slider . . . . $(C''_2 P^L_2)^c$

Besides these special cases, the crossed-slider chain has, lastly, two more special forms, which we can only mention here. These are the forms obtained if, instead of three only, we take all four links of infinite length.

If we make $c = d = \infty$ as before, and then make b and a also infinite but having a finite difference, we get the chain shown in Fig. 254, of which the following is the formula:—

$$\overbrace{C^+\ldots\perp\ldots P^{\pm}_{=}}^{a}\;\overbrace{P^-\ldots+\ldots C^{\pm}_{=}}^{b}\;\overbrace{C^-\ldots\perp\ldots P^{\pm}_{=}}^{c}\;\overbrace{P^-\ldots\perp\ldots C^{-}_{=}}^{d}$$

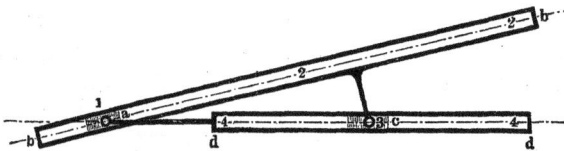

Fig. 254.

We may call this a single crossed-slide chain, and write it shortly as $(CP^+ CP^\perp)$. All its links are dissimilarly placed, it therefore gives us four mechanisms. If the lengths of c and d have a difference as well as those of b and a, but the two differences are unequal, we obtain the chain of Fig. 255, which we may call a double crossed slide chain $(C P^+)_2$. The links a and c are here

similarly placed in the chain, as are also b and d, it gives us therefore only two mechanisms. In all these mechanisms the centroids

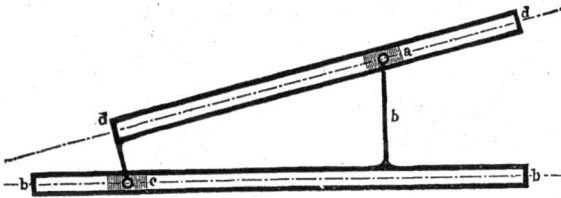

FIG. 255.

have only infinitely distant points. The single crossed slide has
sometimes been used in machinery.

§ 74.

Recapitulation of the Cylindric Crank Trains.

The number of important mechanisms which we have formed or
derived from the chain (C_4'') has been so large that in order that
their mutual relations may be more clearly surveyed it will be
well to place them together in a tabular form. This has been done
in the following pages, with the addition of a small schematic outline of each mechanism, the fixed link being in every case shaded.
The higher pairing, where it occurs, is omitted.

A. Quadric Crank Chain (C_4'').*

1. Lever-crank	$(C_4'')^{d=b}$	
2. Double-crank	$(C_4'')^{a}$	

---

* In this table I have put in brackets words which, although they form an
essential part of the name of the mechanism, might yet very often be omitted
without indistinctness in referring to it.

3. Double-lever	$(C_4'')^c$	
4. Parallel-cranks	$(C_2'' \| C_2'')^{d=b=a=c}$	
5. Reverse anti-parallel cranks .	$(C_2'' \gtrless C_2'')^{d=b}$	
6. Converse anti-parallel cranks .	$(C_2'' \gtrless C_2'')^{a=c}$	
7. Isosceles double-crank . . .	$(C_2'' \lesseqgtr C_2'')^{d=a}$	
8. Isosceles double-lever . . .	$C_2'' \lesseqgtr C_2'')^{b=c}$	

B.   Slider-Crank Chain $(C_3'' P^\perp)$.

9. (Turning) slider-crank . . .	$(C_3'' P^\perp)^d$	
10. Swinging block (slider-crank) .	$(C_3'' P^\perp)^b$	
11. Turning block (slider-crank) .	$(C_3'' P^\perp)^a$	

12. Swinging slider-crank . . $(C_3'' P^{\perp})^c$	
13. Isosceles (turning) slider-crank $(C_2'' \lessgtr C'' P^{\perp})^{d=c}$	
14. Isosceles turning block (slider-crank) . . . . $\Big\}\, (C_2'' \lessgtr C'' P^{\perp})^{a=b}$	

C. Normal double slider-crank chain $(C_2'' P_2^{\perp})$.

15. Turning double slider(-crank) or swinging cross-block . $\Big\}\, (C_2'' P_2^{\perp})^{d=b}$	
16. Turning cross-block . . . $(C_2'' P_2^{\perp})^a$	
17. Swinging double slider(-crank) $(C_2'' P_2^{\perp})^c$	

D. Crossed slider-crank chain $(C_3'' P^+)$.

18. Crossed (turning) slider-crank. $(C_3'' P^+)^d$	
19. Crossed swinging block . . $(C_3'' P^+)^b$	
20. Crossed turning block . . $(C_3'' P^+)^a$	
21. Crossed swinging slider-crank $(C'' P^+)^c$	

E. Skew double slider-crank chain $(C_2''P_2^L)$.

22. (Turning) skew double slider, or swinging skew cross-block . . . . $\Big\}(C_2''P_2^L)^{d=b}$	
23. (Turning) skew cross-block $(C_2''P_2^L)^a$	
24. Swinging skew double slider $(C_2''P_2^L)^c$	

F. Single crossed-slide chain $(CP^+ CP^\perp)$

25. to 28. Four mechanisms.	

G. Double crossed-slide chain $(CP^+)_2$.

29 & 30. Two mechanisms.	

This recapitulation furnishes the best possible proof of the necessity of our previous kinematic analysis to acquaint ourselves even with chains apparently so simple as (C_4'') and those derived from it. We also see how absolutely necessary it was to choose definite names for those of the mechanisms found by our analysis which occur most frequently. These names have been chosen with care and systematically, and they can be easily remembered, especially in connection with their formulæ. The removal of unessentials, which they greatly promote, is an enormous help to the recognition of the real kinematic nature of the constructively complex forms which occur in actual machinery. We shall also see immediately that we have in no way exhausted the list of mechanisms which can be formed from the four cylinder-pairs, notwithstanding its necessary limitations; indeed that we have yet to examine another great family of them, quite different from those we have been considering.

§ 75.

The Conic Quadric Crank Chain (C_4^L).

If the axes of the four cylinder pairs of the chain (C_4'') be not parallel, but have a common point of intersection at a finite distance, the chain remains movable, and (the former conditions being again fulfilled) also closed. The axoids will no longer be cylinders but cones,—as all the instantaneous axes have the point of intersection in common,—and the motion of the links will be determined by their conic rolling, the general nature of which we examined in § 10. If the lengths of the links—measured as arcs of great circles upon a sphere drawn about the point of intersection (M) of the axes—fulfil the conditions laid down for those of (C_4'') in § 65, we have a chain of such a form as is shown in Fig. 256. We may call it a conic quadric crank chain, or

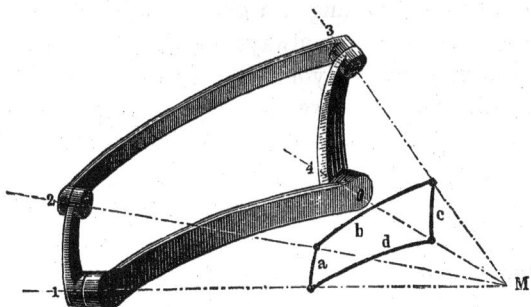

FIG. 256.

four-linked conic crank chain. It stands in a very close relation to the cylindric crank chain, which indeed may be considered as the special case of it when the point of axial intersection, is at an infinite distance.[48] The formula for the chain is

$$\overbrace{C^+ \ldots \angle. \quad C_\pm^\pm}^{a} \overbrace{C^- \ldots \angle \ldots C_=^-}^{b} \overbrace{C^+ \ldots \angle \ldots C_\pm^+}^{c} \overbrace{C^- \ldots \angle \ldots C_=^-.}^{d}$$

It can be contracted into the very simple form (C_4^L), in using which we understand not only that the pairs are oblique to each other, but also that their axes have a common point of intersection, as is shown above in Fig. 256.

The various forms of the cylindric chain repeat themselves with the conic one, but with certain differences in their relations. The principal of these relates to the relative lengths of the links, which would vary if they were measured upon spheric surfaces of different radii,—if they were taken, that is, at different distances from the point of intersection. The ratio, however, between the length of a link and its radius remains constant for all values of the latter, and these ratios are simply the values in circular measure of the angles $1\,M\,2$, $2\,M\,3$, $3\,M\,4$, and $4\,M\,1$, subtended by the links. Instead of the link lengths therefore we must consider the relative magnitudes of these angles, which we can also indicate by the letters a, b, c, and d.

The series of alterations in these lengths which we supposed in the former case, and which we carried on until all the links became infinite, are here represented by corresponding angular changes. The infinitely long link corresponds to an angle of 90°. For the case where two links are infinite but have a finite difference (§ 73) we have now one subtending a right, and the other an obtuse, angle. As however we must always imagine the axis of the links prolonged through and beyond the centre of the sphere, the obtuse angle between two axes gives on the other side also an acute angle between them,—so that no real difference exists between acute and obtuse-angled links. A similar simplification affects the centroids and axoids. The infinitely distant points of the centroids in the chain (C_4'''), of which we had illustrations in § 8, are here represented by the points in which the common normal to the fixed axes cuts the sphere. The axoids here are consequently cones (circular or non-circular) upon some closed base.

Keeping these points in view we may now proceed to examine the mechanisms formed from the conic quadric crank-chain, which we shall do as far as possible in the same order as before.

A. Conic quadric crank-chain (C_4^L) Fig. 257. All links subtend less angles than 90.° We obtain from it, as from (C_4'''), eight mechanisms for its eight principal special cases or positions ; to these we can give the same names as before, only prefixing the word conic in each case. Their formulæ, also, are analogous to the former ones, the form-symbol for oblique replacing that for parallel. I do not know of any applications of these mechanisms, but it is quite possible they may exist, disguised under dissimilar constructive forms.

The parallel and anti-parallel cranks repeat themselves in the conic chain along with the others. The arrangements necessary for passing the dead-points are not, however, those examined before. If we join two conic parallel crank-chains in a way corresponding to Fig. 206, we obtain a mechanism by which it might appear at first sight that a uniform rotation could be transmitted between shafts whose axes are neither coincident nor parallel, a problem for which a solution has often been attempted. The formula of such a train would be $2(C_2^L \parallel C_2^L)^d$. In reality, however, this combination is an impossible one. For the chain $(C_2^L \parallel C_2^L)$ has only four positions—the four cardinal ones—in which its opposite links lie parallel to each other; in all other positions the opposite angles of what was the parallelogram are unequal, and the rotation of the cranks is therefore not uniform. While therefore the chain $(C_2^L \parallel C_2^L)$ has its own special interest, it will be seen that it is not entirely analogous with $(C_2'' \parallel C_2'')$.

B. Conic slider-crank chain $(C_2^L C_2^{\perp})$, Fig. 258. The links d and c are right-angled, that is, the angle between the axes 1 and 4

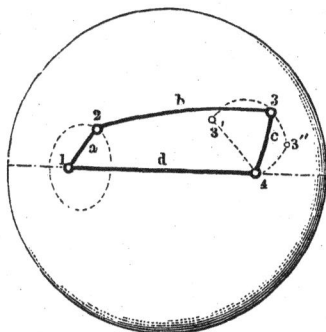 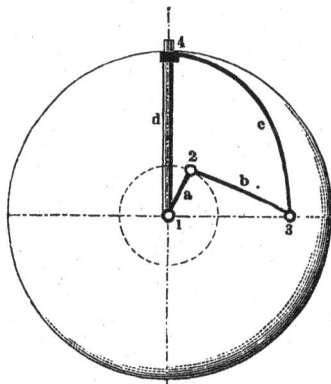

Fig. 257.   Fig. 258.

and between 4 and 3 = 90°. The comprehension of this chain, which may present at first difficulties to some of my readers, may perhaps be made more easy by the help of Fig. 259. Here the principle of pin-expansion is applied to the mechanism. For the arm M 3 (which the figure shows as the projection of a quadrant like c, Fig. 258), turning about an axis at M (corresponding to the

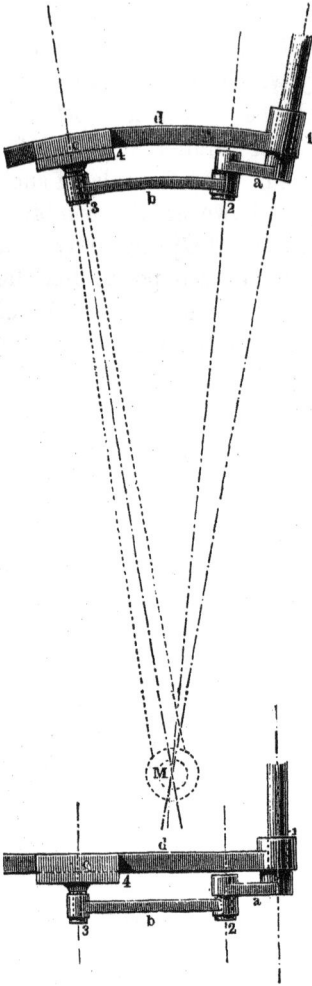

FIG. 260.

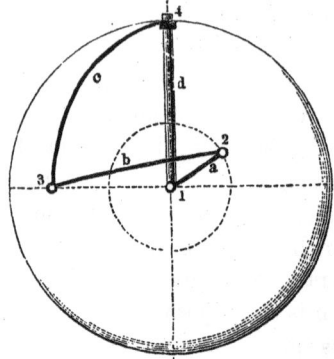

FIG. 261.

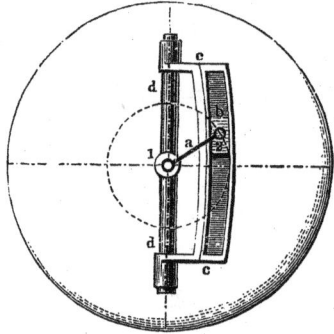

FIG. 259.

Fig. 259.— Conic slider-crank chain, $(C_2^L C_2^\perp)$ compared with cylindric chain $(C_3'' P^\perp)$.

Fig. 260-2. — Normal conic double slider-crank chain $(C_3^\perp C^L)$.

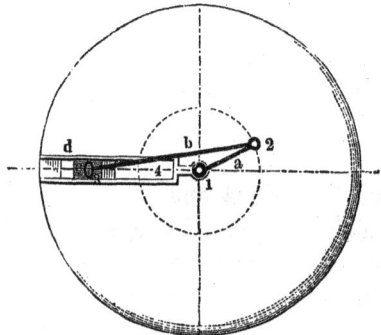

FIG. 262.

rod 4, Fig. 258) perpendicular to the plane of the paper,—we substitute the small section 4 of a cylinder, sliding upon a corresponding section d of another cylinder; c is now the block, d the frame, a the crank and b the coupler as before. Below the conic chain a similar cylindric chain is shown; the juxtaposition of the two makes it very easy to realise that the latter is simply the conic chain with the point M removed to an infinite distance. The conic slider-crank chain, like its cylindric counterpart, gives us six mechanisms, four principal forms and two secondary ones. We shall give them the same names as before with the prefix conic to each. There appears to be very little, if any, use made of them in practice.

C. Conic (normal) double slider chain $(C_3^\perp C^L)$, Fig. 260. Here the links b, c and d are right-angled, and a only acute-angled. This chain corresponds to the one bearing the same name in the cylindric series, and by applying the method of pin-expansion, it can be brought into a very similar form, as in Fig. 261. The mechanism of Fig. 262 is essentially identical with that of the one before it. The slide d is nothing more than a portion of a cross section of the cylinder which in Fig. 261 appears as a round bar, marked with the same letter. The link b subtends an angle of 90°, and is thus identical with the sector b in Fig. 261. This chain, like the cylindric double slider, gives us three mechanisms, which will be called

15.* The conic (turning) double slider, $\left.\right\}$ $(C_3^\perp C^L)^{a=b}$.
    or conic swinging cross-block $\left.\right\}$
16. The conic (turning) cross-block ... $(C_3^\perp C^L)^a$.
17. The conic swinging double-slider ... $(C_3^\perp C^L)^c$.

Considerable use is made of these mechanisms in practice. One well-known application of No. 16 is to be found in the mechanism known as the universal or Hooke's joint.† Writing out the formula $(C_3^\perp C^L)^a$ in full we have

$$\overbrace{C^+\ldots\perp\ldots C_=^\pm}^{b} \overbrace{C^-\ldots\perp\ldots C_=}^{c} \overbrace{C^+\ldots\perp\ldots C_=^\pm}^{d} \overbrace{C^-\ldots\angle\ldots C_=}^{a}$$

and this formula, corresponding to the chain in Fig. 260, we have already found (§ 58) to be that of the universal joint Fig. 263.

---

  * The chains (C_4^L) and $C_2^L C_2^\perp)$ together, as we have seen, give 14 mechanisms.
  † In Germany also as Cardano's coupling.

As either b or d may be the driving link its special formula runs $(C_3^{\perp}C^L)_b^a$ or $(C_3^{\perp}C^L)_d^a$. It may again be noticed that the links b, c and d are completely identical, as indeed becomes visible in Fig.

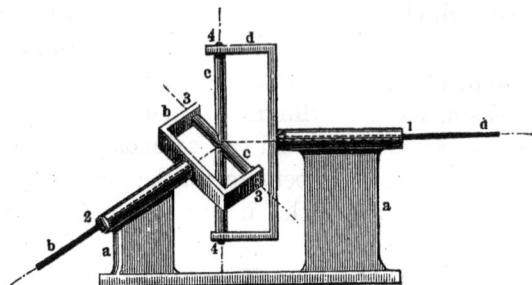

FIG. 263.

260, although in the universal joint they commonly appear so extremely different. We shall shortly have to examine some other very important applications of this chain.

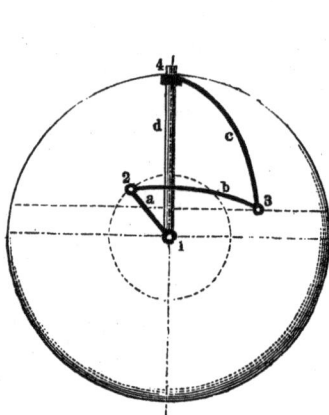

FIG. 264.

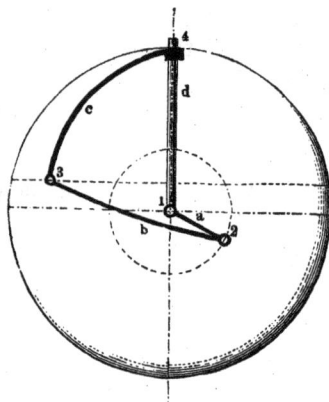

FIG. 265.

D. Crossed conic slider-crank chain $(C_3^L C^{\perp})$. The crossing of the cylindric slider-chain expresses itself here in the altered length of the links, of which one only, in Fig. 264 the link d, remains right-angled. We obtain as before four mechanisms (Nos. 18 to 21), of which very few applications occur.

E. Skew double slider-crank chain $(C^L C^\perp)_2$, Fig. 265, *a* and *c* are acute, *b* and *d* right-angled. This chain corresponds both to the cylindric skew double slider chain and to the cylindric crossed slide chains *F* and *G* (p. 322). It gives us three mechanisms (22 to 24), of which very occasionally we find an application existing.

In all, therefore, this conic crank chain gives us 24 mechanisms, dividing themselves into five different classes. The majority of these have been hitherto unknown; whether they are "practical" or "unpractical" is not a question which concerns us here. Our unerring analysis will allow us further on to obtain very important results from them. Summing up the results of the last ten sections we find that the number of mechanisms formed or essentially derived from the quadric crank chain has been 54, and that they have occurred in 12 distinct classes.

§ 76.

Reduction of a Kinematic Chain.

If we wish to obtain the motion of any particular link in a complete mechanism, without requiring at the same time to use the motions of any other of its links, it is often possible to remove one of these, its place being supplied by a suitable pairing between the two links which it connected. The number of links in the chain can thus be diminished without affecting the particular motion which is required, and it is evident that this may often be very advantageous. We shall examine some examples of it.

Suppose that it be wished to obtain a reciprocating sliding motion by means of the turning slider-crank $(C''_3 P^\perp)^d$ Fig. 266, and that none of the other motions in the chain be required, then the coupler *b* may be removed if we pair the crank *a* to the block *c* direct. This can be done, for example, by attaching to *a* a pin of suitable diameter, and connecting with the block an envelope (§ 3, Fig. 4) for it,—which will in this case take the form of a curved slot touched by the pin upon both sides, as in Fig. 267. The pin and its envelope form together a higher pair of elements. The simplest arrangement will be obtained by using the former crank pin, which will pair with a slot described from the centre of the coupler with a radius equal to its length,—for the slots which would be required as envelopes for pins further from the centre 1 have forms much more troublesome to deal with, such as the path

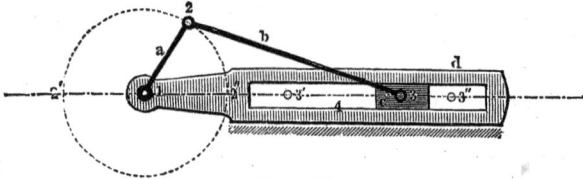

FIG. 266.

(shown dotted) of the point 2'. If the link b be removed and the chain afterwards closed in this simple way, its complete formula, (placed on d), will run :—

$$\overbrace{C^+\ldots\|\ldots C^+,}^{a}\ \overbrace{A^\pm\ldots\perp\ldots P_=^\pm}^{c}\ \overbrace{P^-\ldots\perp\ldots C_=^-.}^{d}$$

We must find means for distinguishing this chain from the former $(C_3'' P^\perp)^d$:—

$$C^+\ldots\|\ldots C_=^\pm\ C^-\ldots\|\ldots C_=^-\ C^+\ldots\perp\ldots P_=^\pm\ P^-\ldots\perp\ldots C_=^-.$$

As it gives us no new motions, but only fewer motions than before, we shall not make a new class for it, but shall treat it as a **derived form of the four-linked chain,** obtained from it by

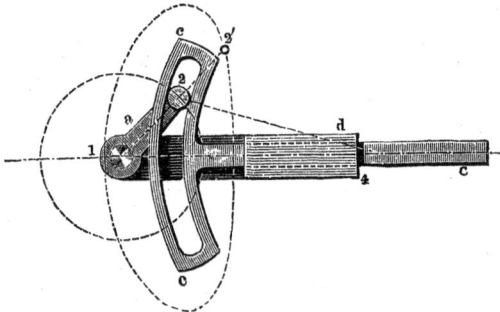

FIG. 267.

the removal of one link, in this case b. We shall call this removal of a link from a complete chain and its replacement by higher pairing a **reduction** of the chain by that link, and shall indicate it in the contracted formula for the new chain by writing the latter :—

$$(C_3'' P^\perp) - b.$$

The chain so reduced has three links;—it can therefore be placed in three ways only, so that only three mechanisms can be formed from it. These are: $(C_3'' P^\perp)^d - b$; the turning slider-crank, $(C_3'' P^\perp)^a - b$ the turning block, and $(C_3'' P^\perp)^c - b$ the swinging slider-crank. All three mechanisms occur in practice.

Instead of removing b any other of the links may be taken away, provided only that the particular motion required can be obtained without it. The two following figures show two methods

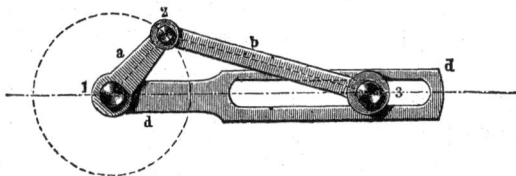

FIG. 268.

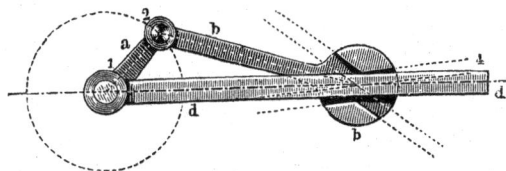

FIG. 269.

of reducing the chain $(C_3'' P^\perp)$ by omitting the block c, they represent therefore $(C_3'' P^\perp) - c$. In the first of them there is added to the coupler, conaxially with the former cylinder pair 3, a solid cylinder, and this is paired with its envelope in the slider, which is a straight slot, or **negative** prism. In Fig. 269, on the other hand, the slider is made a **positive** prism, and paired with its envelope in the end of the coupler. The latter takes the form of an X-shaped recess, which lies closely on the slider only at the points of greatest pressure,—(*i.e.*, when the axes of a and b are at right angles),—but in all other positions has considerable freedom, and works therefore under force-closure. The pair-closure could be complete only if the prism d were infinitely thin. When the arrangement of Fig. 269 occurs in practice it is most frequently with well-rounded corners in the recess in b. The manner in which the true form of these round corners can be obtained is not unimportant;—it may be looked at as follows.

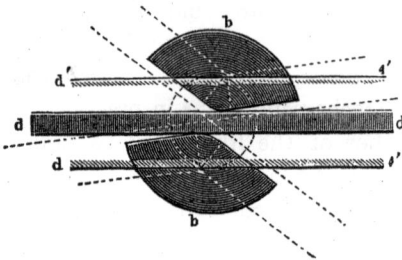

If we have assumed in the first place some considerable breadth for the slider d, and found the required shape of the envelope in the coupler, we can then draw equidistants to the profile thus obtained (§ 35), and may choose such of these as give us for the prism a narrower profile than before. The corresponding

FIG. 270.

equidistants for the recess in b will then give us the rounded corners required, as is shown in the figure above. The pairing is still, however, incomplete (force-closed) for there is some freedom left between b and d in all positions but those of greatest pressure.

The form in which Leonardo's elliptic chuck is most commonly constructed furnishes us with a remarkable example of a reduced crank chain. This form, to which we have already alluded in § 72, is shown in Fig. 271. The reduction here consists in the omission of the coupler b from the turning cross-block,—the form of the mechanism is that, therefore, which would be obtained from Fig. 267 if the radius of the slot were made infinite. Its contracted general and special formulæ are $(C_2'' P\frac{\perp}{2})^a - b$ and $(C_2'' P\frac{\perp}{2})^{\frac{a}{a}} - b$ respectively,—for it is driven by the slider d, turning about the pin 1 (compare the mechanism in § 72, where b is the driving link). a_1 is the end of the headstock, to which the piece a_2 is secured by means of adjusting screws. These two form together the fixed link of the mechanism, the crank a. This carries the bearing for the spindle 1, to which the slide d, carrying the open prism 4, is attached;—d is therefore of the form $C^+...\perp...P^-$. The cross-block c, $(P^+...\perp...P^-)$ consists here of the full prism paired with d at 4, and of the two pieces 3 3, forming the sides of an open prism, which are shown in the lowest of the accompanying figures. These pieces envelop the greatly expanded pin 2 of the fixed link a, and form with it—the link b which connects them in the complete chain having

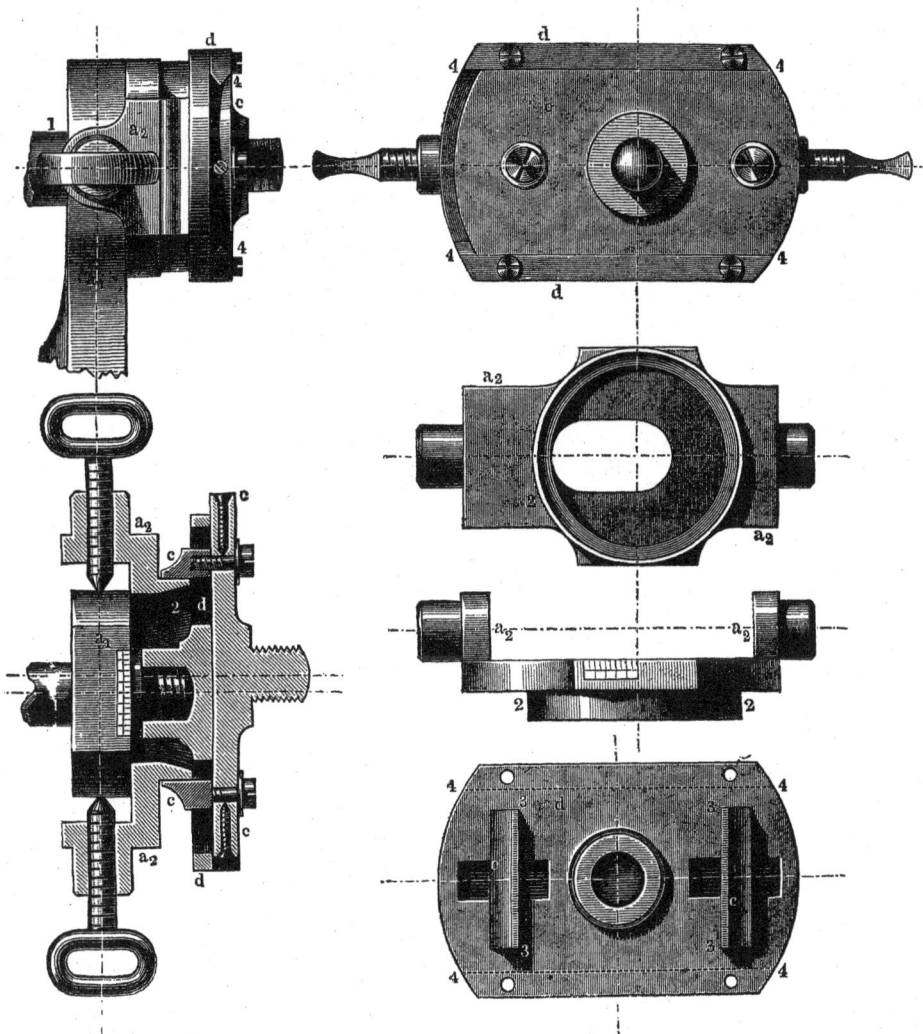

Fig. 271.

been removed—a higher pair. The expansion of an element and the reduction of a chain are here, therefore, used together. By the adjusting screws and scale the relative positions of a_2 and

a_1,—that is the length of the link a,—can be readily adjusted at pleasure,—the piece a_2 being made very open inside to allow of this. On the face of the cross-block c there is a screw to which the chuck can be attached in the usual manner. This apparatus gives us an excellent illustration of the extraordinary way in which the constructive details of a mechanism may hide its real nature. It can easily be seen how this elliptic chuck, although so often used, has been hitherto so little understood. The way in which the very disadvantageous reduction of the chain has perpetuated itself is really extraordinary. The higher pair 2.3 wears rapidly on account of its deficiency in surface, and the motion of the mechanism becomes therefore inexact. The

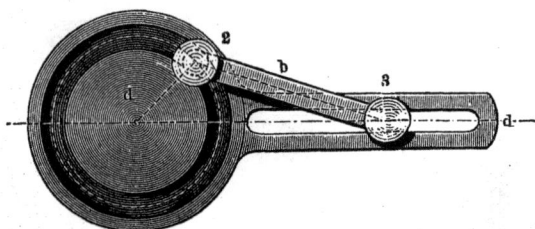

FIG. 272.

arrangement shown in Fig. 247,* in which the chain is used unreduced, and to which our theoretical investigations directly led us, has not this disadvantage. It is also in other ways more convenient than the common plan, especially in its simple arrangement of the prism pairs 3 and 4 upon the back of a disc.

We have seen that it is possible to reduce a four-linked chain to one of three links:—it must in the same way be possible to reduce a three-linked to a two-linked chain under suitable circumstances. This can be done as well with a chain previously reduced as with one containing three links in its complete form, as an illustration will show. Fig. 272 shows the chain represented in Fig. 268 reduced by another link, namely the crank a. For

* Fig. 247 represents the form which Prof. Reuleaux has chosen for the model of this mechanism in the kinematic collection of the Königl. Gewerbe Akademie in Berlin. A machine for cutting elliptical grooves (for man-holes etc.), in which the mechanism was used unreduced, was exhibited at the Vienna Exhibition by a Chemnitz firm. See *Record of the V. Univ. Ex.* (Maw and Dredge), Pl. 195.

this purpose a full cylinder on the end 2 of the coupler is paired with its envelope on the frame, an open annular cylinder having 1 for its centre. Such a chain is written $(C_3'' P^\perp) - a - c$.

It is two-linked:—in other words it has been reduced to a pair of elements, the two pieces b and d. It can be placed therefore in two ways only. Placed upon d it gives us the turning slider-crank $(C_3'' P^\perp)^d - a - c$, in a form which might be used in cases where the only motion required is that of the coupler. Placed upon b we have the swinging block, and can utilise the motion of the slider only.

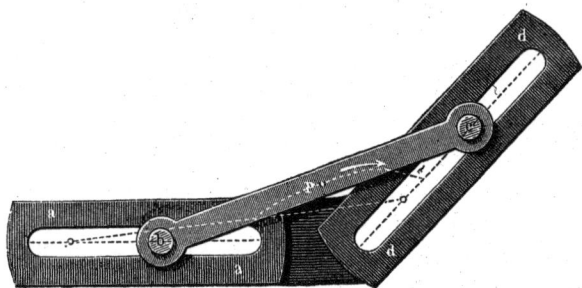

Fig. 273, an arrangement which came before us at the very outset of our work, is another example of a reduced chain. It can easily be seen that it is really a portion of a twice reduced skew double-slider chain such as Fig. 253 (Cf. No. 24 page 326). The piece marked aa dd is the skew cross-block, and $b\,c$ the crank. Its formula is therefore :

$$(C_2'' P_{\frac{L}{2}}) - b - d.$$

Any point p in the crank moves, as we have frequently seen, in an ellipse. The pairing rendered necessary by the omission of the links is here, as in the other cases, higher pairing, the pair of elements obtained by ultimate reduction from a chain is a higher pair. The reduction can be carried no further, for a machinal arrangement cannot be made with less than two bodies.

Practical use is made of this process of reduction in other chains than those which we have been considering. I may just give one example of this. The spur chain (Fig. 274) can be reduced to the higher pair of elements shown in Fig. 275 by omitting the frame. The complete chain has the formula $(C_z^+ C_2'')$ in its reduced form it will therefore be written $(C_z^+ C_2'') - c.$

It must not be supposed that the reduced chains have actually been derived in this way from the complete ones. We have on the contrary (Chapter VI.) seen cause to suppose that the process

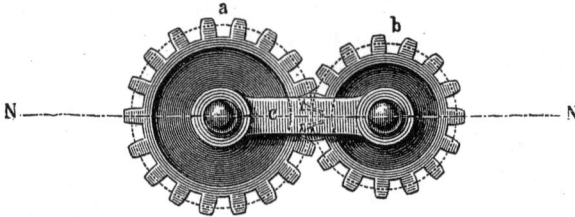

Fig. 274.

of development has moved in the opposite direction. This how-ever need not prevent us from treating the matter deductively when such a treatment greatly facilitates its comprehension, and enables us to avoid endless repetitions. If we had to consider each reduced chain as a separate form of kinematic linkage we should obtain, through the very various forms which the higher

Fig. 275.

pairing can take, an enormous number of additional combinations, without at the same time having added a single new motion to those already existing in the complete chains. The admittance therefore, of the principle of chain-reduction into our work helps us very greatly in simplifying its arrangement. In the case of compound mechanisms it can often be very advantageously employed, and the exact analysis of these combinations forms a very interesting and instructive problem in applied Kinematics.[49]

§ 77.

Augmentation of Kinematic Chains.

The augmentation of a chain stands logically as the contra-positive of its reduction. A chain which has been already reduced can obviously be restored to its original completeness by such a process. But there is no reason why the augmentation should stop here ;—the pairing between any two pieces may be replaced by chaining, *i.e.* by linkage, if the link introduced between them be so arranged as not to alter their relative motions. If the chain already possess the largest number of links which it can have as a simple closed chain, any augmentation must be so arranged as to make it a compound closed chain (§ 3, p. 49). In general therefore the process leads us to such chains, which do not belong to the part of our subject at present under consideration. We may merely mention a few illustrations of it. The (so-called) parallel motions are augmentations of this kind: the parallel motions of Watt or Evans, for instance, replace a prism pair by a kinematic chain having only turning pairs, which therefore essentially is a crank mechanism. A common train of wheelwork again, which is used really as a substitute for a pair of wheels of inconveniently large diametral ratio, may be considered an augmentation in the same way. It will be seen from these examples alone that a very extended use is made in practice of this method of chain augmentation. We shall content ourselves here with having thus stated the general nature of the principle, and shall not go further into the matter. Its further consideration forms indeed a part rather of applied than of theoretical Kinematics.

CHAPTER IX.

ANALYSIS OF CHAMBER-CRANK TRAINS.

§ 78.

Chaining of Crank Mechanisms with Pressure-organs.

HAVING now made ourselves familiar with the application of kinematic analysis to the various forms of the simple crank chain, we may go a step further, and may proceed—limiting ourselves still to the same class of mechanisms—to examine the extent to which this analysis is applicable to actual machines. When we have completed this examination we may hope to have made the road to the practical use of the analysis sufficiently evident and easy.

Among the numberless applications of the crank chain in machinery one special class claims our attention,—that namely in which a pressure-organ,—water, air, steam, gas, &c.,—is used in kinematic combination with the mechanism, so as to produce either a machine for moving the pressure-organ, such as a pump, or its contra-positive, a machine driven by the pressure-organ, an engine or "prime-mover." The combinations employed for these two purposes must obviously have a very close relationship, and a great number of the chain forms which we have considered have been practically used for both of them. At the same time there is perhaps no department of practical work about which more indefinite and certain ideas have been held than this. The method of dealing with it has been little better than a groping in the dark, without the guidance of any principle, without knowledge of its course, without understanding of its materials. Such a multitude of arrangements have been devised—and are still being devised—for the carrying out of one and the same purpose, that it appears almost impossible for a single individual to reduce them all to order, or even to find out their existence. The irresistible tendency towards the invention of "rotary" steam-engines has contributed greatly to increase the number of these arrangements. This tendency has given us many useless, or apparently useless, machines, and has been the means of wasting much thought and capital. Would-be inventors have again and again been warned of it, but the warnings do not seem to have had any effect. From a kinematic stand-point the warning cannot be unconditionally repeated. In the first place the class of inventors here concerned are exactly those whom our warning will never reach, or who will not listen to it. Then again the attempted combinations may not in themselves be bad, although their practical usefulness may not correspond to the hopes of their inventors. And lastly, it does not seem justifiable to check empirical experiments so long as theoretical investigations into the matter fail to furnish the means of doing better or as well, or even to say with certainty what is the real worth or worthlessness of the results obtained.

We have now, however, got a most important auxiliary in the matter, for in kinematic analysis we have the means of determining the real nature even of the most disguised mechanisms, and we shall proceed at once to the consideration of this problem so far as it concerns the machines just mentioned, in which crank trains are formed into pressure-organ machines.

The process of designing such a machine divides itself into two parts. There is, first, (*a*) the making of one of the links into a vessel or chamber, and (*b*) the forming of another (or two others) into a movable diaphragm or piston,—the relative motions of these parts being so arranged that the pressure-organ alternately fills and is driven out from the space between them. The second part of the process consists in the addition of kinematic arrangements for the periodic opening and closing of the passages leading to the chamber. These arrangements may be included under the name of valve-gear, or more generally, as we shall see in § 135, of directing-gear.

The chamber and piston form a kinematic chain with the (liquid or gaseous) fluid, and this is united with the crank mechanism into a compound chain, which, under some circumstances, may require a very extended formula. We may, however, disregard this for the present, keeping before our minds always the forms of piston and chamber, but not considering the fluid itself. This will greatly simplify the matter, while at the same time it will be quite sufficient for our present purposes. Crank mechanisms having their links so formed, (as chamber and piston), that they are suited for the enclosure and motion of a fluid, we shall call chamber-crank trains. As the constitution of the machine itself is independent of that of its valve-gear we need say very little about the latter here; short descriptions of its general nature will always suffice. We shall come later on (Chaps. XI and XII) to the consideration of the real kinematic meaning of the valve-gear, as well as that of the pressure-organs, in such cases as those now before us.

§ 79.

Chamber-crank Trains from the Turning Slider-crank.

(Plates XIV. and XV.)

The arrangement adopted in the table of mechanisms in § 74 gives them in what appears to be their natural order. In order, however, to reduce as far as possible the difficulties unavoidably connected with our analysis I shall here consider them in a different way, commencing with the mechanism which is probably most familiar to the reader,—that used in the ordinary "direct-acting" steam-engine. We therefore begin with the chain $(C''_3 P^\perp)$, and take first the mechanism $(C''_3 P^\perp)^d$, the turning slider-crank. Plates XIV. and XV. show schematically eight forms of chamber-crank gear formed from this mechanism. We shall consider them in order.

Pl. XIV. 1. is the form familiar to us in common direct-acting

Slider-crank $(C_3'' P^\perp)^d$.

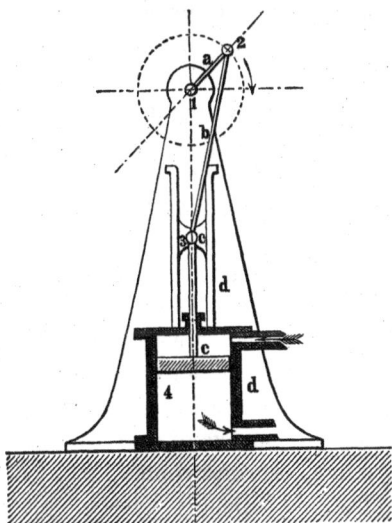

FIG. 1.—Direct Acting Engine or Pump.
$(C_3'' P^\perp)^{\frac{d}{c}}$; $(V\pm) = c, d.$

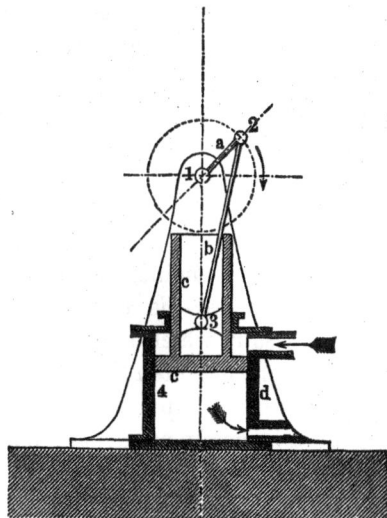

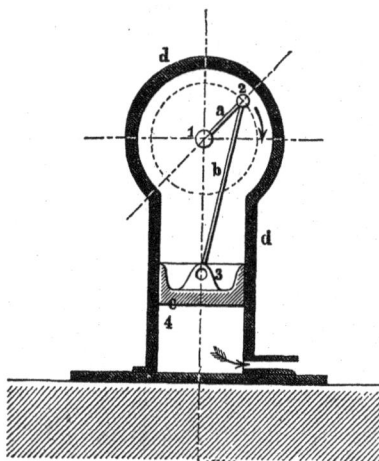

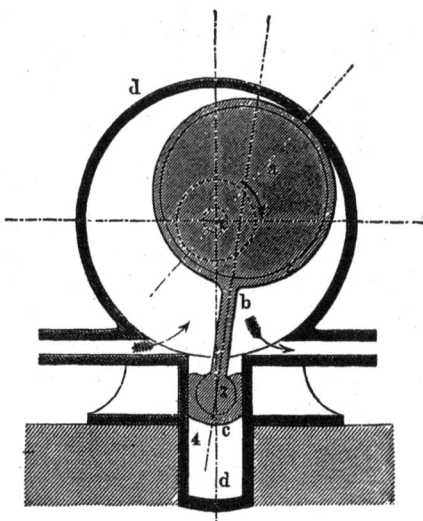

FIG. 2.—Broderip, Humphreys. (Eng.)
$(C_3'' P^\perp)^{\frac{d}{c}}$; $(V\pm) = c, d.$

FIG. 3.—Hastie, Hicks. (Eng.)
$(C_3'' P^\perp)^{\frac{d}{c}}$; $(V\pm) = c, d.$

FIG. 4.—Pattison. (Pp.)
$(C_3'' P^\perp)^{\frac{d}{a}}$; $(V\pm) = b, d.$

engines and pumps. The link taking the form of the chamber is the frame d; its upper part is made the guiding prism for the block c, its lower part being formed as a hollow cylinder, the steam- or pump-cylinder as the case may be. In this moves a tightly-fitting piston or plunger, which is simply an extension downwards of the block c. The feathered arrows show here (as in the following figures) the direction of motion of the in-coming and out-going fluid when the crank a is moving in the direction indicated by the plain arrow. If the machine be a pump the valves almost always act automatically, the pressure of the fluid itself opening and closing them. If the machine be an engine, on the other hand, a special kinematic chain, —the valve gear—is required for working them. There are a few pumps also in which a similar chain or gearing is employed. It must not be forgotten that the kinematic condition for the chamber is that it must be prismatic; the choice of a circular cylinder for its form is merely accidental, and made for the sake of convenience. It has none the less furnished the name,—steam-cylinder, pump-cylinder, etc.,—by which the chamber is commonly known. In some cases where the chambers of ventilating machines have been made of wood, exactly the same considerations of convenience have led to the adoption of a square instead of a circular cross section. The position of the chain relatively to the horizon does not affect it in any way. Fig. 1 therefore serves equally for horizontal engines and pumps, or for those which are inclined, or inverted, or in any other position; whatever practical difference there may be between these various "systems," kinematically they are identical. One other point requires to be noticed,—what is known as the "double-action" of the piston. It is necessary for our purposes to understand clearly what this means. It is that in each period,—that is here in each revolution of the crank,—the piston is paired with the fluid not only on one but on both sides;—and from this it follows that in one whole period the chamber must always be twice filled with fluid. With plunger pumps or single-acting engines this is not the case; the distinction made between them in practice is therefore borne out kinematically.

We have placed below our figure the contracted special formula of the mechanism, from which it will be seen that it contains nothing beyond the four links which we already know,—if we disregard the valve-gear and the fluid organ. The block c appears as the driving link, as would be the case if the mechanism were that of a steam-engine. Were it used as a pump the formula would be $(C_3'' P^\perp)^{\frac{d}{a}}$. The expression $(V\pm) = c$, d is added to show that the piston V^+ and the chamber V^-,—that is the pair $(V\pm)$,—are formed by the links c and d respectively, the block and slider of the turning slider-crank. The principal formula, along with this additional expression for the chambering, is also subjoined in the same way to all the following figures.

Plate XIV. 2 shows a form of steam-engine used by Broderip (1828) and Humphreys (1835)* but specially connected, as the **Trunk engine**, with the name of Penn. The upper part of the prism c is made into a trunk sufficiently large to allow the pin 3 to fall within instead of beyond it (3 thus lying within 4), and is so formed that the coupler b can swing freely although the axis of 3 be placed far back as in the figure. The trunk greatly diminishes the capacity of the upper part of the chamber, but in this it differs only in degree from the former arrangement, where the piston rod also diminished the capacity of the upper chamber in precisely the same way.

Fig. 3, Plate XIV., is an arrangement designed by Hastie† for a steam-engine. It is single-acting only, and the chamber is extended completely over the crank. The piston is made very heavy in order that its weight may to some extent equalize the working of the machine. Hicks has more recently tried to introduce the same arrangement with the frame d placed horizontally; his engine at the Paris Exhibition attracted more attention than it deserved.‡ In order to equalize its action and to carry it over the dead points he used four mechanisms, or rather two pairs of mechanisms with chain-closure as in § 46, and worked the valve-gear of one pair from the pistons of the other.§

* I have in every case done my best to ascertain the name of the inventor or first introducer of each machine, but further than this I cannot answer for the priority of those named, or for the dates given. I have had to content myself with infor-mation at second or third-hand in cases where no more direct sources of information have been accessible to me. I therefore make no pretension to give here an historical account of rotary engines or pumps.—*R.*

† Johnson, *Imperial Cyclopædia,* Steam-engine, p. lx. ; also Bernoulli, *Dampf-maschinen-lehre,* 1854, p. 321.

‡ Offiziellen öster. Ausstellungsbericht, 1868, Motoren und Maschinen der allgem. Mechanik, p. 118. etc. Kittoe and Brotherhood's "Paragon" steam-pump is also an example of this mechanism.

§ Brotherhood's "three cylinder" engine, the general construction of which is

Plate XIV. 4. Pattison's pump.* This arrangement, patented in England in 1857, is simply a mechanism $(C_3''P^\perp)^d$ in which the coupler b is used as the driving link, being shaped as a piston and paired with a chamber forming part of the frame d; $(V^\pm) = b, d$. The chamber is a hollow cylinder coaxial with the cylinder 1, while the piston b, so much of it at least as is required to make a tight joint with d, is formed as a cylinder coaxial with 2. The block c forms, as in the last case, a piston working in the prism 4. Its motions, however, are not utilised, for the necessary valves are omitted. The piston b encircles the expanded pin 2 of the crank a, drawing the water into the chamber upon one side and discharging it from

shown in the annexed figure, is an example of this form of chamber gear which is now very familiar. It is a treble train, having for its special formula 3 $(C_3'' P^\perp)^d_c$.

The form given to the pair 2 is worth noting, it is so arranged that the one crank pin serves as an element in three completely closed pairs. The end of each of the couplers b is formed as a sector of a revolute coaxial with 2 and having both internal and external profiles. The former works upon the crank pin (2) proper,

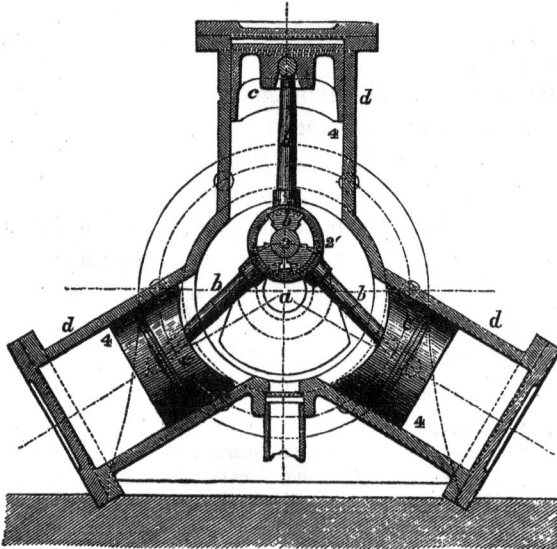

and the latter is always in contact with the ring 2′ which is concentric with 2. The form of the element carried by the crank a at 2 has thus become an open circular channel, while that of its partner element in the coupler b is a corresponding sector. The closed pair (R) or (C) is thus replaced by the pair (A), a change the nature of which was considered at some length in § 71, p. 311. The pair at 3 is a closed pair, although the closure is not seen in the section shown in the figure.

* *Propagation Industrielle,* 1869, p. 178.

the other. The available capacity of the chamber, that is the whole volume of its upper part less the volume of the circular part of the piston, is filled and emptied once in each revolution ; the pump is therefore single-acting. When in its lowest position the piston allows free communication to exist for an instant between the suction and delivery pipes, which, however, with a tolerably quick motion might not greatly injure its action. Between the end sufaces of b and d there is lower pairing, but that between their outer surfaces (those shown in the section) is higher, which of itself renders it difficult to keep the joint tight. The pump requires no valves, none at least for the usual purposes.

Plate XV. Fig. 1. Lamb's steam-engine * (patented 1842). This machine was intended to work with air or gas as well as steam, or to be used as a pump. If the very indistinct description of the inventor can be trusted our figure represents its essential parts. The mechanism is a reduced turning slider-crank, the block c being omitted; its general formula is therefore $(C_3''P^{\perp})^d - c$. The coupler b is again the piston, the frame d the chamber. The special formula is therefore $(C_3''P^{\perp})^{\frac{d}{b}} - c$. On account of the omission of c, the higher pairing between b and d which we have already described in § 76 (Figs. 269 and 270), is used at 3, which almost destroys the possibility of that joint being steam-tight. The inventor rectifies this by the use of some additional closing piece, the nature of which he does not make clear, and which we have therefore omitted. The tightness of this joint is not, however, important if only leakage could be prevented between the circular walls of the chamber and the piston where they are in contact, for Lamb uses the steam both within and without the annular piston. In the position shown there is on the left between the piston and the outer wall of the chamber a considerable space, while between the inner side of the piston and the inner wall of the chamber an opening into which steam is admitted is just beginning to show itself. The action here begins when the outer space is just half full and the crank at the top of its stroke. The steam in the two complementary spaces within and beyond b is allowed to escape freely. As the crank rotates the slot at 3' moves up and down the fixed diaphragm in d, exactly as the pin 3 and block c in the last

---

* See as to this and the following machine, *Repertory of Patent Inventions*, 1843. Enlarged series, vol. 1, p. 98.

example. If no expansion were required only such valve-gear would be wanted as would suffice to prevent escape of steam in the lowest position of the coupler, where (as in Pattison's machine) the piston itself does not hinder communication between the two ports.

Plate XV. 2 represents a second of Lamb's machines, in which two similar mechanisms are united, one being placed within the other. The pistons b_1 and b_2 act quite independently,—they are two couplers of unequal length driving a common crank. The arrows indicating the motion of the steam as well as the ports for the inner chamber are omitted in the figure for simplicity's sake. Mr. Lamb insists most strongly that even more than two of the annular chambers and their ring-shaped pistons should be used together, intending that the second and following chambers should serve for the expansion of the steam. As the available chamber capacity both inside and outside the piston is filled and emptied once in each revolution, the machine may be considered double-acting.

Fig. 3. Pl. XV. shows a form of rotary steam-engine which has very frequently been proposed,—by Bährens (Köln) in 1847,— by D. Napier in 1851, who patented it with certain improvements,[*] and then by Bompard (Piedmont) who re-invented it in 1867.[†] Our figure represents a double application of the train $(C_3''P^\perp)^d$ with the link a as piston and d as chamber. The link b serves merely as a means for obtaining a tight joint, as a link of the chain it is itself incompletely closed. It still consists indeed of two cylinders or portions of two cylinders, one described about the axis of 2, and one about the axis of 3;—but its half-moon-shaped profile is not such as will give a constrained pairing either with a or with c. The kinematic closure is here therefore brought about through the block c, but for this purpose some auxiliary arrangement has to be employed on account of the incompleteness of b. The formula therefore must be so constructed as to express the incompleteness both of b and of c. It may therefore be written as follows; indicating the double application of the same chain by the factor 2 outside the brace :

$$2 \left\{ (C_3''P^\perp)^{\frac{d}{a}} - \frac{b}{2} - \frac{c}{2} \right\}.$$

* *Repertory of Patent Inventions,* vol. xx., 1852, p. 307.
† *Génie Industriel,* vol. xxxiv., 1867, p. 179.

Slider-crank $(C_3'' P^\perp)^{\mathrm{d}}$.

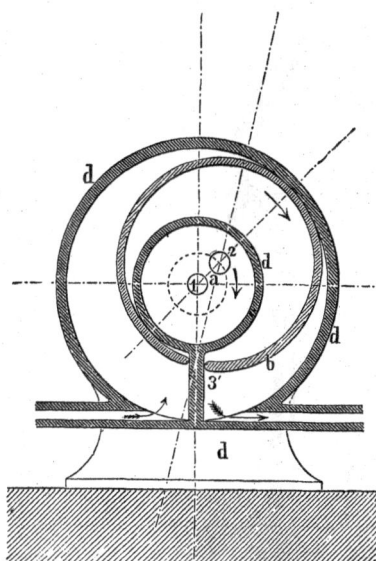

FIG. 1.—Lamb. (Eng.)

$(C_3'' P^\perp)^{\mathrm{d}} - c$; $(V^\pm \mp) = b, d.$

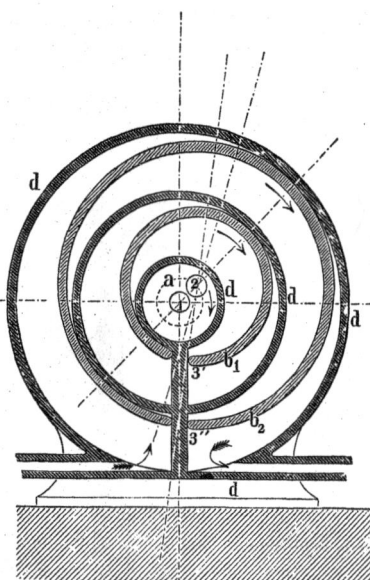

FIG. 2.—Lamb. (Eng.)

$2\{(C_3'' P^\perp)_{b}^{\mathrm{d}} - c\}$; $(V^\pm \mp) = b, d.$

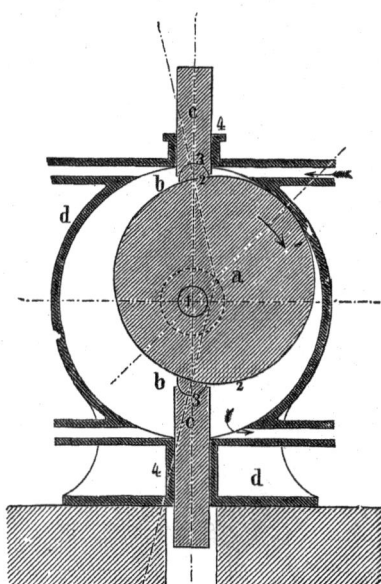

FIG. 3.—Bährens, Napier, Bompard. (Eng.)

$2\{(C_3'' P^\perp)_{a}^{\frac{\mathrm{d}}{a}} - \dfrac{b}{2} - \dfrac{c}{2}\}$; $(V^\pm) = a, d.$

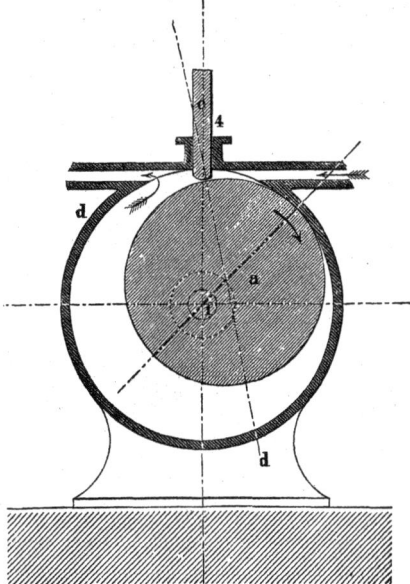

FIG. 4.—Yule, Hall. (Eng.)

$(C_3'' P^\perp)^{\frac{\mathrm{d}}{a}} - b - \dfrac{c}{2}$; $(V^\pm) = a, d.$

PLATE XVI. CHAMBER-CRANK TRAINS.

Isosceles Slider-crank $(C_2'' \leqq C'' P^\perp)^{d=c}$.

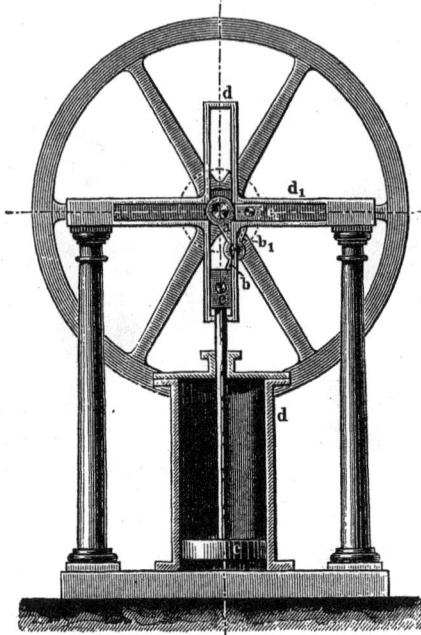

FIG. 1.—Dawes. (Eng.)

$(C_2'' \leqq C'' P^\perp)^{\frac{d}{c}}$; $V\pm = c, d$.

Swinging-Block $(C_3'' P^\perp)^b$.

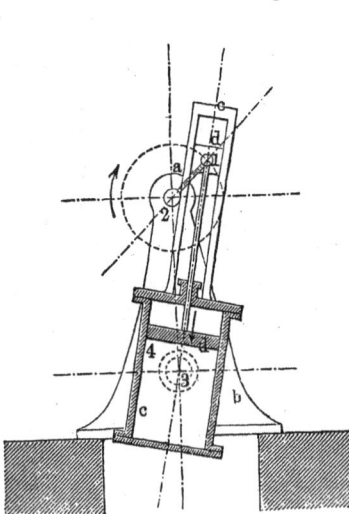

FIG. 2.—Murdock. (Eng.)

$(C_3'' P^\perp)^{\frac{b}{a}}$; $(V\pm) = d, c$.

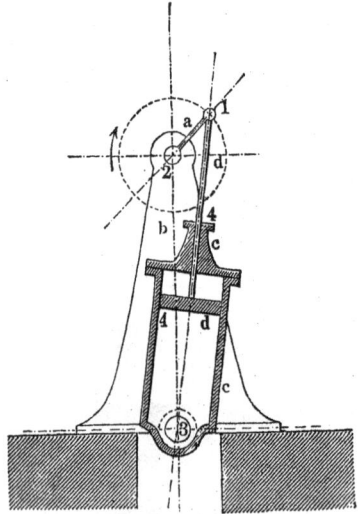

FIG. 3.—Alban, Farcot. (Eng.)

$(C_3'' P^\perp)^{\frac{b}{a}}$; $(V\pm) = d, c$.

If it be desired to show that the same links a and d are common to both chains, that is, that only one of each of these links exists in the machine, we must add $- a - d$ at the end of the formula outside the braces.

Bährens used only *one* block, so that his chain was only single, and he employed a weight to effect the closure of c. Napier, on the other hand, used a perfectly correct chain-closure for that purpose. At first he also used a single chain only, omitting, like Bährens, the lower block and the corresponding ports,—he connected, however, with the block two pins conaxial with 3, one being placed on each side of the chamber d (externally), and he connected each of these by a coupler equal in length to b with eccentrics having their sheaves conaxial with the cylinder 2. In other words, he employed two auxiliary and in every respect equal and similarly acting chains of the form $(C''_3 P^\perp)^d$, which not only had throughout the same link lengths as the chamber-crank chain, but had also the fixed link d in common with it. It is evident without further explanation that such chain-closure must ensure the constrained motion both of the block c and of the crescent-shaped coupler b.

Bompard first employed the double chain shown in our figure, —he used a special chain of several links (which we need not here describe) to effect the closure of the block c, and as it does this only approximately, his engine forms a less perfect solution of the problem than Napier's.

Pl. XV., 4. Rotary steam-engine proposed and constructed both by Yule* (Glasgow, 1836) and by Hall† (before 1869). It is a reduced turning slider-crank with the coupler omitted and with force-closure used for the block c, and will therefore be written

$$(C''_3 P^\perp)^{\frac{d}{a}} - b - \frac{c}{2}.$$

The link a is again the driving link and the piston, while d forms the chamber. Yule arranged his ports as in our figure; Hall made the exhaust as we have shown it, but caused the steam to enter through an opening in the block itself, for which reason he carried up the chamber above the block (Cf. Fig. 4) and connected the

---

* Battaille et Jullien, *Machines à Vapeur* (1847), vol. ɪ., p. 449 ; *Berliner Verhandlungen* (1888), p. 233 ; Bourne, *St. Eng.*, p. 392.

† *Propag. Industrielle*, iv. (1869), p. 340 ; *Génie Industriel*, vol. xxxv. (1868), p. 82.

steam-pipe with it above the latter. In both cases the force-closure of the block was effected by its own weight. The higher pairing between a and c is badly suited for making the steam-tight joint which is there required.

We see now what has not previously been recognised—that these eight chamber-crank trains are all founded upon the familiar mechanism of the first example, and are all derived from it in the same way, by forming two of its links so as to serve as chamber and piston respectively and by the addition, where necessary, of valve gear. We shall not here enter further into the question of the practical value of any of the machines; apart from this, however, all the eight solutions are equally legitimate, and there-fore merit equal attention from our present point of view. The links which are used as piston and chamber are

$$c \text{ and } d, \; b \text{ and } d, \; a \text{ and } d.$$

In every case d is the chamber. This is a very natural choice, for d is the fixed link and the necessary steam or water pipes can therefore be much more easily connected with it than with any of the other links. We shall, however, soon see that in spite of this advantage one of the moving links has not unfrequently been chosen in practice as the chamber. The illustrations which we have given, therefore, do not nearly exhaust the varieties of chamber-crank gear which can be formed from the mechanism $(C_3''P^\perp)^d$,—there would be no difficulty in arranging other combi-nations. There could hardly be machines having less apparent resemblance to the simple crank chain than those of Lamb—they appear at first to have nothing in common with it; it may even be difficult for the reader, until he has become more familiar with the methods of kinematic analysis, to realise their identity. Not-withstanding this our analysis has led us to the result—has shown us distinctly the real nature of the machine—without any forcing or special assumptions whatever. This example serves to show too how necessary our former general investigations were, how important it was for us to become familar with the exchangeability of $+$ and $-$ in the lower pairs, the expansion of elements, the reduction and augmentation of chains and so on, and to be able to recognise the effects of these processes upon the external form of the mechanism.

§ 80.

Chamber-crank Trains from the Isosceles Turning Slider-crank.

(Plate XVI. Fig. 1.)

The special case of the turning slider-crank which occurs when the lengths of the coupler and crank are made equal has been not unfrequently employed in chamber-crank trains. Pl. XVI. Fig. 1, shows a steam-engine formed from it by Dawes[*] in 1816. We called the mechanism the isosceles turning slider-crank. Here again the frame or slide d is made the fixed link, so that its general formula is $(C_2'' \leqq C'' P^{\perp})^d$. The piston is part of c and the chamber of d, so that we have also $V^{\pm} = c$, d. The special formula is therefore $(C_2'' \leqq C''' P^{\perp})^{\frac{d}{c}}$. The coupler is prolonged and carries on its further end a second block c_1, moving in a slide d_1, which forms one piece with d. The chain is thus closed by the addition of a second chain, equal and similar to the first but placed at right angles to it, instead of by the pair-closure described in §§ 47 and 70.

The object of this mechanism is to reduce the distance between the crank-shaft and the cylinder; an examination of it shows, however, that it possesses such constructive difficulties as to leave it little practical value. It appears, however, here and there, as for instance at the London Exhibition of 1851,[†] and in 1868 at the Exhibition at St. Petersburg. At the latter place Prof. Tchebischeff exhibited a working model, in which the second prismatic guide was replaced by an augmentation of the chain (§ 67) in the shape of a "parallel motion" designed by him.[‡]

---

[*] Severin's *Abhandlungen* (*Mittheil. des techn. Dep. für Gewerbe* 1826), p. 64.

[†] See Booth's Machine in the *Official Catalogue*, vol. i., p. 219.

[‡] *Berliner Verhandlungen*, 1870, p. 182. The Imperial Technical School at Moscow exhibited at Vienna, in 1873, a sectional model of this arrangement which was illustrated in *Engineering*, vol. xvi., p. 284.

§ 81.

Chamber-crank Trains from the Swinging Block.

Plate XVI., Figs 1 and 2, and Plate XVII.

The second of the four mechanisms which are obtained from the chain $(C_3''P\perp)$ was the swinging block slider-crank, the mechanism in which the chain was placed on the coupler b, $(C_3''P\perp)^b$. It has been employed as a chamber-crank train in a number of ways.

Probably the oldest form is that of the oscillating steam-engine, Fig. 2, Pl. XVI., invented by Murdock,[*] 1785. The former coupler b has become the fixed link, the crank a turns about the pin which was the crank pin; the block c has taken the shape of the cylindrical chamber and the slide d that of the piston. The chambering can be expressed in the form $(V\pm) = d, c$, while the special formula of the mechanism is $(C_3''P\perp)^{\frac{b}{a}}$. The valve-gear used by Murdock was so arranged that a fixed but somewhat elastic arm caused the rod of a common D slide-valve to move up and down parallel to the piston rod, so that the slide was always in the middle of its stroke when the crank was at a dead point. Later on the steam admission and exhaust were managed by means of suitably formed openings in the hollow pin or trunnion 3, or in some piece conaxial to and connected with it, and the same method is still applied in cases where the mechanism is to be employed as a hydraulic engine or a pump. We may call such an arrangement trunnion-valve gearing. If it be required to use the fluid expansively a more complex gear must be used.

A comparison between Fig. 2, Pl. XVI., and Fig. 1, Pl. XIV. shows that—apart from its valve-gear—the oscillating steam-engine is simply an inversion of the direct-acting engine. The signs of the pair 4 are also reversed, so that the piston of the latter becomes here the "cylinder," while its "cylinder" is here the piston. If the chain had been inverted without this pair-inversion, movable steam and exhaust pipes would have had to be used on account of the motion of the chamber. If, however, the gearing were so altered that the ports could be carried through the piston-rod, it might be possible to arrange the admission and discharge by

---

[*] Muirhead, *Inventions of James Watt*, vol. iii., plate 34.

Swinging Block $(C_3'' P\perp)^b$.

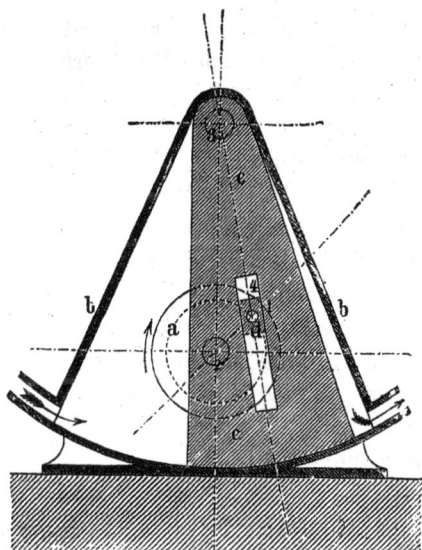

FIG. 1.—Reuleaux. (example.)
$(C_3'' P\perp)^{\frac{b}{c}}; \; (V\pm) = c, b.$

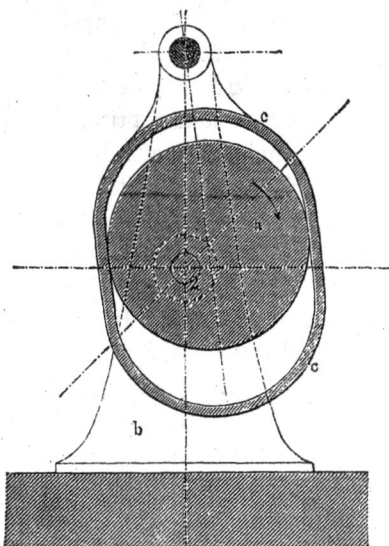

FIG. 2.—Simpson and Shipton. (Eng.)
$(C_3'' P\perp)^{\frac{b}{a}} - d; \; (V\pm) = a, c.$

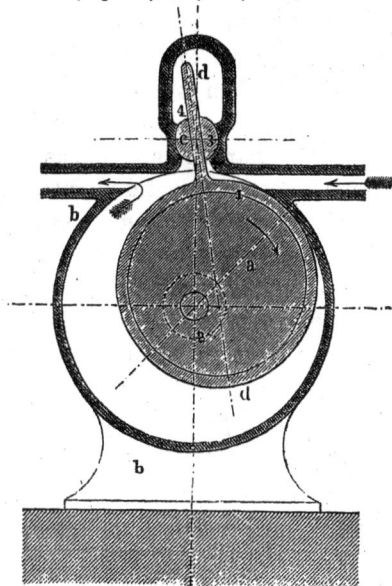

FIG. 3.—Knott. (Pp.)
$(C_3'' P\perp)^{\frac{b}{a}}; \; (V\pm) = d, b.$

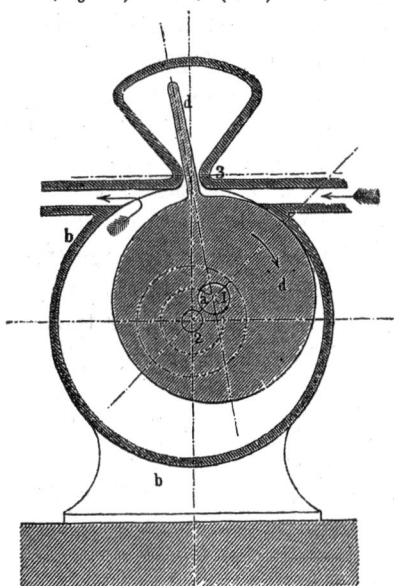

FIG. 4.—Wedding. (Blower.)
$(C_3'' P\perp)^{\frac{b}{a}} - c; \; (V\pm) = d, b.$

trunnion valves in a not altogether unpractical way. I may only mention that this has actually been done in a machine intended either for a steam-engine or a pump, patented in England by Herr E. Jelowicki * in 1836.

No alteration of principle, and very little of detail, is involved in placing the pin 3 below the cylinder as in Fig. 3, Pl. XVI., instead of opposite its centre as in Fig. 2. This arrangement is adopted in the steam-engine of Alban, Farcot, and others, as also in Sir W. Armstrong's well-known hydraulic engines, &c.

The fixed link may be used as the chamber instead of one of the moving ones. In order to show this, and so to generalise somewhat the solution of the problem before us, I add in Fig. 1, Pl. XVII. a construction which might be used as a pump, and in which the link *c*,—the "cylinder" of the last figure,—is made the piston. It swings to and fro from the pin 3 in the chamber *b*. The arrangement cannot pretend to any great practical value, although pumps have been made which are not at all unlike it,— compare, *e.g.*, Fig. 1 in Pl. XXV.

The steam-engine of Simpson and Shipton,† shown schematically in Fig. 2, Pl. XVII., is a more interesting example than the last. A special test of it was made in 1848, which ostensibly proved that it really possessed the advantages claimed for it. Few illustrations, however, show more distinctly than this, on the one hand, how the feeling of interest excited by a novelty may lead even able engineers to see more in it than actually exists there, and, on the other hand, how indistinct and confused our kinematic notions have hitherto been. The inventors themselves, as well as their interpreters, appear completely perplexed in the attempt to analyse the motions going on before them. Ordinary terms will not suffice them for this purpose ; in describing the piston *a* and its motion they speak of an "eccentric revolving in its own diameter." An "eccentric" geometrical fiction, certainly ! Revolving in its own diameter! That this quasi-explanation consisted simply of words without meaning remained unnoticed, as did other matters in which there was an equally suspicious indistinctness of expression. They remained unnoticed simply because

---

* Newton, *London Journal of Arts*, etc. Conjoined series, vol. ix. (1837), p. 34.

† Johnson, *Imperial Cyclopœdia*,—Description of plates, p. 29 ; also Newton, as above, vol. xxxvii. (1850), p. 207.

no way was known of throwing light on the matter, because the methods hitherto used for describing such motions have to a great extent failed in reaching their real underlying principles.

What we have before us is simply a reduced swinging block,— the link d omitted and a and c made respectively piston and chamber, the former being the driving link. The formula for the mechanism is $(C_3''P^{\perp})^{\frac{b}{a}} - d$, and for the chambering $(V^{\pm}) = a, c$. Disregarding the latter and also the reduction, the chain is simply that shown in Fig. 226 (*ante*). Relatively to c, a describes similar oscillations to those of the piston d in Figs. 2 and 3, Pl. XVI., from which it differs only in having rotation as well as oscillation. In the machine itself a suitable valve gear was used to control the admission and exhaust of the steam just as in the former cases. It is obvious, however, that the reduction of the chain, which has necessitated the use of a higher instead of a lower pairing between c and a, has made it extremely difficult to keep the piston steam-tight. I may add that the same mechanism, $(C_3''P^{\perp})^{\frac{b}{a}} - d$, has been employed also by an American—Broughton*—as a pump. He uses trunnion valves arranged upon the shaft 2 of the piston a.

Fig. 3. Pl. XVII. Pump patented by Knott† in England 1863. Here the link d has become piston and b chamber,—the chain is not reduced. The prism 4 of the link d,—which is placed within 3 as in our schematic Fig. 242,—makes a water-tight joint with the block c, as does the periphery of the open cylinder 1 with the cylindrical wall of the chamber. In the upper position of d there is for an instant free communication between the suction and delivery pipes as in Pattison's pump; apart from this, however, the fluid is drawn in continuously, without the necessity of valve gear, upon one side of the piston, and discharged continuously from the other, by the rotation of a. This link is therefore given as the driving link in the formula. According to our former definition Knott's pump is single-acting, while the four other arrangements of the same mechanism which we have mentioned are double-acting.

Fig. 4. Pl. XVII. Blowing machine, patented in Prussia by Wedding in 1868. The chain $(C_3''P^{\perp})$ is here reduced by the

---

* *Propagation Industrielle*, iv. 1869, p. 145. The French patent is dated 1856.

† Newton, *London Journal of Arts*, &c. New series, vol. xix., 1864; also König, *Pumpen* (Jena 1869), p. 103, where, however, there is an inaccuracy in the figure.

link c, otherwise the arrangement is very similar to that of Knott's pump. We have again $(V\pm)=d$, b, and the crank a is again the driving-link. The higher pairing used between d and b on account of the omission of c is that of Fig. 270,—it is exactly the same therefore as that employed by Lamb in the machines shown in Pl. XV. The pressures commonly required in blowing machines are so small that the closure at 3 is quite sufficient, and for the same reason the disadvantage of the momentary communication between the suction and delivery pipes when the crank is at its upper centre is greatly reduced.* The relation between the machine before us and that of Lamb (Pl. XV. Fig. 1) merits closer examination. We find that the inner cylinder d of Lamb's engine, with its prismatic plate $3'$, corresponds to the piston of the machine before us, and the revolving annular piston of the former to the chamber of the latter,—and this again conditions their respective double and single action. Herr Wedding remarks that by suitably forming the upper portion of b and d we could make, them into a second chamber and piston respectively, and that the latter would be in its most advantageous positions when the first piston was in its worst, *i.e.*, at the upper centre, and *vice versâ*. Here again there is an analogy with Lamb's machine, where the actions on the two sides of the annular piston have the same relation to each other.

Once more looking back from the two last machines to Fig. 1, Pl. XIV., we cannot wonder that it has only been after the most careful and searching analysis that we have been able to recognise the turned connecting rod of the common steam-engine in the peculiar looking chamber b, or its framing in the club-shaped piston d.

§ 82.

Chamber-crank Trains from the Turning Block.

Plates XVIII. to XXII.

No one of the crank mechanisms has been turned to account as a rotary engine or pump in so many different ways as the turning

---

* Two 6 horse-power machines of Wedding, used in the Spandau Artillery Works, apparently worked well. (The pistons here were made of wood, the plate at d of thin boards.) Lately, however, both machines have been replaced by common fans, because the diminution of pressure occurring as the crank passed the upper centres was found to affect the smiths' fires.—*R.*

Turning Block $(C_3''P^{\perp})^{a}$.

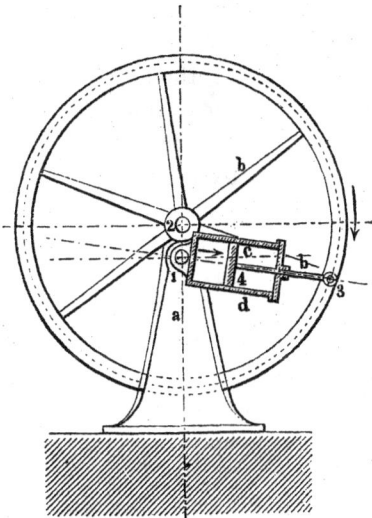

FIG. 1.—Ward, Schneider, Mouline. (Eng.)
$(C_3''P^{\perp})^{\frac{a}{c}}; \; (V\pm) = c, d.$

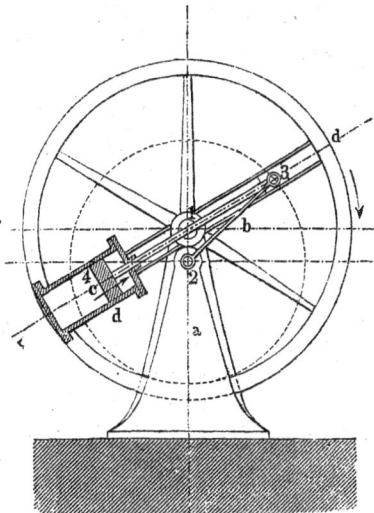

FIG. 2.—Morey, Schneider. (Eng.)
$(C_3''\,P^{\perp})^{\frac{a}{c}}; \; (V\pm) = c, d.$

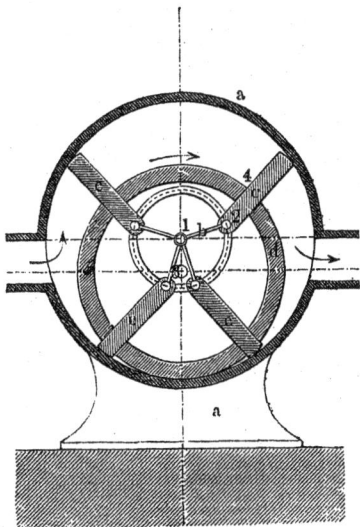

FIG. 3.—Emery. (Pp.)
$4(C_3''P^{\perp})^{\frac{a}{4}}; \; (V\pm) = c, a.$

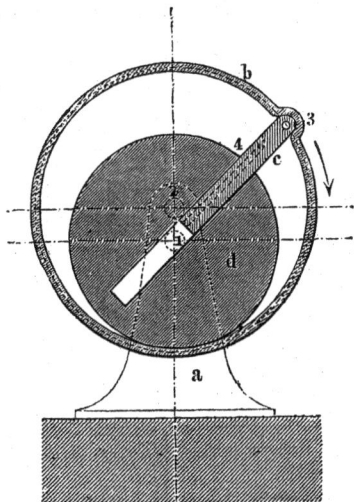

FIG. 4.—Cochrane. (Eng.)
$(C_3''P^{\perp})^{\frac{a}{c}}; \; (V\pm) = c, b.$

block slider-crank, $(C_3'' P^{\perp})^a$. It has two turning links, the slide d and the coupler b, and one link which both turns and swings, the block c, so that all the moving links have rotary motions. In the ceaseless attempts at rotary engine design all three have been used as piston in turn. In the first ten of our examples c is the piston; in the seven next it is d, and in the last three figures b.

Fig. 1, Pl. XVIII. represents a form of steam-engine used by Ward* in 1821 to drive a paddle steamer, and afterwards by Moulinet† (1847), and by Schneider‡ 1862 for factory purposes.

The chamber and piston are formed respectively from d and c, so that the machine is a direct inversion of that of Fig. 1, Pl. XIV. The steam distribution is managed by trunnion valves, which can be conveniently arranged in connection with the centre 1. The fly-wheel is placed upon b, which is the piece whose velocity is required to be as uniform as possible.

Fig. 2, Pl. XVIII. Steam-engine used by S. Morey§ for a paddle-steamer in 1819,—by Cramer, in 1834 as a three cylinder engine,‖—and again in 1862 by Schneider for driving machine-tools. The difference between this arrangement and the last is that here the link d is the one to which the uniform rotary motion is given. Morey, Ward and Schneider are all Americans; the idea once sown in their country seems to have extended itself rapidly. These arrangements have the obvious advantage that all their moving parts have rotary motion; they therefore permit of the use of a number of different constructive forms.

Fig. 3, Pl. XVIII. Emery's Pump¶ (America). This is a combination of four turning-blocks, and so has for its general formula $4(C_3'' P^{\perp})^a$. The fixed link a is the chamber, the four blocks c the pistons. Between the latter and the sides of the chamber there is higher pairing. An arrangement of stops, not shown in the figure, prevents either the formation of a vacuum or the confinement of liquid in the two lower quadrants. The link d is the driving link. In his later machines Emery omits the coupler b,— i.e., reduces each chain by that link, and employs instead of it the

* Severin's *Abhandlungen*, p. 114.—Dingler's *Journal*, vol. ix., p. 291.

† Bataille et Jullien, *Machine à Vapeur*, 1847, vol. ii., p. 241.

‡ *Polytech. Journal*, 1862, p. 401.

§ Severin's *Abhandlungen*, p. 110.

‖ Newton, *London Journal of Arts*, &c. Conjoined series, vol. xx. (1842), p. 454.

¶ *Propagation Industrielle*, vol. iv., 1869, p. 335.

circular groove shown in the figure to pair with the pin 2 of the
link *c.* He has used the mechanism for several purposes, among
others for a water meter.

Fig. 4, Pl. XVIII. Steam-engine designed by Lord Cochrane,*
1834. Here again the block *c* is the piston, this time, however,
the turning link *b* is made the chamber. The steam distribution
takes place through the centre 1 by means of a passage (shown
dotted) in the piston *c.* There is a higher pairing between the link
d and the chamber.

Fig. 1, Pl. XIX., represents the well known gas exhauster of
Beale.† It consists of two turning-blocks, having *d* as their driving-
link. Its special formula is therefore $2(C_3'' P^\perp)\overset{n}{a}$. The fixed link *a*
is the chamber, and *c* the piston ; the coupler *b* appears as a circular
sector (see § 71, Fig. 240, &c.) which works in a corresponding
groove, 2 in *a*, and is made of sufficient length to prevent any
leakage of gas through the groove.

Fig. 2, Pl. XIX. Rotary steam-engine patented by Davies ‡
in England in 1867. There are here three mechanisms of the form
$(C_3'' P^\perp)^a$ united, *c* is the piston and driving-link, *a* the chamber.
The coupler has become a cylindrical bar flattened upon one side,
the flat side being simply a portion of the cylinder 2 which is conaxial with the chamber *a,* it corresponds, that is, to the crank-pin in
the mechanism $(C_3''' P^\perp)^d$. In consequence of the incomplete form in
which the pair 2 is thus used the coupler is not fully constrained.
For its closure Davies used a circular groove (dotted in the figure)
in which a pin connected with the block *c* can work, he employs a
higher pairing, that is, between *c* and *a.* This closure, however, is
not exact, for if the projections upon *c* are cylindrical their envelope
in *a,* that is, the groove, will not be circular, or if the groove be
made circular the projections should be portions of cylinders conaxial with 3.

Fig. 3, Pl. XIX., shows a very old rotary pump, that of

---

* *Propagation Industrielle,* vol. iii., 1868, p. 180.

† Clegg, *Manufacture of Coal Gas,* 5th Ed. 1868, p. 180,—Schilling *Gas-fabrika-
tion,* 1866, p. 204. It will be noticed that the block *c* is in two pieces, which overlap
each other in the middle of the drum *d.* The two pieces are connected separately
to the two sectors *b,* shaded in the figure. A later form of Beale's exhauster is
illustrated in *The Engineer,* May, 27, 1870, p. 329.

‡ *Propagation Industrielle,* vol. iv., 1869, p. 276.

Turning Block $(C_3'' P^\perp)^a$.

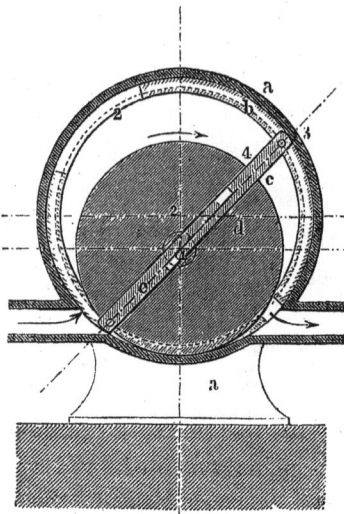

FIG. 1.—Beale. (Gas-Pp.)

$$2\left(C_3'' P^\perp\right)^{\frac{a}{d}};\ (V\pm) = c,\ a.$$

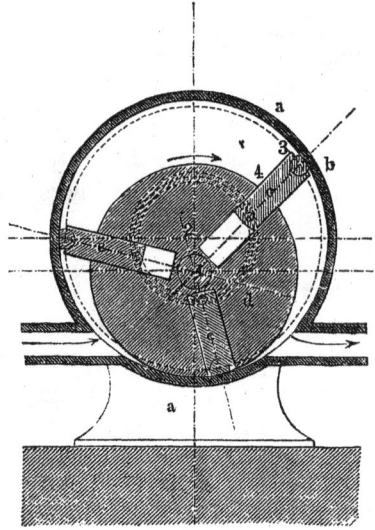

FIG. 2.—Davies. (Eng.)

$$3\left(C_3'' P^\perp\right)^{\frac{a}{c}};\ (V\pm) = c,\ a.$$

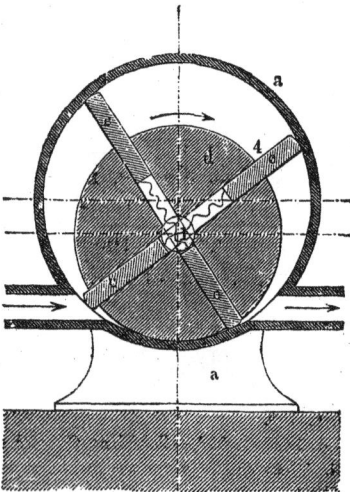

FIG. 3.—Ramelli. (Pp.)

$$4\left\{\left(C_3'' P^\perp\right)^{\frac{a}{d}} - b - \frac{c}{2}\right\};\ (V\pm) = c,\ a.$$

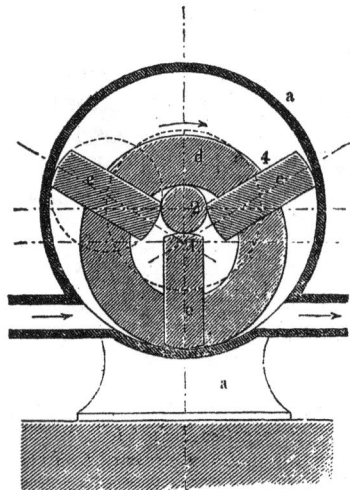

FIG. 4.—Jones, Ortlieb. (Eng., Pp.)

$$3\left\{\left(C_3'' P^\perp\right)^{\frac{a}{c}} - b\right\};\ (V\pm) = c,\ a.$$

Ramelli, described by him in 1588.* It is a combination of four reduced turning blocks, in which d is the driver, the coupler b is omitted and the block c force-closed. Its formula is therefore

$$4\left\{(C_3'' P^\perp)^{\frac{a}{d}} - b - \frac{c}{2}\right\}$$

The springs shown in the figure have often been used for the force-closure. Ramelli himself, however, placed the machine with the centre of the inner drum d above that of the chamber, and contented himself with such closure as he could obtain from the weight of the blocks. The chamber a is made a hollow cylinder, and there is higher pairing between it and the blocks. There have been repeated re-inventions of this pump, among others I may mention the steam-engine of Borrie† which had four blocks and springs, and Caméré's pump, which has only two pistons, but in which the closure is still effected by springs.‡

Fig. 4, Pl. XIX represents a machine very similar to that of Ramelli. It has been proposed in England as a steam-engine or pump § by Jones and Shirreff (1856) and in America as a steam-engine by Ortlieb and White (1867). It is a treble chain $(C_3'' P^\perp)^a$ reduced by b. The higher pairing between a and c rendered necessary by the omission of the coupler is quite correctly made. The first-named inventors leave the cylinder 2 free to revolve, in order to diminish as far as possible the friction between it and the blocks.

Fig. 1, Pl. XX., shows another arrangement of the same mechanism designed by Beale for a steam-engine || and actually applied by him to the propulsion of a vessel. Dalgety and Ledier used it for a pump in 1854.¶ Here the blocks c are made completely cylindrical and left entirely free to roll. That the machine was unsuccessful cannot be wondered at, for the higher pairing used at all the joints rendered it impossible that they should remain steam-tight.

---

* Ramelli, *Arteficiose Machine* (1588), pp. 58 and 167.

† Bataille et Jullien, *Mach. à Vapeur*, 1847, i., p. 445, pl. xi.

‡ *Propagation Industrielle*, vol. iv., 1869, p. 337.

§ Newton, *London Journal of Arts*, &c., new series, vol. vi., 1857, p. 9. *Schweizerische Polytechn. Zeitschrift*, vol. ii., 1857, p. 8.

|| Bataille et Jullien, *Machine à Vapeur*, vol. i., 1847, p. 444. Bourne, *Steam Engine*, p. 392. Beale appears to have employed centrifugal force-closure between the pistons c and the chamber.

¶ *Propagation Industrielle*, vol. iv., 1869, p. 84.

PLATE XX. CHAMBER-CRANK TRAINS.

Turning Block $(C_3'' P^\perp)^a$.

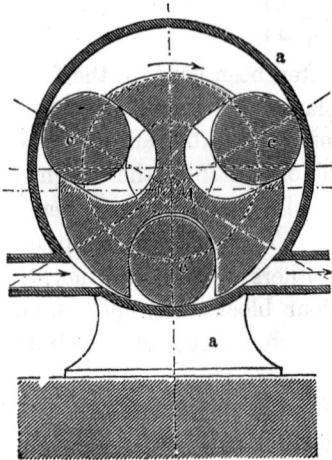

FIG. 1.—Beale, Dalgety. (Eng.)

$$3\{(C_3'' P^\perp)_c^{\frac{a}{c}} - b\}; \ (V\pm) = c, a.$$

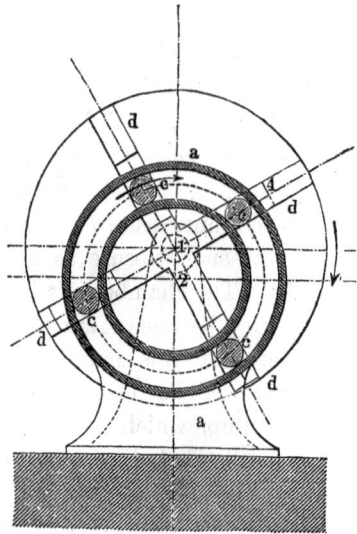

FIG. 2.—Smyth. (Eng.)

$$4\{(C_3'' P^\perp)_c^{\frac{a}{c}} - b\}; \ (V\pm) = c, a.$$

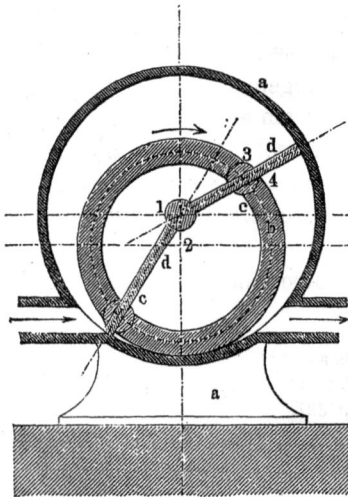

FIG. 3.—Cochrane, Hick, Lechat. (Eng.)

$$2(C_3'' P^\perp)_d^{\frac{a}{d}}; \ (V\pm) = d, a.$$

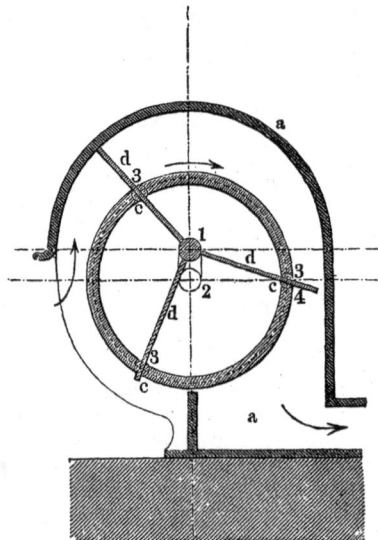

FIG. 4.—Bellford. (Ventilator.)

$$3(C_3'' P^\perp)_b^{\frac{a}{b}}; \ (V\pm) = d, a.$$

Turning Block $(C_3'' P^{\perp})^{\text{a}}$.

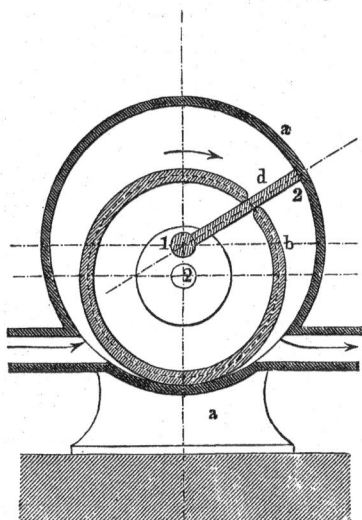

FIG. 1.—Cochrane. (Eng.)
$(C_3'' P^{\perp})^{\frac{\text{a}}{\text{d}}} - c$; $(V\pm) = d, a$.

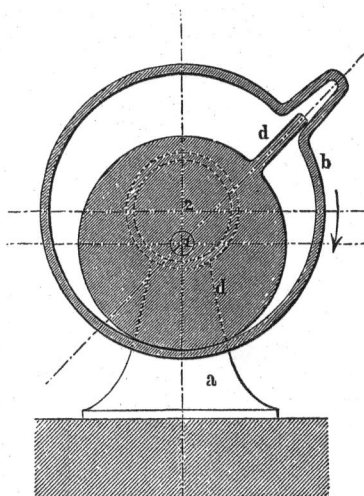

FIG. 2.—Cochrane. (Pp.)
$(C_3'' P^{\perp})^{\frac{\text{a}}{\text{d}}} - c$; $(V\pm) = d, b$.

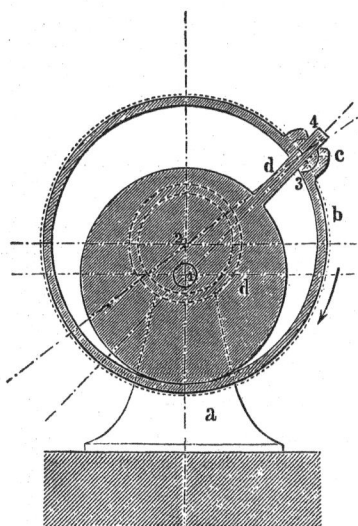

FIG. 3.—Reuleaux. (Example.)
$(C_3'' P^{\perp})^{\frac{\text{a}}{\text{d}}}$; $(V\pm) = d, b$.

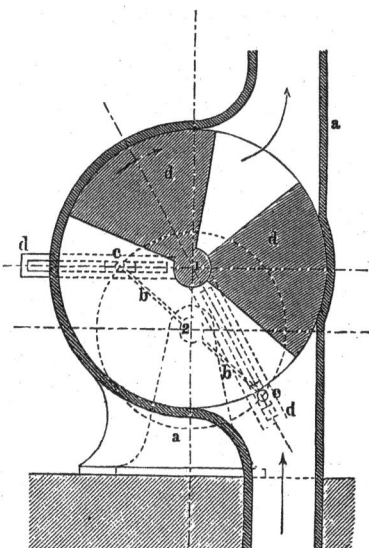

FIG. 4.—Minari, Stocker. (Eng., Pp.)
$2(C_3'' P^{\perp})^{\frac{\text{a}}{\text{d}}}$; $(V\pm) = d, a$.

The form of steam-engine given in Fig. 2, Pl. XX., which was designed by Smyth,[*] shows how far men may be led by the delusive fancy that there must be some special advantage in the rotary steam-engine. We have here a combination of four trains of the class $(C_3'' P^\perp)^a - b$, the blocks c are in their usual form, the higher pairing rendered necessary by the omission of the coupler is supplied by cylindrical bars working in a circular ring, and the steam expands only in the spaces enclosed between the moving cylinders. Its uselessness is obvious; it is only astonishing that it should ever have received any serious treatment at all.

Fig. 3, Pl. XX. Steam-engine consisting of two mechanisms of the class $(C_3'' P^\perp)^a$. Here for the first time we have d used as the piston; the coupler b is a drum, touching the inner wall of the chamber and revolving about the axis 2, the bearings of which are not shown. The block c forms a joint at the place where the slide d passes through the drum b. Lechat has both designed and constructed (1866) a machine on this plan which had one piston only, i.e. which contained only one train $(C_3'' P^\perp)^a$.[†] Hick, of Bolton, had already obtained a patent for it in England in 1843.[‡] Both, however, were long preceded by Lord Cochrane, who in 1831 constructed the machine both with two pistons (as shown in our sketch) and with one only, and at a later date added some small improvements to it.[§] The American Root has recently (1863) re-invented the same machine, using, however, three pistons instead of two.[||]

In Fig. 4 we have the same machine used as a three-armed ventilator. It was patented in this form in England (1855) by Rellford,[¶] and quite recently Root has again brought it to light [**] in very much the same form. We have here a striking example of the waste of inventive energy which has so often occurred in

* *Practical Mechanic's Journal*, vol. xvii., 1864-5, p. 261.

† *Génie Industriel*, vol. xxxii., 1866, p. 27.

‡ *Practical Mechanic's Journal*, vol. xix., 1866-7, p. 249.

§ *Propagation Industrielle*, iii., 1868, p. 181. Newton, *London Journal of Arts*, &c., Conjoined Series, vol. viii., 1836, p. 404, and vol. ix., 1837, p. 216.

|| *Scientific American.* New series, vol. viii., 1863, p. 63. This particular rotary engine, in a form either complete or reduced as in Fig. 1. Pl. XXI, has lately been repeatedly patented, e.g., by Works and Reynolds (1870), Higginson (1872) and Myers (1873). Its original inventor seems to have been Trotter (1805). See *Engineering*, Jan. 1 and June 11, 1875.

¶ Newton, *London Journal of Arts*, &c. New series, vol. v., 1857, p. 112.

** *Scientific American*, Nov. 1872, p. 354.

connection with rotary engines and pumps, and which the hitherto existing notions of Machine Kinematics have been powerless to prevent.

Pl. XXI. Fig. 1 shows another creation of the inventive talent of Lord Cochrane,[*] intended for a rotary steam-engine. It will be seen that this time he has given us a reduced chain. The link c of the arrangement in Fig. 11 is omitted, the mechanism being therefore $(C_3''P^{\perp})^a - c$, and the pairing of Fig. 269 is used, which leaves little hope that the joint will be steam-tight.

The same untiring nobleman, not contented with the series of combinations we have already mentioned, arranged the same mechanism also in the form shown in Fig. 2 of Plate XXI., which was intended for a pump.[†] Here again the chain is reduced by c, but b takes the place of a as the chamber, so that we have $(V^{\pm})=d, b$. If we compare the mechanism shown in Pl. XVII. Fig. 4 with that of the figure before us, we see that, apart from the arrangement of the ports, the one is simply an inversion of the other.

In order to make the matter clearer I add in Fig. 3, Pl. XXI., an arrangement of the same mechanism with the link c restored; it will be seen that it is an inversion of Knott's pump, Fig. 3, Pl. XVII.

We mentioned in considering Pl. XX. Fig. 2, a peculiarity connected with the use of multiple turning-block trains, namely, the enlargement and contraction of the space between the revolving pistons. This principle is applied in the mechanism shown in Fig. 4, Pl. XXI., which is so arranged that the space mentioned alternately almost disappears and becomes a maximum. It is a combination of two chains of the form $(C_3''P^{\perp})^a$, with d as piston and a as chamber. This reduction of the space between the pistons to zero is effected by causing them (the two sectors d), to cover a suitable angle. The slide d with couplers and blocks is placed outside the chamber. Minari,[‡] 1838, applied the chain $(C_3''P^{\perp})^a$ in a three-fold form; Stocker,[§] 1872, used a double train only, as in our figure—both intended their machines as steam-engines. In this case d would be the driving-link, which would be b if the machine

---

[*] *Propagation Industrielle*, as above.
[†] *Propagation Industrielle*, as above.
[‡] *Propagation Industrielle*, vol. iii., 1868, p. 276.
[§] *Baierisches Industrie-und Gewerbeblatt*, 1872, p. 167.

PLATE XXII.   CHAMBER-CRANK TRAINS.

Turning Block $(C''_3 P^\perp)^{\text{a}}$.

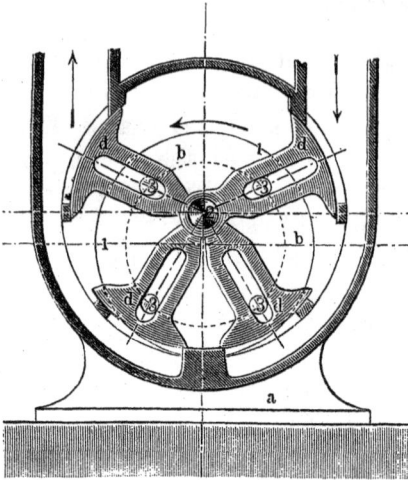

FIG. 1.—Smith. (Pp.)

$$4\lceil (C''_3 P^\perp)^{\frac{\text{a}}{\text{b}}} - c \rceil; \quad (V\pm) = d, a.$$

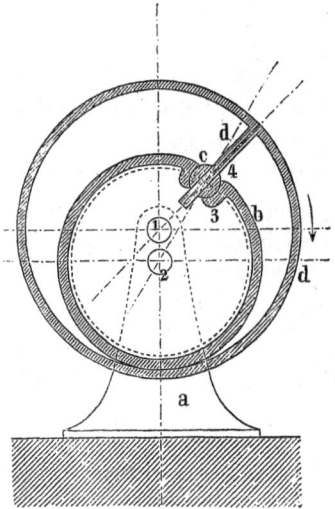

FIG. 2.—Cochrane. (Eng.)

$$(C''_3 P^\perp)^{\frac{\text{a}}{\text{b}}}; \quad (V\pm) = b, d.$$

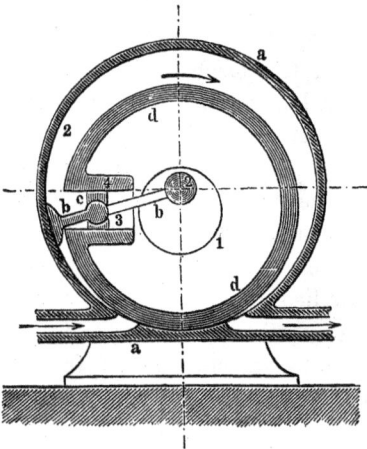

FIG. 3.—Reuleaux. (Example.)

$$(C''_3 P^\perp)^{\frac{\text{a}}{\text{b}}}; \quad (V\pm) = b, a.$$

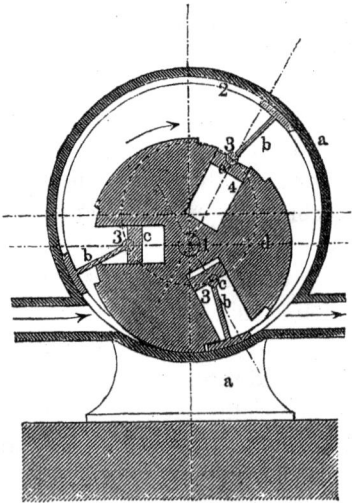

FIG. 4.—Fletcher. (Eng.)

$$3(C''_3 P^\perp)^{\frac{\text{a}}{\text{b}}}; \quad (V\pm) = b, a.$$

were a pump. The joint between piston and chamber can without
difficulty be made steam-tight, as only lower pairing is used.
Geo. Smith * used four connected turning-block trains, and placed
the couplers inside the chamber a. Fig. 1, Pl. XXII. shows his
arrangement adapted for a pump. The block c is omitted, and
the pairing of the reduced chain (Fig. 268) is therefore used. The
formula runs $4[(C_3''P^\perp)_5^a - c]$.

Lord Cochrane gives us still another form of chamber-crank gear
based on $(C_3''P^\perp)_5^a$, Fig. 2, Pl. XXII. Here the link d is the
chamber, and might at the same time be called the piston. We
should, however, perhaps rather call the drum-shaped coupler b the
piston, as has been done in the formula.

The coupler may also be made the piston with lower pair-
closure, to show which I add the arrangement of Fig. 3, Pl. XXII.,
where d is made a drum and c a joint-piece. A similar solution
has also been employed in practice by Fletcher (America ?) in
1843, in a rotary steam-engine which I know only by description
and have represented in Fig. 4. It contains three mechanisms of
the form $(C_3''P^\perp)^a$. The form of the coupler b here is remarkable,
and it was principally to explain it that I added Fig. 3. Instead
of using a circular pin of the common form at 2, the pin is expanded until it reaches the wall of the chamber a, and the pair 2
consists of a ring-shaped channel in a and a portion of a cylindric
ring (A^+) in the piston b. Although this construction disguises
the real nature of the machine in the most extraordinary way, our
analysis places it at once in its proper position without the least
constraint.

§ 83.

Chamber-crank Trains from the Swinging Slider-crank.

(Plate XXIII. Fig. 1.)

The swinging slider-crank $(C_3''P^\perp)^c$ is applied least often as a
mere mechanism of the four trains formed from the chain $(C_3''P^\perp)$,
and it has also least often been used in chamber-crank gear. I
know indeed only a single case in which it has been employed.

---

* *The Engineer*, Jan. 1871, p. 56.

PLATE XXIII.

CHAMBER-CRANK TRAINS.

Swinging Slider-crank $(C_3'' P^\perp)^c$.

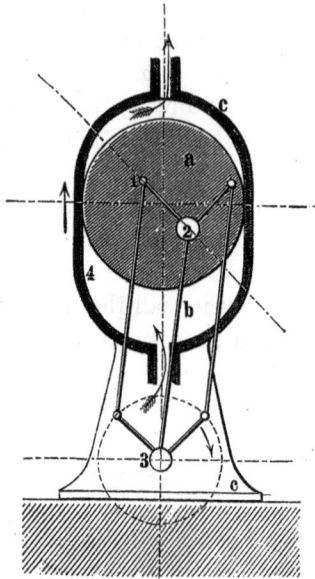

FIG. 1.—Simpson and Shipton. (Eng.)

$(C_3'' P^\perp)_{\frac{c}{a}} - d + (C_4''); \; (V\pm) = a, c.$

Double Slider-crank $(C_2'' P_{\frac{1}{2}}^\perp)^d$.

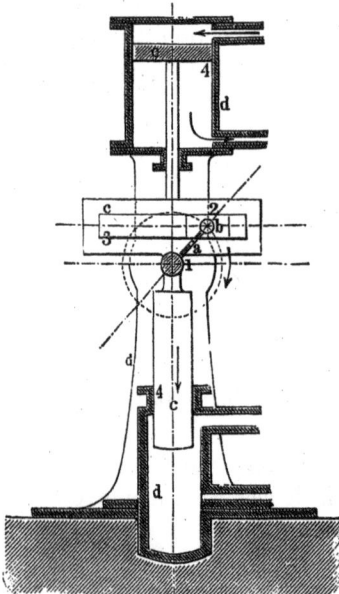

FIG. 2.—Donkey Pumping Eng.

$(C_2'' P_{\frac{1}{2}}^\perp)_{\frac{d}{c}}; \; (V\pm) = c, d.$

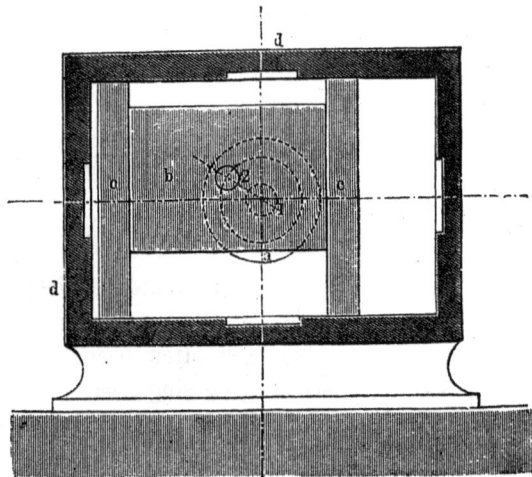

FIG. 3.—Root. (Eng.)

$(C_2'' P_{\frac{1}{2}}^\perp)_{\frac{d}{b+c}}; \; (V\pm) = bc, d.$

Messrs. Simpson and Shipton * employed it in the form of a steam-engine which was shown in action at the London Exhibition of 1851 and attracted no little attention. These inventors are the same as those of the machine shown in Fig. 2, Pl. XVII.; the want of any distinct comprehension of the problem which there showed itself appears here in almost greater measure, and accounts for the employment of a special auxiliary train to transfer the motion of the link a to an axis turning in fixed bearings. This auxiliary mechanism consists of a pair of parallel cranks, attached at the one end to a, at the other to a fly-wheel shaft, coaxial with 3. The couplers of this mechanism are parallel to the principal coupler b, and form with it as it were two pairs of parallel cranks. They are quite rightly adapted to transfer to the axis of 3 the motion of a, which as we know (§ 69) turns completely round during its motion. Apart from this secondary train the machine consists of a chain $(C_3'' P^\perp)$ reduced by d and placed on c, in which the crank a, formed as a cylindric piston, is the driving link, and the block c the chamber. The complete formula for the mechanism is therefore $(C_3'' P^\perp)_a^c - d + (C_4'')$, and for the chambering $(V^\pm) = a, c$. A comparison of this machine with that of Fig. 2, Plate XVII. allows us easily to understand how one and the same inventor devised both of them, for the one, $(C_3'' P^\perp)^c - d$, is simply a direct inversion of the other, $(C_3'' P^\perp)^b - d$.† There is higher pairing between c and a,—the preservation of a steam-tight joint is therefore scarcely possible. There would have been no difficulty, had the chain not been reduced, in employing lower pairing and obtaining a good joint by it; and at the same time the common form of steam-cylinder and piston—arranged in any way in two parts—could have been employed. I may leave it to the reader to investigate this much more practical form,—not advising, however, that even so arranged the machine be again brought forward as a steam-engine. In the form before us, however, the machine is in the highest degree unpractical, in no respect so advantageous as the common direct-acting engine. What can we say therefore to the extraordinary assertion of the inventor before the Mechanical

---

* Johnson, *Imperial Cyclopædia* ; Newton, *London Journal of Arts*, 7. Conjoined series, 27, 1850, p. 207 ; *Repertory of Patent Inventions*, Enlarged series, xiii., 1849, p. 287.

† In his collection of kinematic models at the Gewerbe Akademie in Berlin, Prof. Reuleaux has used the same (invertible) model for both.

Engineers in Birmingham, that "if (as here) the steam could be brought to bear on the crank direct, it would be a more simple and ready means (for utilising steam pressure) than at present in use"?[*] He says this, too, at the very instant at which he has placed between piston and fly-wheel shaft a mechanism constructively so difficult as the parallel cranks!

The particular machine referred to was said to work well;—but while this may be acknowledged, it must be added that it forms only one example of that tendency to aim rather at what we might call the *bravado* in machine construction than at the attainment of practically useful results which has too often proved fatal, in connection with rotary engines, to the judgments of sober and sensible men.

§ 84.

Chamber-crank Trains from the Turning Double Slider-crank.

(Plate XXIII. Figs. 2 and 3.)

The turning double slider-crank (§ 72) allows itself to be used very conveniently as a steam-engine or pump. It is indeed very frequently used in the form of the well known steam ("donkey") pump sketched in Fig. 2, Pl. XXIII. The cross-block c is formed both above and below as the piston, and the fixed link d as chamber;—the steam-cylinder is double-acting, the pump very frequently single-acting only. Its general and special formulæ are therefore $(C_2'' P\frac{\perp}{2})^d$ and $(C_2 P\frac{\perp}{2})_c^d$ respectively.

It will be noticed that the block b moves to and fro in its slot with the same sort of motion as the piston in its chamber, and indeed with the same stroke. If therefore b were to be formed as a piston and c as its chamber we could use with these a pressure organ again, and so cross the dead points by chain-closure (§ 46). It would be extraordinary if no inventive mind had yet hit upon this idea, and indeed we find that the American Root has based a design for a steam-engine upon it.[†] Fig. 3, Pl. XXIII. shows its general form. The two chambers d and c are prismatic,—of

---

[*] *The Engineer and Mechanist* (J. S. Browne), vol. i., 1850, pp. 215 and 234. *Scientific American.* New series, vol. x., 1864, p. 193.

rectangular cross section. All the joints have thus to be made with plane surfaces, a construction which presents some practical difficulties, but (as the surfaces belong to lower pairs) at least allows a steam-tight joint to be made. In its special formula we have to indicate that both b and c are driving links,—it will therefore run $(C_2''P\frac{1}{2})^{\frac{d}{b+c}}$. The sources from which we obtain our information about the machine call it the "very quintessence of simplicity," a judgment in which we can concur only to a limited extent if it be based on considerations as to practical usefulness.

§ 85.

Chamber-crank Trains from the Turning Cross-block.

(Plate XXIV.)

By placing the chain of the last-mentioned mechanism on a we obtain the turning cross-block, $(C_2''P\frac{1}{2})^n$, the motions in which we have already considered in § 72. This mechanism has also been employed in chamber trains intended for use as rotary steam-engines.

Plate XXIV. Fig. 1 shows the steam-engine of Witty, constructed in 1811.* This machine is a simple inversion of the mechanism of Fig. 2, Plate XXIII.—omitting the pump. The crank a is the fixed link or frame, the cylinder d and block b rotate, and the cross-block c makes the cardioidic motions which we now know, for its centroids and that of a are Cardanic circles of which the larger belongs to c. The circles are shown in the figure. The cross-block c is paired to d both by the sliding-pairs at 4 and as a piston. It must therefore be placed in the exponent as the driving-link.

Witty appears to have been interested in the motions of his machine, and to have examined the point-paths of the cross-block, for he constructed a second machine (in Hull) in which the chain was reduced by the block b, and higher pairing consequently used between c and a. (Fig. 2, Pl. XXIV.) The frame a is provided with a rim which has for its profile a curve equi-

* Severin's *Abhandlungen*, 1826, p. 62.

PLATE XXIV.

CHAMBER-CRANK TRAINS.

Turning Cross-block $(C_2'' P_{\frac{1}{2}}^{\perp})^a$.

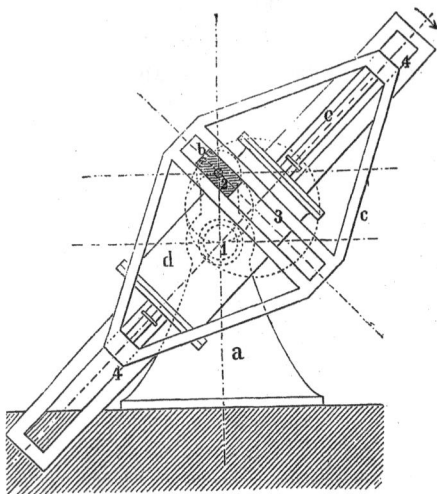

FIG. 1.—Witty. (Eng.)
$(C_2'' P_{\frac{1}{2}}^{\perp})^{\frac{a}{c}} ; \; (V\pm) \, c, d.$

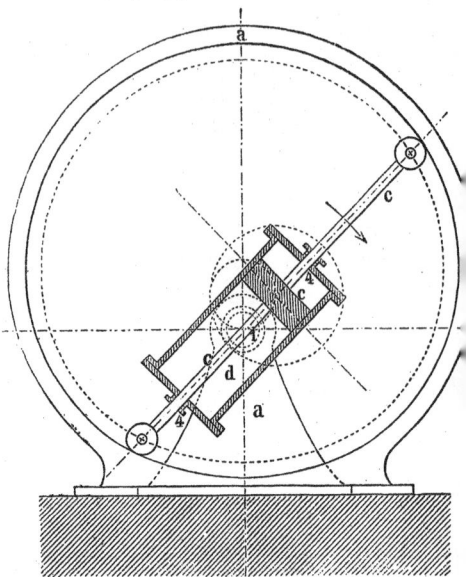

FIG. 2.—Witty, Andrew. (Eng.)
$(C_2'' P_{\frac{1}{2}}^{\perp})^{\frac{a}{c}} - b ; \; (V\pm) = c, d.$

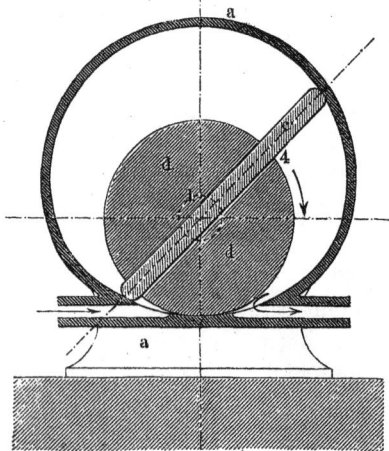

FIG. 3.—Franchot, Serkis-Ballian. (Eng.)
$(C_2'' P_{\frac{1}{2}}^{\perp})^{\frac{a}{c}} - b ; \; (V\pm) = c, a.$

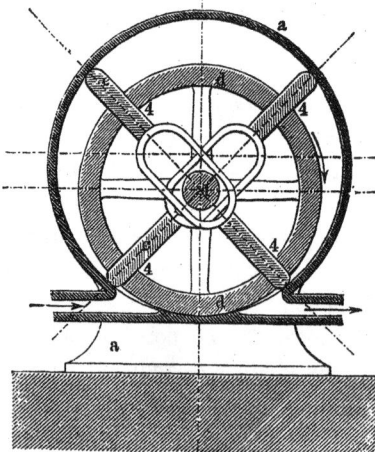

FIG. 4.—Woodcock. (Eng.)
$2[(\dot{C}_2'' P_{\frac{1}{2}}^{\perp})^{\frac{a}{c}} - b] ; \; (V\pm) = c, a.$

distant to a (curtate) peri-trochoid (cf. §§ 22 and 23). All diameters of this trochoid passing through 1 are equal, so that the rollers upon the rod c remain always in contact with the rim, have, that is to say, a closed motion. Witty's second machine can hardly be said to be practically useful, but notwithstanding this it was again proposed, in an almost unaltered form, by Andrew in 1858.[*]

Noticing the eccentric position of the axis 1 in the trochoidal ring, we see that it would not be difficult to use the link c as a piston, working within a suitably formed peri-trochoidal chamber. This has been several times done, as for example by Franchot in Paris. Fig. 3 is an outline of his machine. d is formed as a cylindric drum, c as a piston, its two semi-cylindrical ends in contact with the sides of a. Serkis-Ballian exhibited a very similar machine in Paris in 1867.[†]

Woodcock combined two chains in his machine,[‡] for which the general formula is therefore $2\left[(C_2'' P\frac{1}{2})\frac{a}{c} - b)\right]$, (Pl. XXIV. Fig. 4). The sectional profile of the chamber should again be peri-trochoidal; the block c is of fixed length and has rounded ends. According to the drawings before us, however, Woodcock made the chamber circular. Such an approximation is, of course, defective, but might work tolerably well if somewhat elastic packing pieces were used, and if the eccentricity, *i.e.*, the length of the crank, were made very small. The latter was the case in Woodcock's machine.[§]

Of the four machines mentioned in this section the three last have the least importance, for the peri-trochoidal profile does not possess any special advantages to counterbalance the constructive

---

[*] Newton, *London Journal of Arts*, &c., New Series vol. ix., 1859, p. 335.

[†] He obtained a French patent for its application as a pump,—*Propagation Industrielle*, vol. iv., 1869, p. 241; as to the steam-engine, see further *Génie Industriel*, vol. xxix. 1865, p. 203.

[‡] Newton, *London Journal of Arts.* Conjoined series, vol. xxiii., 1843, p. 93.

[§] Another engine formed from the train $(C_2'' P\frac{1}{2})^a$ in a way very similar to the two last is that of Hyatt, (Bourne, *St. Eng.*, p. 132) or Wilson (Bourne, *St. Eng.*, p. 392). Here, however, the chamber seems to have been made elliptical, or nearly so, and we are told that the working of the piston in it illustrates a "peculiar and unlooked-for characteristic of the elliptical figure," although "the true action is only to be secured when the amount of ellipticity is exceedingly slight!" A couple of force-closed sliding pairs are used to keep the pieces which correspond to the ends of the piston c in contact (with higher pairing) with the sides of the chamber. The arrangement is of course quite worthless.

difficulties connected with it; nor does the first machine, although in it the use of lower pairing allows the joints to be made steam-tight without difficulty, offer any practical advantages.

§ 86.

Chamber-crank Trains from the Lever-crank.

Plates XXV. and XXVI., Figs. 1 and 2.

The quadric crank-chain (C_4'') has also been used in many ways as a prime-mover or pump, by forming some pair of its links into chamber and piston. Our figures represent in the first place several machines formed from the lever-crank (C_4'')$^\text{d}$.

Fig. 1, Pl. XXV. shows a mechanism which has been used by Bramah as a pump,[*] and by Morgan [†] (1830) and then by Ericsson [‡] as a steam-engine. The lever c is in each of the three cases formed as the piston, the frame d as the chamber. The motion of the piston is therefore an oscillation about the axis 4. If used as a pump its special formula is $(C_4'')^{\frac{\text{d}}{\text{a}}}$.

Instead of making the piston form a single sector only, two or more sectors may be employed for the purpose. Gray, in the arrangement shown in Fig. 2, used sectors which had for their outer surface spherical zones,[§]—the chamber being made part of a corresponding hollow sphere,—and on this account he called his machine a spherical steam-engine. This form does not appear to possess any special advantage whatever. Thompson,[||] who used two of his mechanisms as a coupled engine, used the links c and

---

[*] Laboulaye, *Cinématique*, 1864, p. 776.

[†] *Propagation Industrielle*, vol. iii., 1868, p. 151.

[‡] Johnson, *Imperial Cyclopædia*, Ericsson's semi-cylindrical marine engine, Description of the plates, p. 3. See also Bourne's *Modern Examples of Steam, etc. Engines,*—where Ericsson's engines of the *Dictator* are illustrated by plates of which the excellence is worthy of a better subject. Root has also proposed a "double quadrant" engine in which the two pistons swing upon different centres in one chamber, which the crank shaft crosses between them. See Spon's *Dictionary of Engineering*, p. 2444.

[§] *Génie Industriel*, vol. xii., 1856, p. 15. See also vol. xviii., p. 317; and *Schweizerische Polytechn. Zeitschrift*, vol. i., 1856, p. 140.

[||] Newton, *London Journal of Arts*, &c. Conjoined Series, iii., 1834. p. 125.

Lever-crank $(C_4'')^{\mathrm{d}}$.

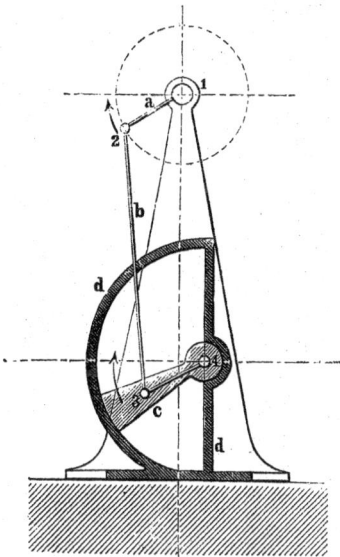

FIG. 1.—Bramah, Morgan, Ericsson. (Eng.)
$(C_4'')_{\mathrm{c}}^{\mathrm{d}}$; $(V\pm) = c, d.$

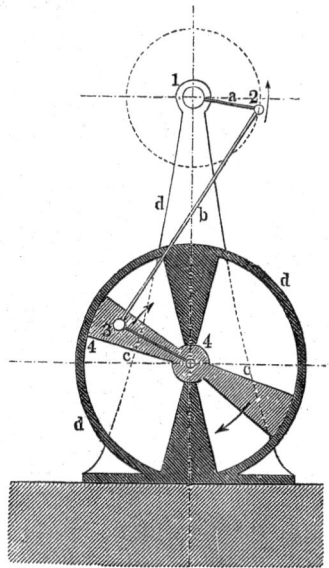

FIG. 2.—Thompson, Gray. (Eng.)
$(C_4'')_{\mathrm{c}}^{\frac{\mathrm{d}}{}}$; $(V\pm) = c, d.$

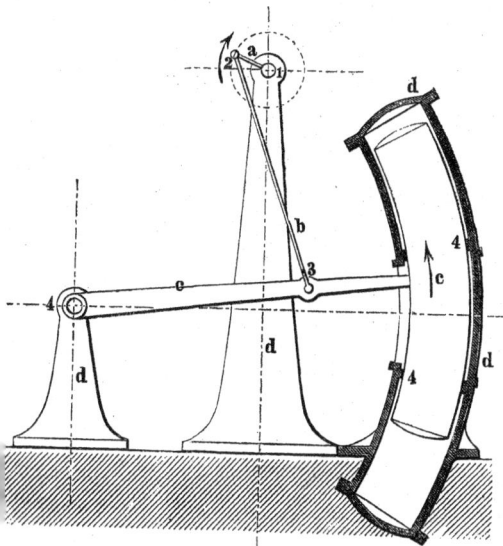

FIG. 3.—Degrand. (Eng.)
$(C_4'')_{\mathrm{c}}^{\frac{\mathrm{d}}{}}$; $(V\pm) = c, d.$

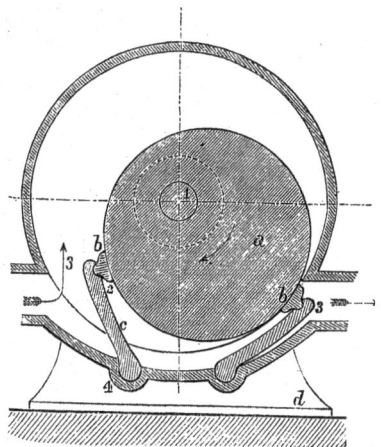

FIG. 4.—Cochrane. (Eng.)
$2\left\{ (C_4'')_{\mathrm{a}}^{\frac{\mathrm{d}}{}} - \frac{b}{2} \right\}$; $(V\pm) = a, d.$

PLATE XXVI.   CHAMBER-CRANK TRAINS.

Lever-crank $(C_4'')^d$.

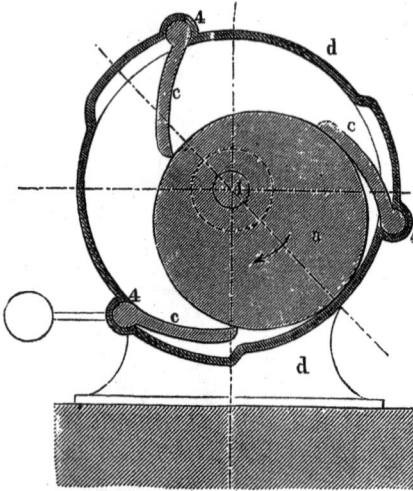

FIG. 1.—Cochrane. (Eng.)

$$3\left\{(C_4'')^{\frac{d}{a}} - b - \frac{c}{2}\right\}; \ (V\pm) = a, d.$$

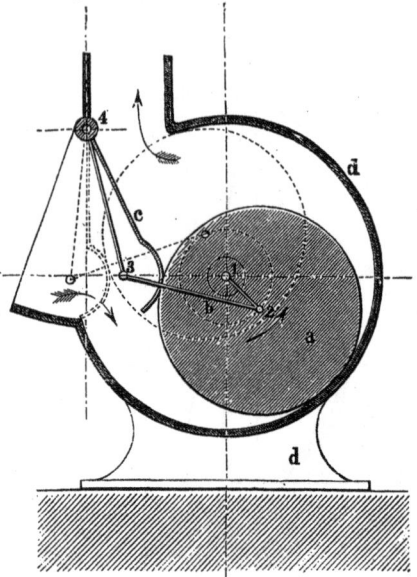

FIG. 2.—Cooke. (Ventilator.)

$$(C_4'')^{\frac{d}{a}}; \ (V\pm) = a, d.$$

Double-crank $(C_4'')^a$.

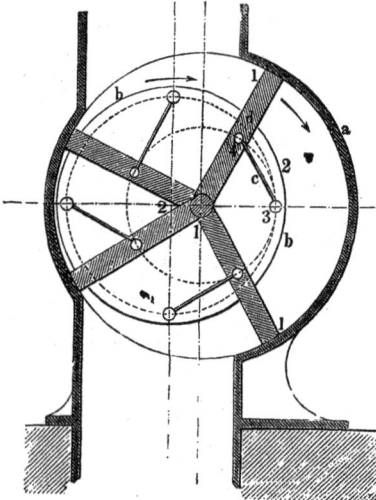

FIG. 3.—Heppel. (Pp.)

$$4(C_4'')^{\frac{a}{b}}; \ (V\pm) = d, a.$$

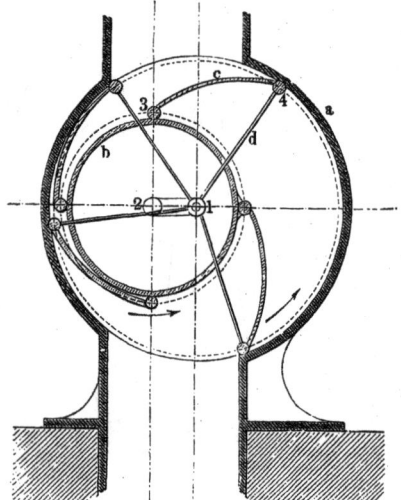

FIG. 4.—Lemielle. (Ventilator.)

$$4(C_4'')^{\frac{a}{b}}; \ (V\pm) = c, a.$$

d as chamber and piston respectively,—$(V^{\pm})$ = d, c,—but fixed the piston and allowed the chamber to oscillate.

Degrand formed the piston c as a sector of a so-called cylindric ring ("globoid ring")* which worked through two stuffing-boxes in the chamber d, as in Fig. 3.† The construction of such a machine prevents obvious difficulties.

Fig. 4, Pl. XXV. represents another rotary steam-engine of Lord Cochrane's.‡ The chain is here used quite differently,—the crank a is made the piston and d the chamber. A doubled mechanism is used in order to prevent the steam passing away at wrong times. The coupler b has become a gun-metal packing piece, and the pins 2 (the crank pin) and 3 are not closed kinematically but only by the steam pressure or by some external load. The lever c forms a diaphragm swinging in the chamber. On account of the force-closure of b the formula will be $2\left\{ (C_4'')^{\mathrm{d}} - \dfrac{b}{2} \right\}$, and we have also $(V^{\pm}) = a, d$.

It will be remembered that the crank a has to work steam-tight both against the sides and the periphery of the chamber. The details of the packing are omitted in our figure.

Lord Cochrane has also used this train reduced, as in the case of Fig. 3. Pl. XX.,—the arrangement he adopted is shown in Fig. 1, Pl. XXVI.§ Three mechanisms are here combined; the coupler b is omitted, the lever c itself being held in contact with a by force-closure. The steam and exhaust ports are formed in the piston, and made to communicate with passages and openings in the shaft 1.

Cooke used the mechanism $(C_4'')^{\mathrm{d}}$ in the first place as a steam-engine ‖ (1868); making a the piston, d the chamber, and forming c as a curved block sliding in and out through the wall of the chamber and having its end force-closed against the piston a, it became therefore $(C_4'')^{\mathrm{d}} - b$. Afterwards, however, he employed it as a blower in the form shown in Fig. 2, Pl. XXVI.¶ The chain

---

* *Cf. Berliner Verhandlungen*, 1872, p. 248, note 3.

† *Propagation Industrielle*, vol. iii., 1868, p. 245. The French patent is dated 1837.

‡ Bataille et Jullien, *Machines à Vapeur*, vol. i., 1847, p. 445; *Repertory of Patent Inventions*. Enlarged series, vol. ii., 1843, p. 193.

§ *Propagation Industrielle*, vol. iii., 1868, p. 182. The patent dates from 1831.

‖ *Propagation Industrielle*, vol. iv., 1869, p. 337.

¶ *Proc. Inst. C.E.*, November, 1875; *Engineering*, vol. viii. (1869) p. 269.

is here complete,—crank, coupler, lever, and frame all appearing in their ordinary forms outside the chamber. d is made the chamber and a the piston, while a piece connected with the lever c remains always in contact with the latter. The contact here is made by forming the end of this piece as a portion of a cylinder having its axis at 3. This machine has been constructed upon a large scale and used as a ventilator in mines. It is the only one of those represented in the last six figures which can really be called practical, for it presents no special constructive difficulties, and with the small pressures necessary in a mine ventilator there should be no difficulty in making the jo'nts sufficiently tight.

§ 87.

Chamber-crank Trains from the Double crank.

Plate XXVI., Figs. 1 and 2, and Plate XXVII.

The double-crank $(C_4'')^a$ has been several times used in chamber-gear ; I illustrate here five cases of its employment.

Fig. 3, Pl. XXVI. represents a pump constructed by Heppel, (Switzerland).[*] It is a combination of four double-cranks. The train of thought which has led up to it may have been somewhat similar to that which produced the machine, Fig. 4. Pl. XXI. The fixed link a is again the chamber, and one of the turning cranks d the piston, while the second turning link b—here made in the form of a disc, having 2 for its centre, and connected by the link c with the piston—allows of the alternate increase and decrease of the velocity of the latter.

Lemielle's ventilator,[†] Fig. 4, Pl. XXVI., which is frequently used, is a machine very similar to the last. Here a is again the chamber, but c is made the piston instead of d, and the crank b is formed as a drum of suitable dimensions. Very much the same may be said of Lemielle's ventilator as of Cooke's ; it has been constructed upon a large scale, and has found many friends in Belgium and in England.

---

[*] *Propagation Industrielle*, vol. iv., 1869, p. 85. French patent dated 1855.

[†] Weisbach, *Mechanik*, vol. iii., part 2, p. 1118 (where a two-armed "piston-wheel" of Lemielle's is described) ; Dingler, *Polytech. Journal*, vol. 150 ; *Civil Engineer and Architect's Journal*, Sept. 1858 ; *Civil Ingénieur*, i., 1854, p. 83.

Double-crank $(C_4'')^a$.

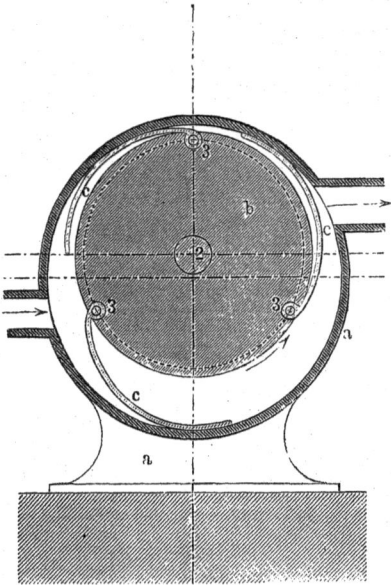

FIG. 1.—Ramelli. (Pp.)

$$3\left\{(C_4'')^{\frac{a}{b}} - d - \frac{c}{2}\right\};\ (V\pm) = c, a.$$

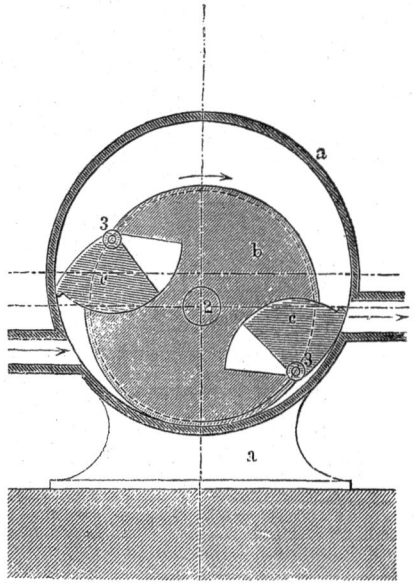

FIG. 2.—Cochrane. (Eng.)

$$2\left\{(C_4'')^{\frac{a}{c}} - d - \frac{c}{2}\right\};\ (V\pm) = c, a.$$

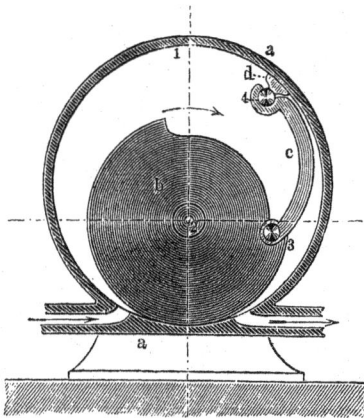

FIG. 3.—Rösky. (Eng.)

$$(C_4'')^{\frac{a}{c}} - \frac{d}{2};\ (V\pm) = c, a.$$

Fig. 1, Pl. XXVII. represents another primitive pump which comes to us from Ramelli.* It is a combination of three mechanisms of the class under consideration, and might be looked at (apart from its date) as formed by reduction from Lemielle's wheel. The guiding crank d, namely, is omitted from the chain, and also the higher pairing which should replace it, so that the piston c is forceclosed only against the chamber. Its general formula, is therefore

$$3\left\{(C_4'')^a - d - \frac{c}{2}\right\}.$$

The relation between this machine and that of Lemielle forms an interesting example of the course which we described in Chapter VI as that taken by machine development,—the newer form of Lemielle is simply the kinematic completion of the older one of Ramelli.

In Fig. 2, Pl. XXVII., we have a machine which is an adaptation of Ramelli's pump to the purposes of a steam-engine,—it is another result of the unwearying activity of Lord Cochrane. The link d is again omitted, and c force-closed. The machine is altogether very incomplete.

Fig. 3, Pl. XXVII., represents another, and much less incomplete form of rotary steam-engine founded upon the mechanism $(C_4'')^a$. It is the invention of Rösky (Elbing) and is known to me only by description, The chain is used singly only, so that a separate valve has to be employed to prevent the steam from passing through the chamber when c is at the lower part of it. The link d is held against the wall of the chamber·by force-closure. Its formula is therefore, c being the driving link, $(C_4'')^{\frac{a}{c}}$, and for the chambering $(V^{\pm}) = c, a.$[50]

§ 88.

Chamber Trains from Conic Crank Mechanisms.

Among the numerous steam-engines and pumps which we have now described and analysed there are many which it has been possible either to explain or to describe by the older methods only with difficulty, and the nature of which, therefore, has remained indistinctly understood by very many mechanicians. Besides these, however, there exists a machine, or rather a small series of machines,

* Ramelli, *Arteficiose Machine*, 1588, p. 60.

which has been hitherto far less understood than any of those which have been mentioned. I mean the so-called "disc-engine," and the rotary engines and pumps constructively connected with it. Since their invention, forty years ago, the real nature of these machines has remained a riddle. It is related that a distinguished philosopher of the last decade said to his students towards the close of his life that all of them had failed to understand him except one, and that he had misunderstood him. The study of these machines reminds us irresistibly of this story, for it may really be said of them that very few have understood them, and these, their inventors included, have understood them wrongly !

There have been a number of these machines. One of them was exhibited at the Paris Exhibition of 1867 ; another spread its work daily all over the world by driving the *Times* printing-press.[*] Wherever we look for an explanatory description of the action of the machine, however, we find, according to my experience, that the whole matter is enveloped in a kind of glorified haziness.[†] The most various turnings and windings are used in the patent specifications in order to make clear to others what a happy inspiration has given to the inventor; while the theorist whose work compels him to treat the machine kinematically obviously hurries over it, sometimes in a very confused manner, and sometimes even with the direct confession of failure to comprehend it.

The cause of all this lies simply in the fact that the methods of investigation hitherto used have attempted to find motions without first looking for the conditions under which they are constrained. What is really before us in the disc-engine and other machines of the same kind is nothing more than a series of chamber-crank trains formed from conic crank mechanisms. For by treating these in the way described in § 78, *i.e.*, by forming suitable links as chamber and piston and adding proper valve gear, we can obtain all those varieties of conic chamber-crank gear which have yet made their appearance.

It is remarkable that the course of empirical invention has hitherto confined itself within the limits of the case where three of the links of the chain (C_4^L) are right-angled, the chain therefore being $(C_3^{\perp} C^L)$. This chain we called above (§ 75) the

---

[*] This engine, which was no success, was thrown aside in 1857.

[†] See for example a passage quoted by R. S. Burn, *Steam-Engine*, p. 137.

(normal) conic double-slider chain. We showed that three mechanisms could be formed by it; two of these appear as chamber-crank gear, and the energy which always seems to accompany the search for solutions of the problem of the rotary engine has found for these two a great variety of forms.

§ 89.

Chamber-gear from the Conic Turning Double-slider.

Plates XXVIII to XXXI.

The rotary steam engine which specially receives the name of disc-engine is represented in Fig. 1, Pl. XXVIII. It is commonly known as Davies' [*] engine, but is sometimes also called after Bishop, who at a later date made some improvements in it. The first inventors are, according to published accounts, the brothers Dakeyne (England), who patented the machine in 1830, and proposed to use it both as a steam-engine and as a pump.[†] There are many descriptions of it; that of Johnson [‡] is very complete, the particular machine he speaks of being the one above mentioned as having been used in the *Times* printing office, which was constructed with Bishop's improvements. It is in reality the mechanism $(C_{\frac{1}{3}}^{\perp}C^{L})^{d}$, the turning conic double-slider, (see § 75. No. 15), in which the fixed link d is made the chamber, and the coupler b the piston. It is therefore the same kinematic chain which, placed upon a, forms the Hooke's or universal joint.

The crank a is easily recognised; it turns upon the pin 1, and is paired by the pin 2 with the coupler b. This carries at 90° from 2 its second pin 3, paired with the block c. The last mentioned link turns about an axis perpendicular to the plane of the paper and therefore also at 90° to the axis of 3. The cylinder-pair or, more generally, the pair of revolutes, of this fourth axis is not fully constructed; it consists of the block c formed as a sector, and

---

[*] In consequence of a misprint in *Bernouilli*, which has now survived five editions, this has been known in Germany for years as D a r r i e s machine.—*R.*

[†] *Repertory of Patent Inventions*, vol. ii., 1831, p. 1; Newton, *London Journal of Arts*, &c., Second Series, vol. ix., 1834, p. 19. The Dakeynes do not seem actually to have constructed any machines under their patent.

[‡] Johnson, *Imperial Cyclopœdia*, Steam-engine, p. 19, plates xii. to xiv.

Conic Double-slider $(C_3^{\perp} C^L)^d$.

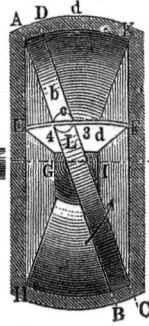

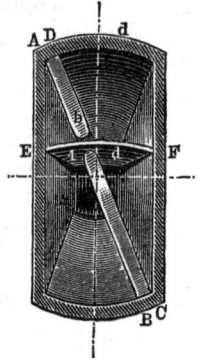

FIG. 2.

FIG. 3.

FIG. 1.—Davies, Bishop. (Eng.)

$$(C_3^{\perp} C^L)_{\overline{b}}^{d} \; ; \; (V\pm) = b, d.$$

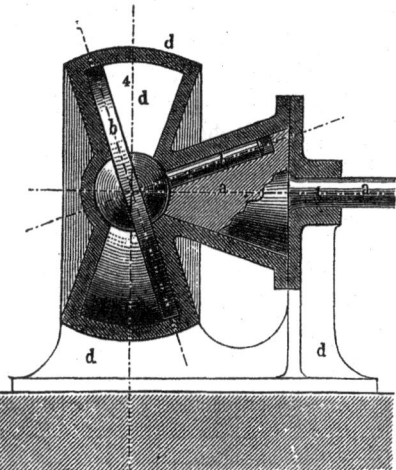

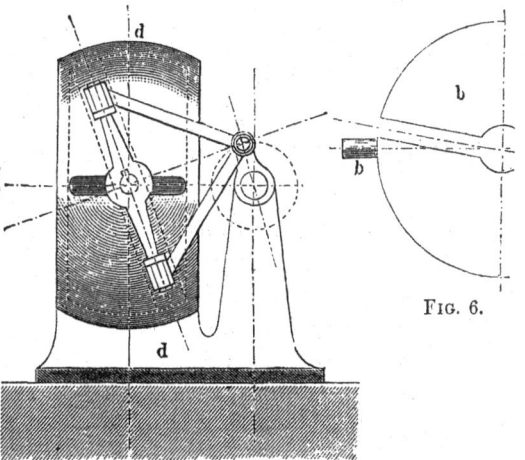

FIG. 6.

FIG. 4.—Bouché, Molard. (Eng.)

$$(C_3^{\perp} C^L)_{\overline{b}}^{d} \; ; \; (V\pm) = b, d.$$

FIG. 5.—Davies. (Pp.)

$$(C_3^{\perp} C^L)^d - c - a + (C_2\, G) \; ; \; (V\pm) = b, d.$$

a channel or groove in which it slides, and which forms a part of
the chamber d. We thus have in the links a, b, c, d, the kinematic
chain

$$C^+ \ldots \angle \ldots (C) \ldots \perp \ldots (C) \ldots \perp \ldots (C) \ldots \perp \ldots C_{=}^-.$$

which is placed upon the right-angled link d adjacent to the acute-
angled link a.

It is most important that the arrangement of the chamber and
piston should be rightly understood. The link b forms the piston.
It has the form of a plane disc provided with spheric surfaces to
work in contact with the chamber both at its centre and periphery.
The form of the interior of the chamber d is that of two circular
cones, which are the envelopes to the motion of the plane sides of
the disc. These cones are indicated in Fig. 2 by the letters $A H G$
and $C I K$. The surfaces $A B$ and $C D$ of the piston touch them
always in one generator, for the axis 2 always makes the same
angle a, the complement of the vertex angle of the cones, with
their axis.

It will be noticed that the geometrical axis of the pin 3 moves
always in one plane, in the figure the plane of the paper. Parallel
to this plane there is a diaphragm 4 fixed in the chamber. It has
plane sides; in reality, however, it is nothing but a portion of a
revolute upon an axis which is perpendicular to the plane of the
paper and passes through the centre of the chamber. In other
words, it is a part of the same figure as the ring channel within
which the block c slides. A corresponding continuation of the
block c itself also exists, as the section in Fig. 2 shows. This
figure is a projection upon a plane somewhat inclined to the horizontal, so as to show a portion of one side of the diaphragm. It
shows at L the continuation of c in the form of two portions of a
cylinder paired internally with the diaphragm d, and externally
with an open cylinder in the disc b. Both pairings are lower, so
that these continuations of the block serve as packing pieces.
Between them and the external sliding-block c there is no kinematic difference; the pair 4 externally takes the form $R_{=}^{\pm}R^-$ (or
$C_{=}^{\pm}C^-$) and internally the form $R_{=}^-R^+$ (or $C_{=}^-C^+$); and the pair 3
takes the forms $C_{=}^-C^+$ and $C_{=}^{\pm}C^-$ respectively.

If we assume that it is possible to obtain a tight joint by the
higher pairing between the surfaces of the disc and the walls of

the chamber we can now see how the periodic sweeping of the chamber by the piston allows us to use the machine with a pressure-organ. The diaphragm $E F$ divides the horse-shoe shaped space on each side of the disc into two parts, the volumes of which change alternately from zero to the whole capacity of the horse-shoe and back again to zero. A suitable valve gear is added and completes the machine as a double acting pump or steam-engine.

I must here mention that I have used the form of a slotted cylinder at L, for the piece which is kinematically identical with the block c, somewhat at hazard, for in the descriptions it is always spoken of simply as a packing-piece and its nature is not always indicated very clearly. In Johnson's beautiful engravings mentioned above this important point is entirely omitted. From several of the patent drawings it might be inferred that the disc was simply provided with a radial slit as shown Fig. 3. This would be the reduction of the chain by the block c and the substitution for the latter of the kind of higher pairing shown in Fig. 270 ; it would be analogous, therefore, to the reduction used in Lamb's steam-engine Fig. 1, Pl. XV. It may be further remarked that in the earliest disc-engines (Dakeyne's and some others) the external semicircular hoop b with the pin 3 and the external block c are omitted, the link c being thus entirely absent. The mechanism shown in our engraving, where b is the driving-link, has for its special formula $(C\frac{1}{3}C^L)_b^d$.

There are considerable difficulties in the way of making a steam-tight joint at the line of contact of the disc and the cones. Bishop attempted to solve this problem by covering each of the conic surfaces with an armour of packing plates, upon which the disc could slide, and which were pressed from behind by adjustable springs. The *bravado* in machine-construction appears to make light of all difficulties! We can only wonder how far men may be led by that fascination of whatever is unusual and singular which once surrounded, and for many persons appears still to surround, everything connected with machinery.

If we look for the counterpart of the disc-engine among the forms of chamber-crank gear obtained from cylindric crank trains, we might take some of those formed from the turning double-slider $(C_2 P\frac{1}{2})^d$, for the three infinite links of this chain correspond to the three right-angled links of the mechanism before us.

The right-angled links in the conic trains do not, however, differ so widely from the others in form as the infinite links in the cylindric chains. We may on this account look for the counterpart of the disc-engine also among the mechanisms $(C_3''P^\perp)^d$, and among them we find in Lamb's engine one which has a great resemblance to it. In both b is the piston and the driving-link and d the chamber; in both also the piston, on account of the omission of c, is fitted with a slot which moves to and fro upon a plate forming part of the frame; in Lamb's engine, too, we might replace c by a slotted cylinder similar to that here shown. The comparison is very instructive.

The disc-engine has been introduced into France by Bouché * and by Molard† in a form somewhat differing from that above described; —the last named engineer, especially, has endeavoured to extend its use as a steam-engine. The form used by both is that shown in Fig. 4, Pl. XXVIII. Here the chamber d is extended over the pin 2 of the link b, and encloses a cone which forms part of a. This cone is simply the revolute 1 properly belonging to a. Bishop's external coupler b and block c are not used. Within the slot of the disc there appears, however, to have been used a metal packing-piece corresponding to the block c.‡ The careful construction of the surfaces of the disc and its enveloping cones is relied upon to give a steam-tight joint between them.

Fig. 5 shows an older form of the disc-machine, proposed by Davies in 1837 and intended for a pump.§ Here again the link c is entirely omitted, the crank a indeed is also omitted, but there is substituted for it an auxiliary mechanism. This consists of a crank with a spheric pin placed upon an axis normal to that of the chamber, and connected by means of a V-shaped coupler to a cross-spindle forming part of b. This spindle, for constructive reasons, is placed as in Fig. 6, and not at right angles to the slot in the disc, an error which might easily have been avoided. The entire machine is a clumsy approximation to $(C_3^\perp C^L)^d$, which could

* *Propagation Industrielle,* vol. iii. 1868, p. 244 (Patent dated 1835).

† *Rapport du Jury International,* 1868, vol. ix., p. 82.

‡ Tresca, *Rapport sur une Machine Locomobile de M. Molard,—Bulletin de la Soc. d'Encouragement,* 2nd Series, vol. xix, 1872, p. 49. Tresca suggests, too, that the " temporary success " of the disc-engine was not unconnected with the novelty of its method of action.

§ Newton, *London Journal of Arts,* &c. Conjoined series, vol. xix., 1842, p. 18.

serve in any case only for pumps working against a very small head. Practically the whole construction is worthless.

§ 90.

Chamber-gear from the Conic swinging Cross-block.

Plate XXIX.

The swinging cross-block $(C\frac{1}{3}C^L)^b$ gives us results which do not differ kinematically from those furnished by the turning double-slider $(C\frac{1}{3}C^L)^d$. This has been already shown in § 75. Here, however, I have separated the two classes of mechanisms, for this enables us to arrive more naturally at the three following machines, which are indeed almost literally inversions of those which we have been considering.

Fig. 1, Pl. XXIX. in the first place, which is given only to make the nature of the others more intelligible, is a simple inversion of Fig. 1, Pl. XXVIII.; b is fixed, the chamber d moves upon it, a turns about 2 instead of about 1; the block c is placed as a packing-cylinder in the slot of the disc b. The figure will help to explain that of Duncan's machine[*] Fig. 2. In it b is again the fixed link, but the piston disc has taken the form of a double cone, and the chamber is made spherical. The revolute 1 of a, which in the last figure is a simple cylinder, is here the double cone; its axis is AA, which is caused by the link d to oscillate exactly as in Fig. 1. The block c might be arranged so as again to form a packing-cylinder, having a perpendicular axis in the centre of the partition in b. Duncan appears, however, to have used hemp packing.

Fig. 3, Pl. XXIX. shows a construction in which the crank of Fig. 2, or rather of Fig. 1, is replaced by a very much less advantageous arrangement. It was proposed by Davies in 1837,—besides the pump we have described (Fig. 5, Pl. XXVIII.),—and the addition consists of a cylinder pair, and four spheric pairs, that is of a chain $(C\ G_4)$, replacing the crank. A patent was taken out in France in 1838, by Gossage,[†] for a very similar arrangement, which was

---

[*] Clark's *Table of Mechanical Motions*, Nos. 61 and 62.

[†] *Propagation Industrielle*, vol. iii., 1868, p. 246.

PLATE XXIX. CHAMBER-CRANK TRAINS.

Conic Swinging Cross-block $(C_3^{\perp} C^L)^b$.

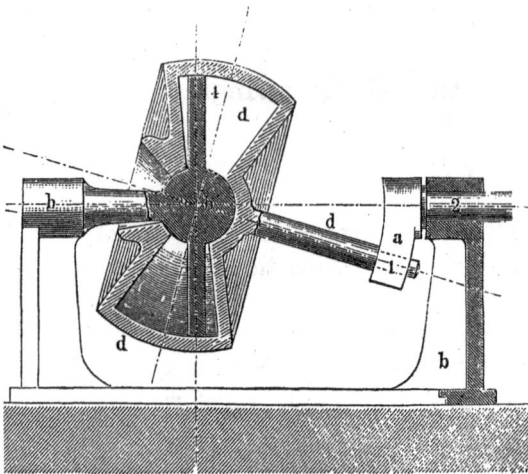

FIG. 1.—Reuleaux. (Ex.)
$(C_3^{\perp} C^L)^{\frac{b}{a}}; (V\pm) = b, d.$

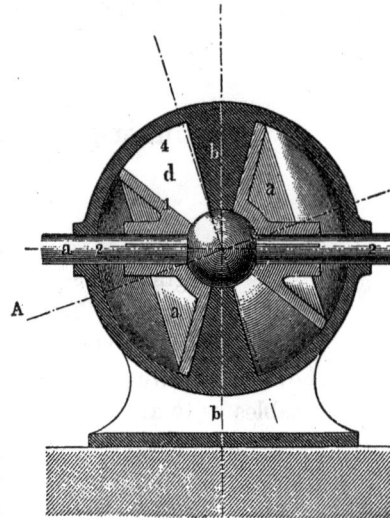

FIG. 2.—Duncan. (Eng.)
$(C_3^{\perp} C^L)^{\frac{b}{a}}; (V\pm) = a, b.$

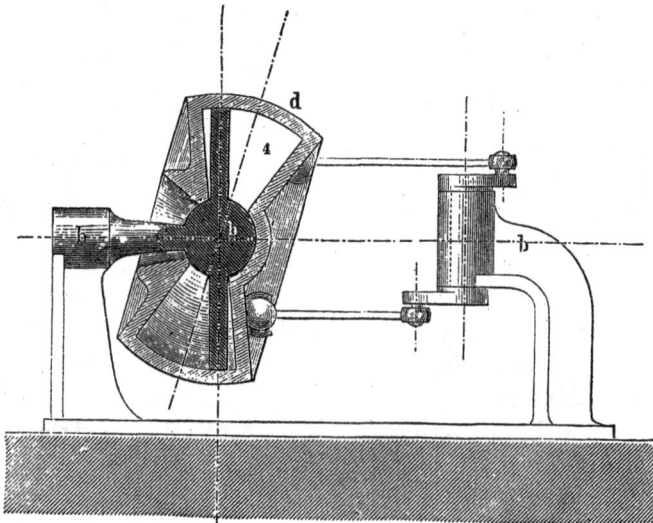

FIG. 3.—Davies. (Eng. and Pp.)
$(C_3^{\perp} C^L)^{\frac{b}{a}} - a + (C G_4); (V\pm) = b, d.$

intended for a steam-engine. If these two inventors had only noticed how much more easily they could attain their object by a mere inversion of the chain such as is shown in Fig. 1, they would have spared themselves the trouble of scheming their complex substitute for the crank..

§ 91.

Chamber-gear from the Conic Turning Cross-block.

Plates XXX. and XXXI.

Of the two remaining positions of the conic double-slider chain, that upon c has not found any favour with the inventors of rotary steam-engines,—while that upon a, the train $(C_3^\perp C^L)^a$, has been very frequently turned to account. This mechanism, which is that of the universal joint, has (unknown to the inventors) formed the basis of the six following machines.

Fig. 1, Pl. XXX. Rotary steam-engine of Taylor and Davies.[*] The chambering is here carried out exactly as in Fig. 1, Pl. XXVIII, and the packing piece c is also added. The crank a, however, is made the frame, so that both the piston b and the chamber d (the latter carrying the block c with it) have pure turning motions. They correspond exactly to the two shafts of the universal joint, while the block c, which we have seen to consist of two revolutes with their axes at right angles, is the cross itself.

Fig. 2, Pl. XXX. Rotary steam-engine of Larivière and Braithwaite.[†] Here a, the fixed link, is the chamber, and is made to enclose the two turning links b and d, and also the block c; d is the only one of the moving links visible externally. The diaphragm in d, corresponding to the revolute 4, is carried across the diameter of the chamber, which makes the machine double-acting without rendering it necessary to make use of the space upon the left of the disc b.

Duclos [‡] also made the link a the chamber (Fig. 3, Pl. XXX), he carried the spindle of b right through it, however, and made the link d merely a rotating blade or wing. It must not be forgotten

---

* Newton, *London Journal of Arts*, &c. Conjoined Series, vol. xviii., 1841, p. 97. Patent dated 1836 ; see also same work, vol. xix., p. 18.

† *Propagation Industrielle*, vol. iii., 1868, p. 211.

‡ *Propagation Industrielle*, vol. iv., 1869. Patent dated 1867.

PLATE XXX.   CHAMBER-CRANK TRAINS.

Conic Cross-block $(C_3^{\perp} C^{L})^{a}$.

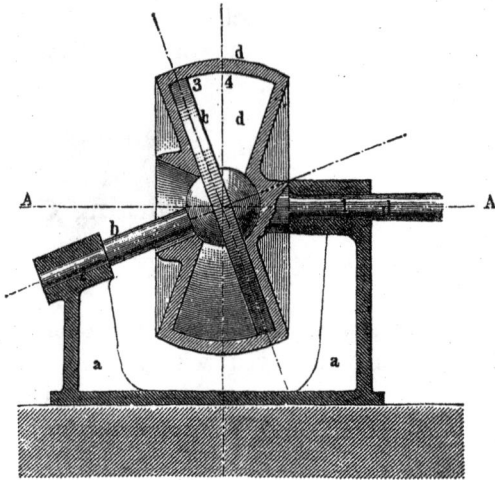

FIG. 1.—Taylor and Davies. (Eng.)

$$(C_3^{\perp} C^{L})_{\bar{d}}^{\frac{a}{d}}; \;\; (V\pm) = b, d.$$

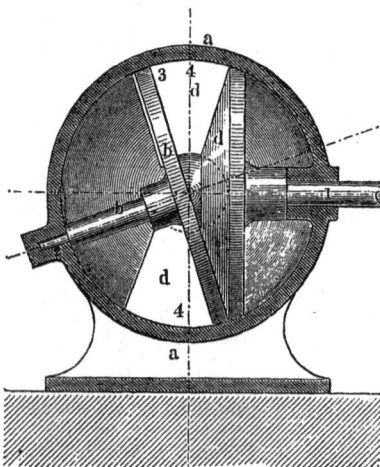

FIG. 2.—Lariviére and Braithwaite. (Eng.)

$$(C_3^{\perp} C^{L})_{\bar{d}}^{\frac{a}{d}}; \;\; (V\pm) = d, a.$$

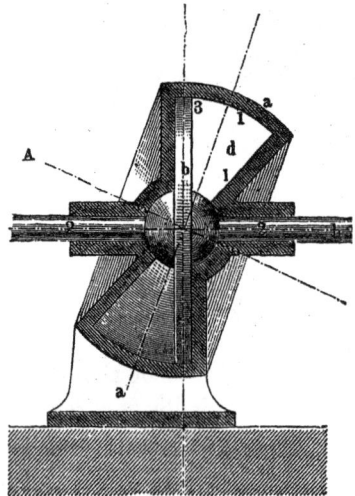

FIG. 3.—Duclos. (Eng.)

$$(C_3^{\perp} C^{L})_{\bar{d}}^{\frac{a}{d}}; \;\; (V\pm) = d, a.$$

Conic Cross-block $(C_3^\perp C^L)^n$.

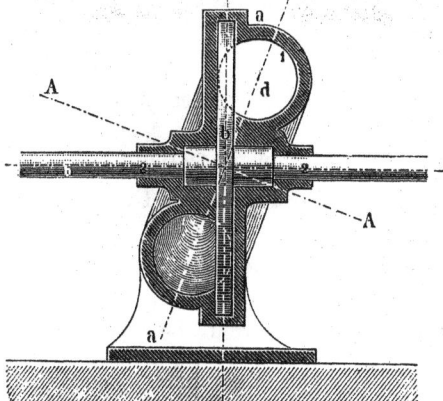

FIG. 1.—Küster. (Eng.)

$$(C_3^\perp C^L)^{\frac{n}{d}} ; \quad (V\pm) = d, a.$$

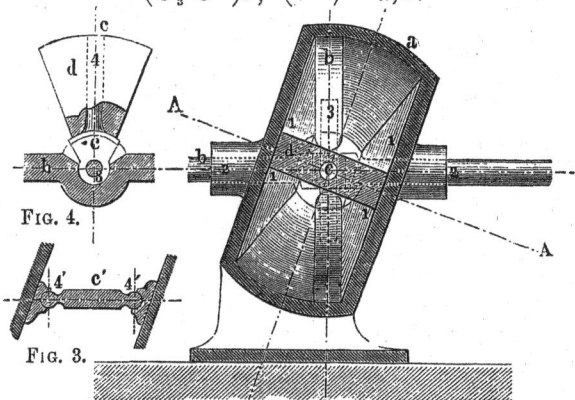

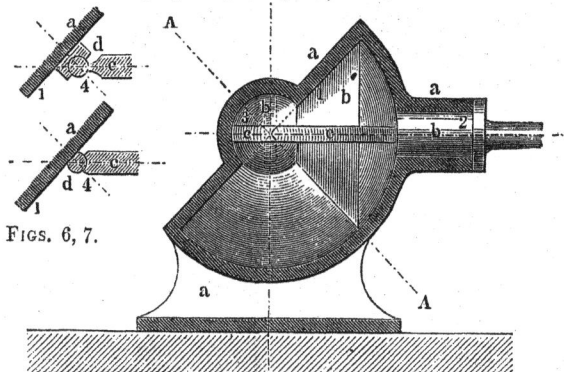

FIG. 4.

FIG. 3.

FIG. 2.—Wood. (Eng.)

$$(C_3^\perp C^L)^{\frac{n}{c}} ; \quad (V\pm) = d, a.$$

FIGS. 6, 7.

FIG. 5.—Geiss. (Eng.)

$$(C_3^\perp C^L)_c^a - d ; \quad (V\pm) = c, a.$$

that the revolutes of which d consists are represented respectively by the double cone 1 of a (having $A\,A$ for its axis) and the plane-pair forming the sides of d itself (and forming a section of a revolute of which the axis is perpendicular to the plane of the paper).

The machine of Küster, Pl. XXXI. Fig. 1, known to me only by description, is very similar to that of Duclos. Here the chamber, instead of being made a double cone with spheric sides, is a globoid or cylindric ring, of which the piston d is a sector. I have no details of the packing used where the piston d passes through the slot in the disc b,—but it appears to be very defective.

Fig. 2, Pl. XXXI. Wood's rotary steam-engine.* This machine is very nearly related to that of Duclos, but here the cross-block c receives a more important position in the train than that of a mere packing-piece. Its formation from two revolutes 3 and 4 crossing each other at right angles is obvious at once. The disc b has a slot profiled as in Fig. 281 for the motions of the piston d. The chamber a has for one of its revolutes the double cone upon the axis $A A$, for the other the bearings 2 of the driving-shaft b. I must mention that I have somewhat altered Wood's drawing. Instead of the cross-arm 4 this shows the two convergent arms 4' as in Fig. 3. With such an arrangement, however, the mechanism cannot move, for the piston, if it were constructed as shown at c', could not revolve in the chamber. For by the spindle 3, which Wood distinctly showed to be rigid, c' is compelled to remain always in the plane of the axis 2, so that the breadth of the piston must vary periodically between the real width of the chamber and the width shown in the drawing. From the existence of this error, and others which may be discovered by a closer examination of the original drawing, it is evident that this machine, in spite of its representation and treatment by Bataille, had never actually been at work.

Our last illustration, Fig. 5, Pl. XXXI., shows one of the latest productions of the inventive spirit which is kept alive by this rotary machine problem. It is the rotary engine of Geiss,† who has it at work at Gebweiler. The chamber a is hemispherical. The link b, the continuation of which serves as a fly-wheel shaft,

* Bataille et Jullien, *Machines à Vapeur*, 1847, vol. i., p. 447.

† *Propagation Industrielle*, vol. v., 1870, p. 132.

terminates in a cone having a spherical head. In this there is made the joint, normal to the axis 2, for the piston disc c, which passes right across the hemisphere. The mechanism is a turning cross-block in which the axes of the revolutes 1 and 2 enclose an angle of 45°. The turning-pair 2 is easily recognisable,—the pair 1 not quite so readily;—for this there indeed exists only one revolute,—the side of the hemispherical chamber—which is a plane cone having its axis in the line $A A$. The link a is therefore C^-... $\angle$ ... C^- or more strictly $K°$... $\angle$ ... C^-. The link b consists of two revolutes 2 and 3 normal to each other, the latter forming part of the spheric head of the link. The link d is omitted, so that the chain is reduced, and has for its general formula $(C_{\frac{1}{3}}C^L)^a - d$. It is on account of this omission of d that we have only the one revolute of the pair 1; and for the same reason there is a higher pairing between c and a, which is effected by rounding the edge of the piston c so that it can work upon the plane cone of the link a. It can easily be seen that only a very defective steam-joint can be made in this way. It would have been easy to have obtained a better joint, using lower pairs only, by retaining the link d and forming it as a packing-piece in some such way as is shown in Figs. 6 and 7. My sources of information about the machine are not, however, so distinct as might be wished, so that it is possible that something of the kind may actually exist in it.

In the last six machines we have had b once for piston, c once, and d four times, as chamber we have had d once and a five times.

Different forms have been given to the chamber; most frequently, however, it has taken the form of a double cone, and by many this particular form has been looked upon as essential. It has certainly given both inventors and improvers much food for thought. They have found it exceedingly difficult to realise distinctly the half rolling, half sliding motion of the disc. Davies must have believed it to be a pure rolling motion like that of spur wheels, for in 1838 he patented a disc-pump in which both the disc and the conic surfaces were toothed like bevel wheels.[*] He included the machine, that is, in the class of chamber-wheel trains which we shall have to consider in the next chapter. The diaphragm

---

[*] Newton, *London Journal of Arts*, &c. Conjoined Series, vol. xix., 1842, p. 153.

formed as it were a tooth common to both cones, the slot in the
plate was the hollow in the double-faced wheel b in which it
geared.* Davies soon noticed, however, that it was impossible
thus to obtain a satisfactory joint between the slot and the
diaphragm, because of the special spheric-cycloidal form of the
teeth, and he seems then to have thrown the whole arrangement
at once aside. The question as to what form the piston and the
surface of the chamber would have if the motion between them were
to be pure rolling is an interesting one. The forms are simply those
of the axoids between the links b and d of the chain $(C_3^\perp C^L)$.

These axoids can be determined without difficulty from the well-
known formula for the relative motions in the universal joint. In
it the link a is fixed, and the links b and d have in turning such a
relative velocity that if the corresponding angles of turning be indicated by ω and ω_1,

$$\frac{\tan \omega_1}{\tan \omega} = \cos a,$$

where a is the angle between the axes of b and d, that is the angle
of the link a.† From this we obtain the ratio of the angular
velocities w and w_1 of the two axes of the expression

$$\frac{w_1}{w} = \frac{\cos a}{1 - \sin^2 \omega \sin^2 a}.$$

This formula expresses at the same time the relative distances of
the instantaneous axis, that is the line of contact of the axoids,
from the axes of the two shafts.‡ If we now suppose that the
latter, instead of being convergent, are parallel, the axoids become
cylinders instead of cones, and the equation gives us the radii of
these cylinders,—and we have for the centroids, or normal sections
of the axoids, curves of the form represented in Fig. 276. In the
position shown the ratio $\frac{w_1}{w}$ is a maximum, when contact occurs
between B and B_1 it is a minimum, after the next turning of 90°
there is again a maximum, and after a third similar motion another
minimum.

* There is a wooden model of this machine in the Patent Museum at Kensington.
The patent is in the names of Taylor and Davies, 1836-8.
† See Rankine, *Machinery and Millwork*, p. 203, &c.
‡ It will be remembered that the instantaneous axis must be in the same plane
as the axes of the revolving shafts.

From these centroids we can, without any great difficulty, obtain the axoids which we require.[51] These are shown in Fig. 277. The two axoids $A\,B\,C$ and $A_1\,B_1\,C_1$ are similar and equal; in the position shown, however, homologous points in them are 90° apart. We do not, of course, propose that the figures thus found should be used as piston and chamber profiles in the disc-engine,—they simply enable us to obtain a representation of the amount of sliding which takes place between b and d when formed in the

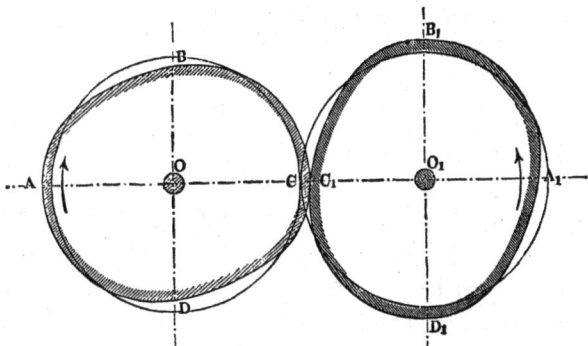

FIG. 276.

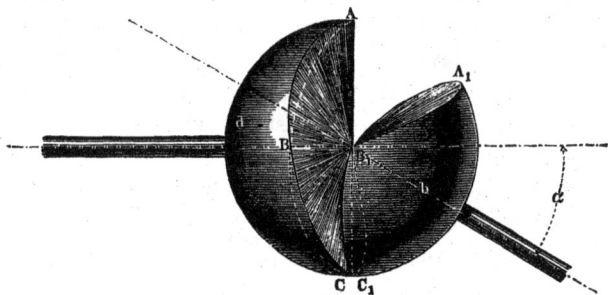

FIG. 277.

usual way. This may be gathered from the figure when it is remembered that in the common disc-engine b is made a flat disc and d a cone whose vertex angle is $90-a$. The determination of the axoids for the links a and c forms a further interesting problem; their profiles are analogous to the pair of Cardanic circles forming the centroids of the double-slider chain $(C_2''P^\perp)$.

§ 92.

Review of the preceding Results.

Chamber-crank gear occurs in such an immense variety of forms that our investigation of it has necessarily been somewhat extended. Even now I cannot for a moment say that I have exhausted all the forms in which it has appeared. I may rather say with Uhland's knight:

> " Wer suchen will im wilden Tann,
> Manch' Waffenstück noch finden kann,
> Ist mir zuviel gewesen."

A comparison of the machines described shows indeed that there are many easily-constructed inversions of existing mechanisms which have not yet been proposed, and many analogies to existing forms which have not yet been tried :—so we may look forward still to the production of whole series of new chamber-crank trains by the never resting empirics.

What our analysis has done for us is by no means unimportant. It has led us in the first place to those general laws which inventors and improvers alike have followed unconsciously and unsuspectingly. The reason for their having so unanimously chosen a crank-train as the foundation for their work must be left for examination in a later chapter. Meantime we have been able to reduce to order and principle the confusion which existed in the material already accumulated. Our arrangement has not been chosen arbitrarily, but has developed itself from the essential nature of the machines ;— on this account it is available also for future forms of the same class of machines, which indeed we have now the means of scientifically designing.[52] At the same time the principles which we have found furnish us with the means of forming a distinct opinion as to the value of the machines individually. A few out of the whole number show themselves to be practically useful ;—the majority, however, are not suited for their proposed purpose, or at best are distinctly inferior to others as practical solutions of the problem. The complete delusiveness of the supposed advantages of many machines has very distinctly revealed itself. Even those, however, which are worthless practically have proved useful to kinematic science, and thus indirectly to practice also. For it is just the great number of the cases existing which has on the one hand so emphatically called for an examination of their general laws, and on the other hand furnished sufficient examples to render such an examination possible.

I believe that it is not too much to hope that, as the recognition of the principles here laid down, becomes more and more general, the aimless search for new solutions of an old problem may gradually cease, and at length disappear; and that our investigations may have materially assisted the scientific comprehension of the rotary steam-engine and pump. In the two preceding chapters, too, we have been able already to gain some insight into the direct application of a synthetic method to the production of new machines. The progress of the analysis furnished us several times with means for applying the converse process. The possibility of replacing "invention," in the old sense of the word, by a scientific method of development—alluded to in the Introduction —has therefore already proved itself to exist. We shall have to return to this part of our subject further on (Chap. XIII.).

CHAPTER X

ANALYSIS OF CHAMBER-WHEEL TRAINS.

§ 93.

Chaining of Spur-Gearing with Pressure-Organs.

THE constrained motion of a pressure-organ, rendered possible by enclosure in chamber-gear, is not limited to the circle of crank mechanisms,—where we have already traced the development of the principle,—but can be obtained in other trains, and has met with frequent applications in them. There are many trains, of various kinds, in which a pressure-organ may be substituted for a rigid link. The method of chambering already described requires again to be carried out, and under certain circumstances gives us very valuable results, and machines really suited for practical use. One interesting series of such inventions, which indeed were not arrived at by our analytical method, but which may none the less be considered under it, is furnished us by machines formed upon the chain $(R_z C_2)$. I proved some time since the mutual relationship—previously unknown—of a number of these machines.[*] We shall give them the common name of chamber-wheel trains or gear.

---

[*] *Berliner Verhandlungen*, 1868, p. 42. I published this investigation on chamber-wheel gear before I was able to avail myself of the kinematic notation. In comparing

A chamber-wheel train consists of a chain (R_zC_2) so formed into a mechanism, by making one of its links a chamber, that it can be chained with a pressure-organ which shall enter the spaces between the teeth, pair with them, move forward with them, and finally be compelled to leave them where the two wheels gear with each other. One or both of the spur-wheels is used as the piston, while the frame $C \ldots \ldots C$ is formed into the chamber. There are necessarily a number of solutions of this general problem. The machines thus obtained may—as we have already found in the case of the crank mechanisms—be used either to cause the motion of a pressure-organ (as in a pump), or as a "prime mover," receiving motion from the pressure-organ, or for other purposes. The general character of the mechanism remains always the same, the special arrangement of it adopted depending upon the particular object in view. We shall here briefly examine a few of the most important forms of chamber-wheel gear.

§ 94.

The Pappenheim Chamber-wheels.

Plate XXXII.

The spur-wheel mechanism $(C_z C_2'')^c$ (Fig. 278), as the geometrically simplest case of $(R_z C_2)$, furnishes our first chamber-wheel train; the form chosen for it being that in which the wheels a and b are made equally large. The frame c becomes a chamber encircling a and b, and furnished with an inlet and outlet passage upon opposite sides of the two shafts. This gives us the oldest form of chamber-wheel gear, the construction of which is shown schematically in Figs. 1 and 2, Pl. XXXII. Two similar spur-wheels, a and b, having their teeth made so as to work without play, are enclosed in a chamber which has two semi-cylindrical wings, with which the points of the teeth remain in contact during their motion. The chamber has two openings, one on each side of the parallel shafts, and has plane end surfaces with which the ends of the

it with the present chapters it will be noticed that the general method and scope of both are very much alike, although here I can go considerably further into the matter than was formerly possible.—*R.*

PLATE XXXII.   CHAMBER-WHEEL TRAINS.

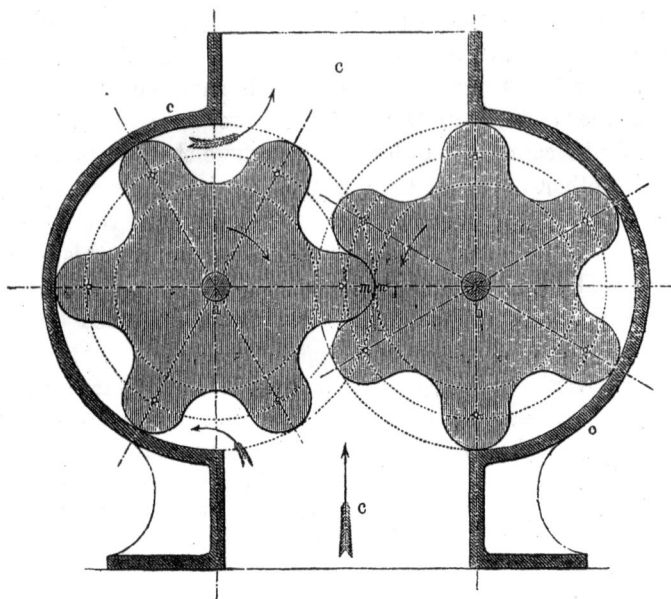

FIG. 1.—Pappenheim.

$$(C_z \, C_2'')^c \, ; \; (V\pm) = ab, c.$$

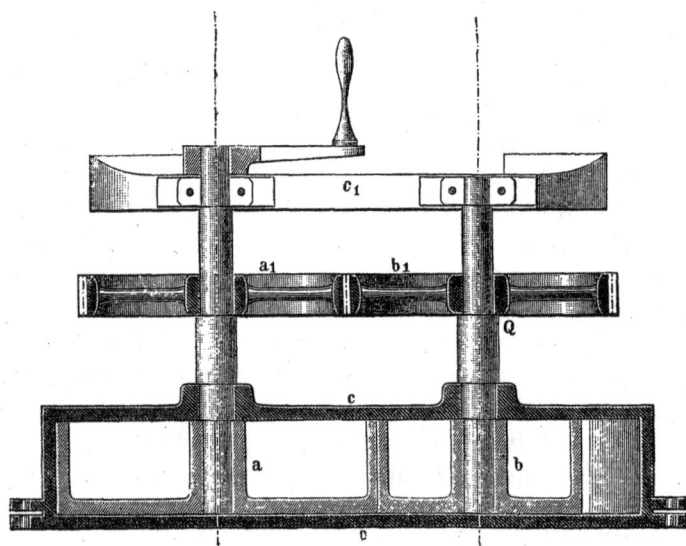

FIG. 2.

wheels must remain in close contact. The shafts of the wheels pass beyond the chamber, and are kinematically connected by two equal spur-wheels, a_1 and b_1. If now one of the axes—that of a, for instance—be caused to revolve, the other will turn with equal angular velocity in the opposite direction. If the turning take place in the direction of the arrows in Fig. 1, and the lower opening be connected with a reservoir of water, then as the wheels revolve the spaces between their teeth will be filled with water, which they will transfer from one side of the chamber to the other. On account of the contact between the teeth where they gear together at $m\,n$ no water can pass backwards, so that it must be driven onwards through the delivery-pipe. The machine may therefore serve as a pump, and for this purpose offers the advantages that it has neither valves nor any motions but rotary ones.

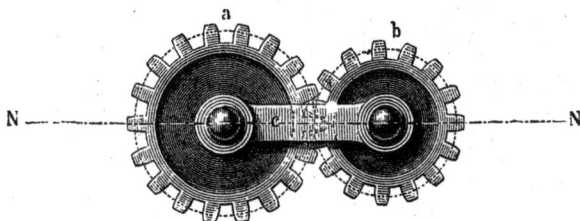

FIG. 278.

There is no difficulty in so forming the teeth of a and b that their profiles are always in contact in at least one point somewhere in the neighbourhood of $m\,n$, and that the point of contact passes continuously through the whole profiles as they move. Under the supposition that this is the case—as it is in Fig. 1—no water can pass back from the upper to the lower part of the chamber between a and b. The amount of water passing through the machine is then directly proportional to the number of revolutions of the wheels. If these move uniformly, the water is delivered in a continuous stream, on which account the machine might be well suited for a fire-pump.

The volume of water delivered by the pump per revolution is equal to the contents of the spaces * of both wheels, or as the volume of the teeth may in this case be considered the same as the volume of the spaces, is approximately equal to the contents of a

---

* A " space " is the volume enclosed between the sides of two consecutive teeth, and limited above and below by the point and root cylinder of the teeth.

cylindric ring whose inner and outer radii are those of the bottom and top of the teeth respectively, the annular space, that is, between the point and root cylinders. This we may call shortly the **tooth-ring.**

If, therefore, it be desired to increase the amount of water delivered per revolution without changing the diameter of any part of the wheel, it is only necessary to lengthen a and b in the direction of their axes. If the head of water be not great, and the angular velocity of the wheels not too small, careful construction may so reduce the loss of water as to make it not worth considering. The arrangement, therefore, is one which in many cases may furnish a really useful water-pump.

As a pump, indeed, the machine is already very old. Weisbach calls it* Bramah's rotary pump, and says that Leclerc improved it (by placing packing wedges in the ends of the teeth); other writers ascribe it to Leclerc himself. This would take back the date of the invention to the end of the last century. But long before this, in 1724, the pump had been described as old by Leupold,† and called "Machina Pappenheimiana"; he headed it "A chamber apparatus with two moving wheels, called by D. Becher Machina Pappenheimiana." Now Becher's work ‡ appeared in the first half of the 17th century. Besides this, however, Kircher, Schott, § Leurechin, and also Schwenter, in his "Mathematischen Erquickstunden" (A.D. 1636, p. 485), have described the same machine with the alteration that the wheels have four teeth instead of six, and that they do not name Pappenheim. The machine is now therefore over 230 years old; it was already known in the time of the Thirty Years' War, and all accounts agree in making it a German invention. Whether Pappenheim was the name of its inventor, or of his city only, remains uncertain; we are quite justified, however, in any case in calling it the Pappenheim pump. In France Grollier de Servières (1719) is often named as its inventor.‖ But this date is that only of the appearance of a description by the younger de Servières of the mechanical collection of his grandfather,

---

* Weisbach, *Mechanik,* iii., p. 843.   † *Theatrum Mach. Hydraul.,* vol. i., p. 123.

‡ *Trifolium Becherianum,* which is unfortunately not to be found either in the Königl. Bibliothek in Berlin or in the Library of the British Museum.

§ Kasper Schott, *Mechanica Hydraulica Pneumatica,* Mainz, 1657. The wheels shown in a little copper-plate engraving have here nineteen teeth.

‖ *Propag. Industr.,* 1868, iii., p. 20.

in which moreover he does not mention the latter as the inventor of the machine.* The collection appears to have been founded about 1630.

I must here remark that the two external spur-wheels, *a* and *b*, are omitted both in Leupold's beautiful copper-plates and in the small woodcuts of Schwenter, as well as by Bramah and Leclerc. There is certainly no absolute necessity for their employment, for the pump-wheels may be used instead; but the oblique action of the latter when in the relative positions shown in Fig. 1 would soon damage the teeth. The use of the wheels *a* and *b* is therefore always to be recommended, and their existence will be assumed in the remaining chamber-wheel trains shown in our plate. We need not here enter into the delineation of the profiles of the wheel-teeth, which have been already mentioned in § 31. We may merely mention that in this case the points of the teeth are semi-circular (as in Leupold's engraving), and the profiles of their flanks are such as work with these curves,— they differ little from circular arcs.

The Pappenheim machine may be used for gaseous bodies as well as for liquids—for a blower, for example, or a gas-pump. Its action also may be reversed, so that the fluid drives the machine instead of being driven by it. The machine thus becomes a prime mover,—a chamber-wheel turbine, if water be the fluid used, or a rotary steam-engine if it be steam. Murdock, a contemporary of Watt,† attempted to apply it for the last-named purpose, using teeth with broader points, so that they could work against the chamber-walls, and fitting them also with packing pieces. The machine thus arranged could only be adapted for very light work, for the closure at the line of contact of the teeth, *m n*, would not suffice if a high pressure were used; Murdock's chamber-wheel steam-engine has therefore never found its way into practical use.‡

* Ewbank, *Hydraulic and other Machines*, Ed. of 1870, p. 285.

† Murdock was for many years an assistant of Watt, and became eventually (practically) a partner in the firm of Boulton and Watt. His engine, made in 1799, is described in Farey's *Treatise on the Steam Engine*, p. 676, and elsewhere.

‡ So far as regards economy in the quantity of steam used Dudgeon's rotary engine, which has been a good deal advertised during the last two or three years, is no doubt superior to Murdock's. In it a return is made to the use of wheels with numerous teeth (thirty-one in wheels twenty inches diameter), for which epicycloidal profiles are

The chamber-wheel gear can also be used in another way, viz., as a measuring instrument. It can, that is to say, if carefully constructed, serve as a water meter; for if a stream of water be allowed to pass through it, driving the wheels, the number of revolutions made by the latter gives the quantity of water passed in terms of the tooth-ring volume. We shall find further on other similar applications of chamber-wheel gear.

Still another application of the machine may be obtained by arranging it with a delivery-pipe of which the sectional area can be varied. By reducing this to a suitable extent the chamber-wheel train, working either with water or oil, forms a brake, which by the use of one or of two valves can be made either single or double acting. If the passages be suitably arranged the same quantity of fluid can be used over and over again; a brake of this kind, too, has no wearing parts, like those of an ordinary block-brake. Such a chamber-wheel brake, acting in the direction of rotation, and not preventing any other motion, may serve as a cataract, and be useful in those cases where it is wished to apply that apparatus in connection with rotary motions.

It will be seen that the chamber-wheel gear has a large range of applications. In its simplest form, without valves, it may be used as a pump (and is suitable for a fire-engine pump), as a steam-engine, or as a fluid meter; a trifling addition makes it available also as a brake or a cataract. It is well suited for working with (driving or being driven by) water or other liquids, and also viscous or merely plastic materials (so that it probably might be used as a clay or pug-mill), as well as for driving gaseous materials, as air

used. The steam is admitted at the side of the wheels into the space between two teeth, and the resulting motion takes place in the one or other direction according to whether the admission opening be placed a little above or below the line of centres. This makes both expansion and reversing possible. The only security against leakage, however, is the higher pairing between the surfaces of the teeth. At the sides of the wheels there is lower pairing, but no means are provided (or at least shown in the engraving) for taking up the wear which must occur there. Altogether I see no reason for supposing that this inventor will be more successful than his predecessors in inducing two bodies to rub upon each other under considerable pressure and at a great velocity without wear taking place, and all the consequences due to that inconvenient action.—See *Engineering*, Nov. 14, 1873. From some correspondence in subsequent numbers of the same journal it seems probable that the first to propose this use of steam from the centre outwards in a chamber-wheel-train was John Hackworth (*circa* 1840-45).

or coal-gas, if their pressure be small. It is indeed capable of a greater variety of useful applications than often exists for one and the same machine.

§ 95.

Fabry's Ventilator.

Plate XXXIII.

This well known machine is a chamber-wheel train used for a " wind pump " or ventilator. The Belgian engineer whose name it bears has introduced it with great success as a suction ventilator for mines, and is still occupied in improving it. Fig. 1 shows the profile of the wheels first used by Fabry.[*] The pump wheels a and b are here three-toothed, the profiles of the teeth at $m\,n$ and $m_1\,n_1$ being epicycloids upon the pitch circles, or their equidistants. At $o\,p$ the profiles touch on both sides of the centre line until m and n or m_1 and n_1 come together. The stream of air is therefore prevented from passing between the wheels, although the point of contact does not, as in the Pappenheim wheels, pass continuously through the whole profiles. The hollowing out of the teeth entails, however, the consequence that as each tooth leaves contact a small quantity of air is carried back to the suction-pipe. If we imagine the teeth to have been first arranged for continuous contact and then hollowed out, the capacity of the hollows thus made would give us exactly the quantity of air returned. The condition therefore remains, that the quantity of air delivered per revolution is very approximately equal in volume to the tooth-ring cylinder. Thus the hollowing of the teeth does not alter the quantity of air delivered; —it prevents, however, the complete uniformity of the delivery,—for the return of air takes place at intervals and not continuously. This want of uniformity might be a serious disadvantage if the machine were working with a considerable water-pressure, but for the purposes of a ventilator, especially where the velocity is small and the pressure low, it has little appreciable influence.

It is not necessary that the recesses in the chamber should be semi-cylindrical in order to insure the joint between them and the points of the teeth being kept for a sufficiently long time; it is

---

[*] Laboulaye, *Cinématique*, Second Edition, p. 793.

PLATE XXXIII. CHAMBER-WHEEL TRAINS.

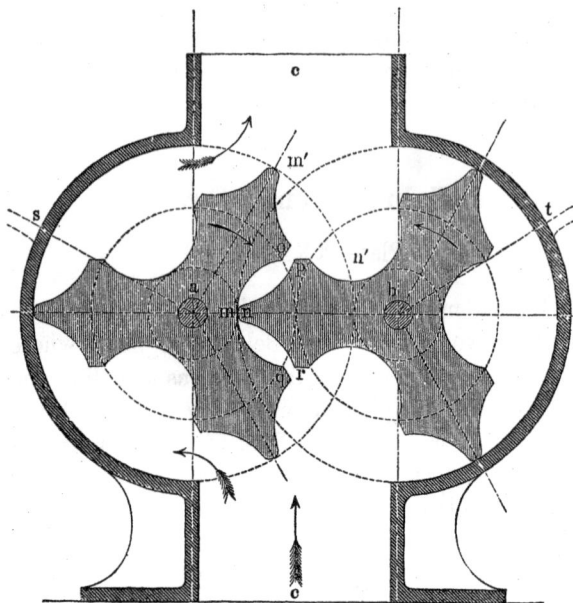

FIG. 1.—Fabry.
$$(C_z \, C_2'')^c \, ; \; (V\pm) = ab, c.$$

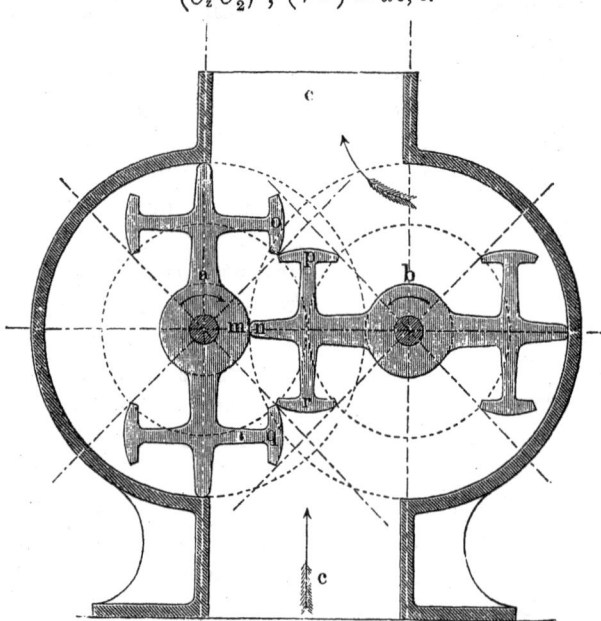

FIG. 2.—Fabry.
$$(C_z \, C_2'')^c \, ; \; (V\pm) = ab, c.$$

sufficient if they subtend the pitch angle, or angle included between one pair of teeth. With such wheels as those of Fig. 1, for instance, they need not extend beyond *s* and *t;* while if they be made semi-cylindrical the wheels need have two teeth only. This form, as shown in Fig. 2, is that adopted in Fabry's later wheels. Epicycloidal profiles are again adopted for the faces *o p, q, r,* etc.; and at *m n* there is contact between the central arm of the wheel *b* and the root-cylinder or boss of *a.* The space between the latter and the cylindrical sides of the chamber is the tooth-ring cylinder, the volume of which again approximates very nearly to the volume of air delivered per revolution. Fabry's ventilators are constructed of from 3 to 4 metres diameter and 2 to 3 metres breadth, and move comparatively slowly, namely, at from 30 to 60 revolutions per minute.* The framework of the wheels is mostly made of wood, tin plates being nailed upon it at the places of contact *m n,* etc., so that the whole construction bears the least possible resemblance to a toothed wheel. We can therefore easily understand how the theoretic connection between the Pappenheim machine and that of Fabry has remained unnoticed by practical men.

§ 96.

Root's Blower.

Plate XXXIV.

The blowing machine of Root represented in Fig. 1, Pl. XXXIV. was exhibited at the Paris Exhibition of 1867.† The wheels were about three feet in diameter and nearly seven feet broad; they were driven at a great velocity, and delivered a large volume of air at a considerable pressure. The profile *p n r* is circular, and works continually in contact ‡ with the profile *q m o* of the other wheel.

---

* Cf. *Zeitschrift des Vereins deutscher Ingenieure,* vol. i., p. 140 ;—Ponson, *Traité de l'Expl. des Mines de Houille; Polyt. Centralblatt,* 1858, p. 506; also *Civil-Ingénieur.*

† *The Engineer,* August, 1867, p. 146.

‡ I believe that now, at all events, the profiles of the wheels are made so as just not to come into contact ; it is considered that the absence of friction thus attained more than compensates for the small leakage of air which occurs at ordinary pressures and velocities.

PLATE XXXIV.   CHAMBER-WHEEL TRAINS.

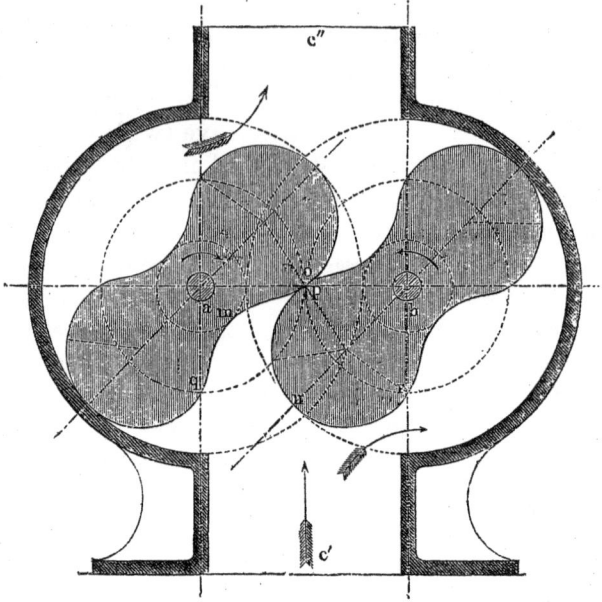

FIG. 1.—Jones, Root.
$(C_z \, C_2''')^c$; $(V\pm) = ab, c.$

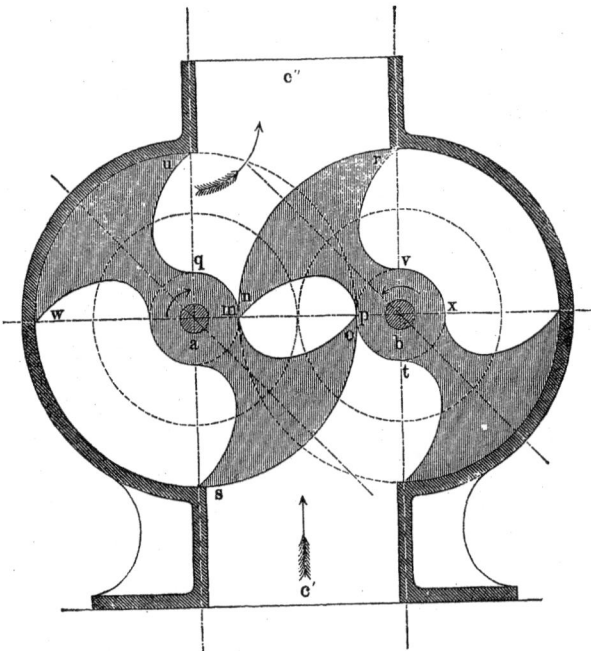

FIG. 2.—Root.
$(C_z \, C_2'')$; $(V\pm) = ab, c.$

It will be seen that the machine is a Pappenheim chamber-wheel train in which each wheel has two teeth. Root made the surfaces of the teeth at first of wood, afterwards of iron. His blower is constructed at various places, and is very extensively used; there were several at the Vienna Exhibition of 1873. Root does not appear, however, to have been the first inventor of this chamber-wheel train, for it was used as a gas-exhauster (made by George Jones of Birmingham) in 1859,* and does not appear to have been new even then.[53]

Fig. 2 is a section of a second form of Root's blower, in which the profiles of the teeth are altered. As in the chamber-wheel engine of Murdock already mentioned, the teeth have here their points made with cylindric profiles, $n\,r$, $s\,o$, $u\,w$, pairing with the walls of the chamber. Each of these profiles extends through a quadrant, *i.e.* through half the pitch, as does also each section $m\,q$, $p\,t$, $v\,x$, of the circular profile of the root cylinder with which the points of the teeth work in contact while crossing the line of centres. $m\,q$ therefore slides upon $n\,r$, $u\,w$ on $v\,x$ and so on. The profiles $p\,n$, $m\,o$, &c. of the flanks of the teeth are here curtate epitrochoids of the rolling pitch circles. The profile $m\,o$ is described by the motion of the point n of the wheel b relatively to the wheel a, these therefore work together as the wheels move as indicated by the arrows. Root does not use exactly the profile thus found, but a profile falling behind it in the wheel, and this is quite justifiable. He sacrifices the second point of contact certainly, but at the same time he avoids the alternate exhaustion and compression of air which would otherwise occur in the space left between the two points of contact. The exact profiles are only shown in the figure for the sake of simplicity. In designing the wheels they have in all cases to be found, in order to determine the limits within which the actual profile can be drawn. Of Root's two arrangements the first is the better, for it delivers a uniform stream of air, which the second, for the reasons mentioned in connection with Fig. 1, Pl. XXXIII., does not. In both of them the volume of air delivered per revolution very nearly approximates to that of the tooth-ring cylinder.

---

* Clegg, *Manufacture of Coal Gas*, 5th. Ed., 1868. p. 181. The engraving given here shows wheels of a profile absolutely identical with that of Root's wheels.

§ 97.

Payton's Water Meter.

Plate XXXV. Fig. 1.

Fig. 1, Pl. XXXV. is a schematic outline of a water meter
exhibited in the English department of the Paris Exhibition of
1867.* It is a two-toothed chamber-wheel train, the profiles of
its wheel teeth being involutes of circles. The line (and normal)
of contact NN makes in our figure an angle of 15° to the line of
centres ; it is necessary to make this angle small in order that the
contact may last sufficiently long, The involutes touching in op
extend from m to q and from r to n; within m and r circular arcs
of any convenient radius (so long as they do not interfere with the
contact) continue the profiles to the bosses of the wheels. The
backs of the teeth have for their profiles curves which are very
nearly parallel to the involutes and which must lie very close to
them in order not to disturb the contact, in order, *i.e.* that the
back of one tooth may not foul the point of the opposite one.
It is for this reason that the teeth have received their peculiar
scoop-shaped form.

Here again a quantity of water, contained in the space behind
each tooth, is returned every revolution, so that the delivery of the
water, as in Fabry's machines and one of Root's, is discontinuous.
This can be seen also from the fact that the point of contact does
not traverse the whole profile continuously. The volume of water
actually passing through the meter per revolution is again very
nearly equal to that of the tooth-ring cylinder.

Whether good workmanship is of itself sufficient so completely
to prevent leakage that the apparatus can make an accurate water
meter can only be determined by experience. The instrument,
unquestionably a very simple one, seems to have been very rapidly
received into favour in England.

* *The Engineer*, Feb. 1868, p. 92. ("Epicycloidal Water-meter.")

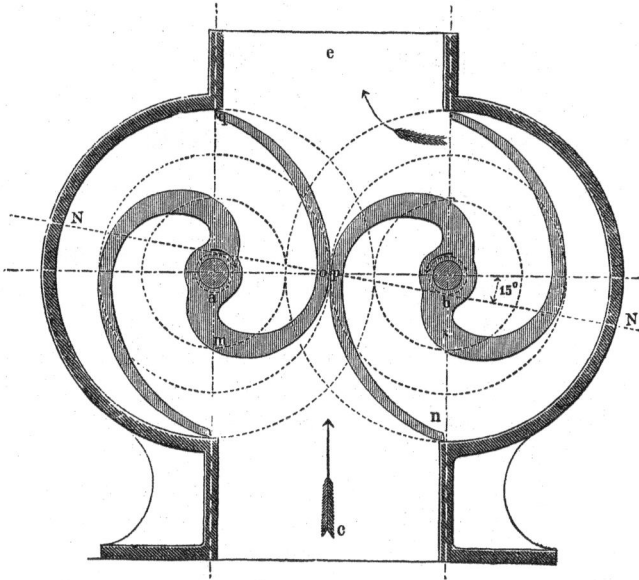

FIG. 1.—Payton.

$(C_z\,C_2'')°\,;\;(V\pm) = ab, c.$

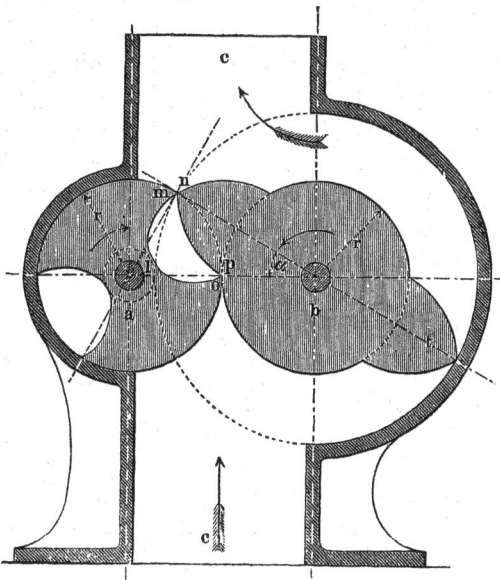

FIG. 2.—Evrard.

$(C_z\,C_2'')°\,;\;(V\pm) = b, c.$

§ 98.

Evrard's Chamber-wheel Gear.

Plate XXXV. Fig. 2.

In the Belgian department of the 1867 Paris Exhibition there was exhibited a ventilator of Evrard's which, although very indifferently constructed, was yet in itself remarkable, and deserves notice here. Fig. 2, Pl. XXXV. shows its general construction.

In the Belgian section of the Vienna Exhibition also, one of the same machines was shown arranged as a water-pump. It is essentially a two-toothed chamber-wheel train in which the wheels are not similar in form, although they still, as in the former cases, revolve with equal angular velocities. The wheel a has two spaces falling entirely with in its pitch circle, while the wheel b has two teeth lying entirely beyond its pitch circle. The teeth of a are very similar in form to those of Root's machine Pl. XXXIV. 2, they lie, however, entirely within the pitch circle, the corresponding spaces of b therefore are entirely outside its pitch circle. The curve $m\,l\,o$ is the curve described by the point n of b relatively to the wheel a, it is therefore a curtate epitrochoid corresponding to the rolling of the two equal circles of radius r. The curve $p\,n$ is the common epicycloid (in this special case a cardioid), described by the point o of the wheel a relatively to the wheel b. Contact ceases at o at the instant that m and n come into gear. In order that this may take place, the angle $m\,l\,o$ must be made equal to the angle subtended by the foot of the tooth on b, that is to double the angle marked a.

The spaces of both wheels carry the fluid from below upwards as they turn in the direction of the arrows. The greater part of the contents of the spaces of a, namely that represented by the opening $m\,l\,o$, is, however, returned again between the wheels. For each revolution, therefore, the volume of fluid passed upwards through the machine is a little less than the volume of the tooth-ring cylinder. The contact of the two wheels has the excellent peculiarity that the tops of the teeth of a roll upon the bottom of the spaces of b without sliding. The ventilator exhibited at Paris, so far as could be ascertained from a somewhat inaccessible machine, was furnished with straight radial blades instead of the epicycloidal teeth of our figure, which gave sufficiently accurate results for practical purposes, and possessed obvious constructive advantages. The return of a portion of the fluid renders the motion un-uniform, but this, as we have seen, is not a serious drawback to the efficiency of a ventilator. On the whole, therefore, it must be said that Evrard's ventilator is a very practical example of chamber-wheel gear. In order to make its delivery uniform,—so as to suit it better for the purposes of a water-pump or a hydraulic engine,— it is necessary only to give the teeth on b circular profiles, and to use the corresponding envelopes for the profiles of the spaces on a.

The special form, however, which Evrard has chosen for his chamber-crank chain was known before his invention,—a much older example of it will be described in § 101. The pump constructed on the same principle which was exhibited at Vienna was shown as the invention of Baron Greindl.[*]

§ 99.

Repsold's Pump.

Plate XXXVI. Fig. 1.

We have seen that the old Pappenheim invention has passed through many changes in the form and number of teeth used. Along with various alterations in the former the latter has been reduced from 6 or more to 4, 3, and even to 2. Only one step more in this really useful reduction could be made, and this has already been taken some years ago in the rotary pump made by

---

[*] In England rotary pumps have been made by Laidlow and Thomson, which are founded upon this chamber-crank train in the form in which Evrard used it. *The Engineer*, May 29, 1868, p. 394.—*R.*

Baker's "Rotary Pressure Blower" is kinematically identical with Evrard's machine, but instead of using such a profile for the spaces of a as corresponds to the relative motion of the point n,—the wheel a is made a hollow drum, with a wide opening along the whole length of one only of its sides. It has therefore to revolve twice for each revolution of the fan-wheel b, while at the same time a second wheel, in every respect similar to it, has to be added in order to effect the necessary closure with the root circle of b when either of the teeth of the latter (which are here also merely thin straight blades) are moving freely across the opening in the drum a.

PLATE XXXVI. CHAMBER-WHEEL TRAINS.

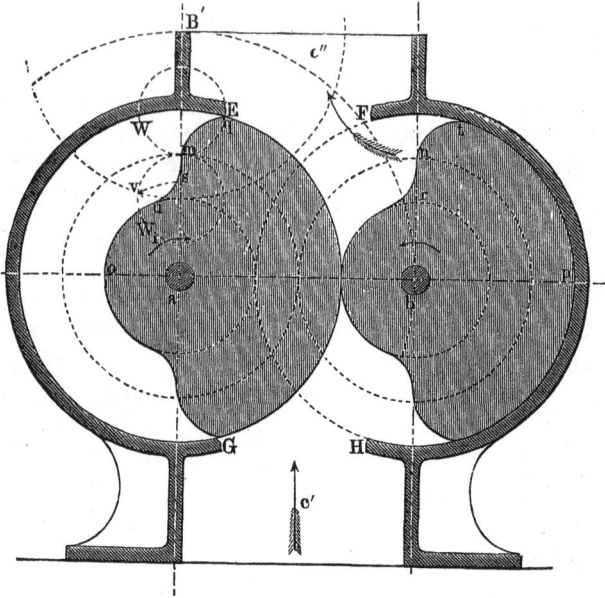

FIG. 1.—Lecocq, Repsold.
$$(C_z C_2'')^c\,;\ (V\pm) = ab, c.$$

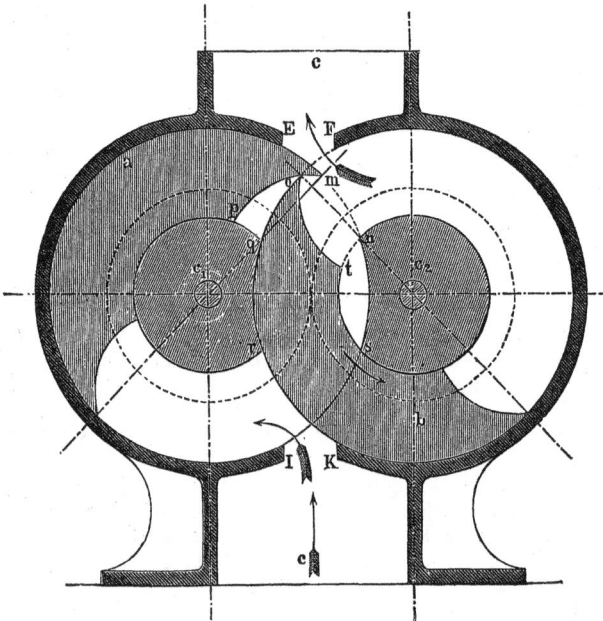

FIG. 2.--Dart, Behrens.
$$(C_z C_2'')^c\,;\ (V\pm) = ab, c.$$

the Hamburg firm of Repsold. This well-known machine, which excited great attention in its time, is a chamber-wheel train, the pump wheels of which have one tooth only. Fig. 1, Pl. XXXVI. is a schematic representation of it. The profiles of the teeth beyond the pitch circles are here epicycloids, as mq and nt, and within them hypocycloids as ms and nr, both obtained, as in ordinary set wheels, by rolling upon and within the pitch circles (primary centroids) the equal describing circles (auxiliary centroids) W and W_1. The portion su of the profile, added at the root of the teeth, is a part of the path of the point t of the wheel b relatively to a; the hypocycloidal arc ms corresponds to the rolling of W_1 through the arc mv. The points of the teeth tp and qG are cylindrical, as are also the corresponding surfaces between their flanks, exactly as in the case of common spur wheels. With the profile forms here described the delivery of the pump is not absolutely uniform, for the whole profile of the wheel does not pass continuously through the point of contact. The want of uniformity is so small that it may fairly be neglected; all that is necessary, however, to prevent it entirely is to use in the tooth faces at mq, nt, &c., such a form as makes the whole profile a continuous curve—as *e.g.* a circular arc—and using for the roots of the teeth the corresponding enveloping form.

The pump-wheels of Repsold's machine are commonly described as " eccentrics of special form " or something of the kind; it is clear, however, from what has just been said, and a glance at the figure makes it still more evident, that each of them is simply a spur-wheel with one tooth. The point and root cylinders of the teeth slide upon one another, so that wear must unavoidably take place at first, as in the Root's blower Fig. 2, Pl. XXXIV. It is therefore difficult to retain a tight joint where these parts of the wheels are in contact, and the pump is therefore most suitable for working with low pressures. The wings EG and FH of the chamber must be greater than semicircles in order to prevent communication between the suction and delivery pipes behind the wheels. Repsold has used packing strips of leather in them.[*] The volume of fluid delivered per revolution is almost exactly equal to that of the tooth-ring.

Repsold's pump is used in mining operations, generally for drain-

[*] *Berliner Verhandlungen*, 1844, p. 208.

ing purposes, and also as a fire-pump; it has also been used in England as a hydraulic motor* (chamber-wheel turbine), and serves often as a gas-pump in gas-works. It has thus been successfully applied to three of the several applications of chamber-wheel gear before enumerated.

So far as the originality of the invention goes—if we may speak at all of the "invention" of what is really only a special form of the Pappenheim chamber-gear—Repsold was not the first to use it, for in France Lecocq obtained a patent for a similar rotary pump in 1832 ; † he called it a "pump with two pistons revolving on one another."

§ 100.

Dart's or Behrens' Chamber-wheel Gear.

Plate XXXVI. Fig. 2

In the American section of the last Paris Exhibition there were two applications of the chamber-wheel train shown in Fig. 2, Pl. XXXVI.; they were invented by Behrens and exhibited by Dart and Co.‡ The two pump-wheels are here again one-toothed, as in the last case. They are fixed at their sides to circular discs (not shown in our engraving) and this renders it possible to remove altogether the portions of them below the teeth, *i.e.*, the root cylinders. The place of the latter is taken by the cylinders c_1 and c_2, which are fixed to the chamber. These have cylindric hollows, $q\,r$ and $n\,s$, the contact of which with the points of the teeth, as the latter revolve, is sufficient to render unnecessary the additional contact of the flanks of the teeth. In our figure these are shown so as also to work together, $m\,p$ being a curtate epitrochoid, described by the point o. In practice the point o is left a little clear of the curve (by rounding it off) in order to prevent the compression of fluid in the triangular space $o\,p\,q$. So soon as the point p reaches q, o has got to the same place, and passes downwards from q to r. The point of the tooth of b, therefore, works closed against $q\,r$, while its root t moves always in contact

---

* *Practical Mech. Journal*, 1855-6, vol. xviii., p. 28.

† *Propagation Industrielle*, vol. iii., 1868, p. 182.

‡ *Propagation Industrielle*, vol. ii., 1867, p. 116. *Engineering*, Apr. 4, 1875, pp. 368-9.

with the cylinder c_2. m soon arrives at n, after which the point of the tooth of a works closed against $n\,s$, while at the same time the portion of the fluid enclosed between the two wheels is delivered back to $I\,K$. Meanwhile the fluid has been passing upwards through $I\,K$ and round c_1, in the left wing of the chamber, while the fluid already above b in the right wing of the chamber has been simultaneously discharged through the delivery opening $E\,F$.

We notice that here a new idea is brought into the chamber-wheel gear, that of the closure of the central passage by lower pairs (here cylinder pairs),—all the other forms of the Pappenheim machine having used a higher pairing for this purpose. The transition from this to the closure before us may be noticed in the lower pair-closure at the teeth points in the machines of Repsold, Evrard and Root, already examined. So far as closure goes the profiles $m\,p$ and $o\,t$ might be omitted; it is, however, well to retain them in order to reduce the quantity of fluid returned, and therefore the un-uniformity of the delivery, as far as possible. The volume delivered for each revolution is again very approximately equal to the tooth-ring volume of one wheel.

On account of the use of lower pairs the prevention of leakage is here more easy than in any of the former cases ; the Behrens' machine is therefore well suited for use as a pump. Its manufacturer, Dart (in whose house in New York the inventor Behrens is a partner), has constructed many for that purpose, and also as hydraulic motors,—indeed he has also applied it as a steam-engine. One of these (of 12 H. P. nominal) was at work at the Paris Exhibition, and drove a Behrens' pump.* It may, however, be doubted whether permanently good results can be obtained in this application of the machine, for it will certainly be very difficult to make its working joints tight against high-pressure steam. At best it is far from reaching the completeness, in this respect, of machines of the ordinary form.

At the Vienna Exhibition there was a steam fire-engine in which engine, fire-pumps and feed-pump were all constructed on Dart's plan.

* *Motoren u. Maschinen auf der Weltausstellung* 1867, Vienna, 1868, p. 124.

§ 101.

Eve's Chamber-wheel Gear.

Plate XXXVII. Fig. 1.

The old chamber-wheel train of an American, Eve, gives us what is really the foundation, as to form, of that of Evrard. In this machine (Fig. 1, Pl. XXXVII.), which was patented in England in 1825,[*] the pump-wheels are essentially two unequal spur-wheels having a diametral ratio 1 : 3. The cylindric axoids of the bodies *a* and *b*, whose shafts are connected beyond the chamber by a pair of common spur-wheels whose diameters are as 3 : 1, roll together at *m n*, while the teeth of the wheel *a* carry the fluid in the direction of the arrow. On the line of centres they pass the space of *b* with higher pair-closure, in precisely the way described in connection with Fig. 2, Pl. XXXV.

In France Ganahl obtained a patent in 1826 for a machine very similar to that of Eve; he intended it both as a motor and a pump.[†] He made, however, the wheel *b* conical, like the plug of a cock. We can see the idea which led to this form of construction,—the inventor looked upon the wheel *a* as a piston-wheel, and *b* as a valve arrangement. Ganahl's machine is strictly a chamber-wheel train formed from a pair of bevel wheels.

§ 102.

Révillion's Chamber-wheel Gear.

Plate XXXVII. Fig. 2.

The general principle enunciated in § 93 that a chamber-wheel train could be made from any form of the mechanism $(R_z C_2)$ includes also the case of screw-wheels. This has been known for a long time, and many attempts have been made to apply it

---

[*] Ewbank, *Hydraulic and other Machines*, 1870, p. 287, also specially Bataille and Jullien, *Machines à Vapeur*, vol. i., 1847-9, p. 440, where other forms are also mentioned.

[†] *Propagation Industrielle*, vol. iii. 1868, p. 55.

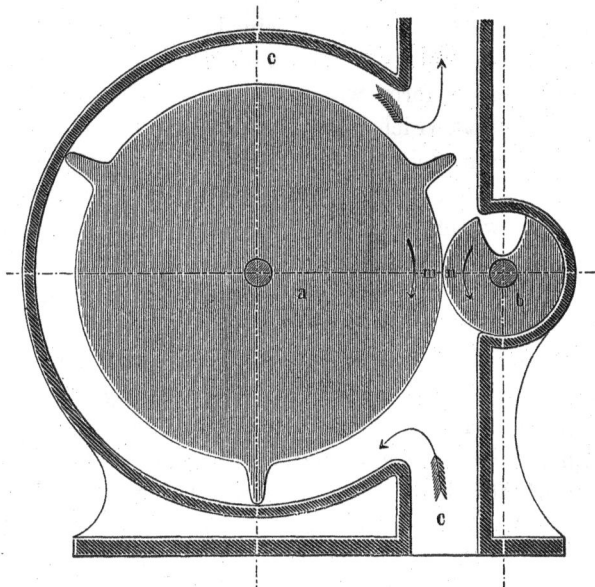

FIG. 1.—Eve, Ganahl.

$(C_z\,C_2''')^c$; $V^{\pm} = a, c.$

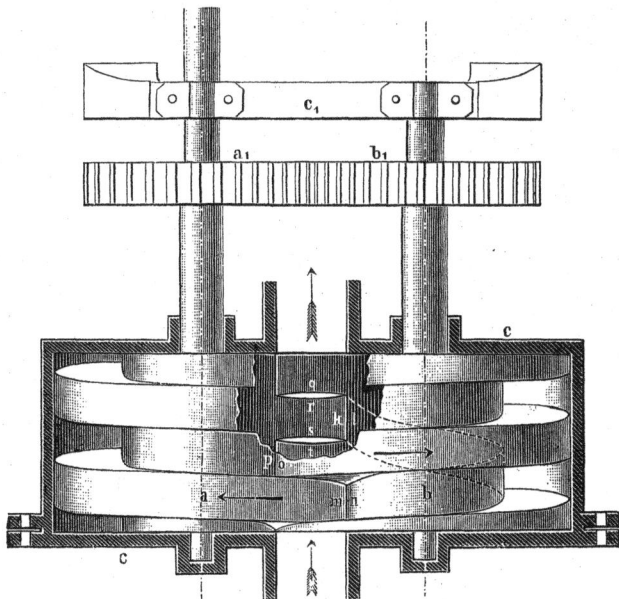

FIG. 2.—Révillion.

$(C_s^+\,C_2''')^c$; $(V^{\pm}) = ab, c.$

practically.　In 1830 Révillion obtained a patent in France* for a
screw-wheel chamber-train.　Such a mechanism is shown in Fig. 2,
Pl. XXXVII., in a form which differs somewhat from that of
Révillion; it is that which I have used for the model in my collection of kinematic models.　The wheels a and b are normal screws
of equal pitch and opposite "hand"; their axes are connected by
the equal spur-wheels a_1 and b_1; the frame c forms the chamber.
The outer surfaces of the threads revolve in (lower) contact with
the chamber, the screws work together with higher pairing at $k\,l$,
$m\,n$, $o\,p$, etc.　I have given them at $q\,r$ and $s\,t$ such a sectional
profile that the outer edges of each thread touch throughout the
sides of the threads between which it is working (see just above
the letters $m\,n$), which has not been done in any former machines
of the kind.　The profiles of the cross sections of the threads are
envelopes of the helix.　The screw cutting lathe makes the
accurate construction of these profiles by no means very difficult.
The fluid fills the spaces between the threads or teeth, and is
pushed forward by the latter just as in the Pappenheim machine.
One of these spaces is, for instance, that included between the
chamber on the one side and $m\,n$ and $k\,l$ on the other, which is
separated from the rest of the chamber by the contact at $q\,r$ and
$s\,t$ and at the similar positions on each side of $m\,n$.　The screw-
wheel chamber gear can hardly be said to have any practical
importance;—I do not think it necessary therefore to consider
here any of the other attempts to adapt it to the purposes of a
steam-engine or a pump.

§ 103.

Other Simple Chamber-wheel Trains.

The various forms in which the simple chamber-wheel gear can
be used have by no means been exhausted by the illustrations we
have given, although these include the best known and more
important of them.　We have seen both equal and unequal spur-
wheels used, as well as cylindric screw-wheels, and a suggestion
(§ 101) of a pair of bevel-wheels.　In this last direction more has
been attempted.　Herr Lüdecke (Dransfeld, near Göttingen)
among others, has constructed a—practically worthless—chamber-

train, in which the pump-wheels are equal bevel-wheels with a very obtuse angle between their axes. The interior of the chamber forms a zone of a sphere, and is separated into suction and delivery spaces by two dividing plates in the plane of the axes. The constructive difficulties are here far greater than in the case of the spur-wheels, and, it may be added, this fact has already made itself felt. Spur-wheels, screw-wheels and bevel-wheels have thus already been used in chamber-trains. One variety only is wanting—hyperbolic-wheels—in which the difficulty of making a tight joint certainly reaches its maximum. It is none the less quite possible that any day we may be startled by the appearance of a "hyperbolic rotary steam-engine."

§ 104.

Compound Chamber-wheel Gear.

We examined in § 61 a specimen of compound spur-gearing in the mechanism $(C_{22}^{+} C_{3}'')^{\circ}$ which is represented in Fig. 279. This mechanism has been used as a chamber-train,—by Justice, among others, who employed it as a steam-engine.[*] Justice, who also constructed a two-wheeled chamber-train, made the four wheels equal, so that b and c were represented by one wheel only gearing with both the others. The frame e was used as a chamber enclosing all the wheels. The design was correct and the construction good, but it is not clear what special advantage could be gained by it.[†] A compound chamber-train, consisting of four bevel-wheels, was constructed by Davies as early as 1838.[‡] It was intended to serve either as a rotary steam-engine or a pump. One of the end wheels, say a, had a large tooth extending across to the opposite wheel d, and the double wheel $b\,c$ had a slot of which the sides moved in very incomplete closure with this tooth. I have already (§ 91) mentioned this machine, which—intelligibly enough—has long ago been forgotten.

---

[*] *Practical Mech. Journal*, vol. xix., 1866-7, p. 360 ; *Propagation Industrielle*, vol. iv., 1869, p. 34.

[†] For an old chamber-train of three wheels see Bataille et Jullien, *Machines à Vapeur*, vol. i., 1847-9, p. 442.

[‡] Newton, *London Journal of Arts, &c.* Conjoined Series, vol. xix., 1842, p. 153.

If the axial distances 1.2 and 2.3 of a compound chain $(C_{z2}^+ C_3'')$ be made equal, the shafts 1 and 3 may be made conaxial. This gives us the chain represented in Fig. 280, where b and c, as before, form parts of the same (ternary) link, while a and d can move

FIG. 279.

independently. We shall call a train of this kind, in which the centres of the last wheel are, as it were, turned back into coincidence with that of the first, a reverted train. This form of train plays no unimportant part in machine practice. Among its other

FIG. 280.

applications it has been used as a chamber-gear, but in all cases with the variation from the forms already described that it has non-circular wheels,—the chain therefore being $(\tilde{C}_{z2}^+ C_3'')$ and the mechanism $(\tilde{C}_{z2}^+ C_3'')^c$.

If we make the mean angular velocity ratio of the two wheels a and $d=1$, then the non-circular axoids will cause any pair of radii of the wheels to have a relative oscillatory motion, while both are turning continuously in the same direction. Then if two sectors connected with the wheels be enclosed in a chamber formed from the frame e, we can use them as pistons; we can pair them, that is to say, with a pressure-organ, either as driven or driving bodies. The relative motions of the two pistons will then be very similar to the motion of those in Fig. 4, Pl. XXI. As a few examples among many, I may mention Smyth's rotary steam-engine, patented in 1838, which had non-circular wheels of complex form;[*] Ramey's high-pressure ventilator with four equal elliptic wheels;[†] and Thomson's steam-engine, with four equal oval wheels, of which two examples were shown at Paris in 1867.[‡] The constructive difficulties connected with these machines, especially when they are intended to serve as steam-engines, are so great as to deprive them of practical importance. The outer surfaces of the pistons at least, however, can be made steam-tight, as they form a cylinder pair with the chamber. Ramey's ventilator is said to have given good results.

§ 105.

Epicyclic Chamber-wheel Gear.

I have still one other kind of chamber-wheel gear to analyse, one of which the nature has never hitherto been understood. Even the inventor himself, Galloway,[§] does not seem to have known it, judging, at least, from his own description of the connection between his machine and others. In order to make our investigation complete it will be necessary to begin somewhat far back.

By placing the simple train of spur-wheels $(C_z C_2'')$ with which we commenced this part of our analysis (§ 94), upon one or other of the wheels instead of upon the frame, we can obtain two mechanisms, $(C_z C_2'')^a$ and $(C_z C_2'')^b$, besides the one $(C_z C_2'')^c$ already

* Newton, *London Journal of Arts, &c.* Second Series, vol. ix. 1834, p. 152.

† *Génie Industriel*, vol. xxx. 1865, p. 254.

‡ *Rapports du Jury International*, vol. ix., p. 81 ; *Propagation Industrielle*, vol. iv. 1869, p. 339 ; *Chambers's Ency.*, 1st. Ed., art. "Steam-Engine."

§ Bataille and Jullien, *Machines à Vapeur*, vol. i., 1848-9, p. 431.

examined. They are similar, so we need only examine one of them, say $(C_z C_2'')^a$; this is represented in Fig. 281, where a is supposed to be fixed to the stand. The frame c becomes a crank

FIG. 281.

turning about the axis 1, while the wheel b rolls upon a. We have first to determine the relation between the angular motions ω' and ω of the wheel b and the arm c respectively. Imagine the wheel a to be movable about 1, move it by the crank c, carrying with it b, through any angle ω, and then leaving c in its new position turn a back into its old one. Then any diameter of the wheel b will have first turned through an angle ω from its original position, and will then (considering the diametral ratio $\frac{a}{b}$ of the wheels) have been further turned through an angle of $\frac{a}{b} \times \omega$, both rotations taking place in the same direction as that of the arm,— so that the whole angular motion of b has been :—

$$\omega' = \omega + \frac{a}{b}\,\omega = \omega \left(1 + \frac{a}{b}\right).$$

If for any given time, as a minute, $\omega = n.2\pi$ and $\omega' = n'.2\pi$, we have for the relative number of revolutions of the wheel and the arm :

$$\frac{n'}{n} = 1 + \frac{a}{b}.$$

If either of the wheels were annular, then the turning back of a into its original position would diminish instead of increasing the angular motion ω' of b, so that we should have

$$\frac{n'}{n} = 1 - \frac{a}{b}.$$

Such a mechanism as that before us is known generally as an epicyclic train. It is frequently applied in practice in the form shown, but more often still in a different shape, that namely of a reverted epicyclic train.

If we place the reverted train ($C_{22}\,C''_3$), already considered in the last section, upon a, as is shown in Fig. 282, we can find the velocity of the turning link $b\,c$ by the foregoing method. It is now necessary however to find the motion of the wheel d (conaxial with a) relatively to that of the arm. Using the same method as before we see that while d is carried forward through ω by the action of the arm, it is caused to turn $\omega \times \frac{ac}{bd}$ in the opposite direction as a is moved back to its original place,—so that the actual total angular motion of d is :

$$\omega_1 = \omega \left(1 - \frac{ac}{bd}\right)$$

and we have for the relative number of turns per unit of time of the wheel *d* and the frame *e*:

$$\frac{n_1}{n} = 1 - \frac{a\,c}{b\,d}.$$

Fɪɢ. 282.

The mechanism $(C_{z2} C_3'')^a$, which we have thus obtained by the simple inversion of a kinematic chain, is frequently called in machinery differential gear. This name has apparently been chosen because of the minus sign in the last formula. We shall not retain it, for it may occasion misunderstanding, but shall call the mechanism a compound (reverted) epicyclic train.

If there be an annular wheel in either of the two pairs of wheels a, b and c, d, the formula for the relative rotations will be:

$$\frac{n_1}{n} = 1 + \frac{ac}{bd}.$$

If each of the two pairs contain an annular wheel* it is again

$$\frac{n_1}{n} = 1 - \frac{ac}{bd}.$$

Or generally, if we indicate the simple velocity ratio of the train of wheel work by a, we obtain the formula

$$\frac{n_1}{n} = 1 - a.$$

Here a itself is positive if there be two annular wheels or none, the minus therefore remains; while if there be one annular wheel only a becomes negative and the sign in the formula is positive.

There are many forms and still more applications of the mechanism before us. It will be noticed that in cases where a is negative in the formula and > 1,—the rotation of d is in the opposite direction to that of e. To simplify the description we shall call a the first and d the second **central wheel** and b the first and c the second **outer wheel**.

The limiting cases which occur when some of the wheels are made infinite are very important. One of these I must specially examine, it is as follows. Let us suppose that either of the wheels a or b be annular, as in the diagrams Fig. 283, in which the pitch circles only are shown; then a is, as we know, negative, and the expression for $n_1 : n$ is

$$\frac{n_1}{n} = 1 + \frac{ac}{bd}.$$

Let, however, the radius of the annular wheel be infinite, then in order to gear with it the other wheel of the pair must be infinite also. The centres of the two infinite wheels lie within the finite

---

* As, for example, in Moore's pulley-block,—illustrated in *Engineering*, Sept. 17, 1875.

ones, namely at 1 and 2, but their points of contact, and indeed the whole of their teeth are beyond our observation; they disappear from the mechanism, and only the two finite wheels *c* and *d* remain. The epicyclic train is thus reduced by two wheels, of which one revolves about 1, while the other turns round the first, carried by the frame *e*. The form at which we have arrived is different from that of Fig. 281, for there the central wheel *a* was fixed, while here the only central wheel left, *d*, turns about its axis. In order, however, that the chain may remain closed, it must contain some representative of the wheels which have disappeared. The use of the latter has taken the point of contact, or instantaneous centre, to an infinite distance; it will be seen therefore that they

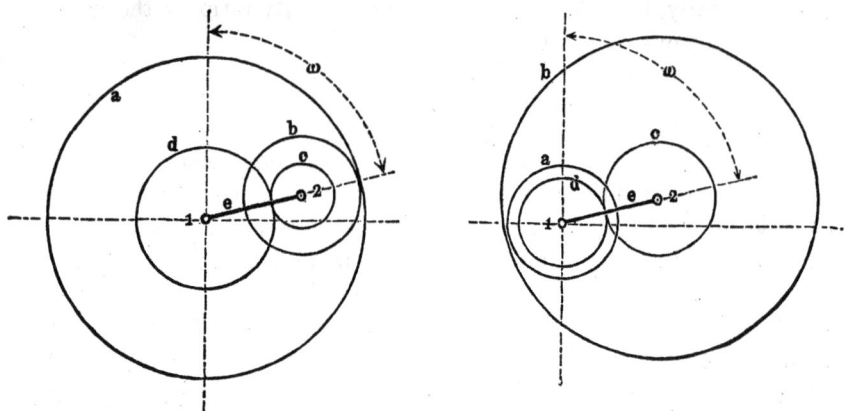

FIG. 283.

may be replaced by any arrangement which will prevent the wheel *c* from making any rotation about its own axis, by means of which, that is, its motion may be restricted to one of translation only, every line upon it moving parallel to itself. This might be done by the addition of a kinematic chain (which we may call accurately a parallel train or motion), so attached to *c* as to constrain it in the required manner.

It is evident, however, that so far as the total motion (in each period) of *d* is concerned, *c* may be allowed to oscillate about its centre 2 to a certain extent, so long as it never completely rotates. This is exactly what occurs in Watt's " Sun and Planet " wheels, Fig. 284, where the upper end of the connecting rod *b* is connected to the beam, and the wheel c fixed to the connecting-rod oscillates with it, but never turns. This mechanism of Watt is therefore a special form of the epicyclic train $(C_{z2}C_3'')^{a}$. Watt usually made $c = d$, from which it follows that $\dfrac{n_1}{n} = 2$.

Fig. 284.

There is nothing here to prevent the use of an annular wheel as one of the pair c d. If this be done a becomes negative. If the outer wheel c be made the annular wheel, the value of a is always greater than unity, and the direction of rotation of d is negative *i.e.*, it turns in the opposite direction to the frame e [*].

Galloway's rotary steam-engine, now to be described, is simply a chamber-wheel mechanism formed upon a planet-train like that of Watt with an annular outer wheel.

The following three figures represent three forms given by Galloway to his engine, which he intended for screw propulsion.

---

[*] Watt himself in his first specification mentioned specially the application of an annular wheel in the sun and planet gear. Muirhead, p. 50. See also Bourne's *Treatise on the Steam-Engine*, p. 21. R. [There is now in the Patent Museum at South Kensington a model of Watt's in which the connecting-rod end is made an annular wheel. To keep this in gear with the wheel on the shaft there is a small roller placed upon a pin attached to its lower part, and this is made to roll upon a suitably shaped cam touching it always on the side furthest from the point of contact of the wheels themselves.]

I have placed on the figures our reference letters to enable the different parts to be more easily distinguished. In the three-cornered
piston of Fig. 285 we have a spur-wheel with three teeth, the
central wheel d of the planet-train Fig. 284, and in c we have
the corresponding four-toothed outer wheel, which is here annular. In order that the motion of the wheel c, which is also the
chamber, might be one of translation only, Galloway carried it
(quite correctly so far as the required motion is concerned) on three
equal and parallel cranks $e\,e\,e$. We can easily recognize in these
the mechanism $2\,(C_2''\parallel C_2'')^a$ which we have already analysed in
§ 66. The proper internal profile of the chamber gave the inventor

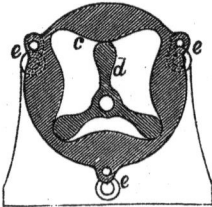 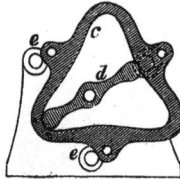 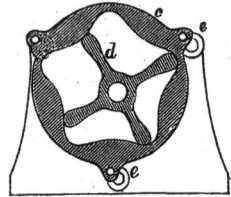

Fig. 285.   Fig. 286.   Fig. 287.

much trouble. It is, in reality, simply the profile of the teeth of
a four-toothed wheel such as can work in gear (and indeed in
steam-tight contact!) with the three-toothed pin wheel d. The
inventor, although he starts from the idea of the annular wheel in
his explanation, does not treat the bodies c and d as toothed
wheels ; but says expressly : " What I propose is to substitute for
toothed wheels, in the majority of cases, the arrangement shown in
the figures, which I shall now explain. . . ." The figures show
distinctly enough that the space between the teeth of c and d varies
periodically from a maximum to a minimum, and is therefore suitable for use with a pressure-organ which can be alternately admitted
to and discharged from it. For the relative rotations of d and e we
have $\dfrac{n_1}{n} = 1 - \tfrac{4}{3} = -\tfrac{1}{3}$, *i.e.* for three revolutions of the centre of the
chamber, or (what comes to the same thing) of the small guiding
cranks e, the piston d revolves once in the opposite direction. Galloway proposed to connect one of the cranks with the screw-shaft of
the vessel, in order that this might rotate three times as fast as the
piston. In Fig. 286 $\dfrac{n_1}{n} = 1 - \tfrac{3}{2}$, in Fig. 287 $1 - \tfrac{5}{4}$. The inventor

points out that the parallel cranks may be connected to the
"piston," and the chamber made to revolve in fixed bearings, which
is equivalent to making c an ordinary spur-wheel and d an annular
wheel, so that $1—a$ again becomes positive.

It is sufficiently obvious that this machine is without practical
value as a steam-engine, although Galloway prepared a design for a
300 H.P. marine engine on his system. Kinematically, however,
it is none the less instructive. In connection with it and the preceding examples our analysis has, I think, shown once more
its capacity for completely solving constructive riddles. These
examples at the same time furnish another illustration of the
remarkable tendency which has so often shown itself in machine-
practice to run through whole circles of solutions for one and the
same kinematic problem by a series of isolated and entirely independent attempts. On account of their very isolation these attempts
have often, as we have seen, led to extraordinary results arrived at by
most roundabout methods. Notwithstanding the simplicity of the
real relation between these mechanisms, which our analysis has
now shown us, we can comprehend in the fullest degree how much
greater the difficulties of the various inventors have been than the
results they have obtained by overcoming them.

CHAPTER XI.

ANALYSIS OF THE CONSTRUCTIVE ELEMENTS OF MACHINERY.

§ 106.

The Machine as a Combination of Constructive Elements.

HAVING in the foregoing chapters considered the nature of the mechanisms of which machines consist, we must now proceed to examine the separate pieces by the combination of which they are actually constructed. Although this may appear at first sight a return to matters already investigated, it is in reality another step forwards upon the road which we have already marked out for ourselves. For it is to a certain extent more difficult to understand the machine in the form in which it actually stands before us than to comprehend the abstract representations by which, so far, we have replaced its constructive complexity. It was necessary that our general notions as to its essential nature should be made distinct, partly indeed re-made, before we could attempt to systematise the complex forms of its single pieces, or distinguish between their fundamental and accidental properties. This problem is indeed by no means a simple one; we shall not wonder, when we have arrived at its full solution, that it has required such long and careful preparation. It was only when chemical science had reached a very advanced stage that it attempted to decompose materials supposed to be elementary; and similarly it has been necessary that kinematic science should be cleared of many erroneous prejudices before it could attempt to analyse the separate pieces from which the machine is formed in the workshop, and make their nature really intelligible.

Wherever the designing of machinery has been made a systematic study, it has to a certain extent been recognised that the machine consists of only a limited number of different parts occurring in it over and over again. Different writers have given to these different names, such as " details," " elements," " simple-parts," etc.; I myself have for many years called them the "constructive elements," (*bauliche elemente*) of machinery.

The constructive elements have formed the subject of many text-books.[54] In these, however, it has not been proposed that this sub-division should be taken absolutely, or indeed without very considerable limitation. It is not assumed, as in the case of the " simple machines," that all machines are simply combinations of these " elements," but only that the latter occur with special frequency in machine construction. Some idea of this sort has always existed below the surface; the want of exact ideas as to the nature of the machine has, however, prevented its clear enunciation, so that as the art of machine construction has advanced there has been a somewhat suspicious uncertainty as to which and what these " elements " were. Neither a very clear enumeration nor satisfactory definitions of them have been given. Only by instinct, as it were, their number has been more or less distinctly limited; or at least they have in general been treated as if some such limitation did exist.

The following enumeration of constructive elements therefore makes no pretension to absolute completeness. It is simply a list of those parts which different writers on machine design have included under the head of constructive elements, or some equivalent title, and fairly represents the details supposed usually to belong to that class. These are considered to be:—

Screws and screwed joints,
Keys, cutters, gibs, and keyed
 joints generally,

Rivets and riveted joints,
Plummer blocks, bearings,
 pedestals,

Pins,

Shafts, axles and spindles,

Couplings,

Framing, bed-plates, brackets &c.

Belts, cords and ropes,

Chains and their connections,

Friction-wheels,

Belt pullies and gear,

Rope pullies and gear,

Toothed-wheels,

Chain-wheels,

Fly-wheels,

Levers,

Cranks,

Connecting-rods, couplers,

Crossheads and guides,

Click and ratchet-wheels and gear,

Brake-wheels and gear,

Pipes and their connections,

Steam and pump cylinders,

Valves,

Pistons and stuffing boxes,

Springs.

In addition to these parts, all of which have very numerous applications, there are others which come into use only in single classes of machinery, spinning and weaving machines, machines for working in metal, etc., but are still employed often enough to have been sometimes included with those above mentioned. By a distinction which appears quite justifiable they have occasionally been called " special " parts as distinct from the above " general " ones. Without giving any illustrations of this second class of constructive elements we shall proceed to consider the first in order. We shall endeavour first to ascertain precisely the kinematic meaning of each, and shall afterwards see how far we can find any general kinematic connection between the whole.

§ 107.

Screws and Screwed Joints.

In the common screw and nut, Fig. 288, we at once recognise the twisting pair (S) or $S \pm S^-$, and we can do the same in some other applications of the screw where, as in the screw-joint, Fig. 289, the nut and screw are themselves parts of the two elements to be united.

The case of the common screw-joint, of which Fig. 290 gives a familiar illustration, is, however, a different one. Here we have a combination of four pieces, a, b, b_1 and c, the object of the whole being the rigid connection of b_1 with c. We see at once that the screw b is prevented from turning relatively to the piece b_1 by the prismatic form given to its neck (cf. § 19), so that as regards rotation b and b_1 form one piece. If the nut a be turned upon the screw, the head of the bolt is brought up to bear upon the piece b_1. This is brought about by the use of the pair (S), that is to say (as we before expressed it, § 47), by pair-closure. Thus the relative turning of b and b_1 is prevented by suitable restraining profiles, and their relative sliding by pair-closure; the two parts therefore form kinematically, a single piece.

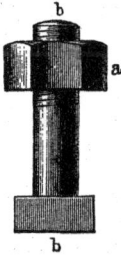

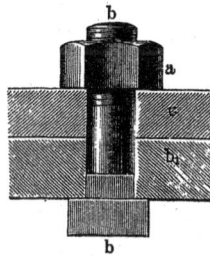

FIG. 288.   FIG. 289.   FIG. 290.

So far as our engraving goes the piece c can turn relatively to b. In the machine itself, however, such a motion is prevented either by the use of a second screw parallel to the first, or by some other means, and the only motion possible to c before the nut is screwed down is translation in the direction of the axis of b. In other words c is paired with $b\,b_1$ by means of a prism parallel to the axis of b; c and $b\,b_1$ form therefore a sliding pair. In reality, therefore, the piece $b\,b_1$ contains two kinematic elements, rigidly connected, a screw S^+ and a prism P^+ parallel to the screw.

The nut a also consists of two kinematic elements, the hollow screw S^- and the plane cone which forms its under surface and rests upon, or more correctly is paired with, c. This surface is not necessarily plane, its general condition is that it must belong to a revolute conaxial with the screw. The pair of revolutes, or turning pair, thus formed by a and c is incomplete, and here pair-closed. This, however, is accidental; essentially the piece a consists of an element S^- having conaxial to it a revolute R, the partner element of which belongs to the piece c.

The last-named piece also consists here of two elements, namely, the above-mentioned prism, paired with that upon $b\,b_1$, and this revolute having its parallel to that of the prism.

·The result of our examination is, therefore, that this screw fastening is a kinematic chain of three links, formed from the pairs (S), (R) and (P). If we write its formula in full, disregarding (for the sake of simplicity) the incompleteness of (R) and further replacing (R) by (C) as we know to be possible from § 57, it will run as follows :—

$$\overbrace{C^- \ldots \mid \ldots S^-_=}^{a}\; \overbrace{S^+ \ldots \parallel \ldots P^-_=}^{b}\; \overbrace{P^+ \ldots \parallel \ldots C^+_=}^{c}.$$

which we may also write, inverting the lower pairs, and noticing that here there is no difference between | and ||;—

$$\overbrace{C^+ \ldots \mid \ldots S^\pm_=}^{a}\; \overbrace{S^- \ldots \mid \ldots P^\pm_=}^{b}\; \overbrace{P^- \ldots \mid \ldots C^-_=}^{c}$$

and in this we recognise a chain, Fig. 291, which we have already examined. We may use $(S'\,P'\,C')$ for its contracted formula. If we consider the link b as fixed, and a as the driving-link, the special formula of the mechanism is $(S'\,P'\,C')^b_a$.

In the applications of the screw-pair to cause rectilinear motion, as in the lathe, or to exert pressure, as in the screw-press, these three links are very distinct, arranged in the first case as $(S'\,P'\,C')^c_a$, in the latter mostly as $(S'\,P'\,C')^b_a$. The form of chain shown in Fig. 291 is also very frequently met with in screwed joints, as, for example, in the "tapped bolt" or "set-screw" of Fig. 292. We also find various methods used in joining b and b_1, as, *e.g.*, the key shown in Fig. 293. In screwed joints, however, of whatever form, we always find that the pair $S^+_\pm S^-$ occurs as part of the chain $(S'\,P'\,C')$.

The action of this chain in different cases varies very much. In the screw-press or the screw-cutting lathe, with which in certain respects the screw-joint might be compared, it is simply used like any other kinematic chain. In the screw-fastening this is also, strictly speaking, the case, but only within such very narrow limits as are allowed by the compressibility of the pieces b and c, beyond these limits it is not used kinematically. When the machine itself is complete, the screw-joint is no longer used as a kinematic chain ; it therefore does not appear in the kinematic formula of the machine. It has been employed as a chain for a temporary purpose only, in order, namely, so to connect two or more pieces that they may be treated as a single body, a purpose for which it is often employed also in structures which are not machines. Such a screw fastening, therefore, as is used for a cylinder cover, or to hold down a plummer-block, has not a machinal but a constructive function in the machine. Its object is to make that connection which we indicate in our formula

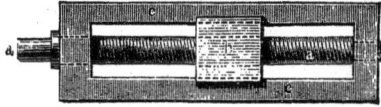

FIG. 291.

by the dotted line ...., in other words it serves to form the links of a kinematic chain.

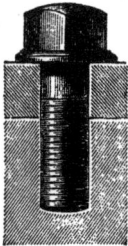

FIG. 292.

FIG. 293.

Kinematically, therefore, its form is indifferent so long as it does not interfere with the required motions of the links ; it is regulated chiefly by considerations of strength. This explains the immense variety of shapes in which screw fastenings occur; the constructive conditions of all are, however, expressed by the formula which we have given above.

We shall have an opportunity of returning to some special forms of screw-joints, safety or locking screws, further on.

§ 108.

Keys, Cutters, &c., and Keyed Joints.

We have already seen (§ 64) that the key * is not a kinematic element in our sense of the word, but that it consists of two prismatic elements, and in its most common application forms a link of a three-linked kinematic chain. This chain, represented by Fig. 294, has the formula :

$$\overset{a}{\overbrace{P^+ \ldots \angle \ldots (P)}} \overset{b}{\ldots \angle \ldots} \overset{c}{\overbrace{(P) \ldots \angle \ldots P_{=}}}.$$

* I use the word universally employed in this connection by engineers, instead of wedge.

for which we have used the contracted expression $(P\frac{L}{3})$. The keyed
joints occurring in machines have indeed always this form, neglect-
ing the occasional force- or pair-closure of incomplete pairs.

The familiar case of the keying of a wheel upon a shaft,
Fig. 295, shows all three links, *a, b* and *c.* The prism pairs 1 and

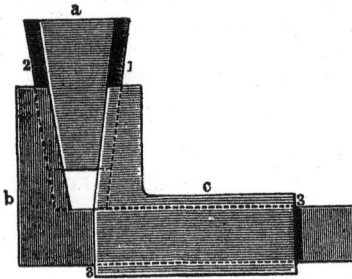

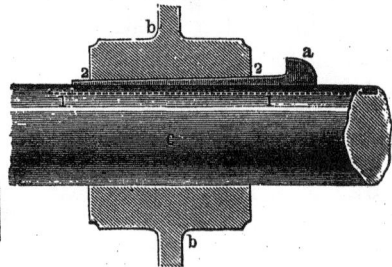

Fig. 294.						Fig. 295.

2 can be at once recognised, each one incomplete in itself but
closed by the other. The pair 3 is omitted, but the wheel, which
is to be moved by the key only in a direction perpendicular to the
axis of *c,* is prevented by force-closure from moving in any other
direction.

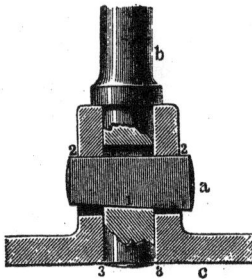

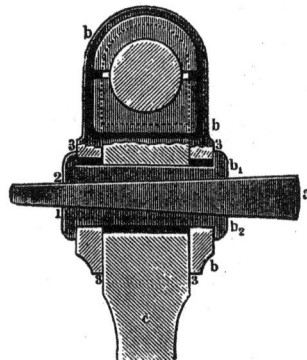

Fig. 296.						Fig. 297.

In the case of a round bar keyed into a socket, Fig. 296, we find
all three links and all three pairs. The pairs 1 and 2 are at the
under and upper surfaces of the key, and the pair 3 appears in the
cylindric surfaces of *b* and *c* as well as the sides of the openings
in *c* through which the key passes. It is these which make the
cylinder into a prism pair. In a " gib and cutter " joint such as is

shown in Fig. 297, the pair 3 is complete, but 1 and 2 are incomplete. The gibs b_1 and b_2 are kinematically parts of the rod c and the strap b respectively.

Keyed joints are therefore in general, as we see from these examples, three-linked kinematic chains, which, however, like those considered in the last section, have not a kinematic function in the machine, but serve simply to form links. We do, however, frequently find the chain $(P_{\overline{3}}^L)$ used in the machine just as the screwjoint chain $(S' F' C')$ is also used, for effecting motion or exerting pressure, but in these cases it is a mechanism, and does not fall to be considered under the head of constructive elements.

§ 109.

Rivets and Riveting, Forced or Strained Joints.

A single rivet joining two plates (Fig. 298) might be regarded as a cylinder-pair $C_-^+ C^-$, the rivet being supposed to be fixed to one of the plates. The latter would then form the elements of a turning pair, and their relative motion would be simple rotation. Rivets are in practice sometimes used in this way, as for instance in flat-linked chains; but such constructions generally come under the head of pins rather than of rivets, and will therefore be considered in the next section. By

FIG. 298.

a riveted fastening we rather understand one in which more than one rivet is used, and in which no relative motion whatever is permitted to the pieces riveted together. The rivets receive their form by hammering while in a more or less plastic condition, and do not in themselves possess the fundamental characteristics of kinematic elements. As constructive elements they serve, like screws and keys, for the formation of kinematic links. They are most frequently used (as my readers know) in boilers and reservoirs of different kinds, in the formation, that is, of the vessels V^- used for the enclosure of liquid or gaseous pressure organs.

A very important part of the action of rivets in pressing together the bodies which they unite is due to the shrinkage or contraction of the rivet as it cools. The same phenomenon is utilized largely in other forms of fastening, and especially in the process of " shrinking " rings of metal over bodies which it is desired to strengthen or unite. The rings are put on their place hot, and of course exert an enormous pressure when they contract in cooling. The same result has of late years been obtained by pressure merely, without previous heating, and in many very important cases this is superseding the older process; in fixing railway carriage wheels on their axles for instance, and in securing the cranks and crank-pins of locomotives, etc. Looked at as a whole the two processes lie very near each other, the latter might almost be called cold riveting. We shall therefore not look at them as distinct, but shall include them both under the name of forced or strained joints.

Kinematically, strained joints represent fastenings of a kind which may be regarded as cylinder or prism pairs, (C) or (P), in which the elements are so closely pressed together that as regards the action of any ordinary forces they form one body only, and which therefore serve for the formation of kinematic links. This close union of the elements is effected essentially by the friction produced by the straining pressure. We shall have occasion once more to return to this point.

§ 110.

Pins, Axles, Shafts, Spindles.

A pin considered kinematically forms one element of the pair $C_\pm^+ C^-$; it is the element C^+, or more strictly R^+ if we use the more general symbol (R) instead of (C). The pin and its bearing, the combinations of elements $R_\pm^+ R^-$, may be considered the most common pair of elements; it occurs in almost every kinematic chain, in large and small dimensions, under light and under heavy pressures, moving slowly and moving rapidly. We shall return to the element R^- in § 112.

Axles are pins joined conaxially; that is, kinematic links of the form $C^+ \ldots | \ldots C^+$. The word axle is used specially in those cases where the forces to be resisted tend chiefly to bend the link.

Shafts are also links of the form $C^+ \ldots | \ldots C^+$. They are therefore kinematically identical with the axles, but the name shaft is used specially in those cases where torsion is the force chiefly acting.

[The word spindle is in many parts of this country used for small shafts].

The kinematic position of these three familiar cónstructive elements in the machine is therefore very distinct.

§ 111.

Couplings.

Under the name of couplings are included a number of constructions by which the motion of one shaft can be transmitted to another. Their kinematic position is not quite such a simple matter as that of the pieces hitherto considered, on account of the very different arrangements which have received and are known by the name of coupling. Toothed-wheels, friction-wheels and wheel-gear generally, although used for the purpose of transmitting the rotation of one shaft to another, do not receive the name of coupling, but frequently enough couplings are trains containing several links. We may perhaps define a shaft coupling as an apparatus which transmits from one shaft to another equal numbers of revolutions in equal times and in similar directions without the use of wheel-gearing. The definition is certainly not a very sharp one, but it seems entirely to cover what is usually meant by a coupling.

Couplings may be divided into fixed, moveable, and loose, the latter being in most cases known as clutches. We shall here con-

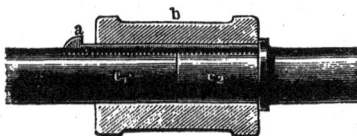

FIG. 299.

sider the first two classes only, returning to the last in § 123.

Fixed couplings join two shafts in such a way that they may be treated as a single body. They are fastened with screws, or with keys, or with both ; indeed there is nothing in principle to prevent their being fastened by rivets. Fig. 299 shows what is known as a muff-coupling, in which the three links and pairs of the chain (P_3^L) will easily be recognized. The flange-coupling, Fig. 300 is a combination of two keyed fastenings with a multiple screw-joint. Other fixed couplings show still further combinations. Their real function in every case is the formation of kinematic links, these links having the form $C^+...\,|\,...C^+$.

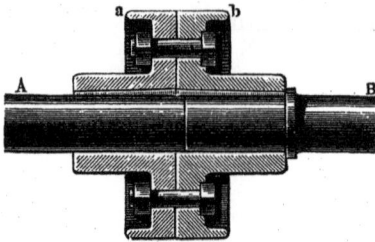

FIG. 300.

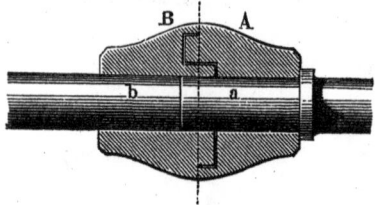

FIG. 301.

Moveable couplings subdivide themselves into those which are moveable axially, radially, and angularly. Sharp's claw coupling, Fig. 301, is an illustration of the first kind. It is formed as a prism pair $P^+_\pm P^-$, for the claws of the piece A and B are prismatic, and are so formed that relative motion can take place between them only in tl e direction of the axis of the shafts a and b. We may suppose the pieces A and B to be connected to a and b by key-fastenings.*

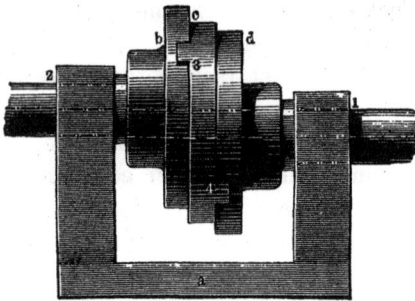

FIG. 302.

Oldham's coupling, Fig. 302, is one which is moveable radially. We have already examined this mechanism fully (§ 72), and have found it to be a turning cross-block, having the special formula $(C''_2 P^\perp_\frac{1}{2})^a_b$.

The universal joint, Fig. 303, is an example of a coupling having angular motion. We have in our earlier investigations repeatedly spoken of this train, and in § 62 pointed out that it was a·conic turning cross-block $(C^\perp_3 C^L)^a_b$. It must not be forgotten that the link $C^-...\angle...C^-$ is omitted from our figure, as is usually the case in representations of the joint.

These examples are sufficient to show that in the moveable

---

* This coupling is not intended for use as a clutch, but for allowing a or b to move axially without disturbing the transmission of rotation.

couplings we have partly pairs of elements and partly complete mechanisms or portions of them, and that it is possible for the

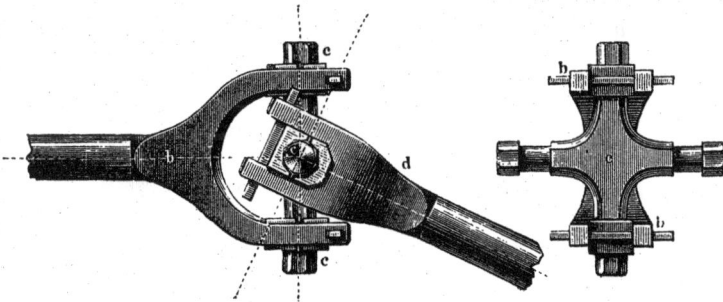

FIG 303.

individual links of the latter themselves to consist of several pieces united by screwed or keyed joints.

§ 112.

Plummer Blocks, Bedplates, Brackets and Framing.

The plummer-block or pedestal makes, along with the spindle or shaft C^+, the pair of elements $C_{\pm}^{+} C^-$; it is therefore itself the single kinematic element C^-. In its actual construction many varieties of screwed and keyed joints and other such auxiliary mechanisms are used, partly to unite the separate pieces of which it consists (*i.e.* to form links), and partly to facilitate lubrication and cleaning. In neither case therefore do they appear in its (principal) kinematic formula.

In the bed-plate which carries the plummer-blocks we have simply the fixed link or frame of a kinematic chain, arranged so that the elements C^- or C^+ may be connected to it by suitable fastenings, or made in one piece with it. Fig. 304 shows a bed-plate for the two parallel shafts A and B. If we imagine the two plummer-blocks to be in their places, we have in the whole simply the constructive form of the frame $C^-...\|... C^-$, Fig. 305.

A bed-plate for two shafts at right angles to each other, Fig. 306, such as is frequently required for turbines, is (with its two plummer-blocks) equivalent kinematically to the frame, $C^-...\perp...C^-$ in

Fig. 307, or, bearing in mind the invertibility of the lower pairs, to the piece $C^+...\perp...C^+$ in Fig. 308. Brasses, screws, cover, bolts and so on serve only to complete constructively the element C^- and to secure it to the floor or building. The compound bed-plate

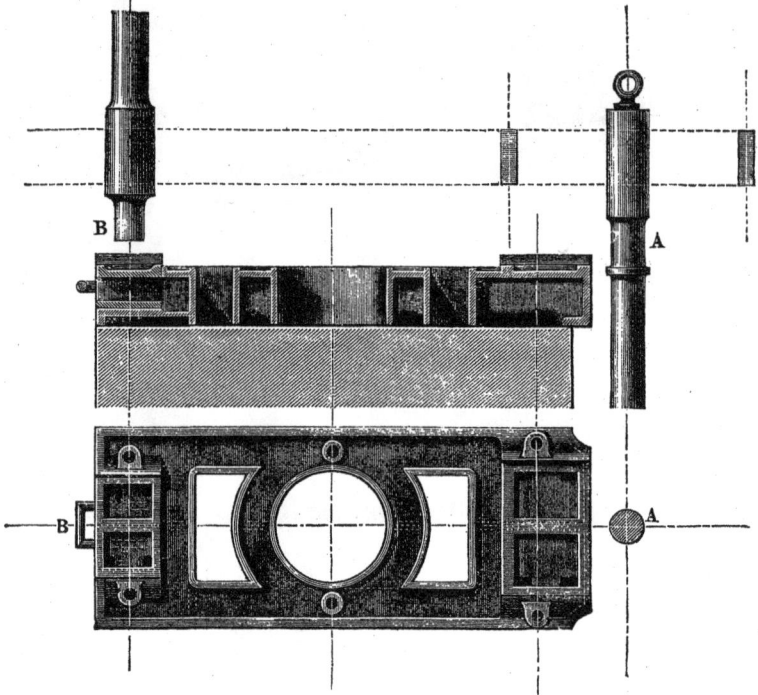

FIG. 304.

Fig. 309 may be represented (always supposing the addition of the plummer blocks) by the four elements C^- of the frame shown in Fig. 310.

FIG. 305.

In the design of any machine it is very advantageous to begin by representing in this way the simple kinematic forms which form the basis of the framing, bed-plates, brackets and structures which

connect them, before proceeding with the design. This will greatly
help the designer in realising his problem in an abstract form, and

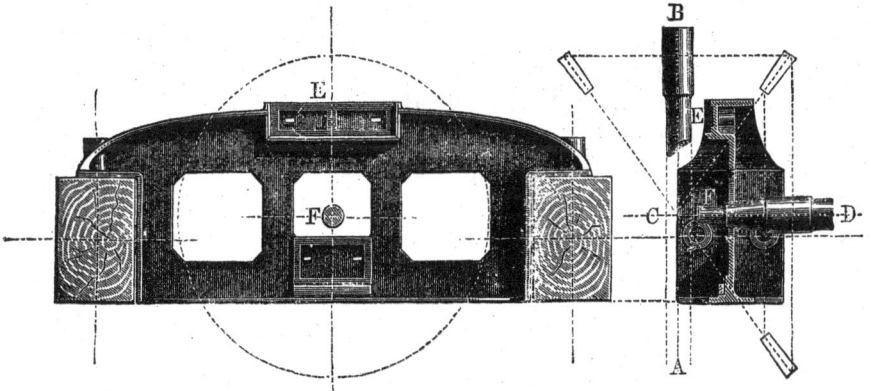

FIG. 306.

FIG. 307.

FIG. 308.

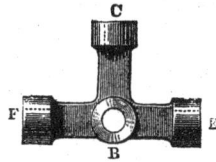

FIG 310.

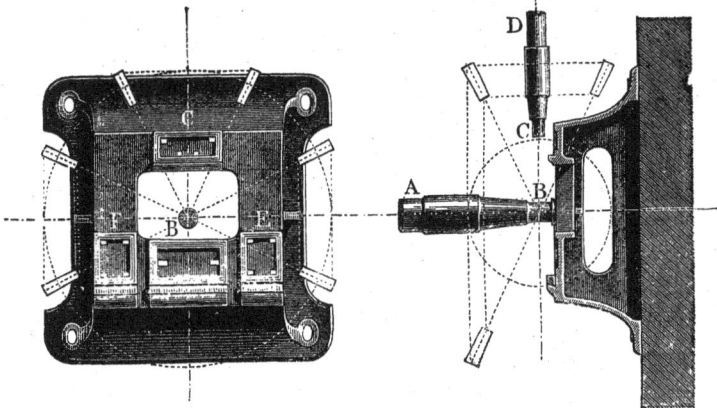

FIG. 309.

the result will be shown in the increased simplicity and excellence
of his work. The first step in this direction is generally to grasp

firmly and carry with us the apparently elementary notion that the
fixed part of the machine is really a portion of its kinematic
linkage. It is only too easy to forget that the masonry, timber,
flooring, and so on, upon which the pedestals, guides or framing
are fixed, have by that very fact become a link of the kinematic
chain of a machine. I have already remarked (§ 58) how often
the fixed link is omitted from engravings. Unquestionably this
omission has arisen from indistinctness as to its nature, and it reacts
in a similar direction upon those for whose
use the engraving has been made. There is
nothing whatever to help the latter to
realise the fact that the important link
omitted is the one which must be fixed. Who
would imagine, for instance, from the accompanying figure of an oscillating engine, taken
from a modern kinematic text-book, that the
bearings A and B must be rigidly connected ?
They are apparently quite without connection
of any kind. The example I have given is,
however, only one among many. We cannot
be surprised, therefore, that this connection

FIG. 311.

has often been carried out incompletely in actual machine construction. Those engineers who are old enough will remember the
noise made about the form in which Penn constructed his oscillating
marine engines, simply because he employed cross-shaped frames
to give special rigidity to the frame of his machine. And yet Penn
did nothing more than carry out the simple requirement which we
recognised at the very beginning of our investigations, as belonging
to that link of the kinematic chain.

We see an exactly similar improvement now being carried out in
the horizontal steam-engine, in the introduction (in America first)
of a straight heavy frame directly connecting the cylinder and the
plummer-block. This bed-plate of the Corliss and Allen engine,
of Tangye's engines and others is nothing else than the frame d of
our turning slider-crank $(C_3'' P^{\perp})^d$ Fig. 312. It is difficult to
believe that this special form of construction has been so recently
introduced that the "improvement" embodied in it is still more
or less a subject of remark, while it seemed to develop itself
as a matter of course from our first propositions. But the way in

which these matters have hitherto been looked at has made many things appear simple and self-explanatory which in reality are complex and require proof, while others have been considered specially remarkable which are only conclusions directly deducible from definite propositions. In the latter circumstance we can recognise the power we possess in having at our command an exact logical system.

Many other examples could be mentioned, which show, like those cited, the want of a distinct perception of the function of

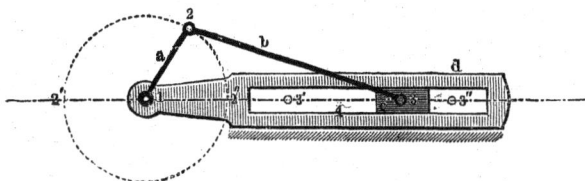

Fig. 312.

those parts of machines and structures of which we have been speaking. Redtenbacher's attempt to treat those machines in which the frames are in one piece as a class by themselves seems to have been due to the same cause.* We have seen that the right treatment of the problem is very simple and intelligible, and does not indicate the existence of any such separation—it will not, therefore, be well to perpetuate it.

§ 113.

Ropes, Belts, and Chains.

We have already found (§ 41) that ropes, belts and chains are kinematic elements. They are the tension-organs T_v, T_p and T_x. If they are so used that by the help of hooks, screws, rivets, etc., they are either made endless (that is, returning upon themselves), or are united with other bodies, they represent links of certain kinematic chains which we shall consider in the next paragraph. The flat-link chains are essentially combinations of numerous kinematic links each of the form $C^+ ... \| ... C^-$, the closure of the whole being effected by the insertion of a frame between the chain-wheels.

* Redtenbacher gave these the name of *Möbel-maschinen.*

§ 114.

Friction-wheels; Belt and Rope-gearing.

Friction-wheels are kinematic elements in force-closed pairs. Two corresponding wheels, such as those of Fig. 313, arranged so as to work in gear with each other, form a higher pair of elements of the form R^+, R^+ or R^+, R^-

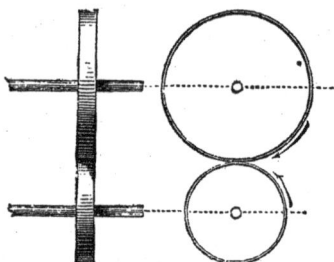 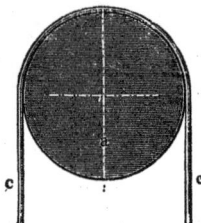

Fig. 313.   Fig. 314.

A pulley which guides a cord or belt, or by its rotation sets such an organ into motion, forms with it the pair R^+, $T^{\pm}$, as in Fig. 314. Two such pairs (which are, as we know, force-closed), when suitably united, give us the belt- or rope-train (as the case may be) shown in

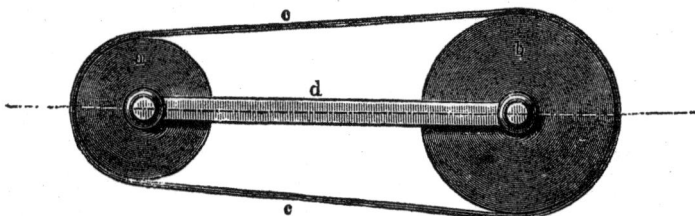

Fig. 315.

Fig. 315, if the shafts of the pullies and the connecting frame be added (cf. § 44). The single rope-wheel or belt-pulley or drum forms with its shaft a link of the chain represented in the figure, the "endless" tension organ being itself another link.

§ 115.

Toothed-wheels, Chain-wheels.

Toothed wheels are links of the chain $(R_z C_2'')$, of which an example is furnished by the spur-train of Fig. 316. The framing carrying the plummer-blocks takes the place (§ 112) of the straight

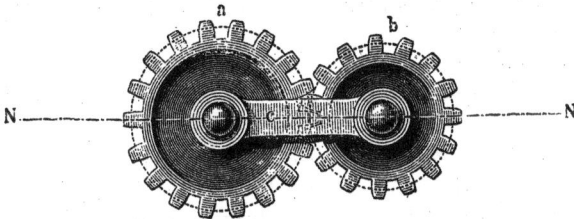

FIG. 316.

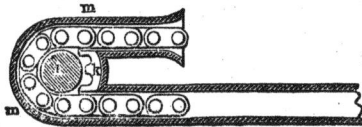

FIG. 317.

link *c*. If a toothed wheel be geared with a chain we obtain the pair R_z, $T^{\pm}_z$ (Fig. 323). A suitable combination of such pairs gives us chain-wheel gearing.

§ 116.

Fly-wheels.

We have already had an opportunity (§ 45) of examining the kinematic meaning of fly-wheels. They are heavy bodies formed as revolutes, and attached to links of the form $C^+ \ldots | \ldots C^+$ in order either to carry the machine over its dead points by their momentum, or to make its motion more uniform. They do not demand any special symbolic indication as links or elements, for our notation is not concerned with the masses of the parts it represents.

§ 117.

Levers, Cranks, Connecting-rods.

Levers, whether simple or compound, are kinematic links furnished with pins about which they can swing (see p. 284). The simple lever is one like the link c of the chain (C_4''), of which the formula is $C^+ \ldots \| \ldots C^+$. The compound lever is a compound link formed from the simple one, such, for example, as is represented by the formula

$$C^+ \ldots \| . \begin{cases} \ldots & C^+. \\ \ldots & C^+ \end{cases}$$

The crank is also a link of the form $C^+ \ldots \| \ldots C^+$, but is so arranged that it can turn completely round its pin or shaft; it corresponds exactly, that is, to the link a of the chain (C_4'') or $(C_3'' P^\perp)$. The connecting-rod, lastly, is also a link formed of two cylindric elements, generally in the form $C^- \ldots \| \ldots C^-$. It corresponds to the coupler b in the trains $(C_4'')^d$ and $(C_3'' P^\perp)^d$. In its kinematic form, therefore, it does not differ from the bedplate Fig. 304 in § 12.

We have here, therefore, a series of links before us, which while they are constructively very different, are kinematically precisely similar, and owe their different characteristics entirely to their position in the chain. The compound lever, too, is exactly similar to the compound bedplate (§ 112) in which the element C is used in precisely the same relative positions.

§ 118.

Crossheads and Guides.

The common crosshead is simply the link c, the "block," of the chain $(C_3'' P^\perp)$. It has the formula $C \ldots \perp \ldots P$. The guides in which it works are formed in many different ways. They constitute the element of the pair 4 which is carried by the frame d in the train $(C_3'' P^\perp)^d$, Fig. 318, and generally have the form P^-, although sometimes they are also made P^+. This prism is shown in the slide-bars $D\,D$ of Fig. 319, where also C is the crosshead,—the block *c* of Fig. 318. As a matter of history the crosshead has passed through an unusually large number of changes of form, which show of what careful study it has been the subject. The

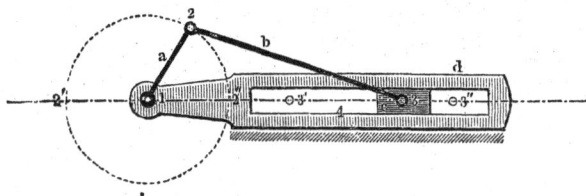

FIG. 318.

production of an exact rectilinear motion in a given mechanism— a matter which at first sight appears so simple—is a problem which for a long period remained without a practical solution.*

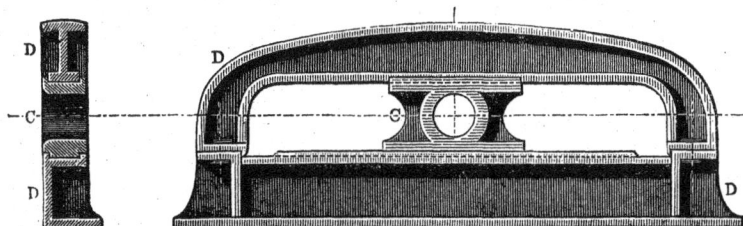

FIG. 319.

We have now seen that the crank, the connecting-rod, the cross-head, the guide bars (including with them the crank shaft bearing), and the lever are in fact all the links of the crank trains $(C_3'' P^\perp)^d$ and $(C_4'')^d$.

§ 119.

Click Wheels † and Gear.

The exact treatment of click- or catch-gear leads to very complex and many-sided problems. We cannot attempt to treat these

---

* Cf. end of § 3.

† For the purposes of Prof. Reuleaux's work it has been necessary to distinguish between two classes of ratchet gear, that, namely, in which the pawl or click acts merely to prevent motion, and that in which it is used to drive the wheel or rack. I propose to call the first class click-gear, and to use the common name of ratchet-gear only for the second and more important division.

exhaustively, but must here content ourselves by looking at a few
of the more important cases which occur.

Among the numerous forms in which click-gear is used the
most common is that of a toothed wheel provided with a click or
pawl, Figs. 320 and 321. The train consists, in both the cases
shown, of three links—viz. the wheel $a = C...|...C_z$, the click
$b = Z...\|...C$, and the frame $c = C...\|...C'$; we shall suppose this

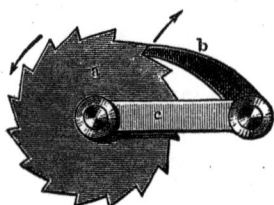

Fig. 320.

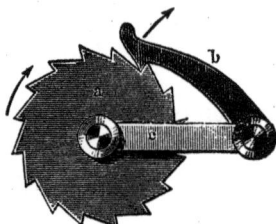

Fig. 321.

last to be the fixed link. The tooth Z, the working end of the
pawl, lies force-closed (as we have already pointed out, p. 180) in
the spaces of the wheel a, the catch being held down either by a
spring or by its own weight. It must also be remarked that b is
kinematically paired with a only for one direction of rotation,
left-handed rotation in Fig. 320 and right-handed in Fig. 321. If
any turning commence in the opposite direction the wheel is at
once held fast by the click, so that the whole mechanism becomes
equivalent to a single piece.

At first sight it might appear that the difference between the
two trains was constructive only, the pawl being formed to resist
pressure in the first case and tension in the second. If, however,
the direction of the arrows shown upon the figures be noticed, it
will be seen that in the first case the turning of the wheel and
click, if the former be set in motion, takes place in opposite
directions, while in the second case it occurs in similar directions.
Between the pressure or push-click and the wheel there is thus
the same relation as between externally toothed wheels, between
the tension or pull-click and the wheel the same as between an
annular-wheel and a spur-wheel. We may therefore use the
symbol Z^+ for the tooth of the pressure-click, and Z^- for that of
the tension-click.

A second property of the gear which must be indicated by our notation is the **single action** of the click-wheel. We may show this by substituting a semi-colon for the comma between C_z and Z. With the addition of the sign for force-closure the pair will therefore be written $C_z ; \dfrac{Z^+}{f}$ or $C_z ; \dfrac{Z^-}{f}$. The point may be taken to denote the immoveability of the chain in one direction, the comma showing that it is moveable in the other.

Placing the chain on c, its complete formula will be therefore

$$\overbrace{C^+ \dots \parallel \dots C_z}^{a} ; \overbrace{Z \dots \parallel \dots C_{\underline{\pm}}}^{b} \overbrace{C^- \dots \parallel \dots C_{\underline{\underline{\cdot}}}}^{c} .$$

The form symbol for Z has been here omitted in order to make the expression more general. The sign for forceclosure is also omitted; it can usually be dispensed with—the unusual nature of the pair being sufficiently pointed out by the semicolon. The latter, indeed, makes it possible for us to use a single element symbol only for the pair $C_z ; Z$, for we shall indicate it quite sufficiently if we take $C_z ; Z, = (C_z ;)$. This contracted form is also justified by the analogy of (C_z) for the spur-wheel pair C_z, C_z —for we may consider the pawl $C \dots \parallel \dots Z$ essentially as a piece of a spur-wheel, carrying a single tooth.

The rack click-gear of Fig. 322, with fixed frame, would have for its extended formula:

$$\overbrace{P^+ \dots \parallel \dots P_z}^{a} ; \overbrace{Z^+ \dots \parallel \dots C_{\underline{\pm}}}^{b} \overbrace{C^- \dots \perp \dots P_{\underline{\underline{\cdot}}}}^{c}$$

for which the contracted form would be $(C\,P\,P_z ;)^c$.

There is another class of click-gear which differs in one very important particular from that which we have been considering;—an example of it is shown in Fig. 323. Here we have click-wheel, pawl and frame exactly as above, but here the pawl so grips the teeth of the wheel as to make its motion in either direction impossible. The click b is therefore, as it were, a combination of those of Figs. 320 and 321, for it acts as a pressure-click against motion in the one direction and as a tension-click against motion in the other. While, therefore, the click-trains just considered were **single-acting**, the one now before us is **double-acting**; we may call them **free** and **fast** click-trains respectively.

We must find a new symbol of relation to enable us to express this double action of the fast click. The tooth may, in the first place, be indicated by $Z^\pm$, and further, following out the same reason for which we chose the semicolon above, we may here use the colon for the sign of pairing. We shall therefore indicate the pair of elements consisting of a click-wheel and double acting pawl by $C_z : Z^\pm$, or by the contracted symbol $(C_z :)$. The fast click train of Fig. 323 will therefore be $(C_2'' C_z:)^\circ$.

This train differs very greatly from the free click train $(C_2'' C_z;)^\circ$. In the latter nothing whatever prevents the free turning of the wheel a in the proper direction; the pawl is lifted by the motion of the wheel itself, and drops again immediately by force-closure. If it be desired that turning should take place in the opposite direction, some special means must be provided for lifting the pawl b, and so throwing the whole train "out of gear." With $(C_2'' C_z :)^\circ$ on the other hand, motion cannot occur in either direction unless the click be first thrown out of gear. If any fast click be thrown out of gear, and then, motion being commenced, be again brought under the action of the closing force (Fig. 324), the rotation lasts only until the next space comes under the tooth of the pawl. The latter then falls instantaneously, and wheel, pawl, and frame become equivalent to a single piece only.

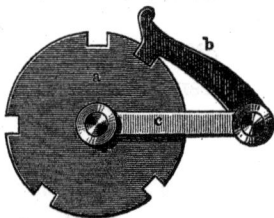

FIG. 322.

With the free click-train (Fig. 325), on the other hand, the pawl falls gradually under the same circumstances (the wheel, Fig. 325, turning to the right), and reaches the bottom of the space even before the moment of closure. It intercepts the wheel teeth therefore with greater safety than in the other case.

FIG. 323.

If after any loaded click train—a train, that is, whose click-piece a is subjected to the action of some continued forward force—be set in motion, its click does not come under the action of the closing force, the wheel a will continue to turn, and will turn the quicker the greater the load be. This motion may be called the reversal of a click train. It is well suited for such a purpose as bringing into action at a given moment mechanical energy which has been stored up in any part of a machine. Click-trains used for this purpose may be called curb-gear; they are employed in many forms, of which an extremely familiar one is the common gun-lock. In this the two "bents" of the tumbler are the teeth of the click-piece or curb, which are released by

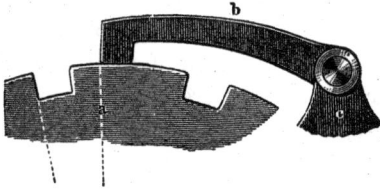 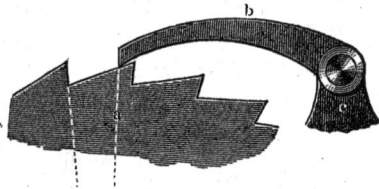

Fig. 324.   Fig. 325.

pulling the trigger. Long ago, in the cross-bow of the middle ages and the catapults and ballistas of the ancients, the principle of the curb-gear was used in mechanisms by which stored-up energy was brought suddenly into action (§ 48). In important modern machinery it serves the same purpose; in the self-acting spinning machine, for example, both free and fast click-trains are used as curb-gear, the special object here being the effecting of a required change of motion at a given instant.*

§ 120.

Reversed Motion in Free Click-trains.

The applications of both forms of click-gear are—as our examples have shown—extremely numerous and important, more important, indeed, than they appear to be at first sight. This makes it necessary to examine somewhat more closely the mechanisms formed from them, several of which will throw considerable light

---

* Click-trains have not unfrequently been turned into chamber-gear, generally in the form $(C_2'' C_z \;)\frac{c}{z}$, $—(V\pm) = b, c$. This is the formula, for instance, for Watt's well-known rotary engine, in which the revolving piston is simply a one-toothed click wheel. This was patented by Watt in 1782, and afterwards by Routledge in 1818, (see Farey, *Steam-Engine*, p. 672, Pl. XV.), but naturally enough it was unsuccessful, the higher pairing $C_z \; ; Z$ could never be steam-tight, and was alone sufficient to destroy its efficiency.

upon the nature of particular classes of constructive elements. We must therefore somewhat overstep the usual limits assigned to the constructive elements themselves.

We have already pointed out that click-trains are employed both for direct and reversed motion. We shall first examine shortly the nature of the reversed motion of the click-piece and the corresponding relative motions of the pawl or click.

If the click-piece a be a wheel, its reversed rotation gives to the pawl $C...\|...Z$ also a motion of rotation, the conditions of which depend upon the particular form used for the back of the teeth of a. This may be considered, in the most general case, as a portion of the curved profile of a disc or cam a, Fig. 326. The motion given by this profile to the pawl will be an oscillation about the axis of the pair connecting it to the frame c. If the radius of the pawl be made infinite it becomes a block b (Fig. 327), which, with suitable force-closure, is caused to slide to and fro in a straight line by the cam a. Here we have, as before, a chain of three links; its formula is

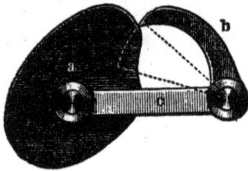

Fig. 326.

$$\overbrace{C^+ \,...\, \| \,...\, \tilde{C},}^{a} \overbrace{\frac{Z}{\tilde{f}} \,...\, \perp \,...\, P^\pm}^{b} \overbrace{P^- \,...\, \perp \,...\, C^-_{\underline{=}}.}^{c}$$

By placing this chain upon the link c we obtain a very numerous series of mechanisms, which we may call slider-cam trains. There are various ways in which the force-closure of Z may be replaced by pair-closure. One of these is the addition to the end of the block or pawl which is paired with the cam of a cylindric pin—an element, that is, of which the profile is a curve equidistant to the end-point,—and the pairing of this pin with a groove in the cam a, formed by drawing equidistants to the original profile, Fig. 328 (cf. § 35). The force-closure of the pawl or block b is therefore not an essential feature in the cam train, but only a separable and accidental property of it. We were therefore fully justified in the omission from our formula, in the last section, of the sign for force-closure. We must not go further into this question here; it is one which must receive extended treatment under Applied Kinematics. It was only necessary to point out that the click-wheel with its sharp teeth belongs strictly to the class of slider-cam trains,[55] which on their

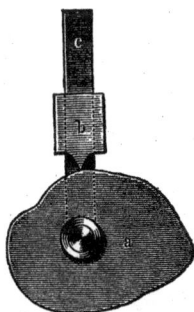 

Fig. 327.   Fig. 328.

part also become, under certain circumstances, spur-wheel trains. We must now turn to some compound mechanisms which are formed from click-gear.*

§ 121.

Ratchet-trains.†

The common forms of ratchet-gear play apparently a somewhat subordinate part in machinery, for which perhaps their force-closed motions may account. None the less do they require our most careful attention, for reasons which I shall show further on, and we must therefore here make ourselves familiar with their principal characteristics.

---

* The slider-cam train—which we might indicate by the contracted formula $(CC,_zP)$ —has, like the chain $(C''_2 C_z;)$, been frequently used as chamber-gear, not unfrequently with a force-closed sliding-block. It will be sufficient to mention, merely as illustrations of the form taken by the chain when chambered, the engines of Davies (Burn, *Steam-Engine*, p. 128), Scheutz (D. K. Clark, *Exhibited Machinery of* 1862, p. 318), or Sudlow (*Engineering*, Ap. 10, 1874, p. 267). The higher pairing in these machines again renders any attempt to form them into satisfactory steam-engines quite useless.

† See note p. 455.

A piece of a machine is said to receive a ratchet motion if it be moved always in the same direction but with an intermittent instead of a continuous motion. The special mechanism by which this motion is given is called a ratchet-train. Such a train requires that the link receiving the intermittent forward motion should be prevented during the longer or shorter pauses that occur from moving backwards,—and for this purpose click-gear is very often employed. A complete ratchet-train is therefore frequently, although not always, a combination of ratchet-gear and click-gear.

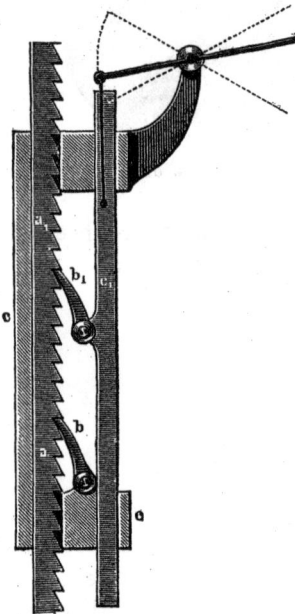

FIG. 329.

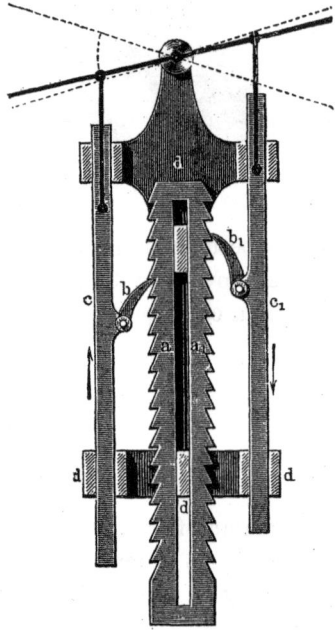

FIG. 330.

The click or other gear thus coming into use we shall call retaining-gear.

A form of ratchet-train which often occurs is sketched in Fig. 329, the object here being the lifting of the rod $a a_1$. The rack click-train $(C P P_2;)$, which we have already examined, is here used as retaining-gear; and the ratchet work consists of an exactly similar chain placed upon a, whose rack a_1 is made one with that of the retaining-gear, and which moves relatively to the fixed frame c. When c_1 is moved downwards the rack $a\,a_1$ is held by the retaining-gear, while the pawl b_1 of the ratchet-train slips over the tops of its teeth; when it is lifted, on the other hand, b_1, c_1, and a_1 behave as a single piece, while b allows the rack to move upwards.

If, retaining the same relative motion of c and c_1 as before, we make half of it into absolute motion, we obtain the double-acting ratchet-train which is represented by Fig. 330. Here the two racks a and a_1 are formed upon opposite sides of the same rod, and guided by an internal prism pair. The pawl rods, c and c_1, are moved by couplers from an equal armed lever above them, in the same way as before. (In practice, where the double-acting ratchet-train occurs not unfrequently, the pawls are generally placed directly upon the working lever, so as to dispense with the couplers and rods.) No retaining-gear is now used, the two sets of ratchet-gear acting alternately. It is to be noticed that the pawls here, although they have only half the stroke of those in Fig. 329, pass over the same distance on the rack in each downward motion as in the former case, supposing the whole travel of the rack to be the same, per period, as in the single-acting ratchet-train.

The "levers of Lagarousse" (Fig. 331) form another double-acting ratchet-train. Here one block carries both a pull- and a push-click, and these act alternately on the ratchet-wheel a. For each upward or downward motion of c, the number of teeth over which the one pawl slips corresponds to twice the distance through which the other (acting) pawl moves the wheel forward. This peculiarity of the motion should be remembered, as we shall have to return to it again.

We notice here that free click-gear is very suitable for use in ratchet work. This is specially the case in one important class of ratchet-trains, the escapements of clocks or watches. These act generally in the manner described in the foregoing section, by the alternate engagement and disengagement of a click with a click-wheel, the latter being continuously driven by some external force in one direction. The engagement and disengagement are caused to take place at intervals of time as nearly uniform as possible, so that the escapement may regulate the

motion of the clockwork by compelling its wheels to move through
equal angles in equal times.

Graham's well-known anchor escapement (Fig. 333) may serve
us for an example of this. In it two free click-trains are united
in such a way that the two clicks b_1 and b_2, the one being a pull-
and the other a push-click, form parts of the same piece, here called

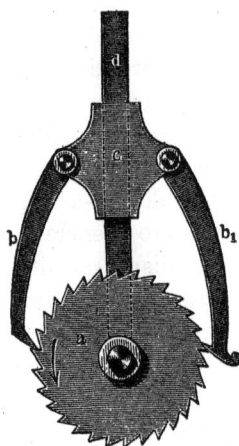

Fig. 331.

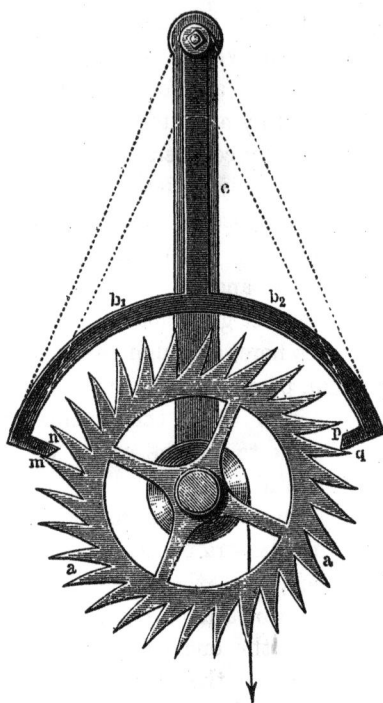

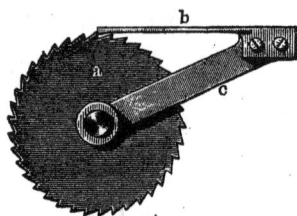

Fig. 332.   Fig. 333.

an anchor. The motion of the pendulum causes the regular
alternate lifting and engagement of the clicks. If the click b_1
be lifted, b_2 falls into one of the spaces and arrests the motion
of the escape-wheel, upon which the driving force acts continually
in the same direction. As the anchor swings back, b_2 is disengaged
and the escape-wheel is held by b_1. Each time the wheel moves
through a distance corresponding to half the pitch. Each tooth
of the wheel, as it slips past the ends q or m of the anchor,
exerts some outward pressure upon it, and slightly accelerates

the motion of the pendulum. This, however, is merely an accidental feature of the escapement, used to adapt it to particular purposes; there are many escapements, especially modern ones, in which it does not exist. In some escapements of specially delicate construction, such as the chronometer escapement, a single click only is used, and is lifted and engaged once in each complete vibration (double swing) of the pendulum, allowing one tooth of the escape-wheel to pass it at a time. In Wheatstone's chronoscope, as improved by Hipp, an escapement of this kind, which acts with extraordinary rapidity, is used (Fig. 333). It is so constructed that it can make 1000 complete vibrations per second. In each of these the escape-wheel moves one tooth forward and is again arrested at the next. We thus see that in the most delicate machines which have been constructed this click-gear, which at first sight appears so rude an appliance as scarcely to be suited for any approach to machinal exactness of motion, is extensively utilized.

Fig. 334 is an example of a ratchet-train with a fast pawl. If the wheel a is to be moved it is necessary first to raise or disengage the pawl b. This is effected by means of a tooth d_1, which forms one piece with the revolving ratchet-tooth d, and lifts the pawl by coming in contact with the face b_1. So soon as this occurs the ratchet d enters one of the spaces of the wheel a, and drives it one tooth forward. At the end of this motion, however, the pawl again drops, the tooth d_1 having passed the projecting piece b_1. As soon therefore as the ratchet motion has occurred the click-train is again fixed. The ratchet $d\,d'$ may revolve in either direction, so that the wheel may be caused to move either forwards or backwards.

If the radius of the wheel a be made infinite, it becomes, as we know (cf. §§ 69 and 71) a straight rack. We should therefore obtain,—if the ratchet $d\,d_1$ were suitably formed,—a train in which a rack could be moved backwards and forwards by ratchet-gear and held during its pauses by a fast pawl.

Without going here into other forms of this kind of ratchet-gear, I must briefly look at one application of the train which occurs with special frequency. This is its application in locks,— where from the common door or box-lock to the most complex

" patent safety " apparatus,—we find this train everywhere applied in the motion of the bolt by the key.

The common door-latch, in the first place, shows itself at once to be a free click-train under our definition. Both the common lifting latch and the ordinary spring-bolt or " sneck," form, with the lock-box or frame, the door-frame and the door itself, click-trains, which belong to the class shown in Figs 320 and 321. They differ from these only so far that after the bolt or latch has fallen into gear by the closing of the door, the socket of the bolt on the one hand, and the frame of the door on the other, prevent any further motion of the door on its hinges, so that the free click has become a fast one.

FIG. 334.

A bolt moved by a key is almost invariably a click-rack a of the form $P...\|...P_z:$, the tumbler is a fixing pawl b, which is made in several pieces in the better classes of locks for security's sake. The key is the ratchet and lifting tooth $d\,d_1$, the frame of the lock the fixed link c. Besides this the bolt a forms with the lock frame and the door a special

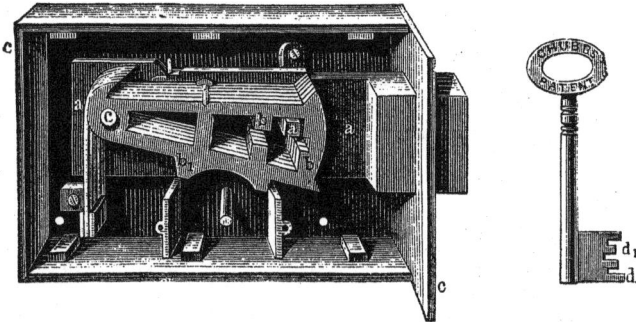

FIG. 335.

fast click-train. In those locks in which more than one turn of the key is required to withdraw, or to shoot the bolt, the rack a has more than one ratchet and click-tooth. In order to prevent any unauthorized opening of the lock, the link of the train which lifts the click and moves the bolt, *i.e.*, the key,—is made separate from the lock itself. The key in many cases serves only to lift the click out of gear, a separate ratchet, connected with a handle, being used to move the rack. Complex forms are given to the key and the tumbler in order to render it impossible, or at least very difficult, to move the bolt by any other key than that specially made for the lock.

The accompanying sketch of a Chubb-lock (Fig. 335), in which the different parts are marked with the letters used for corresponding parts in former figures, may make this matter somewhat clearer. The action of other safety locks, those of Bramah, Hobbs, Yale, &c., are so far the same. The art of lock-making indeed, which has been the parent of so many remarkable and ingenious inventions, has worked in its latest and most refined productions strictly in the spirit of kinematic science,—it has followed its laws throughout with the greatest precision.

§ 122.

Brakes and Brake gear.

Brake drums or wheels are links of kinematic chains,—made usually of the form C ... | ... R,—which serve to control or entirely to stop the motion of the links connected with them, by friction produced upon their surfaces. The blocks or band pressed upon the latter and the mechanisms connected with them form with the drum a complete brake. Brakes are applied both to pieces which move in straight, and to those which move in curved paths.

One fact about brakes which requires to be noticed is that the blocks, slipper or band, form with the drum or rod a pair of kinematic elements so long as the gear is in motion. If a drum be used we have the pair (R), if the block acts on a bar or rail, the pair (P), and so on. Those brakes therefore which are employed completely to stop a motion, are used to prevent the action of a pair of elements, and this is done by uniting the partner elements in such a way that kinematically they may form a single piece only. Under some circumstances brakes act in exactly the same way as click-gear; there is, however, this difference between them that in the case of the brake the two elements are combined by making the motion of the pair gradually more and more difficult, the union of the two elements occurring when this difficulty becomes a maximum.

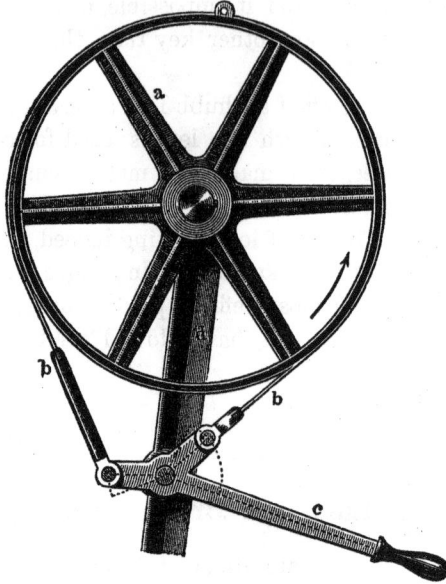

FIG. 336.

Brakes have the further resemblance to click-trains that one class of them act equally well for motion in either direction, as in the fast click-train, while those of another class, like the free click-trains, are either single-acting, or act differently in the different directions, as *e.g.*, in the case of the band-brake in Fig. 336. In the following section we shall be able to investigate more generally these points of similarity.

§ 123.

Engaging and Disengaging Gear.

Among the constructive elements which we have considered there have been several specially arranged and used so as to stop the action of a part of the machine when required, or to set it again free to move. Such arrangements are known as engaging and disengaging gear. It is obviously important that we should have a distinct general idea of the alterations thus occurring in the kinematic chain, in order thoroughly to understand the means by which those alterations are effected. We shall examine some examples of the methods commonly used.

One method very frequently adopted is the separation of the elements of an existing pair, so that the pairing between them may be dissolved. Friction wheels, moved a small distance away from each other,—belt-trains in which the belt can be loosened

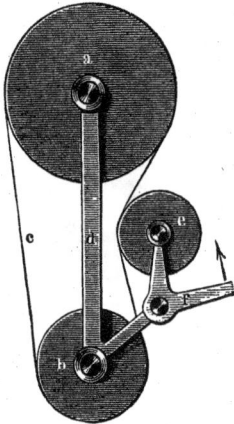

FIG. 337.

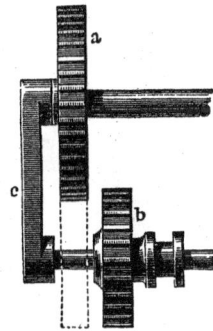

FIG. 338.

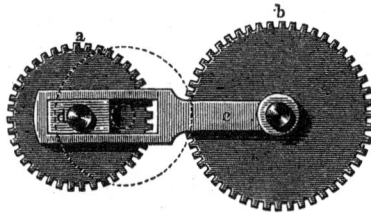

FIG. 339.

by moving a tightening pulley (Fig. 337), or by throwing it off one or both of the drums, spur-wheels which can be moved out of gear either axially (Fig. 338) or radially (Fig. 339) are all examples of this method. When the pairing is dissolved, the motion of the driven link necessarily ceases, no matter whether or not that of the driving-link continues. Engagement, or "throwing into gear," is simply the re-union or re-pairing of the separated elements.

Loose couplings (p. 445) or clutches are another form of disengaging gear. The most common form of this is the claw clutch or crab, the three most important varieties of which are shown in Figs. 340 to 342. The piece a is fixed to the shaft A, while b can slide upon B; the two shafts are then coupled if the clutch teeth be in gear, as shown. These teeth, it will be seen, are formed exactly as those of click-work. The two pieces a and b in fact, form parts of a click-train, a fast-train in the first example and a free-train in the second (§ 119), while in the third case the clutch acts as a free-click if the teeth be engaged to half their depth only, and as a fast click if they be in full gear. These claw clutches are therefore clicks which are thrown out of or into gear when the driven piece is to be stopped or to be again set in motion. The

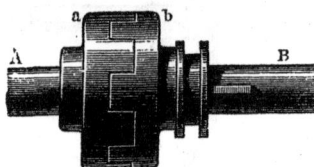 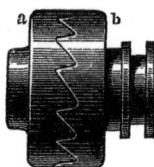 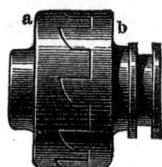

FIG. 340.   FIG. 341.   FIG. 342.

difference between them and the click-trains before described is that here the link formerly fixed is itself in motion. The relative motions in the train, however, are exactly as before.

Such couplings as those of Pouyer-Quertier and Uhlhorn[*] are in principle similar to these,—but in them the driven piece, that corresponding to b in Fig. 341, receives its motion from a second prime mover, and if it be stopped its teeth disengage themselves automatically from those of a, which then slide freely under them.

In friction couplings, of which one is represented in Fig. 343, some arrangement (such as that shown) is employed to press b so closely against a that the friction between them is greater than the resistance to the motion of b, which therefore moves as one piece with the driving shaft A. The coupling is disengaged by the removal of the pressure. Apart from the special purpose for which it is used we have here simply a brake, and this is true also of other friction couplings.

---

[*] These are better known in Germany than with us. They are used where one shaft is driven by two separate prime movers, and are so constructed that the stopping of one of these does not interfere with the continued motion of the shaft, driven then by the other only. Both forms of coupling are illustrated in Reuleaux's *Constructeur*, 3rd. Ed. p. 277, &c. (*Kraftmaschinenkupplungen*).

We thus see that these loose couplings, looked at as mechanisms, are click-trains or brakes,—but that instead of combining a moving with a stationary piece, they are used to unite two moving pieces. As to their action in the mechanism as a whole we notice that when the parts to be coupled are engaged, or put in gear with each other, they become kinematically one piece only. The shafts *A* and *B* become in this way a single shaft *C* ... | ... *C*, while before they are coupled each one separately forms such a shaft. The engagement therefore forms *A* and *B* into one link of a chain, while the disengagement again separates the parts or elements of this link. Disengagement and engagement then, by such methods as

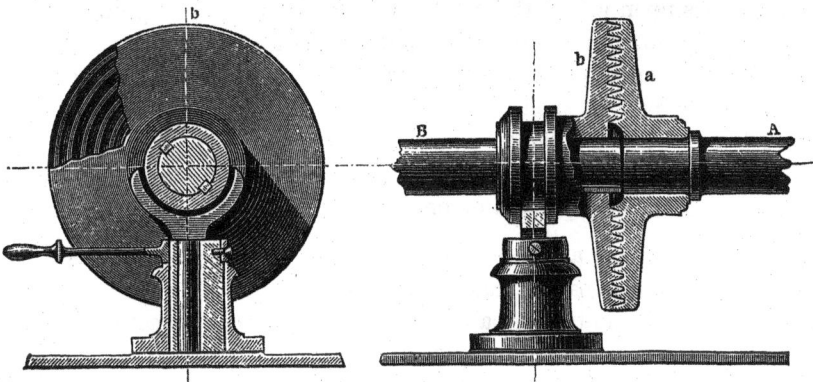

Fig. 343.

have been described, are respectively the separation and re-union of the elements of a link of a kinematic chain.

The various couplings used in this way may be subdivided according to the manner in which the parts of the link move, or are compelled to cease motion, after their separation; as may also the couplings before described. It is the province of Applied Kinematics to examine all these matters, here we must be content with the establishment of the general principles of their construction. I need only further mention that many of the contrivances used to prevent the loosening of nuts or keys are click-trains of the kind here described.

§ 124.

Recapitulation of the Methods used for Stopping and Setting in Motion.

We have seen that disengagement and engagement may be used either in a pair or in a link of a kinematic chain. Its object is in each case to bring to rest, or to set in motion again, some portion of the mechanism. Remembering at the same time that click-trains and brakes often serve the same purpose in connection with the whole mechanism, it will be useful for us to recapitulate here the principles upon which the different methods used for stopping and setting in motion mechanisms or parts of them are founded.

We have already noticed that click and brake act upon the elements of a pair in such a way as to unite them into one body. Such a union of an element with its partner we may call the fixing of a pair. Remembering this, we see that the stoppage of a mechanism or of a portion of one is effected

(*a*) by fixing a pair of elements in the kinematic chain (as in click-gear or brakes) ;

(*b*) by disuniting a pair of elements in the kinematic chain (as in disengaging toothed wheels, throwing off belts, lifting a " gab," etc.) ;

(*c*) by dividing a link in the kinematic chain (as in claw-couplings,—in throwing off pump-rods by removing a key, etc.),—

while the original motion again becomes possible if the chain be restored to its normal constrained condition. We shall see immediately that this classification applies equally to pressure-organs ; it covers therefore the whole ground which we are examining. A general examination of the ways in which it is possible to make a kinematic chain immoveable and moveable at will, without destroying any of its parts, makes it evident that in the classification given above we have exhausted all the means available for this purpose.

§ 125.

Pipes, Steam- and Pump-cylinders, Pistons, and Stuffing-boxes.

Pipes are, as we have already seen in § 41, the indispensable partner-elements of pressure-organs; the connections between them serve to form the links of the kinematic chain in which they occur. In the cylinders of steam-engines and pumps we have the vessel V^- containing the pressure-organ; they are therefore single elements, paired with their pistons or plungers V^+. Piston-rods and stuffing-boxes are partly paired with pressure-organs and partly occur as simple sliding-pairs $P^+_-P^-$. In the tubes therefore we have necessary, and in the four other constructive elements most familiar, forms of pieces which are used as links or as single elements in chains containing pressure-organs. They include, essentially, the chambers of rotary engines and pumps, the channels or races of water-wheels, the housing of turbines, and so on.

§ 126.

Valves.

Valves appear to be the most difficult of all the constructive elements to define kinematically. Their forms are so extremely numerous and varied that they seem to correspond more or less completely with a great number of different cases, without belonging entirely to any one of them. There are clacks, lifting-valves, piston-valves, tapered, cylindric and flat cocks,—the slide-valves of steam-engines, lifting and sliding equilibrium-valves, automatic valves and those which are not self-acting; there is the throttle-valve, the shutters and sluices for water-wheels and turbines, and many others. All these are valves; they serve, that is, to divide the capacity of a vessel containing a pressure-organ in some required manner. They do this in so many ways however, that is, they differ so greatly kinematically, that it appears at first as if it would be impossible to treat them all kinematically as one class. So far as I know, indeed, no attempt has been hitherto made to do this, which remarkable omission we probably owe to the fact that the pressure-organ machines have been almost entirely left without kinematic treatment of any kind. I have elsewhere* attempted a classification of valves according to their constructive characteristics which may be of service to us so far as it goes. It is as follows:

1. **Valves which slide,** including
   a. Cocks and disc-valves,
   b. Slide-valves;
2. **Valves which lift,** including
   (*a*). Clacks, hinged-valves,
   (*b*). Direct lift-valves.

I gave as the essential difference between the two classes that the fluid pressure upon the sliding-valves had no tendency either to open or to close them, while in the lifting-valves it did both, according to the direction in which it acted. The latter, therefore, can be used as self-acting valves, while the former cannot.

There is a good deal to be said for this classification, which does reach to some extent below the surface. It is, however, by no means exhaustive. It is founded on an examination of its subject from without and not from within, and so fails when it is carried to extreme cases; in reference, for instance, to those lifting valves which are completely balanced, and which therefore do not possess the property above named as that characteristic of lifting-valves in general. The division also stands so far upon the same ground as those of the old descriptive school that it does not fully explain its own definitions, and especially that it gives no indication of the position of the valves among kinematic arrangements. Now that we have familiarised ourselves with kinematic ideas by a series of analytical exercises, it is possible to give a definition which really goes to the root of the matter. It is this:—**Valves and their connections form the click-trains, and under certain circumstances the brakes, of the pressure-organs.**

Among these valve-trains also both free and fast clicks exist. The self-acting lifting valves are free clicks, that is, they permit motion past them in one direction and not in the other. The sliding-valves and the balanced-valves above alluded to are fast

* *Constructionslehre für den Maschinenbau,* p. 846, *et seq.*; *Constructeur,* 3rd Ed. p. 583.

clicks, which must be p aced in and removed from their fixing position by external means.

The common pump-clack, Fig. 344, corresponds to the free click-train with the common pawl, Fig. 345. The valve *b* is the pawl, which prevents the click-rack *a*, the water, from moving downwards. The tube *c*, which carries also the joint for the valve, corresponds to the frame *c* of the click-train. It is merely accidental that the hinge of the valve is made of a flectional element, leather; valves with cylinder-pair joints are of course quite usual. The fluidity of the pressure organ which forms the click-rod makes the click teeth unnecessary. The cock, Fig. 346, corresponds to a fast-click, such, for example, as that in Fig. 323. The train shown in Fig. 347, which is used in Thomas' calculating machine, is also exactly analogous to the cock. *a* is the click-piece (here a wheel turning about a fixed centre) corresponding to the water in Fig. 346; if *b* be turned through a certain angle it interposes no obstacle to the

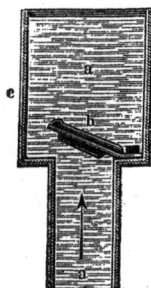

FIG. 344.

motion of *a*, in the position shown it prevents it from turning in either direction. The frame *c* contains two elements, one for carrying *a*, the other for carrying *b*, exactly as in Fig. 346. The sluice-valve used in water or steam-pipes, Fig. 349, corresponds to the fast click in Fig. 349 ;—*a* click rod, *b* valve or click, *c* frame carrying both. Balanced valves of all kinds represent clicks so arranged that the fluid pressure affects their engagement and disengagement to the smallest possible extent. The reader can find a multitude of other analogies,—for we have here an actual correspondence to deal with, and not a mere fanciful parallel.

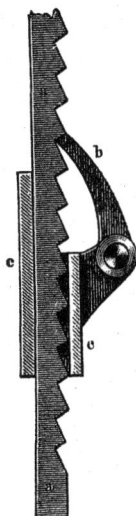

FIG. 345.

If a valve be opened only a small distance, so that tne pressure organ encounters great resistance in passing through it, the click-train acts as a brake. These two classes of mechanisms, therefore, pass here one into the other, exactly as we have already found (§ 122) to be the case with trains composed of rigid elements only.

The analogy between fluid click-trains and those consisting solely of rigid bodies exists equally in those cases where the trains are employed in complete mechanisms or machines. The mechanism Fig. 350, which represents simply a common lift pump, corresponds to the ratchet-train of Fig. 351, which we have already examined. The pump-barrel c, represents the frame c of

FIG. 346.

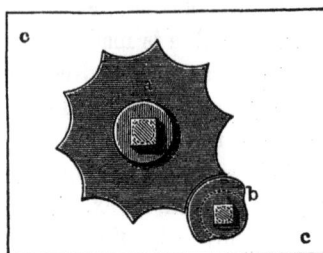

FIG. 347.

the ratchet-gear, the suction-valve the lower pawl b, the bucket-valve the ratchet-pawl b_1, the prismo-cylindric arrangement of bucket and barrel corresponds to the prism pairing between the bar c_1 and the frame c.

The double acting ratchet-train, Fig. 352, which we already know, exactly represents the Stoltz pump, Fig. 353. The pump and

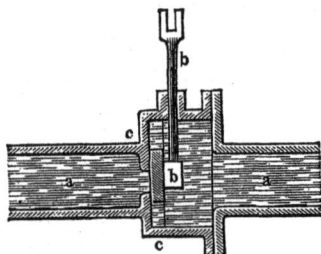

FIG. 348.

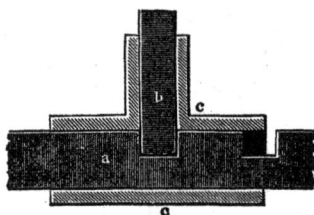

FIG. 349.

the ratchet-train correspond part for part; the two buckets c and c_1 with their rods, are the two bars c and c_1; the valves b and b_1 are the ratchet pawls, the pump-barrels and frame d correspond to the guides and frame d of the ratchet-train. The passing of the one click over twice as many teeth as correspond to the distance through which the rack is lifted has also its parallel in the pump,—the water moving relatively to the descending piston with twice as great a velocity as it is raised by the other. It is not necessary that the pumps should be placed side by side in this arrangement; in some cases they are arranged with the two barrels one above the other and conaxial, one pump being driven from above and one from below.*

The double-acting ratchet-train of Lagarousse, shown once more in Fig. 354, is represented by a double-acting Vose pump, Fig. 355.

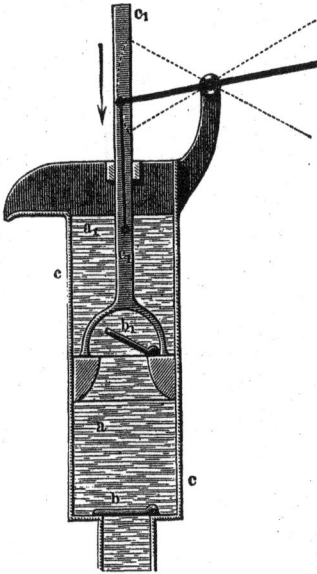

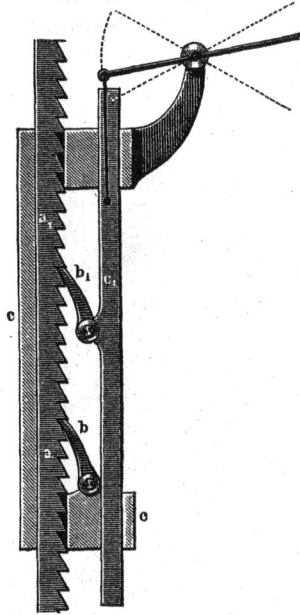

FIG. 350.   FIG. 351.

The analogy can again be followed out through every detail; here again we find, too, that the velocity of the one piston relatively to the water is always twice as great as its velocity relatively to the frame, just as in the case of the clicks b and b_1 and the wheel a.

In a similar way we find in pumps of other constructions a complete analogy with ratchet-trains. The differences are simply those permitted or rendered necessary by the fluidity of the

* Cf. König's *Pumpen*, p. 52. In Prunier's pumps at the Vienna Exhibition, the two barrels are conaxial and the rod of the upper plunger is made hollow, that of the lower one passing through it. They are thus both worked from above. See *Record of Vienna Univ. Ex.* (Maw and Dredge) Pl. lxvii.

pressure-organ.   In other words, piston-pumps with valves are
fluid ratchet-trains.

The view which this proposition affords us of the different
systems of pump construction appears to me to be extremely
instructive, and greatly to simplify the whole matter.   It is
interesting to notice that both free ratchet-gear and piston and
valve pumps were known and had attained some degree of completeness before the introduction of the steam-engine; both click
and ratchet pawls and lifting valves are force-closed arrangements,

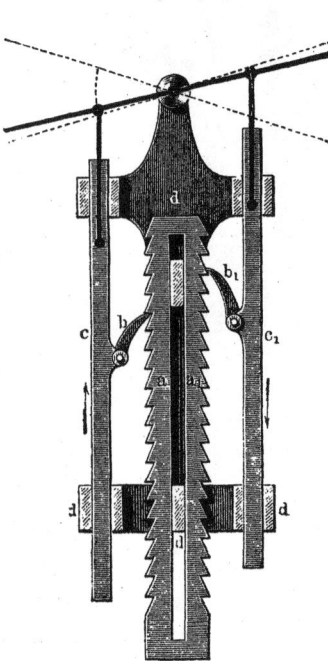

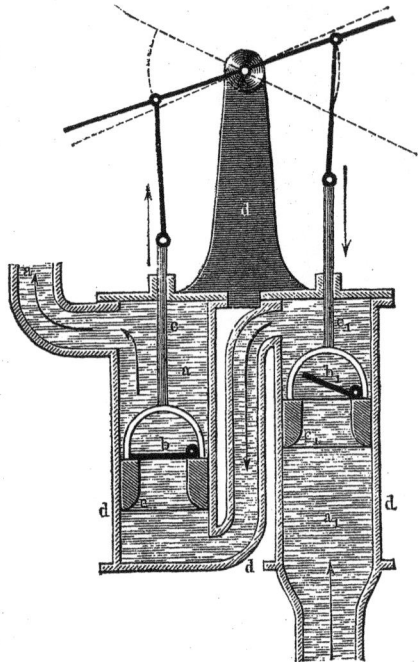

FIG. 352.   FIG. 353.

and so came earlier in the natural course of machine development.
We can observe also a distinct tendency making itself felt among
modern engineers to supersede the force-closed motion of the
valves by a constrained motion—as in pumps with slide
valves, etc.*

* Cf., for instance, Hänel's suction valves in Scholl's *Führer des Maschinisten*,
8th Ed. p. 419.

Hydraulic and steam engines (in their usual forms) have in general the same arrangement as that of piston and valve pumps or fluid ratchet-trains, and are made both single and double-acting. When in motion, the difference between them is simply that in the former the pressure-organ is no longer driven, but is itself the driving link of the mechanism. The valves therefore can no longer be self-acting—but must move with chain-closure. The

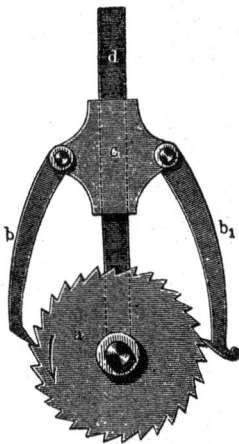

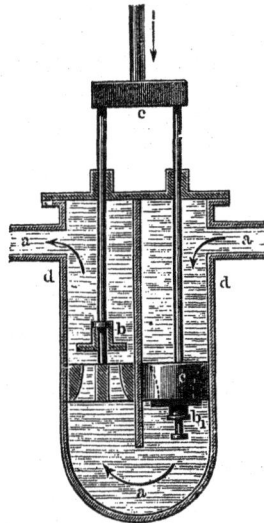

FIG. 354.   FIG. 355.

general relation of their motion to that of the piston is, however, the same as in the case of the pump-valves. In order that this motion may be brought about without any great expenditure of energy, fast-clicks are used, that is, balanced lifting-valves, or (especially in the steam-engine) slides, to which the required motion is given by means of suitable valve-gear. We may therefore say: the common steam-engines, hydraulic engines, etc., are reversed fluid ratchet-trains.* We shall return to this question further on.

* This reversion of the action can easily be seen in Fig. 353, for instance, by supposing the fluid to move in the opposite direction to that shown by the arrows,— the valves being moved at the right time by some separate mechanisms. In the corresponding train of Fig. 352, the same thing would occur if the bar *a* moved down instead of up. It will be noticed that it does not follow in either case that the element *a* should be the driver.

§ 127.

Springs as Constructive Elements.

We have already examined the function of springs in kinematic chains. We found (§ 42) that they were flectional kinematic elements, and might be arranged so as to work under every kind of force-closure, while the tension- and pressure-organs could be used with one force-closure only. Along with the parts attached to or connected with them springs become kinematic links. It follows unquestionably from our earlier examination of the matter that they should be reckoned among the constructive elements.

§ 128.

General Conclusions from the Foregoing Analysis.

The foregoing analysis of the constructive elements of machines has given us some not unimportant results. It has shown us, in the first place, that the parts generally included under this common designation are kinematically of very various descriptions. In part they are really kinematic elements (pins, bearing-blocks, tubes, pistons, stuffing-boxes, cords, belts, chains, springs), in part links of kinematic chains (shafts, axles, frames, levers, cranks, connecting-rods, cross-heads, steam-cylinders, &c.), in part complete pairs of elements (friction-wheels, toothed-wheels) ;—some too are portions of kinematic chains (belt-gear, click-gear, brakes, moveable couplings and disengaging-gear, valves, &c.), and a few complete kinematic chains (screwed and keyed joints). Looking at them as a whole we may draw the general conclusion that the "constructive elements" are really those pairs of elements and kinematic links which are most frequently used. For some of those which are complete chains in themselves, such as the screwed and keyed joints, are not used constructively to obtain the motions of the chains which they represent, but simply as fastenings, that is, for link-formation ; and the belt and cord-gear occurs as a part of a chain, simply because flectional elements can only be used in closed chains under chain-closure. Some moveable couplings, brake-gear, click-gear, &c., are more complex, and seem more distinctly to be complete kinematic chains. These, however, occur so frequently as subordinate parts of larger chains that relatively to the latter they appear elementary, and their appearance among the constructive elements may be justified upon this ground.

The question now presents itself whether we cannot, from the point of view now reached, find some rational classification for the constructive elements, based upon their real kinematic nature. This can certainly be done, and the matter is of sufficient importance to merit a short examination here.

It must be quite understood, in the first place, that no absolutely rigid systematic treatment is here possible. The classification must be based throughout on judicious compromise; we must be content to give and take, that we may accommodate ourselves to the exigencies of the numerous practical questions which refuse to remain within the bounds of a rigid system. This, however, does not of itself involve any error, for it is a consequence of the real nature of the problem before us (p. 437), and in no way interferes with our firm grasp of the scientific kinematic basis upon which the whole matter rests.

We shall in the first place make a more distinct separation than has hitherto been usual between the rigid and the flectional elements. We may begin with the former, placing first the most simple cases which occur. For this purpose, however, we must remember that the simplest things are not always those having fewest parts, and that a combination, therefore, is not to be rejected simply because it contains more than a single element, or a pair of elements. For our purposes those combinations in which no motion occurs—the immoveable fastenings used for forming links— may be considered simpler than the moveable pairs of elements; we may therefore place them first in our list. Next to them come the kinematic elements, pairs and links which give us simple moveable connections. Within these pieces themselves also the immoveable fastenings very frequently occur.

Arranged in this way the following will then be the first series of constructive elements :—

I. Rigid Elements.

a. Joints (for forming links).
$\left\{\begin{array}{l}\text{Rivets and riveted joints,}\\ \text{Keys and keyed joints,}\\ \text{Strained joints,}\\ \text{Screws and screwed joints, pass-}\\ \quad\text{ing into}\end{array}\right.$

b. Elements in pairs or in links.
$\left\{\begin{array}{l}\text{Screw and nut (used for their}\\ \quad\text{motions),}\\ \text{Pins,}\\ \text{Bearing-blocks,}\\ \text{Shafts and axles,}\\ \text{Fixed couplings,}\\ \text{Levers (simple),}\\ \text{Cranks,}\\ \text{Levers (compound),}\\ \text{Connecting-rods,}\\ \text{Crossheads and guides,}\\ \text{Friction-wheels,}\\ \text{Toothed-wheels,}\\ \text{Fly-wheels.}\end{array}\right.$

Whether moveable couplings and clutches should not be treated along with fixed couplings may be questioned, for we have already seen that in general, when their parts are in gear, they simply form parts of rigid links. It must be remembered, however, that these higher couplings contain in themselves numerous subordinate parts, levers, clicks, brake-blocks and so on, and present on this account greater difficulties to the student than the others. For the same reason they require, to a considerable extent, a special treatment dependent upon the nature of their details, and we are therefore justified in placing them rather among the complete mechanisms than here. Let us now go on to the second class of constructive elements, and their simplest arrangement in chains.

II. Flectional Elements.

a. Tension organs by themselves and used with chain-closure.
$\left\{\begin{array}{l}\text{Belts}\\ \text{Cords}\\ \text{Chains}\end{array}\right\}$ and their arrangement in gearing.

b. Partners of pressure-organs.	Pipes, Pistons and plungers, Steam-cylinders and pumpbarrels and chambers, Stuffing-boxes, Valves.
c. Springs.	Tension-springs, Pressure-springs, Bending-springs, Twisting-springs.

We have here another doubtful point, whether, namely, the valves should be included under II. *b*, or whether they should be placed in Class III. along with the click-trains formed from rigid elements, to which, as we have seen, they completely correspond. They fall along with pistons and stuffing-boxes (which also strictly speaking belong to click- and ratchet-trains) so naturally, however, and have been so often treated along with them, that the arrangement adopted will be on the whole the most convenient. The case is one of those in which logical completeness must be sacrificed to considerations of expediency.

Springs are obviously in their right place among the constructive elements in II. *c* above. The calculations connected with them fall to a very great extent, however, into the studies of elasticity and the strength of materials. Whether they be treated there or along with the constructive elements must depend upon the circumstances of each particular case.

We may conclude our list of constructive elements with the few which are more or less nearly complete chains, but which almost always occur in machinery as what may be called elementary groups of parts, and which for that reason may be conveniently treated along with the parts more strictly included under the name of constructive elements.

III. Trains.

Click-gear in its simplest forms,
Brakes,
Moveable couplings and clutches.

These form a kind of transition from the constructive elements to the complete machine. It will be remembered, however, that these three classes of trains are not the only ones occurring among the constructive elements. We had, for example, the screw-train $(S'P'C')^c$ among the rigid elements, while in the chain, rope and belt-trains of Class II we had other complete mechanisms. We know too that the clutches are click- and brake-trains (cf. § 123) while the moveable couplings are mechanisms formed from lower pairs of elements.

Our investigation of the constructive elements from a kinematic point of view has led us rather to a rearrangement of them than to any alteration in their number. It has furnished us, however, with explanations on some points by which, I believe, the treatment of the whole matter will be greatly facilitated.

The analysis has in several cases thrown a new and unexpected light upon very well known and apparently very thoroughly understood constructive elements. This has been specially the case in regard to valves and the machines fitted with them,—pumps, blowing-machines, steam-engines, etc. The conclusions which we reached showed for the first time the close connection existing between many of the characteristics of these machines, and have thus greatly aided their comprehension. They enabled us to define relationships which before had not been proved, even where they had been recognised. In this way we have succeeded in effecting a real simplification of the subject, the advantages of which will be felt specially in the problems of Applied Kinematics.

In reference to locks, too, our analysis has given us an explanation of which the want has often been felt. It has shown us that their kinematic principles are exceedingly simple, and that their treatment falls fairly within the limits of Applied Kinematics. Those arrangements which we have called curb-trains also (the special properties of which have not hitherto been distinctly recognised), and the escapements, we have been able to bring into their proper position among other mechanisms, and to examine from a general point of view instead of from the special one commonly adopted. The same is true also in the case of water-wheels, steam-engines and other complete machines, the more detailed examination of which we shall take up in the next chapter.

Our investigations have, lastly, furnished us with a most important theoretic result connected with the general nature of the closure of a kinematic chain or pair of elements. They have shown us that in every description of kinematic chain, from the most complex to the simplest, we have to distinguish three kinds of closure, namely:

1. Normal constrained closure,
2. Unconstrained closure,
3. Fixed closure.

In all three cases the conditions are fulfilled that the chain returns upon itself, and that proper pairing occurs between each link and its neighbour.

Under constrained closure all the relative motions of the links are perfectly determinate.

Under unconstrained closure these relative motions are made indeterminate by the addition of links to the chain.

Under fixed closure the motions of the links are entirely prevented.

All these kinds of closure are used in practice. The first and most important occurs in every machine, and forms a characteristic feature of it. The second we find in disengaging apparatus, where the action of a portion of the machine is stopped or reversed. The last kind of closure is used both for this purpose and with the object of preventing motions taking place within any single link, or, in other words, for making separate pieces into one link. The common constrained closure lies between the two other cases, and this is sometimes an assistance to us in finding out among the possible closed arrangements of links in any chain the important special case of constrained connection.

The apparent work of the machine designer consists in utilising these three methods of closure in different ways and for different objects. In reality, however, they are to him only means which he employs to solve the problem placed before him in the complete machine. We must now proceed to examine the general propositions which present themselves in connection with this subject.

CHAPTER XII.

THE ANALYSIS OF COMPLETE MACHINES.

"Altes Fundament ehrt man, darf aber nicht das Recht aufgeben, irgendwo einmal
wieder von vorne zu gründen."—Göthe.

§ 129.

Existing Methods and Treatment.

Now that by the help of kinematic analysis we have examined
the machine as a combination of constructive elements, and also
investigated the nature of the latter individually, we have brought
ourselves to our final problem—the examination of the machine as
a whole. This completed, our work will end as it began, with the
machine itself, the dismembering of which, in order that we might
better examine it kinematically from every possible point of view,
we commenced in our first chapter. It might be supposed that we
should by this time have become acquainted with all the essential
characteristics of complete machines. We have indeed done so to
a certain extent, but only by examining these characteristics singly,
as disjecta membra, each one apart from its connection with
others. It still remains necessary to review them and their mutual
relations as a whole, and we shall find that this examination will
explain some things as to the nature and use of the machine which
have not yet come at all under our consideration.

We must first examine the way in which it has been usual
hitherto to treat this subject, in order to see how far it is justified

by facts or can remain useful to us. I alluded in the Introduction to the widely-diffused conception of the nature of complete machines which was specially supported by Poncelet's authority and which has taken such firm root in the French mechanical instruction. This conception is that the complete machine is in general a combination of three parts or groups of parts,—

1. The receiver of energy, or receptor,*
2. The parts transmitting motion, or communicator,
3. The working parts, or tool.

The part or group of parts constituting the receptor is generally understood to be that upon which the natural force driving the machine acts directly ; the tool is the part by means of which the energy received by the machine is directly expended in producing the required change in or in connection with the body to be worked on. As the motions of these two sets of parts are seldom identical, the second group of parts is required to transmit motion from one to the other. The whole conception has something so direct and simple, we might almost say natural, about it as to give a most favourable impression. Poncelet himself spoke of it as well-established—as a matter of which the logical completeness was entirely convincing.† In saying this, too, he rather put in a few words what was previously known, and more or less distinctly recognised, than expressed something entirely new to his time, and now the idea has become to some extent a part of the very foundation on which the study of machinery, in France at least, is based. There is indeed much to be said for its directness, and the ease with which it can be grasped. Its division of its subject into three parts, a beginning, middle and end—two principal parts connected by a third —prepossesses us in its favour by presenting a certain analogy with

---

* A few English authors have mentioned this classification, but none except Moseley have, I think, ever used it to any considerable extent. Both words and ideas will be somewhat unfamiliar to English readers, while Prof. Reuleaux's views will not run counter to any preconceived ideas here as they have done on the continent. The controversial part of the following sections would therefore have been unnecessary had they been originally written in English ;—the conclusions arrived at are, however, none the less valuable.

† *Traité de Mécanique Industrielle,* Pt. III, § 11. "La science des machines, ainsi envisagée, se compose donc de la science des outils, de la science des moteurs, et de la science des communicateurs ou modificateurs du mouvement . . ."

similar ternary divisions in other regions of investigation. In many cases too it corresponds distinctly with the natural divisions of particular machines. It is necessary nevertheless, or rather I should say for this very reason, to submit these apparently fundamental conceptions to the test of the strictest possible investigation.

If we examine more in detail the usual treatment of the subject, we find that those parts of the machine most closely connected with the receptor, for the best possible utilisation of the driving force, are very generally considered by themselves in groups, forming what is known as prime-movers,—steam-engines, water-wheels, turbines, and so on. By a very similar limitation the group of working parts, or tool, has been taken to be more or less exactly co-extensive with a large class of machines, each one adapted for doing certain particular operations, as spinning, weaving, and printing-machines, machine-tools, &c. Collectively we may call them all tools, by a justifiable use of the word in its most general sense.

But if we look more closely at these familiar extensions of the original idea, we shall find that we are treading on very uncertain ground. For if every complete machine must have some motor in the sense used above, must be driven, that is, by steam, air, water, gas, &c.,—then evidently the lathe and planing machine cannot be complete. Scarcely any of the machines in a factory, indeed, could be considered as complete, none of them would be more than portions of complete machines. The common usage therefore, in which they all receive the name machine, must be entirely inaccurate, for it is quite inconsistent with the technical definition of what the machine really is.

Exactly the same difficulty occurs in the case of prime-movers. A very large number of these, considered by themselves, do not possess any part which can be called a tool, for doing work upon other bodies. They also, therefore, cannot be complete machines. Those engineers who devote themselves to the manufacture of steam-engines, turbines, and so on, are makers not of complete machines but only of parts of them. The only really complete steam-engines must be such machines as steam-hammers or crushers, direct-acting rolling-mill engines and so on. All others, no matter how excellent in design or construction, are incomplete machines in themselves and become complete only when combined with other apparatus.

But all this is in direct contradiction to what our natural idea of completeness in a machine would be were we unhampered by theoretical definitions; while the contrast which it presents to our simple direct idea of the machine makes us to a certain extent question the authority of the theoretical conception upon which (logically or not) this popular idea has based itself.

The doubt thus arising is strengthened by another question. If we look at a spinning-machine, we see the thread passing through certain motions which it could not receive were it not itself a transmitter of motion. Is the thread then here the body to be worked upon, or is it a transmitting part, or is it indeed itself the tool? And where does it begin or end to be any one of the three? Similar uncertainties exist in very many other machines. How is it that the spinning-machine, and with it indeed all other machines connected with the manufacture of textile fabrics, will not fit in to the theory? Is it the fault of the machines, the excellence of which every one knows, or of the theory? Let us take another example, the well-known hydraulic (Montgolfier's) ram. The water lifted by the machine is here a portion of the mass of water which works it. The machine is obviously complete, but which part is the receptor, which the tool, which the transmitter of motion? Does the stream of water represent all three itself? And if so, what are the other parts of the machine? Or has Montgolfier bequeathed to us only a tantalising paradox, a machinal will-o'-the-wisp, instead of an orderly and respectable machine which can give a reason for its own existence?

Thus doubt arises upon doubt, question after question, so soon as we seriously attempt to apply the recognised theoretical classification or subdivision to actual machines. But the question is not one to which one answer is as good as another,—it concerns one of the most important factors in modern civilisation, a branch of human activity affecting almost every one more or less directly; while the most strenuous exertions, intellectual and physical, have been made in order that it might be scientifically mastered. It is therefore not without good reason that we shall now commence the examination of the three popular subdivisions of machines.

§ 130.

The Tool.

In commencing with the part of the machine which directly executes the work—the tool—we shall first try to find what portion of a few well-known machines answers to this description. In the lathe, planing-machine, band-saw, etc., this is very easy. The chisel or other cutting tool, the saw-blade and so on, are obviously the pieces directly employed in the work. In the screwing-machine there are generally several pieces, the dies, acting together, and these, together with the frame or stock in which they are placed, may fairly be called the tool, as in practice they actually are. In rolling mills of any kind the two rolls serve as the tool, they share equally in shaping and moving the metal passing beween them. In flour mills the stones act as tools, grinding the corn and passing it from them as meal. In the manufacture of wrought iron nails one tool, a compound one, is used to hold the wire, another, also compound, to cut it, a third to form the head, other parts bring forward a new portion of wire, others remove the nail already made.

In the card-making machine, also, several tools act in succession, some to pierce the leather band, others to cut the wire, to bend it, and to insert it in its place in the band, while others act on band and wire together in order to press home the latter. We have, that is to say, a series of tools working in different ways and for different purposes, and so connected that it is very difficult to say where one ends and the next one begins.

We notice here that the unity of the tool, or indeed of the body to be worked upon, does not appear to be a condition of the machine. This fact has to be kept in mind if it be wished to form accurate definitions, and many writers who have endeavoured to carry scientific exactness through all their work, Redtenbacher among them, have been compelled for this reason to use extended descriptions instead of definitions. Before we consider this question let us look into some further examples.

In hoisting machines the hook from which the body hangs as it is lifted has been called the tool. This is perfectly right, according to the definition, for it is clearly the part of the machine which

directly does the work of lifting. But if we suppose the hook removed, and the rope itself tied round the body to be lifted, the
machine can work precisely as before. The only difference is one
of convenience in securing the load, and this obviously does not
affect the question before us. The hook cannot therefore really be
the tool of the crane or other hoisting machine, for the total removal of any essential part of a machine must necessarily render
it useless. But—it may be suggested—in this case the loop of the
rope, the improvised sling, is really the hook, differing from the
former one only in material and constructive form, not in kind.
Let it be so. But now suppose the load removed entirely and the
rope allowed to hang by itself,—it may hang to such a depth that
its own weight becomes as great as that of the former load,—and
let the crane be set in motion to wind up the empty rope, does it
not work precisely as before? There is still a load to be lifted.
The wheels, drums, shafts, pawls, cranks, all move precisely as before. But neither hook nor sling exists, the only weight to be raised
is that of the rope itself; the body to be lifted has become a link
of the kinematic chain. The tool, in the usual meaning of the
word, has completely disappeared.

Let us look at another example—the locomotive. The coupling-
hooks or other arrangements used for attaching the train to the
engine are here usually said to form the tool. But if this be the
case is it not extraordinary that such an immense number of different coupling arrangements should exist, all intended to serve the
same purpose, and with any one of which any given locomotive
might work? In this case would a change of coupling-gear alter
the machine as a whole? This must certainly be the case according to the commonly received theory, for the tool is an essential
characteristic part of each machine. But in order to look more
thoroughly to the bottom of the matter, suppose the locomotive to
be running with its tender only, or still better, suppose the case of
a tank locomotive carrying also its own fuel and running entirely
by itself. If it come to a steep gradient it may have to exert now
exactly as much power as it would do on a level if it were drawing
the train; the coupling-hooks, however, have now absolutely no
connection with its work. They certainly do not form in any sense
the tool of the machine. It may be said that the latter is now incomplete because it is not carrying goods or passengers,—but this

is obviously a mere accidental condition, and there is no difficulty in supposing a locomotive built like a Fairlie engine on a frame sufficiently long to afford ample room for both. No part corresponding to the tool can, however, be pointed out; it is only certain that the couplings can no longer in any way represent it. The body to be acted upon, indeed, no longer exists beside or outside the machine, but has become a part of it. The one frame supports both carriage and machine.

There are many other machines in which the conditions are exactly the same as in the locomotive,—as, for example, the steamboat; where again we can find nothing corresponding to the tool. In small machines we find the same thing;—it is very difficult to say, for instance, what part of the common clock is the tool. If it be the hands, we ask immediately where the body is upon which the hands work. The hands also are not absolutely necessary to the completeness of the clock; they might be replaced by graduated discs turning relatively to some fixed index; or indeed a mere mark made upon a wheel exposed to view might answer all purposes. The hands therefore are not the tool, and it is not possible to name any other part of the clock which fulfils the functions of that organ.

Our investigation thus leads us to the conclusion that the tool does not form an essential part of the machine. In certain machines only do we find it unmistakably recognisable, in some its distinctness is less and in others it does not exist at all.

Looking at the last class of machines—of which we have given examples in the crane, locomotive, steam-boat and clock—collectively, we find that they have in common the object of effecting some alteration in the position of a body or bodies. The first three examples are machines by which loads are moved, vertically, horizontally, or in both directions. Essentially the same thing is true of the clock, but here, for a special purpose, the alterations of position are so arranged that they enable us to measure the time occupied by the process.

The machines first considered, in which the tool really exists, have, on the other hand, the common object of making some alteration in the form of the body or bodies upon which work is done—such as turning, grinding, dividing, uniting, etc., Lathe, planing-machine, screwing-machine and saw change the form of bodies by removing a portion of them. The nail-making machine and rolling-mill rearrange the molecules or larger portions of bodies worked on, of which at the same time they alter the position. The same is true of the card-making machine. The millstones divide the body into minute pieces, altering its position at the same time. All, however, have one or more tools, and we see that in every case where there has been any indefiniteness about these, it has arisen from the fact that the machine served the double purpose of changing both the form and the position of the bodies worked upon. Apart from this, however, we may now divide machines into two great classes according to the purposes for which they are used, namely :—

I. Machines for altering position, or place-changing machines.

II. Machines for altering form, or form-changing machines.

There is no sharp division line between these two classes, for some form-changes are, as we have seen, necessarily accompanied by changes of position, while some machines, as the corn-mill, seem to belong equally to the two classes. In every case however, those machines which belong wholly or partly to the second class are characterised by the possession of the tool, while this organ is not found in any machine whose object is place-changing alone. The latter are therefore the simpler, and for that reason we have placed them first.

The theory, therefore, which makes the tool an essential part of the machine, is correct only so far as one of these two great subdivisions of machinery is concerned. The tool is not an essential part of the machine; it is accidental to it only, and for this reason cannot form part of the foundation upon which our comprehension of the complete machine is to rest.

§ 131.

Kinematic Nature of the Tool.

Now that we have found what the tool is not, we must turn to the question of what it is, and endeavour to find the kinematic meaning of this organ in the class of form-changing machines, in which we have found it to exist, and the general kinematic principles underlying its use.

Let us first look at the action of the tool in some familiar machine, say a common lathe with a slide-rest in which a bar is being turned. The chisel is held fast in the tool-holder of the rest and moves parallel to the axis of the spindle, the bar to be turned is made to revolve along with the mandril in such a way that the portions of its surface to come in contact with the chisel move always towards its edge. The relative motion of the chisel to the bar is the common screw motion, occurring exactly as if the turning tool were a part of a common nut S^- of which the bar to be turned is the screw spindle S^+. The chisel and bar have therefore the motion of the pair $S_-^- S^+$. This pair does not exist at the commencement of the work, but as the lathe moves, the chisel (being harder than the piece to be worked on) cuts away those portions of the bar which do not belong to its own enveloping form, S^+, upon the spindle. That part of the bar, therefore, over which the chisel has passed, has necessarily taken the form of the element S^+, the chisel itself carrying a small portion of the partner element S^-. Essentially therefore they form a twisting pair, $S_-^- S^+$, as may be recognised more readily, perhaps, if we suppose the lathe worked backwards, and the chisel passing again over the surface it has already formed. The restraint between the two elements of the pair is not complete in itself (§ 18, &c.), but the lathe is so arranged as to supply the want by chain-closure (§ 43). We may notice that the chisel has carried the profile of the nut S^- from the beginning, while the bar only received the form S^+ while the turning was progressing. The pairing therefore of the elements into the form $S_-^- S^+$ is made as the motion of the machine goes on, and at the end of the operation the two bodies are really formed into such a pair.

We have said that the bar being turned takes the form S^+. This is visible enough in roughing out work, where a sharp-pointed chisel is used. In finishing or smoothing work, where the chisel edge is made straight and parallel to the axis of the lathe, the bar becomes in external form a cylinder (cf. § 15); but as regards its pairing with the chisel it is still a screw.

We find in the planing and band-sawing machines the same thing as in the lathe—that the tool and the body to be worked on are combined into a pair of elements, in this case a sliding pair $P^{\pm}P^-$. In the screwing and tapping machines the pairs $S^{\pm}S^-$ and $S^-_{\approx}S^+$ respectively are formed in the same way. We see that in every case the body to be worked upon becomes itself a kinematic element or the part or whole of a kinematic link. In the screwing machine this is specially noticeable, for as soon as the screw is started it itself causes the forward motion of the dies. The body to be worked is therefore not something external to the machine but actually forms a part of it. We shall therefore give this body, to which we shall often have to refer, a special name, calling it the work-piece of the machine.

That the work-piece has formed an element in a lower pair is only accidental to the particular machines chosen for illustration. In other cases we find higher pairs exist. In the rolling-mill, for example, the work-piece forms with the rolls the higher pair R^+, P^+, the work-piece itself forming here a complete link of the chain. In the carding engine the symmetrically placed wires of the cards, compel the fibres of the tangled mass of wool to assume the enveloping forms corresponding to their motion, that is, to lie parallel to each other. In the corn-mill there is a very complex higher pairing, in which force-closure plays an important part, between the grain and the mill-stones.

Our analysis therefore leads us to the following proposition: In form-changing machines the work-piece is a part or the whole of a kinematic link, and is paired or chained with the tool by so arranging the latter that it itself changes the original form of the work-piece into that of the envelope corresponding to the motion in the pair or linkage employed.

This proposition is free from the indistinctness which characterised the older conception of the machine. We see from it, in the first place, that the kinematic chain is not broken at the tool or the working point, but continues through it. It is not the end of the chain, but only a point in it having special importance in reference to the object of the machine. We find here also the answers to several of the questions which came up in § 129. The thread in the spinning-machine, as a link in the kinematic chain, is necessarily a transmitter of force. The spindle, on the upper end of which it first winds and then immediately unwinds itself forms here a higher pair with the thread, and is itself the tool. Relatively to each other, however, the fibres of the thread act as tools. If we imagine for the sake of simplicity simply a pair of such fibres stretched between the spindle S and the draw-frame D,

S. ━━━━━━━━━━━━━━━━━━━━━ D.

and the spindle to make half a revolution, there is, in the first place, a mere crossing of the two fibres,

S. ━━━━━━━━━━━━━━━━━━ D.

but as the turning continues the fibres twist round one another, each fibre acts as a tool in working the other, the screw form of each being simply its envelope relatively to the other. Thus we see that it is not even absolutely necessary that the tool should be harder than the work-piece, and also that occasionally it is not possible to distinguish one from the other. This, however, does not affect our proposition—that the work-piece forms a part, or the whole, of one of the links of the kinematic chain forming the machine.

We notice further that in the work-piece we have a member common to both place- and form-changing machines. We have already noticed that when the so-called tool of the former disappeared, the body on which work was done, the work-piece (where it existed) became a part of the machine. In place-changing machines, therefore, as well as in those which we have been considering in this section, the work-piece is a part or the whole of a kinematic link. In this point the two classes of machines are completely alike.

There follows, lastly, from what we have now found as to the nature of the tool, a proposition which is very important, and which has most numerous applications in mechanical technology. It is the following :—in order that a given form may be given to a body by a machine, we must give to the tool of the latter the envelope of that form. In order to determine this envelope the intended motion of the tool relatively to the work-piece must first be fixed, and as this relative motion may be of many different kinds, not only may the problem admit of several solutions, but as a rule it itself includes numerous other problems. In every case, however, it is a matter of very great importance to be able to include, in a single definite conception, the whole kinematic relations between the tool and the work-piece.

§ 132.

The Receptor.

There has been less variety in the common conception of the receptor than in that of the tool, on account of the limited number of bodies which seem suitable for fulfilling the function assigned to it. These bodies are water, wind, steam and some other gases, weights, springs and living agents. By the receptor of any complete machine has hitherto been generally understood that part to which one or other of these bodies directly imparts the energy by which the machine accomplishes its work. It is important that we should acquaint ourselves with the characteristics of the various ways in which this transference of energy is effected in the cases of the bodies mentioned.

Taking first water-wheels and turbines, we find the receptor at once in their buckets. Our earlier investigations (§ 43) have already shown us that the wheel is not used by itself, but that its buckets are kinematically paired with the water and this again with its channel or pipe. The receptor is here, therefore, unquestionably a link in the kinematic chain. In the various forms of hydraulic engines we note exactly the same thing. Here also the water, paired with the piston, enclosed in the cylinder, guided by the valves, forms a link in the kinematic chain; the whole mechanism is one which we have already examined (§ 126) and found to be a ratchet-train. It is, however, impossible to say certainly whether the piston is the receptor, or the cylinder, or both,—or, indeed, whether the valve gear does not also form a part of it along with both.

The wind is utilized as a source of energy under force-closure of the driving organ, in such a way that a kinematic pairing, in this case a higher screw-pairing, occurs between the wind and the sails of the wheel.

Steam and other gases working expansively are commonly used in piston machines, and occasionally in machines arranged somewhat like turbines, always, however, in such a way that they are kinematically paired or chained with the driving parts. In general this occurs in such a way that it is difficult to point out the receptor in any single piece.

The motors of which we have spoken, water, wind, steam and other gases, are all pressure-organs. Looking generally at the prime movers driven by them so as to classify their leading characteristics, one important fact appears which we must not leave unmentioned. We find two different methods of utilizing the driving energy of the motor, corresponding to two distinct classes of prime-movers. One of these classes, which includes all the machines working with pistons, we have already seen to be ratchet gear, or more fully, reversed ratchet gear. The other class, to which water-wheels, turbines, wind-mills, etc., belong, are characterised by the continuous, or very nearly continuous, motion of their working fluid. This acts no longer periodically or reciprocally, but enters continually at one side and passes away at the other. In water-wheels its action might be imitated by a rack (§ 61), in the wind-mill and in some turbines by a screw, in others by a rope passing down and up round a sheave, etc. The difference between the two classes may be expressed by using names which indicate the principal characteristics of their motion, calling the second running-gear as distinguished from the first, the ratchet-gear. All prime-movers driven by pressure-organs are either ratchet-gear or running-gear.

If we glance again at the chamber-crank and chamber-wheel trains which are described in former chapters, we see that they belong partly to the one and partly to the other class. The chamber-crank machines, both pumps and engines, are partly ratchet-gear and partly running-gear, some of them, indeed, taking a kind of intermediate position between the two classes; the chamber-wheel machines are essentially running-gear. For the purposes of many machines running trains are extremely convenient, because their rotary motion can be so easily and directly utilized. The attempt to design rotary engines and pumps is essentially an endeavour to substitute running-gear for ratchet-gear in machines working with pressure-organs.[56]

The common clock may serve as an illustration of a machine moved by the action of a weight. At first sight it appears scarcely doubtful that the cord or chain from which the weight is suspended is here the receptor, for it is the part of the machine directly connected with the driving weight. Looking at the question somewhat more closely, however, we see that the case is similar to that of the loaded crane (§ 130). We may suppose, that is, the weight to be removed, but the chain or cord to be lengthened sufficiently to make up the load to its former amount; it alone would then be sufficient to work the clock. The weight, therefore, cannot be the driver, for it no longer exists, while the cord has not altered its nature, and cannot therefore have been the receptor in the former case. It is evident, however, that the cord is a link of the kinematic chain, and is paired with the barrel, on which it is coiled when the clock is wound up. Machines driven by weights have then this in common with the prime-movers just described, that the body transmitting the driving effort to them forms itself a part, a link or an element, of the kinematic chain.

The springs used for driving watches and other small machines are, as we have already seen in § 44, kinematic elements or links. Here again it is very difficult, if not impossible, to say which piece of the machine answers to the common definition of the receptor, but we always see distinctly that the piece through which the driving effort is introduced forms a part of the kinematic chainage of the machine.

While, however, all prime-movers possess this common feature, our investigations show us that from another point of view they divide themselves into two general classes. In all those machines which are driven by pressure-organs a change of form takes place in the latter as it traverses the pipes, ports, valves, buckets, etc.; this may be carried to a very great extent, as in the steam-engine, and may be combined with more or less change of place also. With springs the change of form alone exists. With driving weights, on the other hand, the change of place only remains, the change of form has completely disappeared. We have here exactly the difference which we found before to exist in regard to the object of the machine, or as we may now say, in regard to the treatment of the work-piece. We may there-fore again distinguish form-changing and place-changing machines according to the changes undergone by the driving organ in doing its work. This conclusion is not affected by the fact that the number of such place-changing machines is comparatively very small. The difference between the two classes is intrinsically important, for it removes an apparent want of congruity between prime-movers and other machines, and will further serve to explain some remarkable analogies.

Let us turn now, lastly, to the employment of living agents, that is, of the muscular power of men and animals, to drive machines, looking first at the use of human muscles. The common statement here has been that the receptor is some portion of the machine having a form and motion suitable for receiving the action of the driving body, the hand or arm or foot of the worker. In a common grindstone, for instance, the treadle is considered the receptor, the foot or leg of the workman the motor. Although this is certainly what appears on the surface to be the case, the real constitution of the machine must be stated differently. If the crank-pin were made sufficiently long, the grindstone might be turned by the hand of the workman; or if the workman held one end of a cord attached to the crank-pin he could easily, by periodic pulls or jerks, keep the stone in motion.* The treadle is not therefore an indispensable part of the machine. All three methods of driving have, however, this in common, that the body of the worker becomes kinematically chained with the machine. Under certain circumstances this chaining may be very complex, in the case before us, however, it admits of tolerably exact determination. In the first place the crank a with the coupler b, the treadle c, and the frame d (Fig. 356) form a lever-crank $(C_4'')^d$. In this mechanism c is the driving link, so that its special formula runs $(C_4'')_{\sigma}^d$. The workman places his foot on c', the prolongation of c, and (supposing the centre $1'$ of his hip-joint not to move) his leg forms with the treadle c' three links, the crank, coupler, and lever, of another lever-crank $(C_4'')^d$, to the frame d' of which the hip of the worker belongs. His knee-joint forms the pair $2'$ and his ankle joint the pair $3'$. The joint 4 is common to both the trains of this compound mechanism, as is also the frame $d\,d'$ which carries the fixed elements of the pairs 1, 4 and $1'$. The worker exerts muscular force at $1'$ and $2'$, and indeed at $3'$ also,

---

* As the Kalmuck priest drives the prayer-wheel for example, or the Japanese peasant the reel upon which her silk is wound.--*R.*

to give the required motion to the machine. His foot thus forms a part of the link c'. The special formula of the second mechanism is therefore $(C_4'')_{\overline{\text{a.b.o}}}^{\text{d}}$, and it is to be noticed also that the link a swings only, and does not rotate. We find then that the worker makes a portion of his own body into a mechanism, which he brings into combination, that is chains kinematically, with the mechanism to be driven.

A workman applying both hands to turn a crank in a case where there is large expenditure of effort, chains the mechanism formed by his limbs to that of the crank in a very complex

FIG. 356.

manner. He alters at will the action of the force-closure which brings certain joints into, or throws them out of, use, as becomes necessary at each instant.

The motion of the human body in driving a tread-wheel, or still more that of an animal in a "gin," is complicated in the same way. Always, however, we have the same union, by kinematic chaining, of the living mechanism with that of the machine, while no receptor, in the hitherto accepted sense of the word, can be distinctly recognised. Our investigations, then, lead us to the conclusion: that the receptor does not form an essential part of the complete machine.

§ 133.

Kinematic Nature of the Complete Machine.

We have found that the "tool," which has so often been considered an essential member of every machine, occurs only in one half of existing machines. We have seen, too, that the receptor, which has also been considered essential to every machine, is in very many cases quite indeterminate. The prospect that the third member, the communicator, should prove to be essential becomes therefore very small. There are very numerous cases in which it cannot be distinctly identified, although, in some instances, there are large groups of parts which are obviously employed for no other purpose than the transmission of motion. But every link of the kinematic train transmits a greater or less effort from one point of the machine to another ; every link may be looked upon as a communicator between the driving force and the resistance ; and in most instances it is impossible to say where the function of transmission begins and where it ends, so that we must conclude that the communicator also, as a special subdivision of every machine, must be given up. All three, receptor, communicator, and tool, may exist and may be clearly recognisable in one and the same machine ; they are not, however, essential organs of machines in general, they must be reckoned among their accidental members only, for which we shall shortly find another classification.

The fundamental idea to which our investigations have led us, an idea which we have found to be the foundation of, and hidden by, many subsidiary ones, is this : the complete machine is a closed kinematic chain. The driving body and the body on which work is done are equally links or elements in the chain. The laws governing the motion of the motor or driver are the same as those according to which the work-piece is driven and the tool, where it exists, performs its function; they are simply those laws under which the relative motions of any other links or elements take place.

Only one difference appears which tends to impair the simplicity of this conclusion; it is the difference between the form-changing and the place-changing machines. This last remaining distinction deserves somewhat closer examination.

We found that the pairing or chaining of the tool (in the form-changing machines) and the work-piece was such that the former constrained the latter to assume the form of the envelope for its relative motion, the giving of this form being the result aimed at in working the machine. The driver in the form-changing prime-movers passes through an exactly similar process. If we look, however, in this connection at any pair of elements whatever, lower or higher, a pin in its eye, a screw in a nut, a piston in its cylinder, a pair of spur-wheels, we see that in every case form-changes occur in one or the other element, or in both. These changes are of two kinds, viz. (1) temporary changes on account of the unavoidable alterations due to the action of sensible forces even upon the most rigid body, and (2) permanent changes, due to the separation of small portions from the body. In the latter class of changes wear gradually alters the form of the paired elements, and this alteration occurs so that the elements carry the reciprocal envelopes for the motion occurring between them. This law is, however, exactly the same as for the motion between tool and work-piece. In this case we endeavour to carry out the form-change quickly, it is the object of the constrained motion of the machine. In the former case the continuous change interferes with the object of the machine, and we therefore try to limit it in every possible way. In both cases, however, it exists.

The form-changing action which occurs between the tool and the work-piece differs in degree only, and not in kind, from the action taking place between the elements of every other pair in the machine.

We see, therefore, that all complete machines without exception follow the same general laws, and we are now able completely to realise the meaning of the definition of the machine with which we commenced our investigation (§ 1) and which therefore we may conveniently repeat here :—

A machine is a combination of resistant bodies so arranged that by their means the mechanical forces of nature can be compelled to do work accompanied by certain determinate motions.

The "arrangement" of the bodies here mentioned is the kinematic chaining, Motion occurs in the machine when some part of the chain is in a position which cannot be retained under the influence of the natural force acting upon it. The particular motion then occurring is made determinate by the chaining. In the place-changing machines this motion is used for the purpose of altering the position of the work-piece, in the form-changing machines with the object of altering its shape, the nature of both alterations being fixed by the form of the chain. Both results, the motion of a body in given paths and according to given laws, and the alteration (perhaps simultaneous) of its shape, are forms in which the machine has compelled the natural forces " to do work."

A few general illustrations of this may be in place here.

In the common clock, in which we may suppose the chain increased in weight in the way before mentioned, if the chain be brought into such a position that it begins to uncoil itself from the barrel, every part of the mechanism will at once commence its characteristic motion. In clocks of the common construction motion ceases while the chain is being wound up, the process, that is, of bringing the chain into the " unstable" position just mentioned affects every part of the train. In clocks of a better class means are adopted for removing this defect, such as the use of a weighted lever arranged so as to come into action, and drive the mechanism during the operation of winding up. In other words a second kinematic chain is so placed in reference to the first, that when the latter is not acting on account of the winding process, the former is brought into an unstable position, into a position, that is, suited for driving the mechanism.

In the under-shot water-wheel we allow the pressure organ, water, to act on the wheel as soon as the sluice is opened; the two members of the chain are so formed as to pair at once, and motion occurs through the action of gravity upon the water. In the turbine the water forms a screw pair $S^{+}_{-}S^{-}$ with the turbine wheel, it is caused to descend by gravitation, and therefore drives by its motion the element with which it has been paired.

The opening of the stop-valve of a steam-engine allows the column of steam to become a part of the chain (which we have seen to be a ratchet-train, § 126), and to constrain it to perform that particular motion which its form permits. The indefinite length of the driving-link, the steam-column, is obtained by a physical process in the boiler. Similarly in hydraulic machines a meteorologic process furnishes us continually with new portions of the driving link to replace those which have left the machine. It causes the water to move as it were in a circle, always raising it again after it has passed downwards through the machine.

The hydraulic ram mentioned in § 129 no longer presents any difficulties to us. The water in it is kinematically chained with the other parts, the whole forming, in fact, a ratchet-train. So far as the descending water driven by gravity is concerned this train is reversed, but in respect to the portion of the same water which is raised it is direct. The liquidity of the pressure organ allows it to be thus separated into two streams. It is quite indifferent to us, and in no way affects our definition, that the water here both drives, is driven, and communicates motion. In every case we see that the driving body, the driver or motor, forms itself a link in the kinematic chain, instead of being, as in the old conception of the machine, entirely external to it.

§ 134.

Prime-movers and Direct-actors.

We have now reached a position which enables us to give an answer to the question formerly raised (§ 129), whether the steam-engine, the water-wheel and turbine, the lathe and planing-machine, the loom, the crane, and so on, could or could not each by itself be considered a complete machine.

In regard to the three first we can at once say that they are complete machines ; they are moreover place-changing machines, giving to certain of their parts, by suitable kinematic chaining, a determinate motion which may be utilized for any desired end. A steam-engine, for example, may be employed to drive machinery of the most different kinds without in the least altering its own mode of action. The various uses to which portable engines are put gives us a familiar illustration of the way in which advantage is taken of this fact in practice. One prime-mover may always be substituted for another without any alteration in the machines driven by it, if only the effort applied to the shaft and the speed at which it is driven remain unaltered. In other words, the machine fulfils its end in these cases if it give to one link of the chain a uniform rotation, or cause its points to undergo changes of position along circular paths. In treating prime-movers as complete machines, then, popular usage has done not merely what is practically convenient but also what is in theory perfectly correct.

The question whether the lathe, planing-machine, spinning-machine, etc., are complete machines does not appear quite so easy to answer. We may suppose each of them to be arranged so as to be driven by a belt,—for this purpose they only require to be furnished with suitable belt-pulleys. We may then certainly say, if the belt have always the tension necessary to prevent slipping, that they are complete. It is quite indifferent, so far as regards the chaining and the action of the machine, whether the belt be endless or not, whether it be moved by a weight or by muscular force (as in Berthelot's knotted belt Fig. 357,* or Borgnis' "flexible ladder" Fig. 358), or be driven from the shaft of an engine. In each case the belt is as truly the driver of the machine as the steam is of the steam-engine. Just as in that case it is indifferent whether the steam be received direct from the boiler or whether the engine be worked by the exhaust steam from another engine (as has been sometimes the case), so here it is indifferent by what means the belt be set in motion, it is and must in all cases be the driver of the machine.

In mining and tunnelling operations we often find prime-movers worked by air which has been compressed by a hydraulic air-pump, (as in the Mont-Cénis tunnel) or by a pump driven by steam. As prime-movers they are, however, complete, as long as the requisite quantity of driving air is supplied them. Their driver is the column of air in the pipes, and is itself set in motion by another prime-mover. This air-column between the two machines is, however, in precisely the same position as the belt between a steam-engine shaft and a lathe, loom, or any other machine driven from it. Our investigations have already shown us (§ 44) that the cases are not merely analogous but essentially identical. But whether the driven machine be itself a prime-mover or be a machine directly employed in mechanical work is obviously beside the question. The machines considered

---

* Three or more men work beside each other on as many ropes, the pulleys of all being placed on the same shaft *a*. Borgnis, *Mécanique appliquée, Composition des Machines.*

are therefore, taken along with their driving organs, complete machines.

This is equally true of pumps (which are only place-changing machines), of looms, shaping-machines, sawing-machines (which are both form and place-changing machines), and so on; in short, for all machines arranged so as to be driven by a prime-mover,

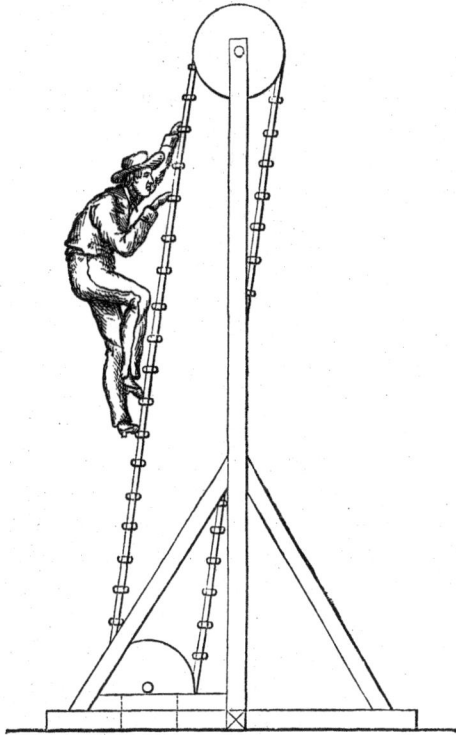

Fig. 357.   Fig. 358.

whatever the nature of the latter be. Such machines we may call collectively direct-actors. By practical machine-makers they have long been considered complete machines, in opposition to the conceptions of the theorists, so that here again we find the popular view to be fully justified upon strictly theoretical grounds.

We have still to consider those direct-actors which are driven
by animal power. We saw above that in these machines the
body of the man or animal combined with the mechanism in a
kinematic chaining sometimes of great complexity. The special
complication, however, lies always in the organic part of the
chain, the links of which receive the necessary constraint by
the action of forces commanded by the will. If we bear in mind
that in the example given—the grindstone worked by the foot—
and equally in the hand-pump, in the tread-wheel, the horse-gin,
etc., the mechanism driven by muscular energy forms in itself
a closed kinematic chain, we see that the relation of the organic
driving parts to the inorganic machine is precisely that of the
prime-mover to the direct-actor driven by it. The man
or animal is to be regarded as a prime-mover, of which the parts
—hands, arms, feet—move so as to drive in the required manner
the given artificial machine. The locomotive has often enough
been called a steam-horse,—we may reverse the comparison and
call the gin-horse, Fig. 359, the locomotive of the machine which
it drives. Its direct work is simply that of moving against a certain resistance. A man working in a tread-wheel, or clambering
Borgnis' endless ladder is in exactly the same position, his work
is that of continually raising the weight of his own body. The
assistance given by the living agents to the process is purely
physical in each case, and not intellectual; it is not in the least
degree necessary that they should know the object of the machine
in order to do their work. This work is precisely that which
would be performed by an inorganic prime-mover in driving the
same mechanism.

So far, therefore, as machines driven by muscular
power are themselves closed kinematic chains, they
may be regarded as complete machines, and do not in
themselves differ from machines driven by any other
than muscular force.

This brings us to another important question, which certainly
has a right to a place in any complete treatment of the theory of
machines, although it has never yet found one. It is the question
of the share taken by living agents, and especially by men, in the
executive portion of the machine's action. If the application of
animal power to machines be considered at all in the study of

machines, as it continually is when that power acts the part of
a prime-mover, it is not consistent to leave unnoticed the share
taken by the same agency in modifying the work produced by
the machine,—in taking, that is, the part of a direct-actor.
This subject is one to which we have been brought by a method
of treatment differing entirely from the old one, under which it
found no place. Here I can only enter into it so far as is necessary to enable us to come to some definite conclusion as to the
completeness of those machines in which human agency is employed in the handling of the work-piece.

In some machines the co-operation of the hand of the worker
in the operations to be performed is, from their nature, essential.
In the spinning-wheel, for instance, which is one of these, the

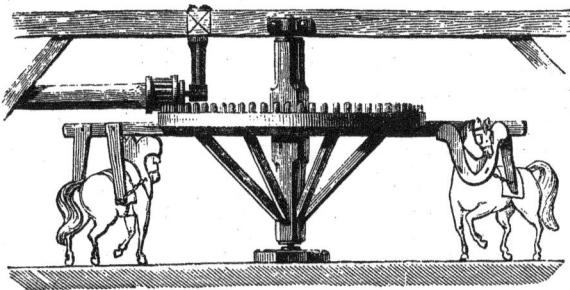

FIG. 359.

spinner has herself to regulate and carry out an important part
of the form-change which the fibres undergo. Her hand becomes
in this way an organ of the machine, in which it forms part of
a very complex chaining controlled by the worker's will. A process
takes place here, therefore, which corresponds entirely to that
described above as occurring, for instance, in the grindstone. The
spinning-wheel is driven by the worker's foot also, so that human
agency has a twofold action in it. The grinder in Fig. 356 is
doubly connected to the machine at which he works in the same
way and for the same purposes.

With the sewing-machine the case is precisely similar. In some
machines the one hand of the worker drives the mechanism while
the other guides the work, in others both hands, often acting in
a very complex manner, are required for the latter purpose.

The needle-grinder works at his stone as a part of a machine to which he does not himself give motion. He holds the needles between his fingers and thumb, moving them to and fro, and making them roll on his fingers at the same time in such a way that each needle receives its conoidal point as the envelope to its motion relatively to the grindstone. In modern factories this machinal work of the grinder is to a great extect done away with. The required motion of the needles is obtained by the use of a special mechanism and by giving a special form to the grindstone itself. For some purposes, too, sewing-machines are made entirely self-acting, being driven by power and having their work mechanically guided, and after long years of study the spinner has found her representative in the spinning-machine. None the less, however, must we regard the grindstone, sewing-machine and spinning-wheel as in themselves complete machines. It is possible to do definite work in all three without the direct intervention of man. The grindstone can polish pieces of cylinders, the sewing-machine can stitch straight strips of material, the spinning-wheel can twist and wind up the loose fibres presented to it. The man adds his own action as that of a machine controlled by will to that of the given mechanism; the living and the lifeless direct-actors together produce, necessarily, a far greater variety of work than was possible for the latter alone.

§ 135.

The Principal Subdivisions of Complete Machines. Descriptive Analysis.

Our examination in the preceding sections of the receptor, communicator and tool has shown us that it is no longer possible to regard these as representing the parts into which complete machines commonly divide themselves. We have found that each of them is absent from some series of cases, so that neither forms a general characteristic of the machine; and in the end our investigations carried us back once more to the closed kinematic chain, which alone we found to belong to every machine. It cannot be denied that when we stand before the machine itself this abstract idea seems bare and unsatisfactory, and does not promise to be of much practical assistance in the work of the machine designer. But in every possible case, however complex, it is of the greatest importance to be able to lay hold clearly of the general underlying principle, and thus to recognise the cause of the non-success of certain combinations which have rested on a defiance of this principle, a destruction of the closure of the chain. The natural wish remains, however, to go beyond this general principle, and to determine at least the more important lines along which the development and application of the principle take place. The three old subdivisions have certainly been of some assistance in this respect, and it was in no way our intention in criticising them to oppose such a desire. But before we could meet it, it was necessary thoroughly to clear the ground, and obtain a rigid, logical basis upon which we could rest when we turned to more detailed matters. Now, however, that we have attained such a position, we may proceed to examine the distinguishing characteristics of certain parts, or groups of parts, which seem to serve very definite functions, within the machine itself.

Our investigation has shown, in the first place, that two parts appear distinctly as forming portions of the great majority of machines, which hitherto have been generally considered to be external to them—the driver and the work-piece. In the steam-engine we recognise the former at once in the driving column of steam;—the latter is less distinct, it may sometimes be the fly-wheel shaft, sometimes a toothed-wheel upon it, sometimes a belt. With the lathe the case is reversed, the work-piece we see directly, the driver is not so obvious. In general the driver is most easily recognised in the prime-mover, the work-piece in the direct-actor. This shows itself very distinctly in the names of the different machines, as steam-engine, gas-engine, water-wheel, &c., among the prime-movers, paper-machine, rivet-machine, &c. among the direct-actors. Thus in name, at least, if not in theory, driver and work-piece have been considered as parts of the machine.

We must also consider that mechanism in the machine which connects the required change of place or form, or both, in the driver with the similar changes in the work-piece, as forming one of its essential parts or groups of parts. We distinguish for instance the piston-engine from the steam re-action-wheel, the bucket-wheel from the turbine, the shingling-hammer from the shingling rolls, and so on. We shall call the mechanism which plays this striking part in every machine its main or leading train. The names we have just cited show that in practice special importance has already attached to the mental separation of this train from the whole machine; indeed the establishment of the foregoing generalisation leads directly to this special distinction.

It is in the design of the leading train of a machine that we meet with those requirements which the old theory has attempted to satisfy by receptor and tool. If either or both of these exist at all they will form part of the main train, and can be treated by themselves if it be wished. I believe I may say that the practical mechanic has very seldom troubled himself about the exact determination of the receptor, while the whole construction of what we have called the main train flashes at once before him so soon as the name of the machine is pronounced. This makes it the more necessary that we should endeavour to ascertain theoretically what is included in this idea.

In our common direct-acting engine the leading-train is a ratchet-train, formed from piston and chamber with valves, and the slider-crank train $(C_3'' P^\perp)^{\frac{d}{\sigma}}$. In a wharf crane of the usual kind the main-train is running-gear (p. 498) formed of the chain, barrel and spur-gearing; in a heckling machine it may be a pair of heckling rollers with their driving mechanism; in the spinning-machine it is the draw-frame and spindles and their driving mechanisms, and so on.

There are many machines in which we find, as in the one last mentioned, that the main-train consists of several parts, or that several main-trains are united, each acting at its own proper time. In some cases this action is periodic, and frequently also, where the leading train is single only, we find a periodic sequence of single changes of motion occurring, governed by special mechanisms. These mechanisms may be considered as forming a group by themselves. In many pressure-organ machines they are represented by the valve-gear, but as they occur in many other cases where there are no valves, we shall include them generally under the name of the director or directing-gear of the machine. The director is therefore the apparatus by which the motions of the machine are caused to succeed each other in their required order.

In the steam-engine above mentioned the directing-gear is simply the familiar mechanism by which the slide-valve is opened and closed at the right instant; in a planing machine driven by a rack the directing-gear determines the periodic reciprocation of the table; in the self-acting spinning machine the gear is a train of no little complexity, comprehending, as Stamm first showed theoretically,[*] four different motions in succession, which he called *Sortie, Torsion, Depointage* and *Renvidage* respectively.

Within the directing-gear there is often an arrangement made to provide for bringing fresh portions of the material which forms the work-piece regularly under the action of the machine. In the carding engine a band of cloth with two feed-rollers is used for this purpose; in the cotton-preparing machine, combs or spiked rollers serve to supply the machine with raw cotton; in the mill feed-rollers are sometimes used to convey the grain regularly to the stones; in the needle-grinding machine a toothed-wheel gives the requisite feed to the needle-frame. In many machines, also, a similar arrangement exists for the purpose of bringing continually into action fresh portions of the driver. All these mechanisms we may include under the one head of feed or supply gear. The arrangements for moving the tool in planing-machines, lathes, drills, &c., as well as those for feeding boilers are examples of them.

For a purpose exactly opposite to that of the supply there is often another special arrangement added to the machine, an apparatus, namely, to remove the finished work-piece from the direct-actor. We may call this the delivery-gear; as examples of it we have the delivery tables in brick-making machines, the delivery drum (or in recent machines a more complicated arrangement) from which the prepared fleece is passed out of the carding-engine, the mechanism for shooting out the finished rivets from the die of the rivet-making machine, and so on. Supply and delivery form as it were the entrance and exit doors of the machine. Through the one the raw material enters the mechanism, through the other the finished manufacture leaves it. It is in connection with direct-actors chiefly that the construction of delivery-gear has been very fully developed.

Along with the director we find in very many complete machines a second mechanism, having special characteristics of its own,

[*] Stamm, *Traité théoretique . . . des Métiers à filer Automates,* &c., Paris, 1861.

and employed to control the passage of the driver or the work-
piece through the machine, to regulate, that is, the quantity of
the material of either of them employed per unit of time
according to the requirements of each instant. We may call
this the regulating-gear or regulator. While the director
determines the sequence of the motions in the machine, the
regulator determines their quantity. The various governors
of prime-movers of course belong to this class of mechanisms;
they regulate the motion or supply of the driving-organ and
consequently the speed of the whole machine. In the Cornish
engine the cataract is the regulator, in the working-gear of clocks
escapements of various kinds fulfil the same function. There is
also regulating-gear in very many direct-actors, as in the regulators
of looms and paper-making machines. Escape-valves placed
upon air, steam or gas-pipes also regulate the supply of the
fluid by preventing its pressure ever increasing beyond a certain
fixed amount.

It is sometimes required, most often in direct-actors, that the
regulator should be able entirely to stop the action of the driver;
as, for instance, when there is danger of any great irregularity
occurring in the work produced by the machine. Regulating-gear
acting in this particular way we may call stop-gear; it is made
in very many forms. As illustrations of it we may mention;—
the arrangements in the loom, which bring the machine to a stand
if the weft thread does not pass; those which stop it if any one
of its numerous threads break; the arrangements for shutting off
the water from a hydraulic lift when the right height has been
reached, and so on.

The regulator and the director are often very closely connected; in modern steam-engines, for example, the former (the
governor) acts directly upon the latter (the valve-gear), and by
means of it controls the motion of the driver (steam); it is always
possible, however, to consider the two separately. When both
of them or other sub-mechanisms exist special parts frequently
(although not always) become necessary simply for the purpose
of transmitting motion. This we may call the transmitting-
gear, or more shortly the gearing of the machine. Gearing is
also frequently inserted between a prime-mover and the direct-
actors which it drives.

In complete machines then, apart from other secondary mechanisms, which can generally be placed without any forcing in one or other of the subdivisions which we have considered, we frequently find that besides their

Driver and Work-piece,

(*a*) the main-train, in which receptor and tool may exist,
(*b*) the director, with its subdivisions supply and delivery,
(*c*) the regulator, with its subdivision stop-gear,
(*d*) the gearing or transmitting-gear,

can be distinguished each as a distinct and separable mechanism. Their separation, which places before us the general purpose of the combination of mechanisms forming the machine, we may call the descriptive analysis of the machine.

The separation of machines into the two classes of form-changing and place-changing is often useful in considering their purpose as a whole, especially in those cases where the change affects the work-piece,—where it concerns the driver (§ 132) it is of less practical importance. It may therefore well find a place wherever machines are systematically treated. It must always be remembered, however, that the distinction is not an essential one, but strictly speaking a difference of degree only and not of kind. This fact makes it sometimes doubtful whether a machine belongs to the one or to the other class. In every case, however, it is according to the nature of its main-train that the machine is classed—an additional reason for examining the characteristics of this mechanism separately.

In the form of analysis which we have indicated in this section, we have rather developed and systematised a method not unfrequently made use of by practical men, and given it distinct form, than introduced a completely new idea. We often find machines explained from a point of view very nearly resembling ours. It appears to me altogether very desirable that this descriptive analysis should form the first part of the description of a machine. The complete or abstract analysis, examining its mechanisms in detail, can be added afterwards. In many cases this may be unnecessary; as, for instance, where the particular mechanisms have already been studied by themselves. Two things must always here be borne in mind;—first, that machines do not invariably divide themselves

into the separate trains we have enumerated, so that the existence
of these is not necessary that a machine may be complete,—and,
secondly, that (as already said) there may sometimes be employed
in the machines arrangements for special purposes which do not
fall exactly under any of the subdivisions spoken of.

§ 136.

Examples of the Descriptive Analysis of Complete Machines.

It will be useful to give here a few examples of descriptive
analysis, in order to show more distinctly by particular instances
what the nature of the problem is, and how far its results extend.
Let us first take a few prime-movers.

A breast-wheel used for driving a manufactory has for its
main-train a mechanism of the formula $(C'C_{z\lambda}V_\lambda)_6^6$,—as we found
from § 62,—that is, a toothed-wheel having a liquid pressure-organ
—guided (in the breast) by a portion of the frame which carries
the wheel,—in place of a rack. The motion is continuous, the
main-train is therefore a running train. The driver is the pressure-
organ, water. Neither director nor supply exists ; the work of the
supply gear is performed in a physical or meteorological operation
which furnishes continually fresh portions of driving material without any action of the part of the machine. A regulator may
exist, as a governor acting on the sluice-valve.

Jonval turbine.—The main-train here also is running gear,—
it is a screw-train with the place of the nut taken by the water
which forms the driver of the machine. There is no director,—a
regulator may be arranged as in the water-wheel. Stop-gear may
be employed, as for instance in the turbines at Schaffhausen, where
in case of one of the driving ropes breaking the regulator suddenly
allows a sluice-valve to fall.

Steam-engine.—Let us examine a high-pressure engine such
as is shown in outline in Fig. 360. Here, besides, the main-train
we have both a director and regulator. The steam is the driver, the
work-piece being the fly-wheel shaft. A is the main-train (in the
form of a cylinder with suitable ports and piston, cross-head, connecting-rod, crank, shaft and frame),—a reversed and double-acting

ratchet-train (§ 126), formed from the train $(C_3''P^\perp)_c^d$ with the addition of a slide-valve. Steam is the ratchet, the piston the click-frame (the links a and c respectively in Figs. 352—355). B, the valve-gear, is the director, and consists of a train $(C_3''P^\perp)_a^d$ formed by the eccentric, eccentric-rod, slide-rod and frame. It drives the slide-valve, which is itself a combination of four separate valves, representing the ratchet-pawls ;—in the Corliss and

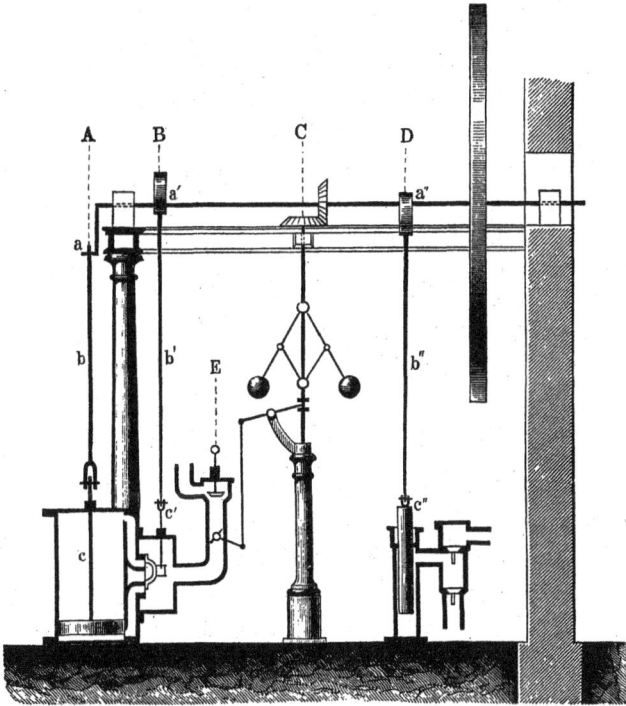

Fig. 360.

similar engines the four valves appear separately. C is the regulator, consisting of a common centrifugal governor (the kinematic constitution of which we must here leave unexamined), with a throttle-valve and gearing.

The feed-pump D requires special notice. It might be considered as a machine by itself, for which the engine was the prime-mover, but if we assume that the engine has a boiler for itself alone we may treat the pump as forming a part of the engine. It would

then form that subdivision of the directing-gear which we have
called the supply. The feeding apparatus is here a direct and
single-acting ratchet train, consisting of the chain $(C''_3P\perp)^d_a$ with
suction and delivery-valves as ratchet- and click-pawls. These
are raised and dropped at the right instants by the fluid ratchet,
the water. The whole arrangement forms therefore a second ratchet
train which differs from the leading train in being single-acting and
in requiring no director. The latter peculiarity, however, we may
neglect; for we could if we chose use a slide-valve worked by an
eccentric instead of the automatic valves. In this case we should
have in our machine two ratchet-trains,—a reversed ratchet-train
for the leading train, a direct one for the supply,—both fitted with
suitable directing-gear. In the case actually before us there are the
additional differences that the one is single- and the other double-
acting, and also that the pressure-organ is gaseous in the one and
liquid in the other. These differences, however, we may suppose
also to be removed, and the question then presents itself, why is
the one ratchet-train reversed and the other direct, although both
work with the same pressure-organ, the water passing through the
boiler? This resolves itself into the general question of the conditions which determine whether a ratchet-train with directing
gear be direct or reversed. The answer is, that it is direct if the
effort in this direction exceed the resistance—reversed in the opposite case. The main train (the piston and its connections) receives
from the column of steam a reversed motion (considered as a
ratchet-train), because the effort of the steam exceeds the resistance
at the crank; while the train constituting the pump is a direct one,
because here the driving effort at the crank (the eccentric a'') is
greater than the resistance at the plunger.* If at any time the mean
resistance at the crank-pin becomes greater than the mean effort on
the piston, the machine runs backwards, and the ratchet-train as
such becomes a direct one, forcing first steam and then air drawn
through the exhaust pipe through what had been the supply pipe.
We have illustrations of this every day in the working of the
locomotive.

The fact that the directing gear of the ratchet-train, as we have
seen, possesses the property of acting either forwards or backwards

---

* I hope to be able to treat this interesting question, and others directly connected with it, elsewhere more fully.—R.

according to the relation between the forces acting on the mechanism, allows the ratchet-gear to have that motion which running-gear has of itself. The director, that is to say, removes the mono-kinetic (cf. § 41) properties of the ratchet-train and so removes the difference existing between it and running-trains.

In order to stop the machine the steam-column is broken by means of the stop-valve *E*. This valve, with its box and fittings, forms a mechanism by itself, arranged to be worked by hand. It is the stop-gear of the engine, and so forms a part of the regulator. Taking the steam-engine as a whole then,—apart from keys, cocks and so on,—we find it to consist of a main-train, a director, a train for feed or supply, a self-acting and a hand-regulator,—five mechanisms in all.

Passing now to the direct-actors, let us take first a common wharf-crane with revolving platform. Such a machine has two leading-trains which can be worked by hand, independently of one another—the chain-drum with its pulleys and spur-gearing for raising the load, and another wheel-train for turning the platform. No director exists, but a regulator is provided in the brake by which weights can be lowered slowly. There is also a click-train used as stop-gear, for preventing any unintended descent of the load. It forms a part of the regulator, but is self-acting, in connection always with the first mentioned main-train.

A common clock with going and striking gear has also two main trains, one for moving the hands and the other for working the striking apparatus. As a rule each train has its own driver in the form of a weighted cord or sometimes a weighted loose pulley ; they stand, however, in close kinematic connection. The going-gear is a compound spur-wheel train. Its motion is determined by a regulator, represented by the escapement and pendulum. We have already seen (§ 121) that escapements are ratchet-trains acting by the periodic disengagement of the clicks. The striking-wheel with its lever-train forms the director of the machine. For each twelfth of a revolution of the hour-wheel this directing gear may cause the hammer (for example) to strike once for the half-hour and to make one of a series of strikings (in arithmetical progression) for the hour. It is driven by the going-train, and its immediate action on the striking-train is the disengaging of a click which allows the hammer-train to work. In order that the latter may act uniformly, a special regulator, in the form of a fly, is attached to the striking gear. The weight drums are connected to their spindles by running clicks, so that they may be moved backwards by hand in order to bring the weights once more into such a position that they can drive the machine. In reality these clicks are simply auxiliary mechanisms for the purpose of supplying the machine anew with its driving organ. There is lastly a lever which can be moved by hand so as to throw the striking-train in or out of gear; this lever with its connections form a directing-train moveable by hand. Summing up the different mechanisms which together form such a clock as we have described, we find them to be eight in number,— two main-trains, a self-acting director, a hand-director, two supply-trains worked by hand, and two regulating-trains;—five of them are automatic and the others arranged for being moved by hand.

A saw-frame, which we may suppose to be driven by a belt from some prime-mover, has its main-train in the crank and connecting-rod mechanism $(C_3'' P^\perp)_a^d$, which drives the frame. The tool (the saw-blade) forms part of the block c; the work-piece is the block or tree-trunk being sawn. The tool forms upon this its envelope, the saw-cut,—removing for this purpose those portions of material which are presented to it at each stroke. The motion of the work-piece by which this is done is effected by a ratchet-train driven from the crank-shaft and moving forward periodically the frame or table upon which the log rests. This train therefore forms the supply. The only regulator is the stop-gear, which enables the workman to place the driving-belt upon a loose pulley.

In the Jacquard loom there are two main-trains, the mechanisms for working the beam and the shuttle; the Jacquard mechanism itself forms a very complex directing-gear. There is also a supply-gear for moving the chain forward and a regulator (already mentioned) for the tension of the warp, and lastly one or more stopping arrangements.

The hydraulic ram, which we have had occasion to mention several times, forms a very easy subject for descriptive analysis. The head water $H\,A\,B$, Fig. 361, is the driver, while the prolongation, $D\,E$, of the same column, forms the work-piece, both being enclosed in suitable vessels. The main-train is a ratchet-train having two pawls in the valves K and D. There is no director, the air-vessel R forms a regulator.

The special feature of the ram, that the work-piece forms a portion of the driver, is to be found also in some other machines, as, for example, in the Chinese scoop-wheel mentioned in § 48 and other wheels resembling it, such as the Noria (§ 49), where the machine consists of a main-train (running-gear), having neither director nor regulator. The chains forming the basis of these machines have three links only, the wheel, the water and the piece forming the frame for the one and the channel for the other.

Where the nature of a prime-mover is already known by analysis it is often possible to include it as a whole in the descriptive analysis without impairing the distinctness of the description. In a paddle-steamer, for instance, we may call the engine with its

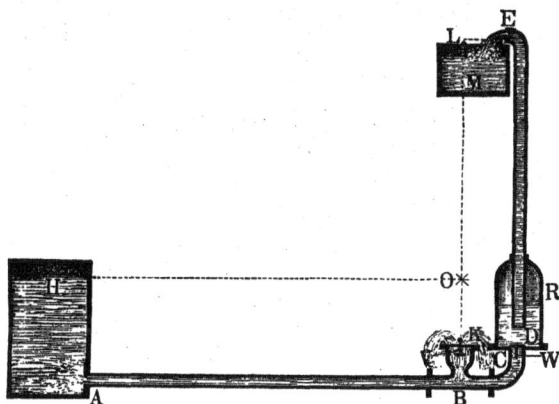

Fig. 361.

paddle-wheels paired with the water the leading train; the rudder and its gearing form here a mechanism for guiding the motion of the whole, that is, a director; a self-acting regulator is seldom applied, there is commonly only an arrangement which can be worked by hand as a stop-gear. The steam-engine itself has its own director and supply, as we have already seen.

The illustrations we have given are sufficient to show how our analysis can be used, and what results it gives. In most of the machines described it would have been useless to apply the old subdivision into receptor, communicator and tool. Any attempt to apply it to (say) the steam-engine, the clock or the loom shows at once that it is absolutely of no help to us. Indeed there has never been any serious attempt made to analyse by it machines having any degree of complexity.

The view which our analyses have given us of the action of the hand in the operation of the machine is remarkable. We see that it occasionally takes part in the directing and regulating gear, and less frequently in the main-train itself,—and also that as each machine develops into more perfect forms both its director and its regulator are made automatic. Looked at historically, from Humphrey Potter, who invented a primitive form of self-acting valve-gear to save himself the trouble of working the valves of Newcomen's engine, to the engineer of an American river steamer, whose business it is to control three polished levers in an elegantly-furnished cabin; from the turner of sixty years ago, whose hand was his tool-holder, to his successor of to-day whose machines, once set, can take five or six cuts off the work-piece simultaneously; we have one phenomenon only, developed in different degrees. This is the reduction of the direct action of the worker with his machine, or, if it be preferred, the increase of automatism in the machine. This process began with the very origin of the machine itself. For between the first timid attempt of men to constrain two external bodies to execute some determinate relative motion, and the most complex production of modern machine-industry, there is an unbroken connection; the lines of development are faintly marked, but are continually increasing in distinctness, while they have always followed and still follow the same fundamental laws.

Those machines must therefore be considered the most nearly perfect or complete in which (as already mentioned in Chap. VI.) human agency is required only to start the machinal process and to cause it to cease. In general the progress in this direction is quite visible, while in some cases existing machines appear to have already arrived fairly within sight of this ultimate perfection.

§ 137.

The Relation of Machinery to Social Life.[57]

From the general point of view to which our special investigations have once more brought us, one question seems so prominent that it is hardly possible to avoid glancing at it. We have traced the growth of the machine from the primitive fire-drill to the Jacquard loom, and have seen to some extent the direction which its future development is likely to take. The growth of the machine has been simultaneous with that of the race; what has been the influence of the former upon the latter? The question is altogether of too wide and too general a nature to be treated here; it may be interesting, however, just to look at a few of the matters suggested by it which seem to be most directly connected with our investigations.

The present form of the industry of civilised nations dates from the introduction of the steam-engine. The ancients certainly carried on important and lucrative manufactures, but the methods of production were then essentially different from those with which we are familiar. They were in general based upon home-industry, each worker doing his own share of the whole at his own house, as is still the case among semi-civilised peoples. The germ of the modern factory appeared when the home-worker took assistants to work along with him. In the middle ages this system had already attained considerable proportions; and since the close of the last century, it has grown with increasing rapidity, until we now have huge factories, full of busy workmen and workwomen, in every part of the country. It is the steam-engine which supplies the driving energy in these factories; had we been still dependent upon the older motors—upon muscular force, or wind, or falling water—they would never have existed. It is easy to see how this prime-mover, once introduced, brought with it the rapid growth of machinery in general. Its influence made itself felt in both directions in which, as we have seen, the machine naturally grows (cf. § 51). It increased, on the one side, the force at our command; not only did it react upon itself, so that engines were made larger and more powerful, but the older hydraulic prime-movers also received new development from the ease with which they could now be constructed. It increased, on the other hand, the attainable variety of motions by removing all difficulties as to want of sufficient power for their execution. In this way it has become the parent of an immense number of direct-actors, and we owe to it, in very great measure, both the advantages and the drawbacks of our modern industrial life.

In the great majority of cases the change from the old to the new industry has taken the form of a concentration of isolated workers and work-places. This is naturally noticed most of all in connection with those productions which possessed an importance before the time of the steam-engine which they have since retained, and in none more than in textile industry. Here the results of the change cannot be said to be in every respect advantageous. The home-worker, the small master, has all but disappeared. This in itself may be in many instances a cause for regret. But with him has also disappeared much of his individual skill. The work in the factory does not call for the possession nor allow of the employment of that personal skill which was required and shown by the old home and hand-worker, and the skill therefore no longer exists, at least in these industries. The breaking up of home life, too, which is involved in the factory-system is a matter having many possible draw-backs; it has already called for public attention more than once, and may do so still more pressingly in the future.

It is in connection with the future of these industries that the construction of small, cheap prime-movers becomes a matter of special importance. The direct-actors are every day being made better and less expensive,—but it is at the same time found that the prime-mover works the more economically the larger it is. For factories, therefore, one huge, expensive but economical engine drives an immense number of small direct-actors, and in this way only can the goods be made cheaply enough for the market. I believe that in many places and circumstances it would be an advantage if the home-industry could be placed in a position to compete with the factory-work. This can only be brought about when it becomes possible for the workman who has a little money at his disposal to buy a small and cheap prime-mover which is at the same time economical, to drive the two or three direct-actors which he may be able to possess. It is in this direction that I look for a future for the gas-engine, which has lately been brought into practical shape, and perhaps also for small hydraulic motors and hot-air engines.

But there are many industries,—the manufacture of engines and machinery for instance,—in which the drawbacks I have mentioned do not appear, or appear in a less distinct form, while in others the influence of the introduction of machinery has been almost altogether good. This is the case especially in those industries where the work in its own nature is disagreeable, unhealthy or even degrading. In mining operations, for instance, we can look forward with unmixed pleasure to the substitution of machine labour for much of the work of the colliers, and to the consequent amelioration of the sad social conditions so often associated with their work.

It is remarkable, also, that the place-changing machines, as distinguished from the form-changing machines, have had an influence upon social life which is almost entirely favourable. Railways, steamers, cranes have only to be named to make this recognised. The work of those connected with them may be hard, but it is as a rule healthy, while it demands every day more and not less skill and knowledge on the part of the workman. Of the good which the State as a whole derives from them it is unnecessary to speak.

There is one other characteristic of modern industry as affected by the growth of the machine to which I may direct attention. This is the substitution for manufacture of what I have elsewhere called machinofacture.* The two differ essentially in degree only, but the difference is so marked as to thrust itself upon our notice. This machinofacture appears especially in those cases where many machines have to be made from the same model, or from a limited number of forms in different combinations. In gun-making or in waggon-making, for instance, it has done extraordinary things, and in many other departments, locomotive building among them, it is making rapid progress. We have to thank it, too, for the spread everywhere of cheap and well-made sewing-machines. Reacting upon its source, the growth of machinofacture is accompanied by an increase in a variety and capacity of direct-actors, that is, especially, in the varieties of constrained motion at our command.

The wonderfully quick development of machinofacture which has taken place within the last few years must be ascribed in great measure to the particular direction in which the ideas of inventors have turned, and especially to the fact that they have given up the attempt to copy the operations of the hand or of nature in

* *Offizielle Bericht über die Pariser Welt-ausstellung,* 1867. p. 401, *et seq.*

the machine, and have tried to mak the latter solve each problem
in its own way, a way often very different from that of nature
Attempts were made for many years to construct a sewing-machine
which should produce work exactly the same as hand-stitching,
but they always resulted in failure. As soon as this idea was
completely discarded, and a new form of stitch specially adapted
for the machine was looked for, the spell was broken, and very
shortly the sewing-machine appeared. It is the rolling-mill, which
works in a way so greatly differing from the time-honoured operations of the smithy, that has been the special means of developing
the manufacture of malleable iron. The attempt to imitate nature
in the machine rests upon an altogether mistaken idea, and it was
when this was entirely thrown overboard that machine development received the impetus under which it is still making such
rapid progress.

CHAPTER XIII.

KINEMATIC SYNTHESIS.

"In magnis et voluisse satest."—PROPERTIUS.

---

§ 138.

General Nature of Kinematic Synthesis.

HAVING now examined at some length and through a great variety of cases the problems of kinematic analysis, we come to the consideration of the reversed operation—kinematic synthesis. While the former showed us the nature of the constrained motions obtained by the use of given combinations of elements, links, or chains, the province of the latter (which has already been mentioned in § 3) is the determination of the pairs, chains, or mechanisms necessary to produce a given constrained motion.

This problem is the highest of those which here come before us, and perhaps the most important of the whole series, for it has for its immediate object the creation of new machines. For this reason, and also because its solution presupposes an acquaintance with kinematic analysis, it fitly forms the conclusion of our investigations. The reader who has followed these so far, however, cannot failed to have noticed that various synthetic propositions have presented themselves in the course of our work, both in the general view of machinery to which the history of its development led us, and also in our special examination of single elements and of several series of mechanisms and complete machines.

These propositions have more and more limited the road to the solution of the problem, so that we already know something of the general nature of the results to which we may expect kinematic synthesis to lead us. The methods, however, in which the problem may be treated differ very greatly, and we must in the first place endeavour to determine which form of application of the synthesis promises the best results in the treatment of its problems.

There are, I think, two principal methods by which the desired end can be attained;—these may be called direct and indirect synthesis respectively. Each of these again divides itself into two branches, according as its treatment is general or special. We shall attempt to determine, *a priori*, the usefulness of these different forms of synthesis.

§ 139.

Direct Kinematic Synthesis.

The general direct synthesis should give us immediately the mechanisms which are required in each machine to effect a given change of place or form in its work-piece, or to utilise in it a given natural force. It is evident at once, however, that we cannot hope to arrive at useful results by this method. For our analytical investigations have shown us that one and the same motion can be obtained in different and often very many different ways. The synthesis must therefore either give us a great number of different simultaneous solutions to the one problem, or must be able to furnish us with that one out of them all which is the best. The latter is, however, impossible,—for the practical merits or defects of each single result lie to a considerable extent beyond the sphere of kinematics (§ 3). Two steam-engines, for instance, in different circumstances may be equally good, equally useful, equally "practical," although they may be kinematically very different. We cannot expect, therefore, to build up any system of general direct analysis which can be useful to us.

The function of special direct synthesis would be to furnish us directly with a pair of elements suited for any required place- or form-change. This is generally possible, for if we know the required motion we can determine (Chap. II.) its axoids, and these themselves may be used (as was shown in Chap. III.) as the profiles of the elements. In the cases, however, when the centroids are infinite (§ 9) this cannot be done, and the problem requires a wider treatment, which presents the same difficulties as those of general direct synthesis. But it is not necessary to consider this further, for we saw long ago that the practical value of solutions of these problems by pairs of elements was far less than that of solutions based upon kinematic chains. This second method of synthesis, therefore, cannot furnish us with results of any practical value.

§ 140.

Indirect Kinematic Synthesis.

It is the province of the indirect synthetic method to give us beforehand the solutions of all those problems under which it is possible for the given problem to fall; to solve, that is, all the problems of machine-kinematics in advance. At first sight this problem may seem so extended, in fact so measureless, that any attempt to solve it may appear to be nothing more than a mere theoretic proposition. We must not forget, however, that some of the investigations we have already made point to the conclusion that the region covered by the problems of machine-kinematics is not unlimited. We may remember, for instance, the remarkable smallness of the number of lower pairs (§ 15), or the definite number of mechanisms which can be formed from a given chain (§ 3). So here also we shall find on closer examination that all these kinematic problems lie within a region which it is possible for us at least to survey. If the requirements of the case be not made too high, the difficulties attending the use of this form of synthesis, although great, are by no means insurmountable, especially within the limits ordinarily covered by machine-construction.

That the special indirect synthesis is really practicable is proved by the results of our analytical investigations, Its function is to show us what kinematic pairs actually exist. Now we know

(§ 56) that the number of elements is not very large, for we have been able to express them all by a very moderate number of symbols. It follows necessarily that the pairs built up from these elements can only vary within tolerably narrow limits. This is really the case; we see at once therefore that a field for the application of kinematic synthesis is becoming visible.

The object of general indirect synthesis is to do for the kinematic chain what the special synthesis does for the pair of elements. The great number of possible cases at once presents itself as a difficulty. On examination, however, it will be found that these fall together very much. The number of simple chains especially—that is, of chains in which no link contains more than two elements—is by no means as large as might be at first sight imagined. The determination of these possible simple kinematic chains alone, however, forms no inconsiderable part of the whole problem.

There is, of course, no limit to the number of compound chains which can be formed, so that in this direction the solution of the problem can never be complete, and these compound chains demand investigation on just the same terms as the simple ones. In actual machinery, however, the compounding of chains is not carried very far. In those cases which appear most complicated it is almost always possible to subdivide the whole according to the purpose of each of its groups of parts, and to treat it as a series of separate mechanisms, no one of which is in itself very complex. The method of descriptive analysis (§ 135) has given us very extensive and satisfactory illustration of this, and we shall further on find a more exact method of distinguishing between different cases of compound chains. There are certainly, however, many compound chains which cannot be subdivided in this way. Some of the more important of these we can already treat fully synthetically without extending their investigation to any excessive length ; others will no doubt yield in time to synthetic methods.

We see, therefore, that we may apply the method of indirect synthesis to our subject with every prospect of obtaining by its means results which are really of practical value.

§ 141.

Diagram of the Synthetic Processes.

The importance of this part of the subject is so great that I have thought it worth while to add the accompanying diagram (Fig. 362) in the hope that it may make the connection between the different synthetic methods somewhat more clear to the reader. Kinematic synthesis as a whole divides itself into direct and indirect, and each of these classes again subdivides into general and

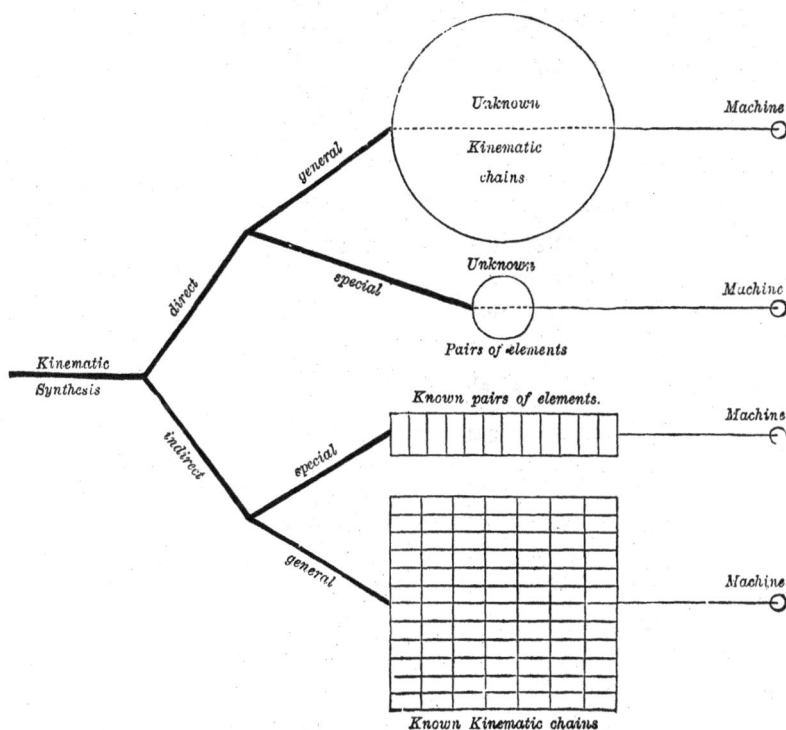

Fig. 362.

special. The direct synthesis should combine the kinematic elements at its command into the required pairs or chains according to the laws of pair- or chain-formation. In part it strikes upon insoluble difficulties, in part it furnishes results which have no practical value. The indirect synthesis first (as special synthesis) forms and arranges all the possible pairs of elements, and then (as general synthesis) finds all the combinations of these pairs into chains. From this systematised arrangement of pairs and chains the special combinations best suited to each particular case can then be chosen by an inductive process. When the required chains have thus been found the remaining processes of forming them into mechanisms and machines present no difficulties.

We must now examine the results to which this indirect synthetical method leads us.

§ 142.

Synthesis of the Lower Pairs of Elements.

In § 55 we chose twelve class symbols for the kinematic elements, of which ten were for rigid elements :—

S Screw,	H Hyperboloid,
R Revolute,	G Sphere,
P Prism,	A Sector (portion of a revolute),
C Cylinder,	Z Tooth,
K Cone,	V Vessel,

and two for the flectional elements,

T Tension-organ. Q Pressure-organ.

We shall first look at the combination of the rigid elements into pairs. We may therefore omit the element V, which is always paired with the pressure-organ Q. G, also, stands only for a particular revolute, and A for a portion of the same element, so that these symbols may both be included under R. Seven elements therefore,

$$S, R, P, H, K, C, Z$$

remain for synthetic treatment. Three of the pairs which can be formed from these elements are already well-known to us; the three common lower pairs :—

$S^+_-S^-$ or (S) the screw- or twisting-pair,
$R^+_-R^-$ or (R) the revolute- or turning-pair,
$P^+_-P^-$ or (P) the prism- or sliding-pair.

Strictly speaking the word "lower" should be added in speaking of the first two pairs, but we have seen that in most cases it may be omitted without fear of misunderstanding. For (R) we commonly write (C), calling the pair often a cylinder-pair; we can, however, always return to the more general symbol when necessary.

The two pairs (R) and (P) may, as we saw in § 3, be treated as special cases of the form (S). If we place the tangent of the pitch angle as an exponent to the symbol S (as was done, for instance, in the case of the plane hyperboloid, p. 254), we have at once $(R) = (S^o)$ and $(P) = (S^\infty)$. We may also, in places where we require only to distinguish between classes of elements, include all the lower pairs under the symbol (S). We require, in other words, in such a general classification as is required for the synthesis, to consider only the one lower pair (S).

§ 143.

The Simpler Higher Pairs.

The element C not only forms the closed pair (C), but is used also in higher pairs, such, for example, as the cylindric friction-wheels, Fig. 363, which would be written C, C or more generally

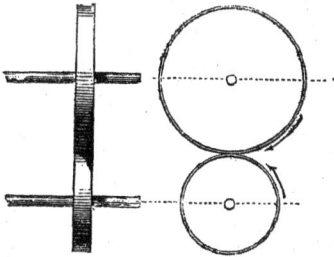 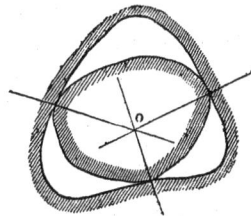

Fig. 363.   Fig. 364.

$\bar{C}, \bar{C}$. The class of pairs of which this one forms a special case is the pair of general hyperboloids, $\bar{H}, \bar{H}.$[58] Non-circular cones, $\bar{K}, \bar{K}$, and non-circular cylinders, $\tilde{C}, \tilde{C}$, are also special cases of the same class. We had numerous examples of the last-named elements in the higher pairs of § 21 *et seq.*, of which Fig. 364 represents a general case.

The pair C, C results from a further simplification of form in the same class, and it again takes several definite but general forms.

We may use for the symbol of these pairs, as we have done for former ones, a single letter. This presents no difficulty, as the

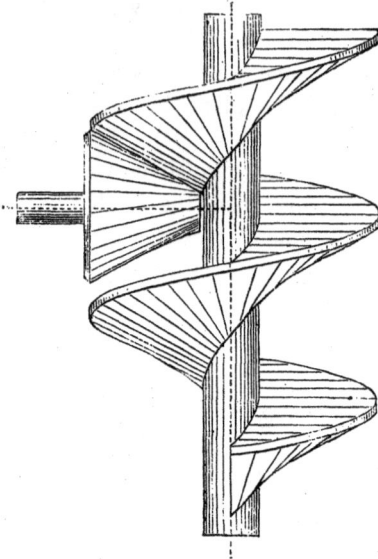

Fig. 365.

two elements always have the same name-symbol. We have only to provide means for distinguishing them from the lower pairs, (the name-symbol being sometimes the same), and this can easily be done by adding a comma to the name-symbol. We have therefore the class

$$\tilde{H}, \tilde{H} \text{ or } (\tilde{H},) \text{ the general pair of hyperboloids,}$$

and within this the following subdivisions:

$$\tilde{K}, \tilde{K} \text{ or } (\tilde{K},) \text{ the pair of non-circular cones,}$$
$$\tilde{C}, \tilde{C} \text{ or } (\tilde{C},) \text{ the pair of non-circular cylinders,}$$

as well as the three special cases of greatest simplicity :—

$$H, H \text{ or } (H,) \text{ the pair of hyperboloids of revolution,}$$
$$K, K \text{ or } (K,) \text{ the pair of circular cones,}$$
$$C, C \text{ or } (C,) \text{ the pair of circular cylinders.}$$

Forms intermediate between the general and the special also occur, such as the pair H, S, which is represented in Fig. 365, and the pair $H°, K$, the plane-hyperboloid and cone, both of which pairs we have mentioned before (pp. 81 and 83). These forms would be indicated by the symbols $(\breve{H},)$ and $(H,)$ respectively, for the screw in Fig. 365 enters the pair because it is a **ruled surface**, *i.e.* $\breve{H}$, and not because it is a screw S.

The pair C, C is to be distinguished from the closed cylinder-pair; this is easily done by using the symbol $(C,)$ for the higher pair. The pair C, P is the special case of $(C,)$ when one of the cylinders is of infinite radius. The sign (C_p) may be used for it, the comma indicating sufficiently that the pairing between C and P is higher.

§ 144.

Synthesis of Toothed-wheel Pairs.

We come now to toothed-wheels. These might also be included under the class $(\breve{H},)$, but the repetitions of the same profile occur

in them so characteristically that it appears well to treat them as a class by themselves, as it has hitherto been usual to do. It must not be forgotten here that chain-closure does not necessarily accompany the use of toothed-wheels. We have already seen (§§ 43 to 50) that they may be employed with pair-closure alone, in such a way, for instance, as is represented in Fig. 366.

Taking first toothed-wheels for which the centroids are circular, we may include them all under the symbol H_z, H_z or (H_z), which stands for a pair of hyperboloidal toothed-wheels, and consider

$$K_v, K_z \text{ or } (K_z) \text{ the pair of bevel wheels, and}$$
$$C_v, C_z \text{ or } (C_z) \qquad \text{do.} \qquad \text{spur wheels}$$

as special cases under this general class.

The teeth of these wheels are in general formed as ruled surfaces of the same character as the axoids for the motion which they transmit; they may, however, be made helical, and in that case H_z becomes H_v. The most general class formed in this way has the higher screw $\tilde{S}$ for its tooth form, it would therefore be written H_v^-, H_v^- or (H_v^-). As special cases under this class we have

(K_s) conic or bevel screw-wheels, and
(C_s) cylindric screw-wheels.

The pair of elements S, S or (S_s), represented in Fig. 367, is included in the last subdivision. For this pair the symbol (C_s) is generally preferable. The symbol (S_s) is, however, valuable, in relation to its higher form $(\tilde{S}_s)$ as pointing out that the general closed screw-pair (S) may be looked at as a subdivision of the general class of reciprocally enveloping screws.

We have also to distinguish other and higher classes of toothed-wheels, those namely which have non-circular centroids. Of these we have the general cases

$$(\tilde{H}_z) \text{ and } (\tilde{H}_v^-),$$

which include as special cases $(\tilde{K}_z)$ and $(\tilde{K}_v^-)$, $(\tilde{C}_z)$ and $(\tilde{C}_v^-)$. The forms $(\tilde{H}_s)$, $(\tilde{K}_s)$, etc., may also be considered as subdivisions under $(\tilde{H}_v^-)$, $(\tilde{K}_v^-)$, etc.

The intermediate forms H_z^0, K_z—(hyperboloidal face-wheel with bevel wheel (Fig. 36) and H_z, S_z (Fig. 365)—may be included under these class signs. This holds good also for those pairs in which a spur-wheel becomes a rack P_z, in which P may be treated as a special case of C.

§ 145.

Cam Pairs.

The pairing between the links a and b in the cam trains Fig. 368 and 369 may be considered a special case falling under the class $(\tilde{H}_{,})$. In the cases shown we have a non-circular cylinder $\tilde{C}$

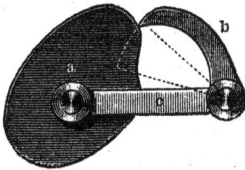

FIG. 368.   FIG. 369.

paired with a tooth Z; the general case would be a non-circular hyperboloid $\tilde{H}$ paired with a tooth of general form $\tilde{Z}$. The pair may be called the **general cam pair** and written

$$\tilde{H}, \tilde{Z} \text{ or } (\tilde{H}_{,z}).$$

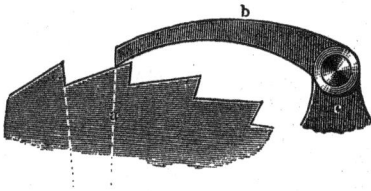

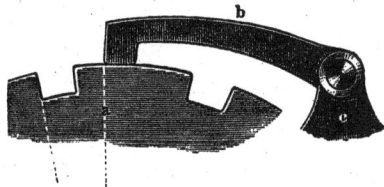

FIG. 370.   FIG. 371.

As subdivisions we have, besides $(H_{,z})$

$$(\tilde{K}_{,z}) \text{ and } (K_{,z})$$
$$(\tilde{C}_{,z}) \text{ and } (C_{,z})$$

Among cam pairs we have also to include (as we know from § 120) the click-pairings in such trains as are shown in Figs. 370 and 371. These may be written in the general case

$$(\tilde{H}_z\text{;}) \text{ and } (\tilde{H}_{z'}\text{·}),$$

respectively (see § 119), and these classes subdivide themselves as in the case of toothed-wheels. The case of the toothed-rack occurs here also, as the limiting case of $C = P$.

<h2 style="text-align:center">§ 146.</h2>

<h1 style="text-align:center">Recapitulation of the Pairs of Rigid Elements.</h1>

We have seen in the foregoing sections that the pairs which are obtained from the rigid elements can be systematically arranged in divisions and subdivisions so that each special form may fall under the more general case to which it naturally belongs. Of the divisions obtained in this way we may call the highest and most general the **order**, and the next lower the **class**, while special subdivisions of the latter we have treated as **groups**. The following table gives a general view of the pairs of elements which we have considered, arranged in this way.[59]

Pairs of Rigid Elements.

Orders.		Classes.			Groups.		
I. $(\tilde{S},)$		(S)			(S)	, (R)	, (P)
II. $(\tilde{H},)$ ...	$(\tilde{H},)$	, $(\tilde{K},)$	, $(\tilde{C},)$	... $(H,)$	, $(K,)$	, $(C,)$	
III. $(\tilde{H}_z)$ ...	$(\tilde{H}_z)$	, $(\tilde{K}_z)$	, $(\tilde{C}_z)$	... (H_z)	, (K_z)	, (C_z)	
IV. $(\tilde{H}_s)$ ...	$(\tilde{H}_s)$	, $(\tilde{K}_s)$	, $(\tilde{C}_s)$	... (H_s)	, (K_s)	, (C_s)	
V. $(\tilde{H}_{,z})$ ...	$(\tilde{H}_{,z})$	, $(\tilde{K}_{,z})$	, $(\tilde{C}_{,z})$	... $(H_{,z})$	, $(K_{,z})$	, $(C_{,z})$	
VI. $(\tilde{H}_{z;})$ ...	$(\tilde{H}_{z;})$	, $(\tilde{K}_{z;})$	, $(\tilde{C}_{z;})$	... $(H_{z;})$	, $(K_{z;})$	, $(C_{z;})$	
VII. $(\tilde{H}_{z'})$ ...	$(\tilde{H}_{z'})$	, $(\tilde{K}_{z'})$	, $(\tilde{C}_{z'})$	... $(H_{z'})$	, $(K_{z'})$	, $(C_{z'})$	

We have here all the pairs already considered, in their numerous varieties, included in seven orders. Those special classes and groups which are obtained by the use of the limiting case $C = P$,

$H = H^\circ$ and $K = K^\circ$ may always be considered as falling into the divisions given above. Some cases which are, apparently, extremely complicated as, for example, the pairings occurring in Rose-engines, which are often constructed in the freest form, belong to the second order. Even the first order includes necessarily many quite free-forms. Although the number of primary divisions which we have employed is so small, we include among them, so as to make the classification as useful practically as possible, several which in strictness are only varieties of other orders, as, for instance, Nos. VI. and VII.

§ 147.

Pairs of Elements containing Tension-organs.

The characteristics of the tension-organs, band T_p, rope T_s, wire, T_c, chain T_z or T_r, so far as their manner of pairing goes, may be sufficiently indicated by the use of the symbol T for the whole of them, and this accordingly will suffice us here. We have seen that the pairing of T with other elements can take place only

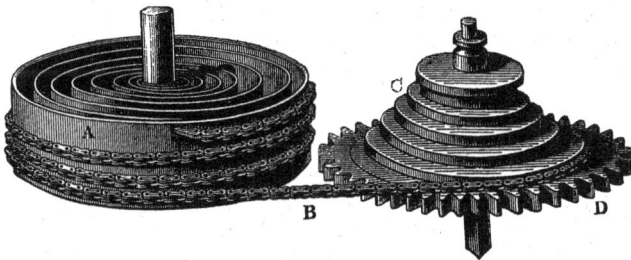

Fig. 372.

under a tensile force-closure. It can be laid upon, or wound round, rigid elements, but obviously only upon "positive" (see § 56) elements, so that its pairing with rigid elements is restricted to a certain class of forms.

Beginning with the screw, we find both higher and lower forms of the pairing of T with S frequently in use. The common chain drum of a crane forms an element of the pair S,T, which we may shortly write (S_{st}); and exactly the same formula represents the cylindric rope-drum, on which the rope is spirally coiled, the cylinder itself here becoming a screw (§ 15). Higher screws $\tilde{S}$, are not unfrequently paired with T, the fusee of a clock or watch, for example, Fig. 372, or the conic rope-drum which has lately been much used in winding engines, or the fusee in Robert's mule, Fig. 373. The conoid with a rope enveloping it (Fig. 374) belongs also to this class. The order to which these pairs belong is therefore $(\tilde{S}_{,t})$, under which the class $(S_{,t})$ is a lower form. At

FIG. 373.

FIG. 374.

the limit $S = R$ we obtain the group $(R_{,t})$ of which Fig. 375 shows two examples.

The next order is furnished us by the pairing of T with $\dot{H}$, the pair being therefore $(\tilde{H}_{,t})$. As a representative of the class $(\tilde{C}_{,t})$

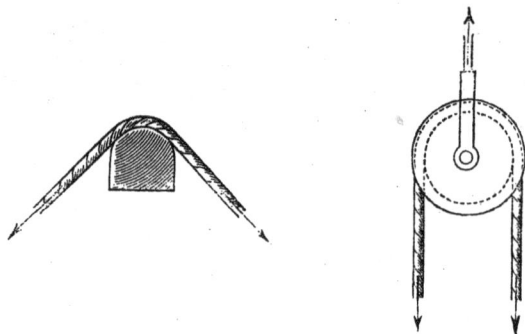

FIG. 375.

we have the pair formed by a rope-drum and a flat belt laid spirally round it.

With $\tilde{H}_{z}$ we can pair the chain T_{z}. It gives us the order $(\tilde{H}_{zt})$, of which there are many applications.

The combinations of the element T with $\tilde{H}_{\mathfrak{r}}$ and $\mathring{H}_{\mathfrak{s}}$ may be included in the order $(\tilde{S}_{rt})$; it does not require, therefore, special examination here.

Click-trains containing tension-organs exist, and their number has increased of late years. They are both single acting (free-clicks) as in pulley tackle, and double-acting (fast-clicks). Fowler's well-known clip-drum (Fig. 376), which does such excellent service in agricultural and in towing operations, is a case of the latter. We have thus the two orders $(\tilde{H}_{rt};)$, and $(\mathring{H}_{rt};)$ not merely theoretically possible, but actually in practical use. If in a click-train

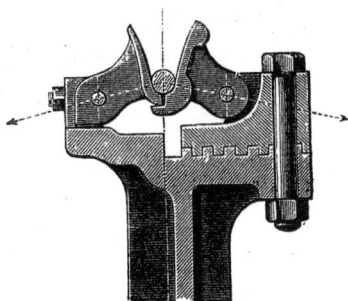

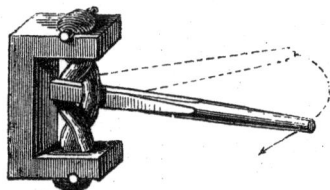

FIG. 377.

FIG. 376.

of this kind the element T be used in the form of a chain, T_{z}. (as *e.g.* in Bernier's pulley) the element H_{z} takes the place of H. We thus may have both the orders $(\tilde{H}_{z;t};)$ and $(H_{z;t})$.

We find therefore that it is possible to employ a tension-organ in every one of the seven orders into which we divided the pairs of rigid elements. One other pairing may also be carried out, that, namely, of two tension-organs. We have already mentioned one such case in speaking of the spinning process (§ 131). In the wrenching-spring, Fig. 377, we have another example of it, so that the pair T, T was one which came very early into use. The symbol for this order of pairs is (T_{r}).

§ 148.

Pairs of Elements containing Pressure-Organs.

We have seen (§ 56) that the pressure-organ Q takes several special forms : the liquid Q_λ, the gaseous Q_γ and the globular or grained $Q_\ddot{z}$ or Q . Although these forms have very important points of difference, yet so far as their kinematic pairing is concerned they may all be denoted by the single symbol Q.

The characteristic of the element Q that it has no resistance except to compression, allows it to be paired with rigid elements in the most various ways. It can be used as one of the elements in all the first five orders of pairs mentioned in § 146, as a substitute for a rigid element. The turbine, the screw-propeller, the water-wheel, the chamber-wheel train, the pug-mill and so on, give us numerous and various examples of such pairs. We have therefore the orders

$$(\tilde{S}_{,q}) \quad , \quad (\tilde{H}_{,q}) \quad , \quad (\tilde{H}_{z,q}).$$

The order $(\tilde{H}_{\bar{z},q})$ and its lower form $(\tilde{H}_{z,q})$ may also be distinguished, but it is more convenient to include both in the order $(\tilde{S}_{,q})$.

We have already found, further, that pressure-organs are paired both in free and in fast click-trains,—in those namely, in which the clicks are valves. If we consider the valve in such pairs as a tooth, Z, (which the analogy of the rigid click-trains allows us readily to do), we can indicate the two orders of pairs thus obtained by the symbols

$$(Q_{,z}) \quad \text{and} \quad (Q_{,z}).$$

In none of the cases mentioned is the pairing possible without another pairing taking place at the same time, that namely of the pressure-organ with its vessel or chamber, V^-, Fig. 378. Besides this we have also the pairing with the piston V^+ (Fig. 379), so that we may write this order of pairs in general as

$$(V_{,q}).$$

We have noticed before (§ 41) that this kind of pairing can be and has been also extended to the tension-organs, as in the link-chain guided in a pipe Fig. 380 and the spring brake Fig. 381. The same principle is also utilised in certain machines for the manufacture of wire-work. It is unnecessary, however, to treat these pairs as an order (V_{jt}), for in all cases the flectional element of the pair may be considered as a pressure-organ. They therefore are all included in the order (V_{jq}).

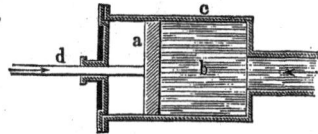

FIG. 378.   FIG. 379.

Another and very remarkable pairing of Q is that with T. This can be very easily carried out if the latter have the form T_z or T_r, if it be made, that is, as some form of chain. The "Paternoster" pump, whether with buckets or piston discs, the grain carriers of mills, the ladders of dredging machines, etc. all furnish illustrations

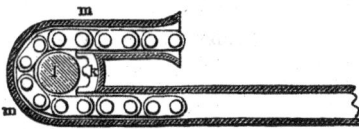

FIG. 380.   FIG. 381.

of this pairing. It would be possible to indicate by a special sign the peculiar construction of the chain, calling a tension-organ provided with scoops T_v for instance. This particularity is not necessary, however, for if a symbol show a pairing between T_z and Q it may be held to indicate that the tension organ is arranged in such a way that the pairing is possible. In some cases also no special alteration of the form of the tension-organ has to be made for this purpose, as in Vera's "rope-pump," where a mere rope T_s lifts water by adhesion.* All these pairs together form there-

---

* The rope-pump, or water-rope machine, has been frequently ascribed to the elder Brunel. It is certainly older than his time, however. See Langsdorf, *Maschinenkunde*, ii., p. 226 ; Hâchette, *Traité Élémentaire*, p. 134.—*R.*

fore an order for which we may use the symbol $(T_{,q})$ and one of which very frequent use is made in machine practice.

We have, lastly, pairings between two pressure organs, analogous to the pairing of T with T which we have already examined. These occur somewhat frequently in a form which we may express by the symbol $(Q_{\gamma,\lambda})$, and of which we have illustrations in the air vessels of pumps and various hydraulic machines, in spiral pumps, hydraulic blowers and so on. We may indicate this order of pairs generally by the symbol $(Q_{,})$.

§ 149.

Recapitulation of the Pairs containing Flectional Elements.

In summarising the pairs of elements considered in the last two sections it will be sufficient to tabulate the symbols for the orders alone, those for the classes and groups can be formed from these as in the case of the pairs of rigid elements. We have to add to the seven orders already tabulated :—

(*a*) The six following orders containing tension-organs :—

VIII.	$(\tilde{S}_{,t})$	XI.	$(\bar{H}_{,t;})$
IX.	$(\bar{H}_{,t})$	XII.	$(\bar{H}_{,t:})$
X.	$(\bar{H}_{z,t})$	XIII.	$(\tilde{T}_{,})$

(*b*) The eight following orders containing pressure-organs :—

XIV.	$(\tilde{S}_{,q})$	XVIII.	$(Q_{,z:})$
XV.	$(\bar{H}_{,q})$	XIX.	$(V_{,q})$
XVI.	$(\bar{H}_{z,q})$	XX.	$(T_{,q})$
XVII.	$(Q_{,z:})$	XXI.	$(Q_{,})$

If it be required to indicate that the pressure-organ is a fluid, the symbol λ or γ (as the case may be) can be used instead of the q.

Examples of the majority of these twenty-one orders of pairs already exist in machinery, others of them have not yet found practical application. Our object here does not allow us to treat them further in detail. Our investigation has, however, gone far enough to show us the notable fact that the number of possible pairs of elements is limited, and that the whole can be determined collectively by a synthetic treatment. We may now assume this to have been done, and proceed to the synthetic determination of kinematic chains.

§ 150.

The Simple Chains.

We cannot adopt so direct and definite a treatment in the case of kinematic chains as was possible with pairs of elements. In treating the latter we could build directly upon the definite and limited series of axoidal forms discussed in Chapter II ; here, however, we can make but accidental use of these, for the relative motions of the links of very different chains may be identical, and have, therefore, similar axoids. We might indeed treat chains by working through every possible combination of two, three, four, &c. pairs, and take systematically all the relative positions for the pairs in each combination. But the extreme unwieldiness of such a method, and the certainty that very many of the combinations so found would prove useless, unpractical, or altogether impracticable makes it very desirable that some other treatment should be adopted, even at the expense of external uniformity in our methods.

We shall adopt in general an inductive method, as is so frequently done in mathematical investigations, and choose for each series of problems the treatment which seems best to suit the special conditions of the case. It must be remembered, at the same time, that our object here is not to complete the synthesis of the chain, but merely to note its general direction. On these grounds we shall not begin with the general case of compound chains, but with the simple ones, the essential characteristics of which we have already examined somewhat closely. The conclusions arrived at in § 128, where we found that the constrained chain took its place in the series of possible combinations of links between the unconstrained and the fixed chain, will be of great service to us here. If, that is to say, we find a chain to be fixed, we can convert it into a constrained closed chain by inductive addition of links to it, or can obtain the same result from an unconstrained chain by the removal of links from it. We shall follow in general the same order in treating the simple chains that we have adopted in classifying the pairs of elements, without adhering to it rigidly.

§ 151.

The Screw Chain.

If we combine three conaxial screw pairs into a simple chain we obtain such an arrangement as is shown in Fig. 382, for which we see at once that we may use (S_3') as a contracted symbol. This chain forms an order by itself. The three mechanisms which can be obtained from it are essentially the same.

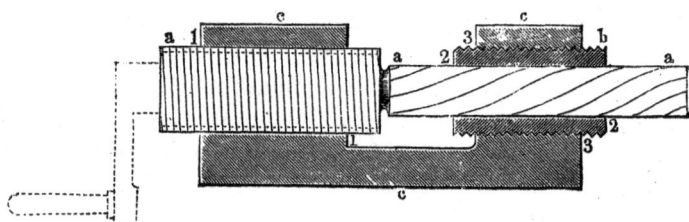

Fig. 382.

Besides the class which the order (S_3') itself represents we obtain special classes from the limiting cases in which one or other of the screws becomes $S^\infty = P$, or $S^0 = R$ or C.

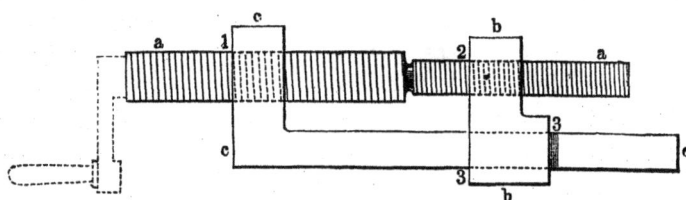

Fig. 383.

In the chain shown in Fig. 383 the pairs 1 and 2 remain (S) as before, while 3 has been made $(S^\infty) = (P)$; its formula runs $(S_2'P')$. From this chain two different mechanisms can be formed, the trains $(S_2'P')^c$ and $(S_2'P')^b$ being similar. They are the well-

known " differential screws," whose invention has been ascribed
both to Prony and to White. Our previous investigations have
put the reader in a position from which he will recognise in
Hunter's press,* the differential screw vice † and so on, only such
alterations of $(S_2'P')^c$ as are due to the reversal of pairs or to
external differences in constructive form. The train $(S_2'P')^a$ does
not appear to have been hitherto applied.

If we make the pairs 2 and 3 = (S), and pair $1=(S^0) = (C)$, we
obtain the chain shown in Fig. 384, which gives the two mechan-

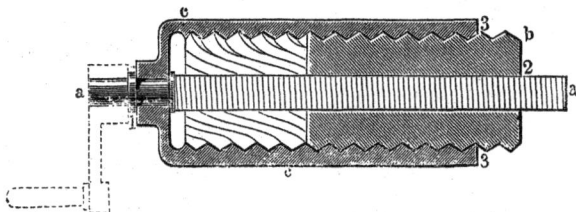

FIG. 384.

isms $(S_2'C')^c = (S_2'C')^a$ and $(S_2'C')^b$. The latter appears new,
the first has been more than once applied, as for instance, very
happily by Skinner ‡ in his steering-gear, which, however, is a
compound train.

If the pair 1 be made = $(S^0) = C$, the pair 3 = $(S^\infty) = (P)$, the pair 2
alone remaining (S), we get the chain $(S'P'C')$, Fig. 385, which

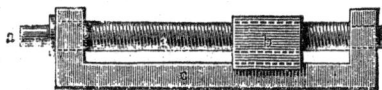

FIG. 385.

has already several times come under our notice. Of the three
mechanisms which it gives us $(S'P'C')^c$ especially has found, as
we know, numerous applications. (Cf. §§ 43 and 107.)

The synthesis has therefore given us here three classes of chains
furnishing seven mechanisms, and three of the latter appear new.
Let us now apply our method to the cases in which one of the
rigid elements is replaced by a flectional one.

* *Moseley, Mechanical Principles, &c.,* vol. i.
† *Ibid,* also Weisbach, *Mechanik,* iii., p. 288.
‡ *The Engineer,* 1868, vol. xvi., p. 182.

The use of a tension-organ in this way does not give us any useful results. It is otherwise, however, with the pressure-organs. If, in the first place, we replace the link b in the chain $(S' P' C')$ by a fluid we can form several practical mechanisms from it. The complete formula would in this case run :

$$\overset{a}{C^+ \ldots \mid \ldots S,Q} \overset{b}{\ldots\ldots Q,P} \overset{c}{\ldots \mid \ldots C^{\equiv}_{-}}$$

which would be contracted, the link c being supposed fixed, into $(S'_{,q}P'_{,q}C')^\circ$. If we now take a as driving-link we get the machine $(S'_{,q}P'_{,q}C')^\circ_{\overline{a}}$. This formula represents the screw-pump, the Archimedian water-lifting screw, the screw-ventilator, the main-train of Schlickeysen's clay-press, &c.

If with the same mechanism the pressure organ b be made the driver, which would give us $(S'_{,q}P'_{,q}C')^\circ_{\overline{b}}$,—we have the simple screw turbine.*

If we place the chain on b and make a the driver we obtain the mechanisms $(S'_{,q}P'_{,q}C')^b_{\overline{a}}$ which is the leading train of the screw-steamer. a is the propeller, c the vessel and b the water.

The train of Fig. 384 is also applied in a well-known machine. If we replace b once more by a pressure organ (here specially by a liquid) we obtain a chain of which the complete formula is :—

$$\overset{a}{C^+ \ldots \parallel \ldots S,Q_\lambda} \overset{b}{\ldots\ldots Q_\lambda,S} \overset{c}{\ldots \mid \ldots C^{\equiv}_{-}.}$$

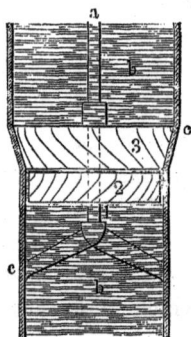

FIG. 386.

Placing the chain on c and making the fluid link b the driver we obtain a mechanism $(S'_{,\lambda_2}C')^\circ_{\overline{b}}$ which is that of the Jonval or Henschel turbine, Fig. 386. To improve the working of the machine the screws 2 and 3 are higher screws, so that strictly the formula should contain $\tilde{S}_{,\lambda}$ instead of $S_{,\lambda}$.

Once more we have a series of machines placed together which differ immensely in their objects and in their constructive form, but which are formed upon one and the same kinematic chain. We shall have to mention other screw chains in § 154.

---

* Such for example as the turbines at the mill of St. Maur described by Leblanc.

§ 152.

Cylinder-Chains.

We have already (Chap. VIII.) studied the chains (C_4'') and (C_4^L), and found that they divided themselves into twelve classes containing fifty-four mechanisms. Our investigation resembled so very much a synthetic treatment of these chains that it is not necessary here to repeat the investigation. Let us look what other simple chains containing none but cylinder pairs can be formed.

If we attempt to form a chain from three cylinder pairs we see at once that its closure is fixed (Fig. 387), we need not, therefore, examine it further.

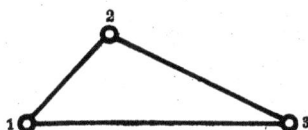

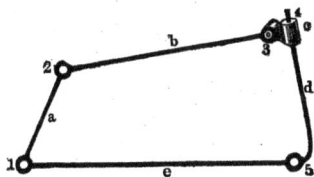

FIG. 387.   FIG. 388.

Five parallel or conic cylinder pairs give us a simple chain which is unconstrained, so that this combination also is of no importance to us. If we put a normal or normally crossed pair in place of one of the parallel pairs we obtain a chain which is constrained, and which contains five cylinder pairs, but here no motion can take place in the normal pair. It might thus be altogether omitted without affecting the motion of the chain, which is therefore really one of four links only. If the pair instead of being normal be oblique or obliquely crossed, as in Fig. 388, the chain becomes fixed, no motion whatever can take place in it. We have not, however, reached the limits of the cylinder chain, the two last mentioned may be considered to be only special cases of one containing a larger number of links. Working upwards to this from Fig. 388 we obtain first, by the addition of another cross-joint, and by destroying the parallelism of 1 and 6, the six-linked chain of Fig. 389, which again, like Fig. 388 is immoveable. If, however, we divide the link joining the cross-blocks into two parts, as in Fig. 390 for example, where a cylinder pair whose axis passes through 3 and 5 has been inserted in it, the chain becomes moveable. The links a and f move in different planes, but the linkage $b c d e$ allows them to be constrainedly connected. The chain now consists of seven turning-pairs, and may be written generally (C_7^+). Crossed-crank-trains formed from this chain are used in machinery; as *e.g.* in planing machines for giving motion to the fork which moves the belt from one pulley to another. The chain is indeed rich in

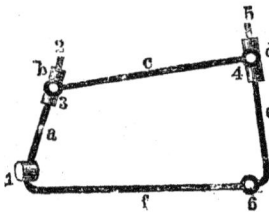

FIG. 389.

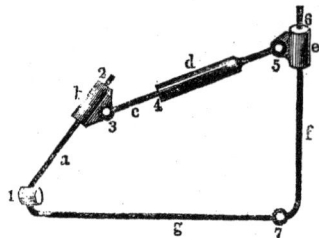

FIG. 390.

special cases, and out of many of these mechanisms can be formed. It deserves a complete and systematic examination; a glance at the subdivisions of the much simpler chain (C_4'') gives some idea of the number of special cases to which such an examination would lead us.

Under certain conditions the six-linked chain, which we may call (C_6^+) can also be made moveable and constrainedly closed. Applications of this occur in practice, but not in an easily recognisable form. Indeed it is specially noticeable that in these applications the principle of chain reduction (§ 76) is almost always employed. In order to fully understand them it becomes necessary to add or

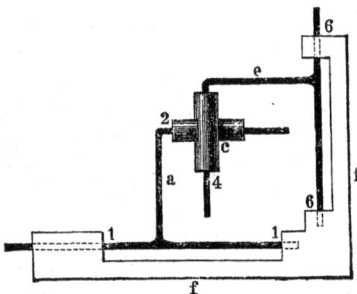

FIG. 391.

suppose added the omitted links.

The mechanism represented by Fig. 391 gives us an illustration of this. It serves here and there for working shunting signals which require to be turned through 90°,—the signal is connected with the link *e*. Of the links in this reduced train *c* and *f* are normal crossed links, *a* and *e* have parallel cylinders; the links *b* and *d* are omitted. The two open cylinders 2 and 4 of the link *c* have therefore higher screw motions relatively to the full cylinders of *a* and *e* upon which they work. In order to complete the chain it would be necessary to add the two links *b* and *d*, (which are each in the form $C^-...\|...P^+$, and to make the elements in *c* a pair of open prisms.

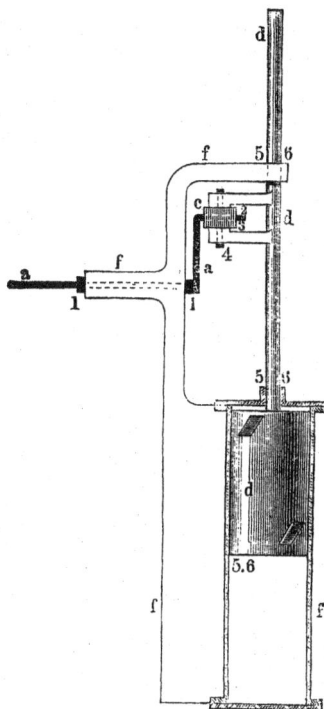 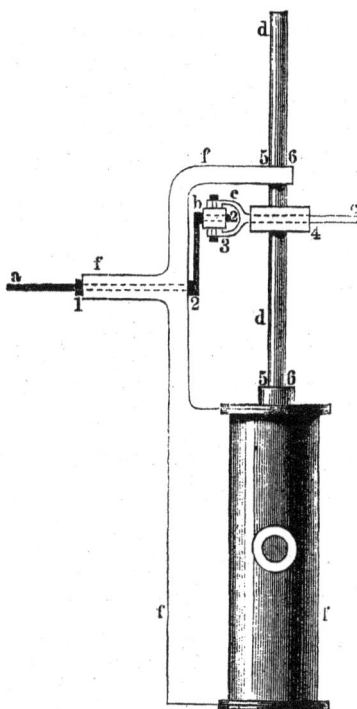

Fig. 392.   Fig 393.

Robertson's steam-engine [*] is another mechanism of the same kind, its leading train is represented by Fig. 392. The link *a* has the form $C^+...\|...C^+$; the link *b*, $C^-...\|...P^+$, is omitted, but *c* is still left. On account of the omission of *b* the latter has no longer

[*] *Artisan*, vol. xxix., 1871, p. 2 ; *Revue Industrielle*, June, 1874, p. 192 ; *Dingler's Journal*, 1874, vol. 213, p. 183.

the form $P...\perp...C$, but becomes $C...\perp...C$. It is carried by the link $d = C ... \parallel ... C$. This link Robertson uses as the piston-rod, that is the driving-link of the chain, he omits, however, the link $e = C ... \parallel ... P$, so that the piston d is constrained to oscillate about its axis at the same time that it moves to and fro in the cylinder. The contracted formula of the train, unreduced, is $(C''C''P\perp C''C''P\perp)^{\frac{t}{a}}$, and in the form used by Robertson, omitting b and e (putting together consecutive similar symbols),

$$(C_2''P\perp C_2''P\perp)^{\frac{t}{a}} - b - e.$$

We need not examine here whether the machine be practically useful or not; it serves equally well in either case as an example of empiric synthesis, which seems to be carried over all constructive difficulties in the delight of originating new mechanisms.

Robertson has used also another form of the chain for his machine, as is shown in Fig. 393. Here e only is omitted, but the arrangement of the links is at the same time somewhat altered. The formula of this chain unreduced is $(C''C\perp C\perp P\perp C''P\perp)^{\frac{t}{a}}$, or, more shortly, and with the reduction, $(C'' C\perp P\perp C''P\perp)^{\frac{t}{a}} - e$. The reader will have no difficulty in finding still other forms in which the chain (C_6^+) can be employed. Some of these may find useful practical applications.

The seven-linked cylinder chain may be arranged constrainedly in other ways than that above mentioned, in the way, for example, shown in Fig. 394. By making one or more of the links of infinite

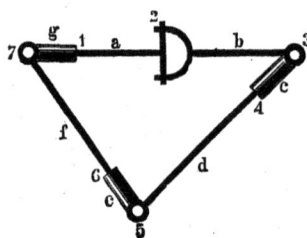

FIG. 394.

length we can of course obtain very numerous forms of the chain. A very interesting example of it is shown in Fig. 395, a mechanism which has been applied by Brown* as the leading train of a steam engine. If I am not mistaken it had been used earlier for the same purpose by Maudslay.[60] Here the link a has the form $C^+...\angle...C^+$, the link b being $C^-......G$, it consists, that is, of a cylinder and a sphere, the centre of the latter lying upon a normal to the axis of 2 drawn from the point of insection of 2 and 1. The sphere is paired with the cross slide e.

* *Engineering*, Feb. 1867, p. 158.

The " ball and socket " joint which we have here is simply the result of the omission of one of three cylinder pairs, the axes of which intersect at right angles in one point. It is therefore marked 3.4.5. The block $f = P^-...\perp...P^+$ serves as the piston-rod, the driving-link, of the steam-engine. The contracted formula of the mechanism is therefore $(C^L C_4^\perp P^\perp P'')_f^g - c - d$. In order to make

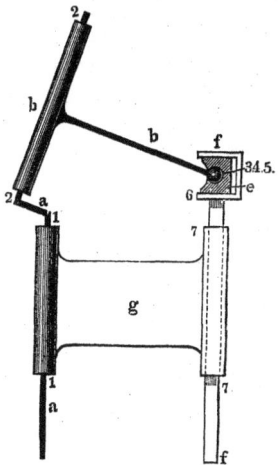

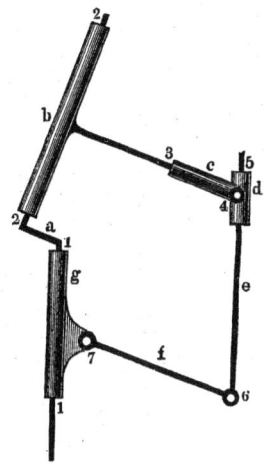

Fig. 395.   Fig. 396.

a comparison with the general form (C_7^+) more easy, I have represented the chain in Fig. 396 so that the two prism pairs, which we may consider as R^∞ or C^∞, are replaced by the cylinder pairs 6 and 7. The formula for the mechanism in this form, placed on g, runs $(C^L C_4^\perp C'' C^+)^g$.

The foregoing examples are sufficient to show the importance of the chains (C_7^+) and (C_6^+) and to serve as an introduction to their complete synthesis. The former appears to have the largest number of links that can be used in any constrained simple chain formed from the lower pairs of elements.

§ 153.

Prism Chains.

We have already more than once examined (§§ 64 and 108) the three-linked prism chain, or wedge-chain, (P_3^L). Fig. 397 represents it in a form with which we are now familiar, Fig. 398 shows it in another form, in which all three prism pairs are closed. If instead of three we attempt to combine two prism pairs into a chain we

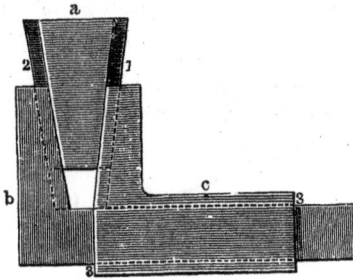

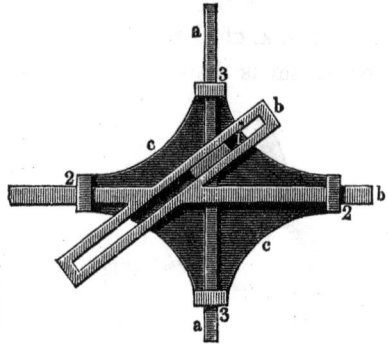

Fig. 397.   Fig. 398.

obtain either a single pair or a fixed chain. From four prism pairs, however, we can very easily form a simple chain, as is shown in Fig. 399, of which the formula is (P_3^L). We might treat the chain

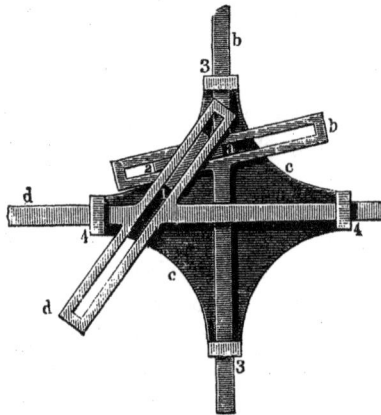

Fig. 399.

(P_3^L) as derived from this one by making the angle between the pairs 3 and 4 infinitely small. The chain (P_4^L) itself, however, might be considered as obtained from (C_4'') by making all four links infinitely long. But closer examination shows that the chain (P_4^L) is not constrainedly closed. If we suppose, for example, that the links b and c be fixed together,—the pair 3, that is, made immoveable,—we see at once that the chain still remains moveable, and indeed has become simply that of Fig. 397.

§ 154.

The Crossed and Skew Screw Chains.

In our development of the cylinder chain, (§ 152), we found seven to be the limiting number of links in a simple chain. We were not there dealing, however, with the most general case, for the cylinder pair is not the highest form of the lower or closed pairs. This position is occupied, as we know, by the screw-pair (S). We shall therefore obtain the most general form of chain containing

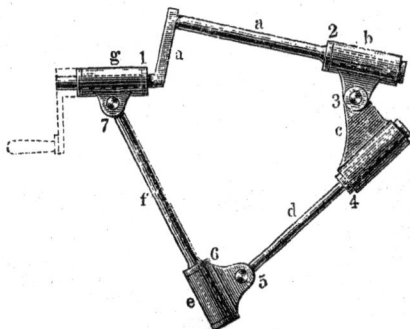

Fig. 400.

only lower pairs if we place (S) instead of (C) in the most extended cylinder chain. The highest chain formed from closed pairs will therefore be the chain (S_7^t). The complete synthetic examination of this chain, of which those already treated in §§ 151 to 153 are really special cases, is part of the work still before synthetic kinematics. Further on we shall have to return to the question, it would suffice here merely to indicate the existence of the chain were it not necessary to look at some of the forms in which it is practically applied in machinery.

Referring again to the chain (C_7^+) which was shown in Fig. 394, it will be noticed that we can make the cylinder 2 oblique or crossed instead of normal, as for instance in Fig. 400, where an exceedingly complex motion can be obtained. Leaving this unexamined, however, we may go a step further, and make the cylinders 1 and 2 conaxial. We obtain in this way the chain shown in Fig. 401. It is no longer, however, constrainedly closed, for the link *a* can be turned about its axis, (the coinciding axes of 1 and 2), without any motion occurring in the other links, while in these

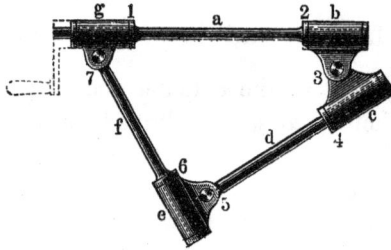

Fig. 401.

remaining links the chain is fixed, they move together like a single link, or rather element, relatively to *a*.

These conditions can be entirely altered, however. If either of

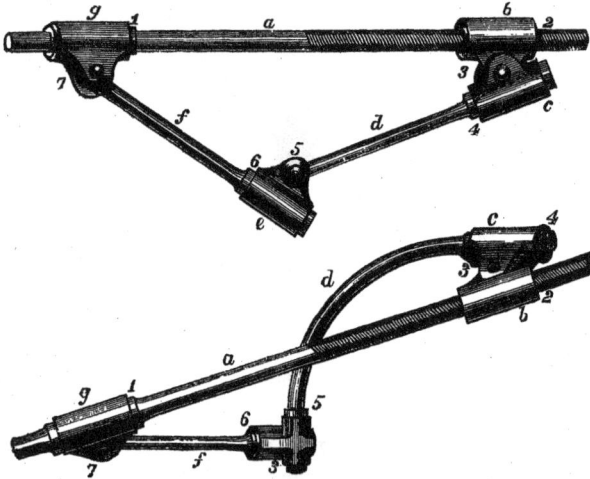

Fig. 402.

the pairs 1 or 2, let us say 2, be changed from $(C) = (S)$ to (S), and at the same time the chain be so arranged that the axes of 6 and 4 are not in the same plane, we obtain such a chain as is shown in

Fig. (402). In order to make the relative positions of the axes 2, 4 and 6 more distinct we have here given a plan as well as an elevation of the chain, which is now constrainedly closed, and which we may call a crossed screw-chain. Its formula is (beginning with the pair 1); $(C'S^+ C_{\frac{1}{2}}^{\perp} C^+ C^{\perp} C^+)$. A more general

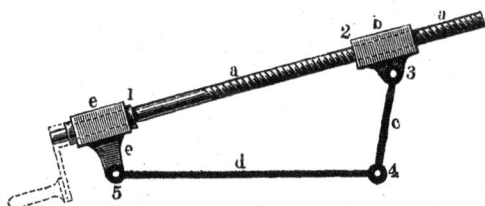

FIG. 403.

case of the same chain could be formed from the chain shown in Fig. 400. It must be noticed as a condition of the moveability of the chain that the axes of the pairs 4 and 2 must always intersect each other, and also those of the pairs 6 and 1.

If in the chain Fig. 401 we replace 2 by a screw pair but leave the axes 4 and 6 still con-plane, there is no longer any motion in the pairs 4 and 6, and they may be altogether omitted. The chain therefore takes the form shown in Fig 403.

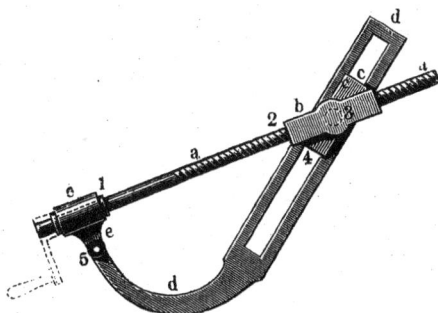

FIG. 404.

Here there are five links only instead of seven; *f* and *e* have been united and *d* and *c* have also become a single link. The formula for this chain is $(C'S^+ C_{\frac{2}{2}}'' C^+)$. This five-linked chain has very many practical applications. By placing it on *d*, for example, we get a screw-reversing gear which has been used for locomotives, by placing it on *e* we have a train which has been used as steering gear, knuckle lever presses, &c.

If the link c in such a chain be made infinitely long we obtain
the chain shown in Fig. 404, which also finds a number of applications. Its formula is $(C'S^{\perp}C^{\perp}P^{+}C^{+})$.

If now the length of the link e be also made infinite, or in other

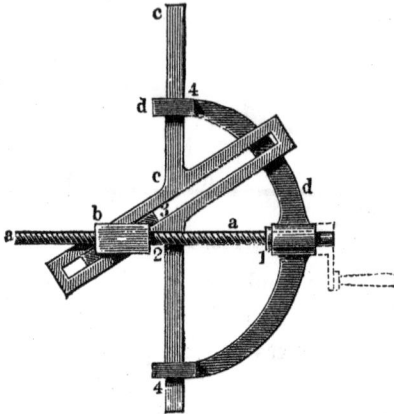

FIG. 405.

words the axis 5 removed to an infinite distance, the chain takes
the form shown in Fig. 405. The varying angle between the links
b and c in Fig. 404 has here become constant, the cylinder pairing
at 3 is therefore superfluous, and we obtain the four-linked chain
$(C'S^{L}P^{L}_{\frac{1}{2}})$ which we may call a skew screw-chain. Placed upon

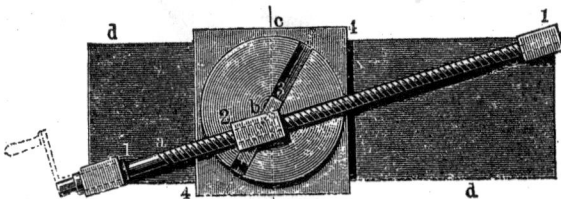

FIG. 406.

d it has received a very neat application by Nasmyth in his dividing
machine.* Fig. 406 shows the arrangement adopted by him.
The frame d is here the bed of the dividing machine and c the
slide. The angle between the pair 3 and 4 is made variable, so
that the motion of the slide for each revolution of the screw can
be altered with great nicety within very wide limits.

We see that chains derived from (S_7^+) take in themselves many forms which are of practical value ; they can and do also receive very numerous useful applications in compound chains.

§ 155.

Substitution of Higher Pairs for Pairs of Revolutes.

We have seen that we can regard the chains (C_7), (C_6), (C_5) and (C_4) with all their special cases as derived from the chain, (S_7), as the highest form of the chains formed from closed pairs. This, however, we are not obliged to do. For the element C may be considered as a special case not only of S but of other higher forms, namely the general cylinders and cones $\tilde{C}$ and $\tilde{K}$, with which (as we saw in § 21, etc.), we can form higher pairs of elements. Considering, then, the circular cylinder C as a particular case of the general cylinder $\tilde{C}$, we can substitute the pair $(\tilde{C},)$, formed from the latter, for the pair (C) where it occurs, and thus obtain entirely new motions in the cylinder chains. In this way an immense series of chains can be formed, and an almost inexhaustible series of constrained motions obtained.

The substitution of $(\tilde{C},)$ for (C) in the general chain (C_7) is not, however, possible in every case. It cannot, for instance, be carried out in the chain (C_4^L) or (generally) in those cases where oblique cylinder-pairs are applied, for here the universal condition of the closed pairs, the coincidence of the axes of the two elements, becomes essential, and this is not fulfilled by the elements of the higher pair. In these cases, however, we can still use the higher pairing if we substitute $(\tilde{K},)$ for $(\tilde{C},)$,—the higher cone for the higher cylinder. The instantaneous axis of the pair then always passes through the same point.

It must not be supposed that this use of $(\tilde{C},)$ or $(\tilde{K},)$ instead of (C) is mere speculation. We find many instances of it in practice, especially in chains of the class (C_4''). As an example very often met with I may give Hornblower's curve-triangle train (Fig. 407), beside which is placed (in Fig. 408) the slider crank-train from which it is derived. The latter is a reduced turning double slider-crank

---

* *Civil Ingénieur*, 1863, p. 215, 1864, p. 21.

($\S\S$ 72 and 76), its formula runs $(C''_8 P^\perp)^\mathrm{d} - b$. In Hornblower's train the curve-triangle $\breve{C}$, (which we have already examined in

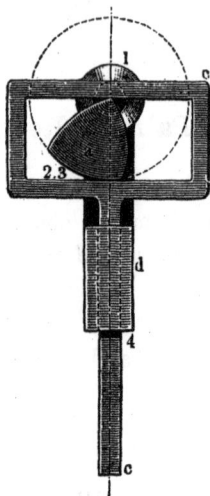

FIG. 407.

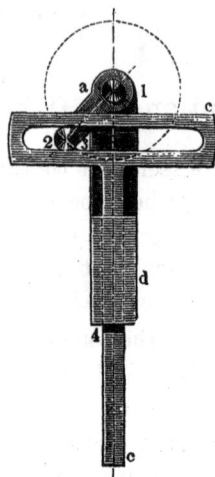

FIG. 408.

§ 26) takes the place of the pin 2. The chain being reduced by the link b, the curve-triangle appears without its rectangular partner

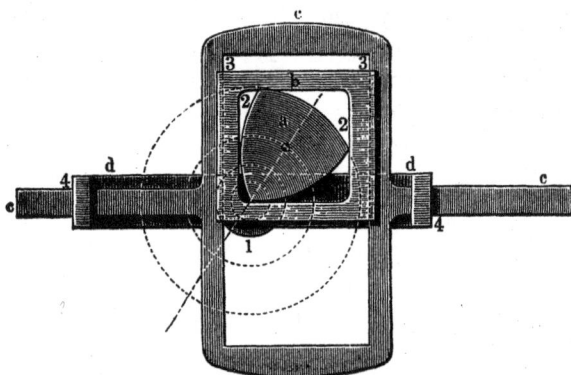

FIG. 409.

element. The formula of the chain runs therefore $(C''' \breve{C}, P\frac{\perp}{2})^\mathrm{d} - b$. If we bring both trains by augmentation into their complete condition we obtain the mechanisms shown in Figs. 409 and 410. The centroids of a and c become somewhat complicated, we cannot here examine them.* This curve-triangle train has also sometimes been

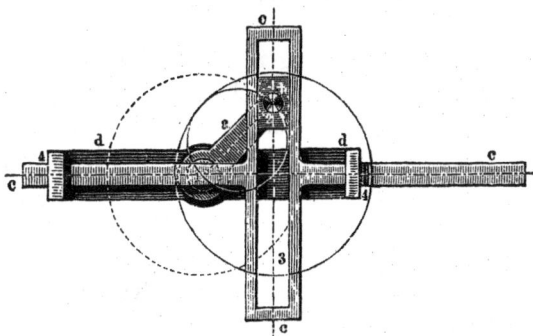

Fig. 410.

employed in practice unreduced; I know of one case, at least, in which the pair 2 has been used complete. Fig. 411 shows this

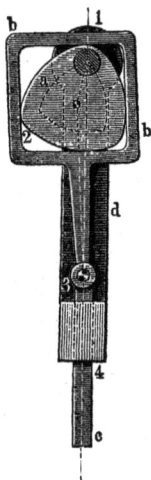

Fig. 411.

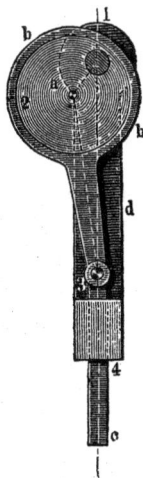

Fig. 412.

mechanism in our schematic form and with the addition of our symbols; its formula is $(C''C''',C''P^{\perp})^{d}_{a}$. It is used for driving the

---

* In the kinematic collection of models at Berlin I have shown these for various mechanisms of this class.—*R.*

slide-valve of a 100 HP. Woolf steam-engine.* I place beside it
in Fig. 412 the analogous mechanism $(C_3''P^\perp)_{\frac{a}{a}}^d$, which we already
know, in order to make the comparison between them more easy.
In both trains the pair 2 is expanded.

I may just note here in passing that the whole series of forms
obtained by pin-expansion from the chain (C_4''), &c., (see § 71), can
be used directly in the higher chains which we have been considering. This has scarcely been noticed as yet by machinists,
and many forms possessing considerable constructive advantages
have consequently not been utilised. There are many cases
indeed, as the foregoing example shows, in which these cam-
trains may be employed as easily and advantageously as the
common eccentric-train of a steam-engine.

Chains of the class (C_7) containing more than one higher pair
have not, to my knowledge, ever been practically applied. It is
probable enough that really useful applications may be found for
some of the numerous cases which we see here to be possible. It
must suffice here to have noticed the general case.

§ 156.

Simple Wheel-chains.

Among the simple chains which consist of wheels with their shafts
and bearings (cf. § 43) the friction-wheel chains naturally come
first. The circular wheels with the frame which pair-closes them
give us the chain $(C\frac{1}{2}H_,)$, with the special forms $(C\frac{1}{2}-K_,)$ and $(C_2''C_,)$.
Hyperboloidal wheels seldom occur in this way, but are occasionally
employed. Still higher forms, indeed, have ocasionally found application, as *e.g.* the spiral friction wheels of Dick's cotton press.
Friction-wheels generally occur in compound chains, I merely refer
to them here because of their importance in some industries,
in particular in rolling-mills, where the rolls themselves are really
friction-wheels.

We need not non-examine the series of special forms which are
taken by the simple toothed-wheel chain $(C_2\bar{H}_,)$, for we have
already (§ 144) investigated the various forms which the pairing $\bar{H}$·

---

* By Ad. Hirn in the Logelbach Works.

can take. We must mention here, however, the use of a pressure-organ in the chain. The chamber-wheel trains of Chap. XI. belong to the compound chains; we have, however, simple chains in which a fluid—that is a pressure-organ—takes the place of one of the wheels, and the pairing of the fluid with its chamber takes the place of one of the cylinder pairs. Among these are the common water-wheel, the lift-wheel, and the paddle-wheel (cf. §§ 61 and 62), and also some turbines and centrifugal pumps.

§ 157.

Cam Chains.

We noticed the cam trains very briefly in § 120, and recognized the desirability of their separate treatment. Fig. 413 represents one of these trains. Its formula is $(C_2'' C_{,z})$;—we have already examined (p. 537) the nature of the pair-ing between the cam and the click or tooth. The special forms which this chain can take are very numerous. The cam chain, however, is not here represented in anything like its highest form. The latter, so far as is conditioned by the form of the cam and the tooth, would be the chain

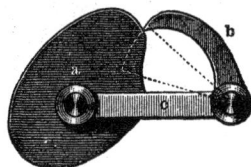

FIG. 413.

$(C_{\frac{1}{2}}^t \tilde{H},\!)$, which is formed from two pairs $(R) = (C)$ of the first order, and the highest forms of the pairs of the fourth order (§ 146). The most general form of all will be obtained if we substitute (as in § 155) the higher pairs $(\bar{C},)$ or $(\bar{K},)$ for (C), a method which also leads to the highest forms of the simple spur-wheel chains.

Under this most general form of the cam chain there come those important special cases which we have called click-trains. We obtain these by using in the chain pairs of orders VI. or VII. (§ 146). We thus obtain the chains:

$$(C_2 \tilde{H}_z;) \text{ and } (C_2 \tilde{H}_z:)$$

with their numberless simpler forms. As we have studied several typical cases of these in Chapter XI., we might now leave them without further examination. There is one other click-train, however, the one represented by Figs. 414 and 415, which deserves a little notice here. In both these forms the train can also be used as a ratchet-train. The train of Fig. 414 was called by Redtenbacher the "one-toothed-wheel;" it is somewhat widely known by the names of Maltese cross or Geneva ratchet, the one being taken from the form of the wheel b, the other from the employment of this click-train in Geneva musical-boxes. Fig. 415 shows that the essential condition of the train is not that the wheel a should have one tooth only; in the majority of cases, however, the wheel b is more or less star-shaped, on which ground it has been proposed *

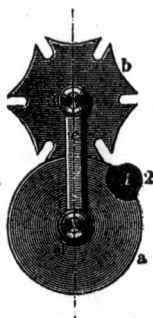

FIG. 414.

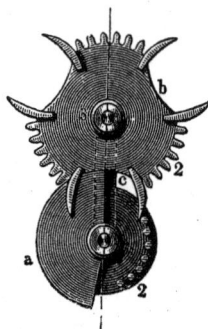

FIG. 415.

to call the wheels Star-wheels. It is evident that we have here a special form of the train $(C_2 \bar{C}_{n})$ or if it be preferred, of $(C_2 \bar{H}_{n})$, and it is desirable to indicate by a special formula its relation to the other wheel-chains. The special characteristic of the wheels is their segmental arrangement. We may therefore use here the symbol A, and will indicate chains of the kind before us by the formula $(C_2 A_x\cdot)$, or more generally $(C_2 \bar{A}_x\cdot)$.

If the radius of the wheel b in the click-train $(C_2''' A_x\cdot)$ be made infinite, b becomes a rod or bar, carrying upon it the curved recesses and the hollows for receiving the teeth. Its formula then becomes $(C^{\perp} P^{\perp} A_x''\cdot)$. The bolt-train in the Bramah lock, Fig. 416 furnishes us with an interesting example of this. The piece $A B C D$ belongs to the bolt b; the opening 2 in it is the hollow for the tooth 2 of a, $A B$ and $C D$ are adjacent curved recesses in the piece

* *Polytech. Zentralblatt*, 1864, A s t e r, *Sternräder.*

b corresponding to the circular recesses which are made in *b* in
Fig. 415. There is no longer any segment of *a* made to fit these
curved recesses (as in Figs. 414 and 415), the pin 2 sufficiently
answers the purpose of such a segment. Similar arrangements
are to be found in other locks, although not often in such a

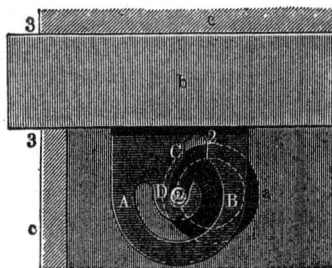

FIG. 416.

distinct form. In the Bramah lock the piece *a* is further connected with a fast click-train, the ingenious nature of which is
well known.

In cam and slider-cam- trains, and also in their special forms—
click-trains—pressure-organs are sometimes used. This always
occurs, however, in compound chains.

§ 158.

Pulley Chains.

The mono-kinetic properties of the tension-organs greatly increase
the difficulty of arranging them in simple chains along with rigid
elements (cf. § 41 *et seq.*). Such simple chains do exist, however,
in belt-chains or rope-chains, as Fig. 417, and also in chain-gearing,
where T takes the form T_x. The form taken by these chains,
apart from the special form of the tension-organ, is $(C_2 R_{,t2})$. Of
such higher forms as $(\bar{C}_2 \bar{H}_{,t2})$ or $(C_2 \bar{H}_{t2})$ but few applications
exist. One special and limiting form of the chain, however,
demands special mention. If we imagine a belt-train with crossed
belts (Fig. 418) to have its two pulleys *a* and *c* brought into
contact, and then the pulley *c* made infinite in radius, the pair

4 necessarily becomes a prism-pair, and the organ *T* is fixed at both ends to the prism into which *c* has been changed. Fig. 419 shows

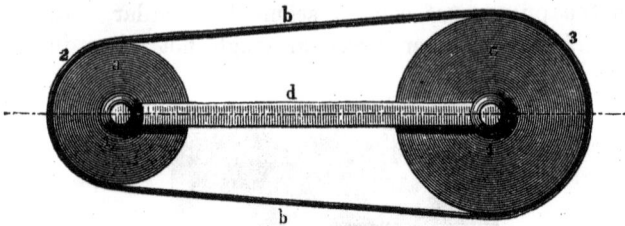

FIG. 417.

this chain, in which it is specially noteworthy that on account of the fixing of the organ *T* we have a chain of three links only instead of four. Its complete formula runs

$$C^+ \dots | \dots R, T^{\pm} \dots\dots P_{=}^{\pm} P^- \dots \perp \dots C_{=}^{-}.$$

In contracted form this would be $(C' R_{vt} P^{\perp})$. The chain is used here and there, placed both upon *c* and upon *d*.

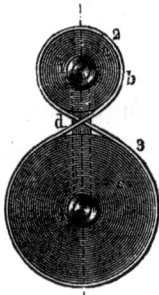

FIG. 418.

A common pulley-tackle is an unconstrained closed chain, or at least is constrained only by force-closure; here therefore we need not consider it. If it be made complete by pair-closure, it becomes (as we saw also in § 43, with the simplified roller arrangement) a compound chain. The cases thus obtained are very interesting

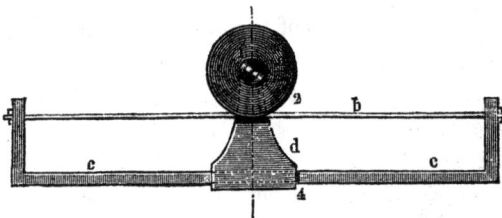

FIG. 419.

in themselves, but do not come into this part of our subject (cf. p. 568). This is true also of a number of other important applications of tension-organs.

§ 159.

Chains with Pressure-organs.

In the foregoing treatment of the simple chains we have repeat-edly had to consider the replacement of a rigid element by a pres-sure-organ, and in doing this have also examined the chains of which the pressure-organ became a part. The mono-kinetic properties of the pressure-organ have in all these cases shown themselves to be most important, and in the majority of cases force-closure has been necessary. If we wish to avoid this we come at once

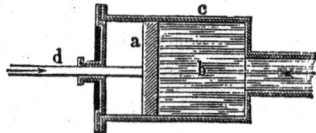

Fig. 420.

into compound chains. Complete simple constrained closed chains containing pressure-organs do not appear to be possible. Such a simple chain, for example, as that shown in Fig. 420, which would be written

$$\overbrace{P^+ \dots \mid \dots}^{a} \overbrace{V,^+ Q_\lambda \dots\dots\dots}^{b} \overbrace{Q_\lambda, V^-_. \dots \mid \dots}^{c} P^-_=$$

or in a contracted form $(PV,^+_q V,^-_q)$, is essentially force-closed. If the force-closure did not exist, for instance, the water would at once

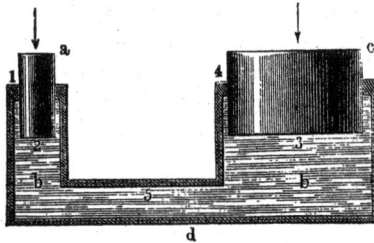

Fig. 421.

leave the chamber. The arrangement has none the less its own value, it is simply that of the common squirt.

If we arrange a second piston in the delivery pipe we obtain the arrangement of Fig. 421, already known to us. The chain has now four links instead of three; it is, however, a compound chain for two of its links are ternary; the water is paired with the three elements 2, 3 and 5, and the frame d with the three elements 1, 4 and 5. Its formula is

$$\overset{d}{\overbrace{\qquad}} \overset{a}{\overbrace{\qquad}} \overset{b}{\overbrace{\qquad}}$$
$$V- \;\;\underset{c}{\underbrace{\dots\dots \begin{cases} \dots (P) \dots\dots \;(V^{+}_{,q}) \dots \\ \dots (P) \dots\dots \;(V^{+}_{,q}) \dots \end{cases} \dots\dots}} \;\; Q_{\lambda}.$$

By itself, however, the chain is still not constrainedly closed,—force-closure is always a condition of its working.

One means of making such chains, as well as others containing flectional elements, pair-closed, consists in doubling them. We have found already that this method has been applied in the case of the belt-trains (cf. § 44), and have also seen the nature of the doubling of the chain now before us. Fig. 422 represents such a doubled chain. Its motions are now completely pair-closed; but at the same time, although it contains only five links, it has become a compound chain even more completely than before. If we imagine the chain of Fig. 420, also, to be made pair-closed by any means,—the addition for instance of a suitable lever-train, which could make the motions of the piston dependent upon those of another,—we

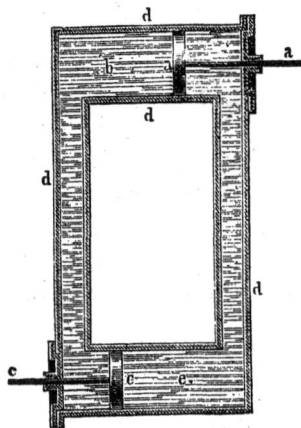

FIG. 422.

see at once that we are again brought into the region of compound chains. It may just be noticed that this method of doubling chains can be applied also to ordinary pulley-tackle in such a way as to transform it into a constrained closed chain.*

* I have illustrated this by a model in the kinematic collection of the König. Gewerb. Akademie in Berlin.—*R.*

§ 160.

Compound Chains.

Our synthetic investigation has now brought us, in a number of different directions, to the limits of the simple chains and to the ground covered by the compound ones. We see at once that the latter include so many important practical cases that it would be impossible to leave their synthetic treatment untouched. But there is no end to the possible combinations which can be made by joining one kinematic chain to another, and it is therefore very necessary to inquire if every problem arising in this way must necessarily fall within the region of kinematic synthesis, or if some distinction which may simplify our work does not exist between different classes of problems.

A distinction of this kind is, fortunately, furnished by the way in which the compounding has been carried out. A compounding may be a mere placing in sequence of known motions or chains, giving us therefore nothing new, or it may be so arranged as to give us some result in itself quite different from before. It is evident that these two methods of compounding may be treated in quite different ways. Let us first examine a few examples of them.

To take first a very simple case ; we obviously obtain nothing kinematically new by placing one belt-train or one wheel-train behind another. The relative velocities of rotation of the different parts may be altered, the nature of these rotations is, however, exactly the same as in the simple chain, and the advantage of the compounding in such cases is connected simply with the repeated use of one and the same form of train.

Fig. 423 is a sketch of the leading train of a beam-engine. The chain here used, which consists of the seven links a, b, c, d, b_1, a_1 and d_1, is clearly compound. It consists of a lever-crank $a, b, c, d = (C_4'')^d$ and a crossed swinging slider-crank $(C_3''P+)^c$. The latter is shown separately, in a form already known to us, in Fig. 424. The compounding has been carried out by combining the fixed links d and c_1 of the two chains into one frame, and the links c and b_1 into a ternary link, the "beam" of the engine. The angle of swing of the lever c and the coupler b_1 becomes therefore equal, and the stroke of the slide d_1 is made dependent upon the length of the crank a.

Each train, however, might have precisely the motion which it now has were it entirely separated from the other.

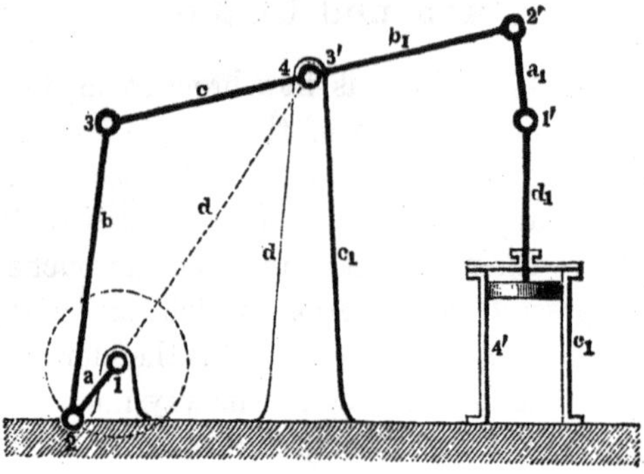

Fig. 423.

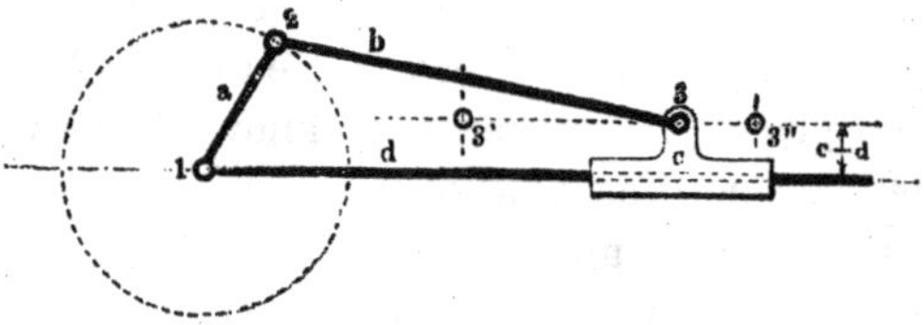

Fig. 424.

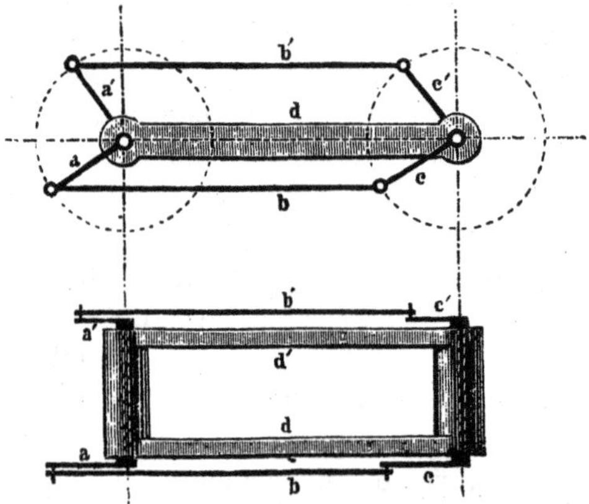

Fig. 425.

The case is quite different with the compound train shown in Fig. 425, which we have already studied. This mechanism is a pair of parallel crank-trains,—it consists therefore of two chains of the form ($C_2'' \parallel C_2''$) having their d links equal and common, and their links a and c combined into ternary links. We know that this chain 2 ($C_2'' \parallel C_2''$) has the property that both its parallel cranks can pass their dead points, which are also change-points (cf. §§ 46 and 66) without stoppage or change of motion. This property, however, is characteristic of the combination, neither chain by itself

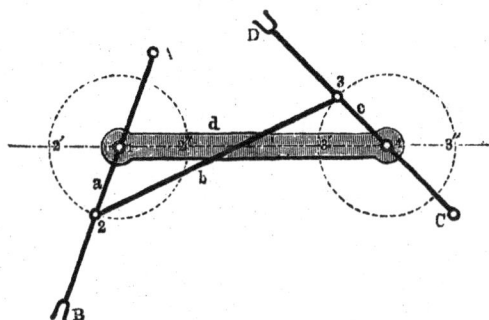

FIG. 426.

possesses it,—the compounding has therefore given us in this case something new.

The anti-parallel cranks, Fig. 428, give us another illustration of the same thing. The object of our examination of this train in §§ 47 and 67 did not lead us to notice that here also, although the number of links is not increased, we have a compound chain. It consists of our well-known four links a, b, c, d and a second chain having for its links, $A1\,B$, $C4\,D$, and d. The latter may be written in full

$$\overset{a}{C^+} \dots \parallel \dots (Z) \dots \overset{c}{\parallel} \dots (C) \dots \parallel \dots \overset{d}{C^-_=}.$$

Its frame d is identical with the link d of the chain (C_4''); the two links $C\dots\parallel\dots Z$ coincide with the links a and c,—both are therefore made ternary links. By themselves neither of the chains could move continuously, (C_4'') would be stopped at the dead points, ($C_2''Z''$) is unclosed in every other position.

These illustrations will suffice to show the difference between the two classes of compound chains. We shall call the class last considered combined chains,* and the former class mixed chains. The consideration of the combined chains forms an essential part of kinematic synthesis, while that of the mixed chains is not in every case necessary.

§ 161.

Examples of Combined Chains.

The compound chains having a larger number of links than the simple ones, the mechanisms formed from them have a proportionately greater number of applications than those from the former. Their investigation, therefore, to be in any degree complete, would far exceed the space here at our command. Our object here, too, is rather to point out the existence and nature of problems than to attempt any complete treatment of them. I must therefore limit myself to a few examples.

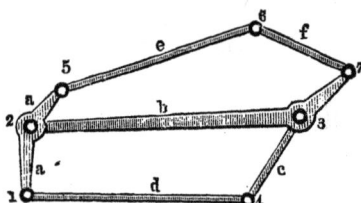

Fig. 427.

An immense number of compounds can be formed from chains of the class (C_4'') and its modifications. Among these compounds some of the mixed chains also give us something new if they be placed upon certain links. The chain shown in Fig. 427 is a combined chain. It consists of two chains of the form (C_4''); the first is $a\,b\,c\,d$, the second $a\,e\,f\,b$.

When it is remembered that the lengths of the different links can all be changed, and also that they can be increased to infinity, it will be recognised what an enormous number of special cases arise out of the general one shown in the figure. If, for example, we

---

* Prof. Reuleaux uses the expressions *ächt* and *unächt zusammengesetzt,* —real and apparent compound,—for what I have called combined and mixed chains. I think I am justified in using the latter much shorter terms, especially as a very closely analogous use of them is familiar in chemical terminology.

make the links e and f infinite, and further make the axes of the pairs in each of the ternary links (1, 2, 5 and 2, 3, 7,) conplane, we obtain the combination shown in Fig. 428. If we make the original chain (C_4''), a, b, c, d, a parallelogram (as here shown) we obtain a combined chain which has some remarkable properties, although they have not yet been utilised. The line 7·4′ parallel to a is always=2·1, and the length 1·4′ is constant, the lines 5·6 and 1·4 therefore always intersect in the same point O. If we place the chain on d we obtain a mechanism in which the bar e will

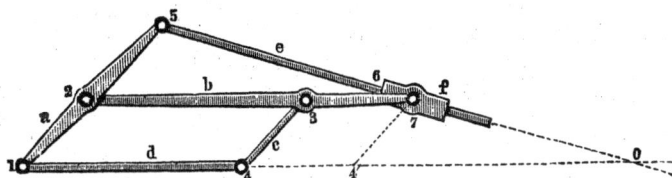

FIG. 428.

move so that its axis passes always through a fixed point beyond the mechanism, and which therefore may be itself inaccessible.

If we make e finite and therefore f and 3·7 infinitely long, we obtain the chain shown in Fig. 429, which is essentially different from the last.

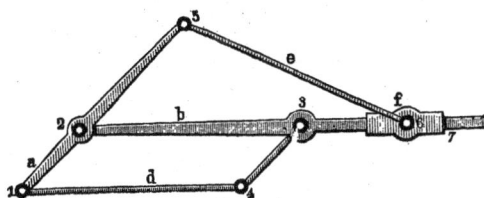

FIG. 429.

The combination of cylinder-pairs already described in § 60, which is again represented in Fig. 430, is a combined chain. The closure of the links 4·7 and 3·6 of the (C_4'') chain makes the otherwise incompletely closed five-linked chain 1. 2. 3. 4. 5 constrained. This chain finds several useful applications in "parallel motions," trains in which one or more points move in (accurately or approximately) straight paths. One of these, for instance, given by Tchebyscheff,[*] and another by Harvey,[†] are formed on this

---

[*] *Dingler's Journal*, 1862, vol. 163, p. 403.

[†] *Practical Mechanic's Journal*, 1850, vol ii., p. 174.

chain. Both are placed upon the link 6·7, and give a very near approximation to the required motion. Compound (C''_4) chains are also employed in numerous modifications as weighing machines.

The skew screw-chain $(C' S^L P^L_2)$, which we examined in § 154, has also come lately into use in the "dogs" used upon the face plates of lathes,* &c., in several forms.

Another example which is in place here is that of the reverted wheel-chain $(C_{z2} C''_3)$, which we examined in § 105. I must content myself here with merely mentioning this: we have already

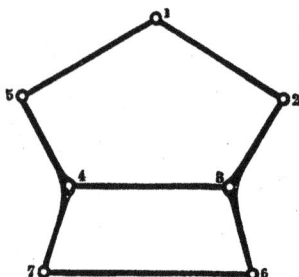

seen what an immense number of mechanisms are formed from the chain $(C_{z2} C''_3)$.

As a fifth and last example we shall take the chain shown in a general form in Fig. 431, which gives us some very notable mechanisms. It is a combined chain consisting of the simple spur-wheel chain $(C_z C''_2)$ with two links, each containing two parallel cylinder pairs, added to the two wheels. The chain has

FIG. 430.

therefore five links and six pairs, the latter being the cylinder pairs

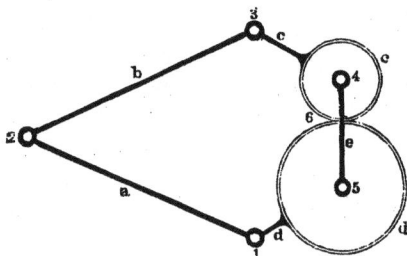

FIG. 431.

1, 2, 3, 4, 5 and the pair 6 of the form (C_z). We may write it shortly as $(C''_5 C_z)$ and in full :—

$$
C_z \dots \left\{ \begin{array}{ccccccc} 6 & d & 1 & a & 2 & b & 3 & c & 6 \\ \cdot \parallel \dots (C) \dots \parallel \dots (C) \dots \parallel \dots (C) \dots \parallel \cdot \\ \dots\dots \mid \dots (C) \dots \parallel \dots (C) \dots \mid \dots\dots \\ d & 5 & e & 4 & c \end{array} \right\} \dots C_z.
$$

---

* See for example Danbury's drill-chuck, *Scientific American*, vol. xix. (1873), p. 215.

For distinctness' sake I have added the names of the links in the formula, and also the numbers of the pairs. The formula makes the symmetrical construction of the chain very distinct. The lengths of the links can be altered within the widest limits ; so

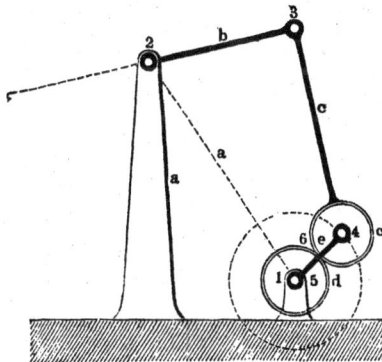

Fig. 432.

long as the closure be not made either fixed or unconstrained, any link may be fixed and any other made at the same time as the driving link, and in this way we can obtain from the chain most

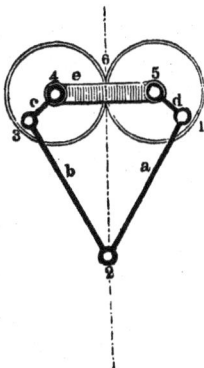

Fig. 433.

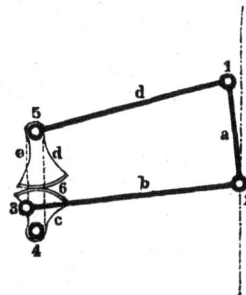

Fig. 434.

various mechanisms. We shall examine briefly a few important cases. For simplicity's sake I have omitted the teeth of the wheels and shown only their pitch circles in the figures.

(1.) Let the length 1·5 be made =0,—the pairs 1 and 5 then

becoming conaxial. If the chain be then placed on a we obtain
Watt's planet-train, Fig. 432, the motions in which we have
already examined (§ 105). Following the name which Watt gave
to the mechanism $(C_s''C_z)^a$* we may call the chain itself $(C_s''C_z)$ the
planet-wheel chain.*

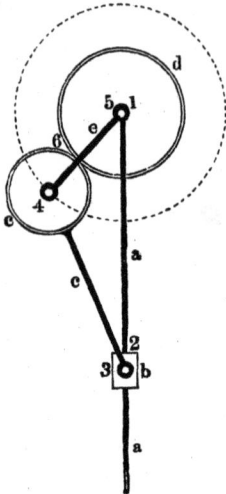

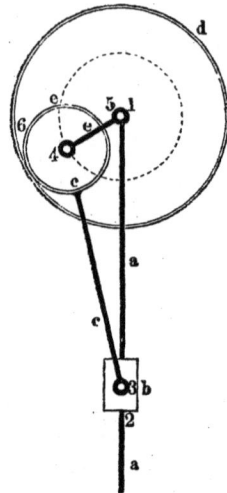

FIG. 435.   FIG. 436.

(2.) If we make $1\cdot5 = 3\cdot4$, $a=b$, and the two wheels also equal, so
that the links c and d are equal and similarly placed, the whole
chain becomes symmetrical about the line $6\cdot2$, Fig. 433, and placed
on e it gives us Cartwright's parallel motion.

(3.) We can make the lengths $5\cdot1$, $1\cdot2$, $2\cdot3$ and $3\cdot4$ un-symmetri-
cal, but by suitably proportioning them, and giving the wheels a
particular diametral ratio, we obtain, by placing the chain on e, a
mechanism in which 2 moves approximately in a straight line,
Fig. 434. This arrangement is that proposed by Maudslay. The
path of 2 is very nearly straight if the link d be not allowed to
swing through too large an angle.

(4.) We obtain important special cases by making single links
infinite. Let us do this first with b and a, using at the same time
the simplification employed in Watt's planet-train, namely, making
the length $1\cdot5 = 0$. We obtain in this way such a chain as is

---

* The planet-wheel train used by Galloway was more complex than the chain
before us, and so does not come into consideration here.

shown in Fig. 435, of which the contracted formula, beginning with 1, is $(C^{\perp}P^{\perp}C''_3C'_z)$. If it be placed upon a it gives a planet-wheel train with a straight slider, a combination which has found numerous applications.

We obtain a special form of this by making d an annular wheel as in Fig. 436. In this form the chain, without recognition of its nature, has recently found several applications. Among others it has been used in an arrangement of steering-gear by Caird and Robertson.* They place the chain on a and use e as the driving-link, formula $(C^{\perp}P^{\perp}C''_3\,C_z^-)^a_o$. The diametral ratio of the wheels is made very nearly equal to unity, so that the rudder moving

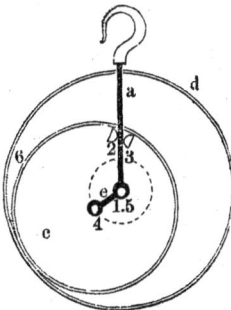

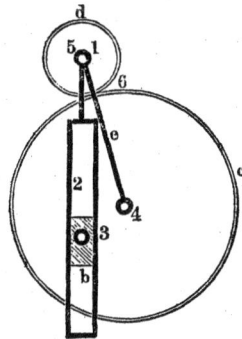

FIG. 437.   FIG. 438.

slowly is well under control. The rudder shaft is conaxial with d. With the wheel as the driving-link this mechanism is sometimes used in sewing-machines.

Eade's pulley-block,† schematically represented in Fig. 437, is another application of the same mechanism. It is again placed on a and driven by e. The link $b = C...\perp...P$ is omitted, and the higher pairing described in § 76, Figs. 269 and 270, is employed in its place. The formula of the train is therefore $(C^{\perp}P^{\perp}C''_3C_z^-)^a_o - b$.

The same mechanism, with the same reduction, has been used by Wilcox‡ and also by Taylor§ in counters or numbering machines.

---

* *Génie Industriel*, 1869, vol. xxxvii., p. 29. Caird and Robertson have applied the same mechanism also in capstans.

† *The Engineer*, 1867, p. 135.   ‡ *Engineering*, January, 1869, p. 38.

§ *Engineering*, July, 1869, p. 1.

(5.) By making the length 3·4 less instead of greater than 4·5 we obtain in the chain a motion differing very greatly from any occurring in either of the former cases. Fig 438 shows this arrangement. While in Figs. 436 and 437 the whole motion of *b* relatively to *a* was equal to twice the distance 4·5, that is twice the length *e*, it is now equal to twice the distance 4·3. I have formerly suggested this mechanism as a leading train for punching, riveting or stamping-machines, and given it the name of **toothed-eccentric** * (cf. also p. 300).

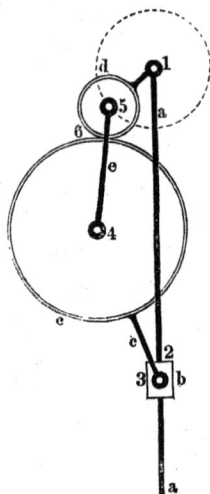 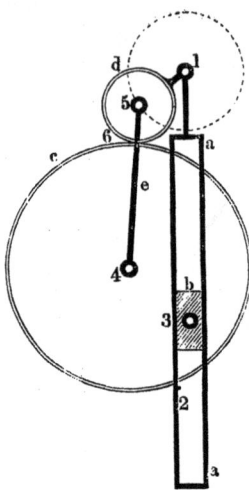 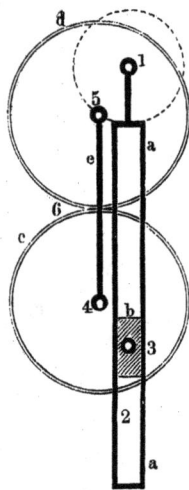

Fig. 439.	Fig. 440.	Fig. 441.

(6.) Leaving still the links *a* and *b* infinite, but giving to 1·5 some finite value, we obtain a chain represented in a general form in Fig 439. If we here make 3·4 <*e*, we have the chain represented in Fig. 440, which I formerly called the general case of the toothed-eccentric. Placing this chain on *a* we obtain a mechanism which may serve to give to a link (*b*) reciprocations of varying stroke. We obtain an interesting case if we make the toothed-wheels equal and the lengths 1·5 and 3·4 also equal, and at the same time place the latter symmetrically to *a*, as in Fig. 441. I have

---

* See *Civil Ingénieur*, 1858, p. 4; "Das Zahnexzentrik, ein neuer Bewegungs-mechanismus." In this article I examined the whole series of these mechanisms. I had not then recognised their connection, above explained, with the planet-wheel trains. —*R.*

called this mechanism the **symmetric toothed-eccentric**. The centroids of c and a and also those of d and b are Cardanic circles.

(7.) By making b and c infinite instead of b and a we obtain an altogether different mechanism, as is shown in Fig. 442. Its formula is $(C''C^{\perp}P^{\perp}C_2''C_2)$. Placed on c and driven by e we obtain a somewhat complicated reciprocation of b. Among other applications of the train is one by Whitehill for the motion of the needle in a sewing-machine; he makes the two wheels equal.

If we make c an annular wheel we obtain the chain shown in Fig. 443. If here the diameter of d be made half that of c, and

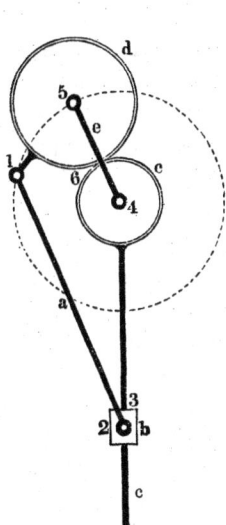
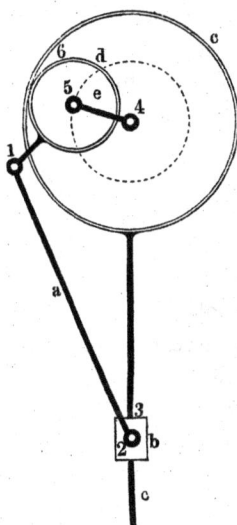
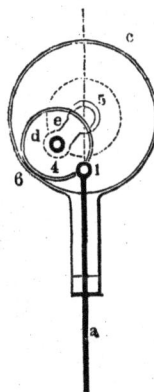

FIG. 442.   FIG. 443.   FIG. 444.

1·5 be made equal to 5·4, the chain takes the form shown in Fig. 444. The point 1, upon the circumference of a smaller Cardanic circle, moves along a diameter of c. Placing the chain on c, therefore, we obtain the well-known hypocycloidal "parallel-motion." This is a very old mechanism, called both after Lahire and after White, and is often used in printing-presses. There is no longer any motion in the turning pair 2, so that the link b may be altogether omitted. If the same chain be placed upon a instead of c we obtain again a parallel-motion, this time for the link c. So far as I know this mechanism is new.[*]

[*] There is a model of it in the Berlin kinematic collection.

I can here only mention further, that if the prism pairs in the chain $(C_5'' C_z)$ be made crossed instead of normal to the axes of the cylinder pairs, a number of special cases occur : these must be here left altogether unexamined.

§ 162.

Closing Remarks.

The sketch of the synthesis of machines which we have now ended has given us several results differing greatly from those which have hitherto been deduced from a general and apparently scientific treatment of the subject. The most important discovery which we have made is undoubtedly that the region within which kinematic combinations are formed is much more narrowly limited than has usually been supposed. This is apart, I think, from the inexactness of the treatment with which so many former writers have been satisfied, for even the more accurate ideas as to combinations of elements with which we commenced our study of the problem did not in themselves indicate that the synthesis could be successfully used over so large a field as that in which we have found it available.

It is very noteworthy also, in regard especially both to practice and to instruction, that all the principal problems of machinery are connected with a comparatively very small number of kinematic chains. These are : —

> the screw-chain,
> the wheel-chain,
> the crank-chain,
> the cam-chain,
> the ratchet-chain,
> the pulley-chain,—

in all of which flectional elements may take the place of rigid ones. The problems not covered by these chains are all more or less inferior in importance.

In § 92 I directed attention to the extraordinary unanimity with which the inventors of " rotary " engines and pumps have chosen crank-trains as the foundation for their chamber-gear. This now explains itself. Among all the kinematic chains just mentioned as those most generally and easily applied, the crank-chain is that which contains the pairs of elements,—the cylinder and the prism-pair, most suitable for chambering and for the making of fluid-tight joints. Invention has thus, unconsciously, fallen generally upon this chain.

We have seen at the same time how extremely important it is that the synthetic treatment should be carried out to the fullest extent possible, for it is full of promise of new and valuable results. The question is, what form this treatment should take; for what we have here been able to accomplish in this direction has brought us only to the outer limits of the subject. It might appear at first sight that the best plan would be to make " Synthetic Kinematics " a special subject of study and instruction, treating it in separate books, and working completely through it, pair by pair and chain by chain. I do not think, however, that this is the best method. It appears to me far more advisable that under "Applied Kinematics" we should treat mechanisms, which might then be arranged according to their practical applications, both analytically and synthetically. Synthesis should be here simply one of the aids in the investigation, not its governing idea ; it must be used with and beside other methods, the whole being combined for the most advantageous treatment of each particular branch of the subject.

Another remark, however, must be made here. After the satisfactory consciousness which our investigation has given us that we are not working in a field of which we can never see the boundaries, there may arise a doubt whether the material now placed at our command may not too soon be exhausted, whether our scientific treatment of it may not speedily work the mine altogether out. The doubt is made all the stronger by the stress which we have laid upon the simplifications of the matter to which we have been able to make our way. It is not one, however, about which we need to trouble ourselves.

We have carried the synthesis far enough to allow us to look round, forwards and backwards, and to compare the ground which has been explored with that which still lies untouched before us. And in the latter we can see an immense, indeed, an inexhaustible series of problems awaiting the earnest investigator. The short sketch which we have given of the planet-wheel chains gives some indication of one of the thousand points in which the region of the compound chains awaits investigation. And here, after all, we considered only the abstract mechanism as formed from rigid elements. If we substitute for some of these flectional elements and use for all of them the materials actually employed in construction, each with its special natural characteristics, we find a multitude of new demands upon us which must be met before the abstract scheme is suitable for working under its altered conditions. Before these, that is before the never-ending demands of practical work, the doubter may well make himself once more happy in the knowledge of the essential simplicity of the means with which we have to work. We are encouraged by the conviction that the many things which have to be done can be done with but few means, and that the principles underlying them all lie clearly before us.

And now, finally, I have reached a matter upon which I touched long ago in the Introduction, and with which this whole chapter has been, without directly mentioning it, indirectly connected. This matter is the invention of mechanisms. What I meant in saying that the process of invention might become a scientific one, and might especially be performed synthetically, has now been made clear, and the truth of my assertion has, I believe, been proved. The kinematic synthesis, however, makes the finding of mechanisms easier only to those who have scientifically grasped their subject, while at the same time it places the goal which they attempt to reach ever higher and higher. It does not decrease, but rather raises, the intellectual work of the inventor, while it enables him to see more clearly, not only the object he wishes to attain, but also the means at his disposal for attaining it, and the best method of employing those means.

NOTES.

[1] (P. 5.) This letter is dated June 30, 1784, and others, written before the specification was drawn up, extend as far as the 22nd July of the same year. The specification is, however, dated April 28th, 1784, an inconsistency in Muirhead which seems to require explanation.

[2] (P. 9.) Through 113 chapters he is compelled to employ long descriptions for what we express by the word pump. For example : Chap. 1:— "Cette cy est une sorte de machine, par laquelle facilement et sans point de bruit l'on peut faire monter l'eau d'une fontaine ou d'un fleuve à une proportionnée haulteur . . . . Chap. xvii. : Ceste autre façon de machine, par laquelle l'on faict pareillement monter l'eau d'un lieu des en hault . . . . Chap. lvii. : L'effect de ceste autre façon de machine est de faire monter l'eau d'un canal à une juste haulteur," etc.

[3] (P. 11.) *Calcul de l'Effet des Machines.*

[4] (P. 11.) *Introduction à la Mécanique Industrielle.*—Compare also § 129, Chap xii. of this work.

[5] (P. 12.) From a philological point of view, "Kinematics" is incorrect,— Ampère should rather have said "Kinetics," (*Cinétique*). [It is of course impossible to make any alteration in this direction now in this country, where the word Kinetics has already obtained extended use in another sense, and there are good reasons for not making any change on the Continent, although the word Kinetics has not yet come into general use there.] It is in every way better to use the word with a K, as in the language from which it has been derived, than with the C which it owes to its transmission through the Latin and French.

[6] (P. 21.) Parerga ii., Chap. iii., § 41, — also Wille und Vorst. ii., Chap. xiv.

[7] (P. 35.) It is very remarkable, and has long ago given occasion for reflection, that we so seldom find definitions of the machine which agree with each other. The following examples show how uncertain, and often how altogether indefinite, have been the attempts made to define the machine even by those who must have known the thing itself.

Weisbach. "Machines are all those artificial arrangements, by means of which forces are made to act in a way differing from that in which they would otherwise have acted." Any tool whatever ; a needle, a pencil, &c., is therefore by itself a machine.

Poncelet. "The industrial . . . . machines have for their purpose the performing of certain work by the help of motors or moving forces provided for us by nature." An explanation full of restrictions, which gives us only one of the purposes of the machine.

Bresson. "A machine is a tool of which the general purpose is the transference of a force from its point of application to a position where it can act so as to overcome a resistance, and execute work which it would be difficult, and sometimes impossible for the same force to execute if applied directly." What then is a "tool?" And how does "sometimes" find its way into a scientific definition?

Rühlmann. *Geostatik,* 3rd Edit., 1860. "By the name machine, we indicate a combination of rigid bodies, moveable and immoveable, into a rigid unalterable 'free' (*lose*) system, by means of which forces, through changes in their direction and magnitude, may be made to balance each other." He has explained in an earlier part of the work what a "free" system is. According to this definition, a suspended iron chain would be a machine ; an hydraulic press, however, could not receive that name, for the water is not a rigid body.

Rühlmann. *Geostatik,* 2nd Edit., 1845. Almost exactly as Weisbach.

Rühlmann. *Allgemeine Maschinenlehre,* i. (1862). "The machine is a combination of moveable and immoveable rigid bodies which serves to receive physical forces and to transmit them, changing if required their direction and magnitude, in a manner suitable for the performance of definite mechanical work." Here are three definitions coming to us from the same pen : to which shall we trust?

Kayser. "Machines are arrangements which transmit the action of forces in order to balance or to overcome other forces, and to produce motions for definite purposes." This covers, *e.g.,* the tow-line of a ship.

Schrader. "A machine is an arrangement for the alteration of a given force." Very concise, but somewhat difficult to understand. What is it to "alter a given force"?

Wernicke. "A machine is a combination of bodies, of which the purpose is the accomplishment of any work by some disposable force." The first few words sound like a definition, the conclusion, however, becomes altogether indefinite.

Poppe. "By machines, we mean those artificial arrangements by which motions may with advantage be produced, prevented, or transmitted in definite directions." What has "advantage" to do with science? Motions also cannot be produced by "arrangements ; " and so on.

Delaunay. (*Analytische Mechanik,* 1868.) "A machine is an apparatus which serves to transmit mechanical energy, or also, to make a force to act at a point which does not lie in its own direction." Again, only characteristics, no explanation, no rigid definition ; and that fatal "or also ! "

Willis. "An instrument, by means of which we may produce any relations of motion between two pieces." We might call this definition an equation with two unknowns.

[I think Prof. Reuleaux is mistaken in giving this as Willis's d e f i n i t i o n of a machine. Willis merely says (Preface, 1st Ed., p. xiii.)—" . . . . instead of considering a machine to be an instrument by means of which we may change the direction and velocity of a given motion, I have treated it as an instrument by means of which . . . .," etc. He nowhere gives a formal definition of a machine, but in one place (Preface, p. iv.) describes it in a sentence which more nearly agrees with Reuleaux's definition than any other I have seen, although it certainly is still incomplete :—"Every machine will be found to consist of a train of pieces connected together in various ways, so that if one be made to move, they all receive a motion, the relation of which to that of the first is governed by the nature of the connection."]

G i u l i o. "An arrangement which is designed to receive motion from the action of a motor, to alter this motion, and to transmit it to an instrument formed so as to execute any kind of work." This gives certain characteristics of the machine, not, however, what it actually is.

L a b o u l a y e. "We give the name machine to a system of bodies which is intended to transmit the work of forces, and consequently both to alter the intensity of the forces, and to make the velocity and direction of the motion produced such as is suitable for the purpose in view." What is a "system of bodies ? " is it sufficient that it should be "intended" for all this ? and so on.

B e l a n g e r. "A machine is a body (o n e body ?) or a number of bodies intended to receive at one point certain forces, and at other points to exert certain forces, the latter being in general different from the former, both in intensity and direction, and in the velocity of the point at which they are exerted." Again the "intention." The whole also is a description, not a definition.

H a t o n. "A machine is an apparatus which is intended to connect a motor with the material to be worked upon." Apparatus, motor, material to be worked on, connection ? From a logical point of view, how many riddles ! The reader may fairly exclaim *Davus sum, non Oedipus !*

Lastly, P i e r e r ' s *Universal Lexikon* (? Hulsse). "Machine—an arrangement by which a motion, *i.e.,* a change of place or of form, may be given to a body, by which, that is, in general, some work may be done or mechanical effect obtained." This is only descriptive, and suits a multitude of things which are not machines.

[The definition I have quoted from Willis is certainly more accurate than any of those which have followed it, it is indeed so obviously better than most of them that it is matter for wonder that it has not been generally adopted. In order simply to shew to what extent indefinite definitions have passed current here also, and not from any desire to criticise, I may add the following to Prof. Reuleaux's catalogue :—

Hart (1844). "A machine is defined to be an instrument, by means of which a given force is caused to make equilibrium with a resistance to which it is either unequal, or not directly opposed."

Goodwin (1851). " Any contrivance by which force is transmitted from one point to another, or by means of which force is modified with respect to direction or intensity, is called a machine." Goodwin expressly includes as machines an oar, a poker, etc.

Galbraith and Haughton. " A machine is an instrument, by means of which pressure or motion may be transmitted from one point to another, and altered both in magnitude and direction."

Goodeve (1860). "A machine may be defined to be an assemblage of moving parts, constructed for the purpose of transmitting motion or force, and of modifying, in various ways, the motion or force so transmitted."

Rankine. "Machines are bodies, or assemblages of bodies, which transmit and modify motion and force. The word 'machine,' in its widest sense, may be applied to every material substance and system, and to the material universe itself; but it is usually restricted to works of human art, and in that restricted sense it is used in this treatise."

Todhunter. "Machines are instruments used for communicating motion to bodies, for changing the motion of bodies, or for preventing the motion of bodies."

Magnus (1875). "A machine is an instrument by means of which a force applied at one point is able to exert, at some other point, a force differing in direction and intensity."

I conclude that the authors, some of them very ⸲⸲⸲inguished men, whom I have cited, can hardly have been satisfied with ⸲⸲⸲ ⸲wn definitions, which, indeed, ought not strictly speaking to receive that name at all, so far as the machine is concerned. Almost without exception the machine is an "instrument," a "contrivance," an "assemblage of bodies," by means of which something is done or can be done. This something is the thing defined in each case, the machine itself is quite left out. It is instructive to note also how completely the idea of the "fixed link," which will be found presently to be of absolutely vital importance, is absent from most of them. Prof. Rankine recognised it,[*] (although it is excluded by his definition, if "a body" can be a machine), but it is expressly or tacitly excluded by most of the other writers.]

The reader must not be surprised that I have placed here beside each other names of such very various degrees of importance, nor that I have omitted others which are so well known, e.g., Moseley, Redtenbacher, Jolly, Karmarsch, Holzmann, and among the older authors Langsdorf, Eytelwein and others. These authors give no definition of the machine. They consistently avoid it, going at once into classification and description. I have given so many examples in order to show the more clearly that no authoritative definition has ever yet been arrived at.

The older definitions, much more naïve than the modern attempts to grasp a multitude of phenomena, are by no means uninteresting. Leupold, for example, says (*Theatr. Mach.*, 1724) :—"A machine or engine[†] is an artificial work, by means of which some advantageous motion can be obtained, and something moved with a saving, either in time or in force, which would not be otherwise possible." [The word Rüstzeug, which occurs in this old definition, and which I have represented (I cannot say translated) by engine, continued in use in Germany until well on in the present century. Prof. Reuleaux traces it back to an adaptation by Zeising, 1607, of an old definition of Vitruvius.]

---

* See p. 24.
† "*Rüstzeug*," a word covering machinal arrangements of all kinds, and used very much as the word *engine* was used among us in Leupold's time ;—*e.g.* "Let all the dreadful engines of war," &c.

All the definitions which I have given have a common property ; they are entirely or chiefly d e s c r i p t i v e, they do not go down to the essentials of the matter. In criticising them in this way I do not wish to be misunderstood ; the question is not one of mere criticism, but of the relative import- ance of the fundamental propositions upon which a scientific study is built up, for the whole matter begins with the definition. It may be objected to this that the definition can in this case only be determined after the object defined has been thoroughly investigated,—its position at the beginning is therefore artificial. The statement is perfectly true : it is, and must be, true however for every definition. The beginning m u s t presuppose the end. A scientific study is not a chronological account of its own investigations. It is not necessary, however, that the learner should know at the very com- mencement the full meaning of the definitions. The explanation and develop- ment of the subject, as they are carried on, refer him back always to these first propositions, which he finds, as he proceeds, to be a mirror in which he can see a reflection of each later one,—until at the end of his work he discovers the full and complete justification of the definitions with which he started. It has seemed to grow more and more full of meaning as he advanced, until finally it has shown itself to be a true representation of the whole matter in a form of the greatest possible conciseness.

An incomplete or merely descriptive definition forming the key-stone of any branch of science reflects the condition in which that study must be. We shall have an opportunity of showing in the course of this work how the machine has gradually shaken itself free from the problems of Pure Mechanics. The student only of this science, in whose work the machine is merely an accident, does not feel the drawbacks of its incorrect definition. Nor do these present themselves even to the engineering student, so long as his science has not yet been put in the form of rigidly logical propositions. I cannot do better than quote here an excellent sentence from Mill (*Logic* I., bk. i., chap. viii., § 4) bearing upon the subject. He says :—" What is true of the definition of any term of Science, is of course true of the definition of a science itself ; and accordingly the definition of a science must necessarily be progressive and provisional. Any extension of knowledge or alteration in the current opinions respecting the subject matter, may lead to a change more or less extensive in the particulars included in the science ; and its composition being thus altered, it may easily happen that a different set of characteristics will be found better adapted as differentiæ for defining its name."

[8] (P. 47.)   The immense importance of this proposition will appear greater to the reader presently than it can do at present. Its principle seems to have been hitherto entirely unrecognised. I have found but a single trace of it. This is in Chasles (*Aperçu historique sur l'Origine* . . . *des Méthodes en Géométrie*,1837, note xxxiv., p. 408, *et seq.*, and later on), where he speaks of the elliptic chuck of Leonardo da Vinci. The consequence in that case of fixing another link of the kinematic chain Chasles has taken for an indication of the great law of d u a l i t y, upon which he enlarges at great length. His reason- ing, however, is not well grounded, and goes seriously wrong ; we have here not d u a l i t y, but a most characteristic p l u r a l i t y, which contains naturally

all the consequences which Chasles obtained artificially from the duality, and an immense number of others.

[I have used the word t r a i n as synonymous with m e c h a n i s m through this work. I regret that it was omitted, by an oversight, in line 8 of p. 47. I would venture to suggest, in connection with this subject, that it would be really more accurate, and in every way better, if the three-bar, five-bar, &c., linkworks and cells (about which so much that is interesting has recently been written and done by Peaucellier, Sylvester, Hart, Kempe, and others) should rather be called four-bar, six-bar, &c., linkworks respectively. The numbers hitherto attached to them are those of the moving links only of the kinematic chain. I need hardly point out that the plane in which the two fixed points are supposed to be, and relatively to which some other point describes some particular curve, is in every sense as much a " bar " as any of the other links. In view especially of this most interesting subject finding its way from the hands of the mathematicians to those of the students, it seems a pity that this most important fact should not be recognised in the names given to them. Prof. Sylvester distinguishes between the chain and the mechanism or train in the special cases mentioned by calling them linkage and linkwork respectively.]

⁹ (P. 72.) It seems suitable to call here particular attention to the fact that the two centroids are absolutely reciprocal, that is, that neither of them possesses, as a centroid, any property which the other has not. This may a p p e a r to be the case when (as in Fig. 21) one of the two curves is fixed. We see, however, from the above problem, that this difference of appearance may always be removed, and both the centroids made to move, if another link of the chain be fixed. The conditions of both pairs of centroids are identical, the fixing of one curve is merely accidental. The difference made by Poinsot between Poloid and Serpoloid is precisely the difference due to one of the curves being fixed,—it cannot, however, at least in the study of machines, be justified. It must be given up in other investigations also, I think, for there is no real difference between the two curves ; the particular distinction made is indeed more apt to confuse than to explain, for it seems to point to the existence of only t w o centroids in a moveable system, while there is no such limit to their number ; we have already noticed one case in which there were six pairs in one mechanism. Mechanisms such as those of Fig. 135 also—the spur-wheel train—give further illustrations of the undesirability of making this distinction. There the centroids of $a : c$ become a point, as do also those of $b : c$; they are therefore no longer visible as curves. The centroids of $d : b$, on the other hand, the only ones remaining as curves in the chain, b o t h m o v e if the latter be placed on c, and are therefore absolutely indistinguishable. The difference in name cannot therefore be logically justified, and in a Science, especially a young one, whatever cannot be logically justified should be carefully kept at a distance, and by no means taken up on mere grounds of convenience. I should not have spoken of the matter if it had not been for Prof. Aronhold's proposal to call the stationary and moving centroids by different names (*Polbahn* and *Polkurve* respectively), which has been too hastily accepted by the younger forces. [I believe Aronhold has now given up the use of the word Polkurve.] My experience has shown me a

NOTES.   591

hundred times that the students,—in spite of the reiterations of their teachers that no logical difference between the things exist,—do make such a difference. I hope it may still not be too late to return to a correct nomenclature. If at any time it be necessary to distinguish between the two curves, it will be both correct and sufficient to do so by calling one the stationary and the other the moving centroid.

¹⁰ (P. 75.) The base circles used in drawing the profiles of involute teeth are secondary centroids of this kind. The third centroid is the straight line, rolling on these centroids, of which a point generates the profiles.

¹¹ (P. 80.) [The theorem is contained in Poinsot's celebrated memoir *Théorie nouvelle de la Rotation des Corps*, presented to the French Institute in 1834, and afterwards published in an extended form in *Liouville's Journal*, vol. xvi. pp. 9-129 and 289-336 (March, 1851). A translation (by Whitley) of the original paper was published at Cambridge in 1834. The theorem referred to is stated as follows : " The most general motion of which a body is capable is . . . . that of a certain external screw which turns in the corresponding internal screw." The same paper contains also the first enunciation and proof of the theorem : " the rotatory motion of a body about an axis which incessantly varies its position round a fixed point is identical with the motion of a certain cone whose vertex coincides with this point, and which rolls, without sliding, on the surface of a fixed cone having the same vertex." The simple treatment of such a problem which we now adopt was not possible forty years ago ; Poinsot's reasoning included the ideas of force and velocity, instead of merely the notion of change of position.]

¹² (P. 80.) [*Traité de Cinématique*, 1864. Belanger here re-states Poinsot's theorems of motions in a plane and about a point (pp. 55 and 58), and gives the following theorems respecting general motion in space (p. 59) :—" Tout mouvement continu d'un système invariable équivaut au roulement d'un cône lié au système, sur un autre cône qui aurait un mouvement de translation dans l'espace, le premier cône étant le lieu géométrique dans le corps en mouvement, des axes instantanés de sa rotation autour du point choisi ; le second cône étant le lieu des mêmes axes instantanés dans le système de comparaison en translation " . . . . (p. 79) " Il faut ajouter ici que ce même mouvement équivaut à celui d'une surface réglée qui, étant liée au système, toucherait continuellement, suivant une génératrice, une autre surface réglée fixe dans l'espace, sur laquelle elle roulerait en glissant à chaque instant le long de la génératrice de contact des deux surfaces. En effet, les positions successives dans l'espace, de l'axe instantané de rotation et de glissement, forment une surface réglée, c'est le surface immobile ; et les positions de ce même axe instantané, relativement au système ou corps mobile, forment une autre surface réglée mobile, emportant le corps avec elle."

Prof. Ball points out (*Theory of Screws*, p. xix., etc.) that the discovery of the theorem given in § 12, the "canonical form " of the displacement of a rigid body, is due to Chasles. In view of Ball's most interesting investigations, as well as of the more elementary treatment of the subject now adopted in other books, I regret that Prof. Reuleaux adheres to the view taken on page 83, where twisting is expressed as a rolling about two axes, one at an infinite distance. This conception seems rather to add difficulty to, than to simplify, that of the twist pure and simple, which for every reason it appears to me better to treat as the ultimate and general case.]

[13] (pp. 64, 119 and 121). [It must, I think, be admitted that it is very desirable to have a somewhat clearer understanding as to the use of these names than has hitherto existed. I shall be very glad if the following table assist in any way in promoting this understanding. It shows the nomenclature I have myself used throughout the book, which differs somewhat from Professor Reuleaux's, but which agrees with that adopted by our best writers on the subject, so far as I have been able to make out a system from their references to these curves. The last column requires a word of explanation, the rest of the table, I think, explains itself. The case referred to there will be understood by a reference to Fig. 97. Two circles are there in internal contact, of which the larger rolls and the smaller is stationary. It may seem at first sight unnecessary to separate this case, for what I have called the pericycloid can always be described as an epicycloid. The necessity for having a separate name arises, however, from the fact that the curves described by points without and within the large rolling circle, that is the curtate and prolate peritrochoids, cannot be described as curtate or prolate epi-trochoids. The cardioid can be generated as a roulette as a special case both of the epi- and the pericycloid. I have retained "curtate" and "prolate" for want of better words, but they are sometimes singularly inappropriate to the external form of the curve, in Fig. 96 for example, where the larger ellipse is the curtate curve. Professor Cayley's kru-nodal and ac-nodal hardly seem adapted for popular use, and an excellent suggestion of Professor Clifford's, "looped" and "wavy," fails also in (external) suitability in the case of Fig. 96.

I use trochoid as the general name for the whole class of curves.]

DESCRIBING POINT.	TROCHOID.			
	EXTERNAL CONTACT.		INTERNAL CONTACT.	
	CIRCLE ROLLING UPON STRAIGHT LINE (LINEAR TROCHOIDS.)	CIRCLE ROLLING UPON CIRCLE. (EPI-TROCHOIDS.)	SMALLER CIRCLE, ROLLING. (HYPO-TROCHOIDS.)	GREATER CIRCLE, ROLLING (PERI-TROCHOIDS.)
On circle	Cycloid.	Epicycloid.	Hypocycloid.	Pericycloid.
Beyond circle	Curtate trochoid	Curtate epitrochoid.	Curtate hypotrochoid.	Curtate peritrochoid.
Within circle	Prolate trochoid.	Prolate epitrochoid.	Prolate hypotrochoid.	Prolate peritrochoid.

[14] (P. 119.) Chasles (*Aperçu historique sur l'Origine* . . . . . . *des Méthodes en Géométrie*, 1837), cites Cardano's *Opus Novum de Proportionibus Numerorum Motuum*, etc.

[15] (P. 121.*) Without the use of a model it is very difficult for anyone unaccustomed to this class of problems to realize fully the nature of the motion which occurs here. In my model the curve-triangle, UTQ, is etched upon a glass disc, the duangle, PVQW, is engraved upon the plate RPSQ. [Professor

* The reference is to the top line of p. 121, the 5 of the reference number has unfortunately been omitted in printing.

Reuleaux's beautifully made models are now (May 1876) at the Exhibition of Scientific Apparatus at South Kensington, where they will remain during the summer. By drawing the triangle ABC with its centroid upon paper, and the duangle with its centroid upon tracing paper, the rolling of the centroids and the relative motion of the elements can both be tolerably well followed. By this method I have found it very easy not only to examine the motion, but also to draw series of point-paths for these higher pairs of elements.]

[16] (P. 125.) For drawing these and similar roulettes I use a special three-legged compass, made for me by Herr J. Kern of Aarau (Switzerland). The third leg is jointed and its length also can be altered, so that obtuse as well as acute-angled triangles can be taken up, which could not be done with the old form of three-legged compass.

[17] (P. 125.) What a strong hold this idea has obtained is shown, for example, in the following passage from Weissenborn's *Cyclischen Kurven,* (Eisenach 1856) p. 3 :—" If the circle described about m_0 roll upon that described about M, and if the describing point B^0 describe the curve $B_0 P_1 P_2$ as the inner circle rolls upon the arc $B_0 b$, then evidently, if the smaller circle be fixed and the larger one rolled upon it in a direction opposite to that of the former rotation, the point of the great circle which at the beginning of the operation coincided with B_0 describes the same line $B_0 P_1 P_2$. This " evidently" expresses the usual notion, and the one which is suggested by a hasty pre-judgment of the case. In point of fact B_0 describes the pericy-

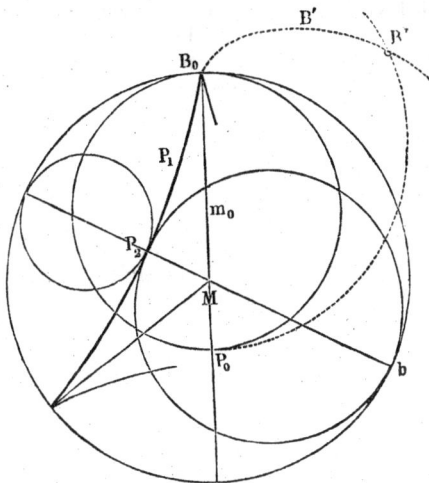

FIG. 445.

cloid $B_0 B' B''$, which certainly differs sufficiently from the hypocycloid $B_0 P_1 P_2$.

Centroids are also of very great assistance in understanding planetary movements, and are well suited to remove the difficulties which commonly occur in connection with their "real" and "apparent" motions. (I have a very instructive model for this purpose in the kinematic collection at Berlin.) If we test the ideas held by the majority of people upon this matter, we shall perceive how true a remark of Poinsot's (in the Memoir already referred to) still is,—" Mais s'il s'agit du mouvement d'un corps de grandeur sensible et de figure quelconque, il faut convenir qu'on ne s'en fait qu'une idée très-obscure."

[18] (P. 129.) These curves have often engaged the attention of mathe-maticians. Cf. Schlömilch's *Zeitschrift,* vol. ix., p. 209, Durège, *Ueber einige besondere Arten cyklischer Kurven.*

[19] (P. 145.) Open square figures, like ABCD, give square point paths if the centroid of the curved disc be a duangle (compare Pl. XII. Fig. 1). This occurs here, however, if QT be normal to PS. For we have, by similar triangles, $\angle 1PQ = \angle m_1\, SP$, and hence $m_1\, P = \dfrac{PR}{2}$ cos $1PQ = PS$ sin $1PQ$, or (as $PR = PS$), tan $1PQ = \frac{1}{2}$. The form of the disc must therefore be so chosen that the tangent of the angle $m_1\, SP = 0\cdot 5$, or that $m_1\, P = 0\cdot 5\, m_1\, S$. [I regret that the Figures on Pl. XII. have been transposed by mistake. Fig. 1 is referred to in the text as Fig. 2, and *vice versa*.]

[20] (P. 154). Willis (*Elements of Mechanism*, 2nd Edition, p. 90) gives Camus' theorem as follows :—

" If the pinion is to turn the wheel with a uniform force, the curve of its leaf, and that of the tooth of the wheel must be generated in the manner of epicycloids by one and the same describing curve, which must be rolled within the circle of the pinion to describe the inner form of the leaf, and on the outside of the circle of the wheel to describe the outer form of the tooth," etc. [Willis gives the quotation from Camus also in the original.]

[It is very interesting to compare the " solutions " given in Willis's third chapter with this portion of Reuleaux's work. Willis gives the method of § 32 as the " general solution," his other solutions are all for the special cases of circular centroids. With this limitation the first solution corresponds with § 34, the second and third are special cases of § 32, and the fourth corresponds with § 33. In a subsequent section he uses the method of § 35 for pin teeth.]

[21] (P. 154.) *Ann. Ph.* 1706, p. 379. De la Hire enunciates the proposition as follows :—" It is always possible to find a curve which, by revolving upon a given base-curve, shall generate by some describing point, in the manner of a trochoid, a second given curve ; provided that the normals from all points of the second curve meet the first." He gives as an illustration the production of a straight line by rolling a curve upon a second straight line cutting the first. The describing curve, as can be easily seen, is a logarithmic spiral. If the lines are parallel the spiral becomes a circle.

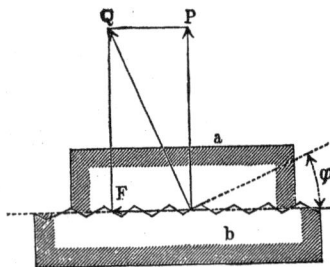

Fig. 446.

[22] (P. 171.) Suppose the two pieces a and b, Fig. 446, to be pressed together by some force, P, normal to their surfaces, and that these are covered with regularly formed teeth whose sides are inclined at a uniform angle, ϕ, in the way shown in the figure ; the resultant, Q, of the pressures upon the sides of the teeth opposed to any intended motion resolves itself into the resistance, P to the load and the resistance, F, to the force tending to produce sliding. F therefore is $= P$ tan ϕ. We could in this way determine, by experiments on the "friction of rest," the mean angle of the roughnesses upon the two surfaces in contact.

I may be permitted to take this opportunity of pointing out the inexactness of the treatment of friction which is customary in text-books of elementary Mechanics, and which, in my opinion, is very unfavourable to its proper comprehension.

There is first the notion that friction prevents motion, but that it does not produce it. This is the view commonly taken in theoretical text-books. Weisbach, for example (*Theor. Mech.* i, 4th Edition, § 167) uses the heading, "the resistances of friction and rigidity," and says in the text, "In relation to the motion of bodies friction is a passive force or resistance, for it only prevents or retards motion, and never produces or aids it." Kayser (*Statik* § 161) says, "Friction may be regarded as a passive force, merely hindering motion, and never causing or aiding it." All writers do not express themselves so frankly, but the prevailing idea is certainly that friction is a "resistance," and substantially this is said by all. Rühlmann is an instance, and Wernicke, Mosele, too, and Poncelet, and even Duhamel. Among modern writers even the clear-headed Ritter has adopted the common mode of expression, indeed, I have found in text-books no exception whatever to the rule. And yet the idea is not founded upon any investigation as to its accuracy, and it stands in manifest contradiction to the known laws of motion, or more generally to the law of the conservation of energy. For friction is a force and must itself be treated as one, no matter whether or how it be derived from other forces. A multitude of other forces are derived in the same sense as it is. There is no real reason why this force (and the stiffness of cords, which is generally treated in the same way as friction) should be suddenly thrown out from all systematic connection with others, why it should be absolutely asserted that this force does not possess the essential characteristic of other forces, the capacity, namely, of producing or aiding motion,—why it should always appear with the negative sign. We have here a survival from the ancient Mechanics, from which the modern scientific treatment of the subject has in some places shaken itself entirely free, and against which, in others, it is still struggling.

It is necessary, of course, to prove my position, and it is at the same time exceedingly easy. Both in nature and in machinery there are numerous illustrations of the falsity of the statement that friction does not produce motion. The wind sets the surface of water in motion by friction ; the wind itself is retarded by its friction upon the water, the latter being at the same time accelerated. The violin-bow sets the string in vibration by friction, in a way which Helmholtz describes as follows :—(*Sensations of Tone*, trans. by Ellis, p. 133)—"During the greater part of each vibration the string clings to the bow, and is carried on by it ; then it suddenly detaches itself and rebounds, whereupon it is seized by other points in the bow and again carried forward." If a fast-running belt or cord be slid on to a stationary pulley, it slides upon it at first and then, by a more or less gradual operation, sets it in motion. The force which acts as a driving force from a velocity $= 0$ to a velocity equal to that of the band is friction. It resists the motion of the band, but accelerates that of the pulley. If we look into the matter more closely we shall find that in every single case friction both causes and hinders motion,—although the former may occur only as small alterations of form in the body acted upon. It is not even necessary to fall back upon pure mathematics, as might easily be done, and show that the retardation is itself a production of motion, motion, that is, with a changed sign. Neither upon practical nor upon scientific grounds, therefore, is there any justification for the popular view, the correction of which is much to be wished, although a number of connected errors may make such correction difficult.

" Every friction causes the disappearance of actual energy," says Helmholtz, in one of his excellent *Vorträge* (Hft. ii. p. 129). From this undeniable proposition it is only too easy to infer the other, and false one, " in every case friction causes only the disappearance of actual energy." That friction always causes energy to disappear, in no way contradicts the fact that it also produces energy. It is therefore not unnecessary to warn students that the proposition quoted, although in itself true, may lead to some very erroneous conclusions if it be blindly followed. Let us take the piston of the steam-engine for an example. The piston fits the cylinder closely, and loses not inconsiderably in actual energy by friction against its sides. Even the keen experimenter Hirn, however, did not succeed in detecting the slightest loss of energy from this cause. He himself gives the reason for this quite rightly in saying that the energy lost by friction appears again in the correspondingly raised temperature of the steam. Here friction causes the simultaneous disappearance and re-appearance of energy in such a way that at the end we can perceive nothing of the process. The proposition, " Every friction causes actual energy to disappear," by itself, would not be apparently substantiated by mere measurement ; if it be given in an explanation of the nature of friction (for which purpose Helmholtz did not use it), it must be completed by the addition of the clause, " and also to reappear in another form."

I wish that writers upon Mechanics could be persuaded to give this matter a proper logical treatment, or at least to say something about it in elementary text-books. The further down the defect in logic be mended, the fewer corrections have to be made in the higher part of the work.

Another point to be mentioned is the statement of the l a w s o f f r i c t i o n. Taking up almost any text-book of Mechanics whatever, we find the three following important propositions given :—(1) Friction is p r o p o r t i o n a l to the normal pressure between the rubbing surfaces ; (2) it is i n d e p e n d e n t of their extent ; (3) it is i n d e p e n d e n t of the velocity with which sliding takes place between them. These are, as a whole, the propositions of Coulomb and Morin. Later, although now by no means recent, experiments have shown that they express the real phenomena of the case only within very narrow limits ; that for those surfaces and velocities which are commonly to be found in machinery they do not apply ; that in the latter, in fact, we must really read : n o t p r o p o r t i o n a l and n o t i n d e p e n d e n t. Indeed we know that machine-design has fallen into a hundred errors through its adhesion to the propositions of Coulomb and Morin, and in fact that in recent practice they have been entirely disregarded, and dimensions adopted which are altogether at variance with them. Is it not time that the experiments of Rennie, Hirn, Sella, Bochet, and others, were transferred from the notes to the text ? It is to be desired equally for the sake of mechanical science itself and for its many practical applications.

[I am glad to be able to point to at least one excellent elementary text-book, in which the results of experiment are " transferred to the text,"—Prof. Ball's *Experimental Mechanics.* The extent to which inaccurate statements on this subject are made by writers of ability, who are perfectly well acquainted with the physical facts in question, is really extraordinary. The following sentences, for instance, occur on the first page of a well-known treatise on Friction :—(the italics are mine) " . . . The second class of forces, comprising the resistance of fixed obstacles, the resistance which fluid media oppose to the passage of bodies through them, and friction, are *not capable* of producing sensible motion, and *show their effects only* by the apparent destruction of the motion which has been caused by forces of the former class. These may be termed ' resisting ' forces. . . . Resisting forces are *incapable of producing sensible motion,* and if they do not actually destroy it they do convert it into a species of motion which is, to sight, insensible."]

[23] (P. 203.) The series of phenomena adduced by Lubbock (*Origin of Civilization,* etc. Lond. 1870) in support of the theory of the unity of the human race (which is not to be confused with their growth from a single pair) is really extraordinary.

[24] (P. 205.) Chamisso (iv. 244) gives the following description of this and other methods :—

" In the Caroline Islands a piece of wood is fixed to the ground, and over it is held perpendicularly a second piece, about a foot and a half long, tolerably round, and of the thickness of one's thumb. This is caused to twirl by the palms of the hands, its lower and roughly pointed end being pressed against the fixed piece. The first slow regular motion is quickened and the pressure increased as the wood-dust, which is formed by the friction and collects round the borer, begins to carbonise. This dust is the tinder, and soon catches fire. The women of Eap possess wonderful facility in executing this process.

" In Radack and the Sandwich Islands they hold over the fixed piece of wood another which is about a span long and roughly pointed, and slope its upper end away from them at an angle of about 30°. It is held with both hands—the thumbs being placed below and the fingers above to improve the grip, and moved backwards and forwards in the plane of its slope through a distance of two or three inches. When the dust which collects in the groove formed by the friction begins to carbonise, the pressure and velocity are doubled.

" It is remarkable that in both methods the two pieces of wood used are of the same kind. They are best when of equally fine grain, not too hard and not too pliable. Both methods require practice, skill, and patience.

" The method of the Aleutians is the first of those above given, mechanically improved. They use the twirling-stick as they do the drill which they employ for other purposes. They hold and pull one end of the cord, which is twisted twice round it, with both hands ; its upper end is passed through a piece of wood prepared for the purpose, which they hold in the mouth. We have seen two pieces of fir-wood thus used give fire in a few seconds,—a result which otherwise would have taken a much longer time.

" The same people also produce fire by striking together two stones rubbed with sulphur over dry moss strewn with the same material."

These accounts are clear and intelligible—peculiarities with which, unfortunately, we cannot always credit the corresponding descriptions of other travellers. It is much to be desired that expeditions to remote parts of the world should attach greater importance to objective observation of the technical industries of the natives, and should make their descriptions of these as full and accurate as possible, and unmixed with subjective additions. The fire-producing apparatus of these people is one of their most interesting possessions ; it is in most cases of extremely great antiquity, and has formed the first step towards their other industrial operations. Many methods of fire producing have indeed been observed which have escaped the notice of writers on technical subjects. I may mention briefly three which have been orally described to me by esteemed friends.

Herr Jagor found the Malays employing the following method. A piece of dry bamboo, a foot long, is split up lengthways, and the tender inside bark which forms its inner coating is scraped together into a little ball in the middle of one of the halves. This half is then placed on the ground with the hollow side (and the little ball) downwards. The worker then splits so much away from the other half as to make it into a sharp-edged straight piece like a knife-blade. This he draws like a saw or file across the middle of the first piece, in which he has perhaps previously cut a little notch. This notch is widened and deepened by the sawing, and its edges get so hot that when at last a hole is made through the cane, the little ball of pith within catches fire.

Prof. Neumeyer saw a similar process used in New Holland. Instead of bamboo, wood was there used, and wherever it was possible a split log was used for the fixed piece. Some easily-kindled pith or other material was placed in the crack, and the process went on as above described.

Consul Lindau witnessed the following method of making fire in the Sandwich Islands. Some little stones of a kind which give sparks when struck together were placed, along with easily ignitable leaves, in a box formed from a large dry leaf, and then fastened to the end of a switch. This was twirled round in the air in a particular manner with great skill, so that the stones rattled against each other and the leaves caught fire.

The question of the invention or discovery of fire is not yet cleared up. Peschel in his excellent *Völkerkunde* (1874) deprecates premature conclusions on account of the scarcity of available material. Caspari (*Urgeschichte der Menschheit,* 1873) develops fully and carefully the hypothesis that the use of the borer may have led to its discovery, and his hypothesis is repeated and treated at great length in Baer-Hellwald's *Vorgeschichtliche Mensch* (p. 554 *et seq.*). Here some very remarkable survivals of primitive customs are pointed out as having been observed in Germany and England in the kindling of beacon-fires, and in Appenzell (Switzerland) as a child's play. (*See* Kuhn, *Herabkunft des Feuers,* Berlin, 1859 ; Caspari, as above, vol. i. p. 37 ; Schwarz, *Ursprung der Mythologie,* 1860, p. 142.) Herr Kuhn says that the case at Essede, among the Hanoverians, described by him, is not the only one which he has seen. The fire was lit by means of a horizontal pole, the ends of which were pivoted in hollows in two upright posts. In one of these hollows tow was placed, and this was kindled by twirling the pole rapidly in both directions. This was done by a cord twisted round it, pulled at both ends by men.

[25] (P. 206.) See Rau, *Drilling in Stone without Metal*, Smithsonian Report, 1868. In the cause of archæological science Rau has made the tremendous sacrifice of completing such a boring with his own hands. With a wooden borer such as we have described he pierced a hand plate of Diorite, 45 mm. thick, by making two hollows on opposite sides and meeting in the centre. He succeeded only after two years' (more or less intermittent) work. The form of the hole made is exactly that of the holes in numerous rude axes found in various places in Europe.

In the Ethnographic department of the Berlin Museum there are several excellent specimens of American work in rock crystal, among others a specially characteristic carved horse's head of something like 70 mm. long.

Mr. A. R. Wallace, in his *Narrative of Travels on the Amazon and the Rio Negro* (p. 278), says :—" I now saw several of the Indians with their most peculiar and valued ornament—a cylindrical, opaque, white stone, looking like marble, but which is really quartz imperfectly crystallized. These stones are from four to eight inches long, and about an inch in diameter. They are ground round, and flat at the ends, a work of great labour, and are each pierced with a hole at one end, through which a string is inserted, to suspend it round the neck. It appears almost incredible that they should make this hole in so hard a substance without any iron instrument for the purpose. What they are said to use is the pointed flexible leaf shoot of the large wild plantain, triturating with fine sand and a little water ; and I have no doubt it is, as it is said to be, a labour of years. Yet it must take a much longer time to pierce that which the Tushuaúa wears as the symbol of his authority, for it is generally of the largest size, and is worn transversely across the breast, for which purpose a hole is bored lengthways, from one end to the other, an operation which I was informed sometimes occupies two lives. The stones themselves are procured from a great distance up the river, probably from near its sources at the base of the Andes ; they are, therefore, highly valued, and it is seldom the owner can be induced to part with them, the chiefs scarcely ever."

[26] (P. 208.) This subject is most fully treated by Ginzroth, *Wagen und Fahrwerke der Griechen, Römer und anderer alter Völker*, Munich, 1817 ; also Weiss in several places. The four-wheeled vehicle was also in use, especially for carrying heavy loads. It had fixed axles, and was therefore much less easily guided than the two-wheeled one. In India there still exist, in the use of the natives, four-wheeled vehicles with a kind of moveable fore-carriage, an arrangement which must therefore be considered somewhat ancient. We know that the war-chariots of Porus were drawn into the immediate neighbourhood of the battle-field by draught oxen and not by horses. This may have occurred by no means unfrequently. It may well have happened that two of the empty chariots were then sometimes fastened together, the pole of the one to the frame of the other, and in this way a vehicle would be formed which had separate fore and hind carriages. The great ease with which such a compound vehicle could be guided must have struck its possessors, and may have led to the deliberate use of the revolving fore-carriage in regular vehicles.

[27] (P. 208.) Wooden war-chariots would be burnt by the conqueror for lack of horses to carry them away, iron vehicles he would render useless by

breaking some essential part of them, very much as we spike cannon. In 2 Sam. viii. 4, we have : " And David took from him a thousand and seven hundred horsemen, and twenty thousand footmen : and David houghed [rendered useless] all the chariots, and reserved of them an hundred chariots ; " also Joshua xi. 6 . . . " Thou shalt hough their horses and burn their chariots with fire . . . (v. 9) Joshua . . . houghed their horses and burnt their chariots with fire." That the Jews had long known wheeled vehicles is evident from a passage in Numbers (vi. ; 3—8), where six (wooden) wagons, each drawn by two oxen, are spoken of. The wheels of Solomon's laver carriages (about 1000 B.C.) were cast of bronze (1 Kings vi. 33), " their axle-trees and their naves and their felloes and their spokes were all molten."

²⁸ (p. 209.) The museum in Toulouse contains two remarkably well preserved antique bronze chariot-wheels of 54 cm. diameter, with naves 40 cm. long and 7 cm. diameter ; casts of them are in the Romano-Germanic museum at Mainz. These wheels have five round spokes and deeply-recessed felloes ; in the latter the rivets for fastening on the wooden rims still remain. The Esterhazy collection at Vienna, and the national museum at Pesth contain similar excellent specimens ; an existing Egyptian carriage-wheel of wood is described and illustrated in Wilkinson, *The Ancient Egyptian,* vol. i. p. 383.

²⁹ (P. 209.) In the gradual development of the chariot-wheel in constructive form, the tire plays a very important part. It was evidently the metal rim which first made the wheel durable when used for quick motion along heavy roads. It was long, however, before the iron tire made in one piece was reached. Homer mentions in his celebrated description of the chariot of Juno, tires of copper (*Iliad,* v. 722, *et seq.*) :—

> " Quickly Hebe fixed on the chariot the rounded wheels
> Of copper, eight-spoked, around an iron axle ;
> Their felloes, indeed, were of gold, imperishable, but around
> Tires of copper were firmly fitted, a wonder to behold."

The last words show that there must have been great difficulty in completing the " firmly-fitted " tires (in segments ?). That these were made of copper does not exclude the possibility that iron rims were also in use. Assyrian and ancient Persian relievos show chariots of many forms, most of them with smooth tires, scarcely distinguishable from the rest of the wheel. A few wheels are specially remarkable as showing rings of little projections (Fig. 447), like strings of beads all round the tire. Prof. Lindenschmidt, of Mainz, who called my attention to this peculiarity, solved the riddle at once. The projections are intended for the heads of nails. The whole wheel is covered with nails, driven into the wooden rim in close rows, their heads overlapping each other like scales. Among the discoveries in ancient burial-places in South Germany there are not a few iron tires about a metre diameter. They are always found in pairs, and are evidently the remains (after the rotting away of the wood), of the chariot wheels of the dead warrior, which had been buried along with him. These tires are covered with radial spikes on the inside, and their outer surfaces show the scale-like overlappings just mentioned. Close inspection

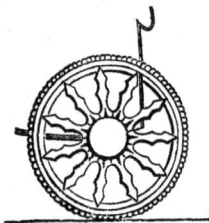

FIG. 447.

shows them to be simply the nails which had been driven into the wooden rim rusted together! The collection at Sigmaringen contains some beautiful specimens of them. They form evidently a very early step in the direction of making tires out of one piece.

I have found what appears to be a confirmation of Lindenschmidt's view in a model of a two-wheeled Chinese cart which was sent to the Vienna Exhibition of 1873. The tires are here made of iron, drawn out under the hammer ; they are, however, very narrow, and on their outer surfaces are deeply stamped into forms resembling strings of beads. This appears to be simply a transference of the traditional outward form to the solid tire. The stamping is a fashion only, use and wont give value to this external form, although the new construction has made it worthless—a process which fashions of every kind experience. It may be noted here, also, that in the great Pompeian mosaic, the "Alexander-schlacht," the Persian chariot in the centre is represented as having a tire of nails made in the way we have described.

[The ancient wheels in the British Museum form a very interesting study. The Egyptian paintings show but few chariots ; their wheels have always six spokes, and only one drawing (so far as I have noticed) shows any constructive details. This belongs to the 18th or 19th dynasty. The nave is made in one piece, and has sockets for the (round) spokes ; at the end of each spoke is a tee-piece, which forms a socket both for the spoke and for the segments of the tire. Spokes and tire segments are coloured red, the nave and tee-pieces are left white.

The ancient Greek vases (*circa* 800—500 B.C.) show numerous racing chariots, sometimes in very great detail. A great number of their wheels have four spokes only, but the way in which the wheel is put together is not shown. The tires are very narrow and apparently (from some of the end views) made in segments. A number of the drawings seem to indicate that the spokes are flat, and very much wider (in the plane of the axle) at the nave than at the rim. Strengthening pieces are always used at the junction of the spokes with the rim. Upon one vase, a prize at an Athenian chariot-race about 700 B.C., chariot-wheels are very distinctly shown as having one pair of radial spokes only, these being crossed at right angles by two bars passing at a considerable distance on each side of the nave. In the bronze room there is a four-wheeled bronze brazier from Vulci, and two others from Eschara, none of them probably later than 600 B.C. The wheels are four or five inches diameter.

The most interesting wheels are, however, those of the Assyrian sculptures. Here the uses of three different forms of wheel can be distinctly noticed. The vehicles used for heavy carriages and drawn by oxen have four spokes only. The sculptor has not thought it worth while to show their constructive details, but the spokes are very broad and clumsy, and probably square in section, the tires or rims are also very heavy. The ordinary war chariot-wheels have commonly eight spokes. These are apparently round and fit in sockets formed upon the nave. The rim of the wheel is very deep, and is always shown as consisting of three concentric rings, of which the outer one, the tire, is much the deeper, and is made in segments. The rims are generally strengthened by two pairs of clips slipped on from the inside and reaching partly across the tire,—Prof. Reuleaux tells me that in some instances at least these have proved to be pieces of leather. The royal chariots, at least those of Sennacherib and Sardanapalus, not only have ornamented spokes and naves, but have also the nail tires mentioned by Prof. Reuleaux. These occur nowhere but on the royal chariots, so that they must have been the " latest improvement " in Assyria in the eighth century B.C. I have noticed wheels with more than eight spokes only upon one slab. One has sixteen, one thirteen and several twelve spokes. Some of these, however, are certainly intended for wheels belonging to the chariots of the people with whom the Assyrian are fighting.]

[30] (P. 209.) According to Herr Detring's own observation, so that the disc-wheel of the plaustrum forms a step in the growth of wheeled vehicles all the world over.

[31] (P. 210.) In Sanscrit the chariot is called *ratha*.

[32] (P. 210.) We may remember, for instance, the method of transporting the pillars of the Temple of Artemis in Ephesus, described by Vitruvius (x. chap 2). The master-builder Chersiphron fastened iron pins to the ends of the enormous cylindrical blocks of hewn stone, and laid upon these a wooden frame fitted with proper bearings for them. To this frame the draught oxen were yoked, and with their aid the pillars were dragged, after the manner of our street rollers, from the quarry to the site of the building—the same site upon which excavations have recently enabled us to appreciate the magnitude of the work, and the advantages of the method used in carrying it out.

[33] (P. 211.) It has been successfully shown by experiment that apparently blunt fragments, if they have crystalline edges, are specially suited for boring harder stones.

[34] (P. 212.) The rare form *tornator* is to be found in Jul. Firmicus (Mathesis, iv. 7) :—*facit quoque tornatores, aut simulacrorum sculptores.*

[35] (P. 212.) Among the specimens at the Berlin Museum which undoubtedly belong to the old kingdom, there are several which have certainly been turned in the lathe, and the latter must therefore have been used by the Egyptians between 2,000 and 3,000 years before our era. These are again vessels, partly of alabaster and serpentine (as Nos. 93 and 88), partly of marble and even granite (Nos. 62 and 100). The hypothesis of the connection between the lathe and the potter's wheel (which turned out most excellent work, as we see from the museum collection, even at that early date) seems to be supported by this.

[36] (P. 215.) Cf. Böckler, *Theatrum Mechanicum Novum*, Nuremberg, 1762, Plates 35, 36, 80. Neither in this whole work of 154 plates, nor in Rosberg's *Kunstlichem Abriss* &c. Nuremberg, 1610, do we find any apparent trace of our present belt-train. Arrangements for driving by a cord or rope twisted two, three, or four times round a pulley, are given by Ramelli, *Arteficiose Machine*, Paris, 1588, Plates 171, 175, 183.

[37] (P. 217.) There are specimens of Ancient Egyptian spindles in the Berlin Museum. Wilkinson, who mentions the Berlin specimens in his *Ancient Egyptians*, places beside them (Fig. 385, 1 to 5, vol. ii.) three illustrations of distaffs or portions of them, which he erroneously takes to be spindles also. He had apparently been misled by a note in an older catalogue. [Nos. 1 and 2, Fig. 168 have unfortunately been printed upside down.]

[38] (P. 220.) Dr. Wetzstein writes to me : " The word schaduff or shadoof

comes from the root *schadf*, which means to hang down to one side. This is very appropriate to the irrigating machine in question, because its lever, when not in action, always slopes downwards towards that side which is weighted with stone. The machine is not found in Syria, I have seen it only in Egypt." In the *Descr. de l'Egypte* (xviii., 2, p. 539, *et seq.*) the shadoof is also called *delú (delou);* at the junctions of water channels from thirty to fifty shadoofs are not unfrequently to be seen together.

[39] (P. 225.) Endeavours in this direction are even now to be met with among a few cultivated nations. Baron Von Korff saw, as he told me, in Egypt, a gunsmith who, while both hands were busy with his iron work, used his feet in working a saw to cut the wood for his gun-stock. The Tartars, both men and women, although engaged in their domestic duties, seldom lay aside their great curved embroidering frames. We need only look at the European stocking-knitter, too, to see the connection between these customs and our own.

[40] (P. 229.) A Spanish word, from the Arabic *nā-'ūrah*, so called from the snorting noise made by the emptying of the buckets :—*na'ara*, to snort (Heyse). Vitruvius also knew these wheels, which even in his time must have been of great antiquity (x. chap. v. [Vulgo x.]) : . . . "Circa eorum frontes affiguntur pinnæ, quæ cum percutiuntur ab impetu fluminis, cogunt progredientes versari rotam, et ita modiolis aquam haurientes et in summum referentes sine operarum calcatura, ipsius fluminis impulsu versatæ, præstant quod opus est ad usum."

[41] (P. 230.) A splendid example of this kind of machine stands in Zürich in the immediate neighbourhood of the Polytechnic School, a contrast which is humorous enough. It would be worth while to preserve at least drawings of this mammoth among machines, this doomed representative of a past epoch for the benefit of the coming race.

[42] (P. 235.) [In the Patent Museum at South Kensington there are to be seen several wooden models of Watt's, of his proposed arrangements for obtaining rotary motion from the beam, as well as the sun-and-planet engine which for so long drove the machinery at his Soho factory. See also Note, p. 433.]

[43] (P. 237.) It appears not to be well known, and may therefore be mentioned here, that the Greeks were perfectly well acquainted with the pulley. The Romans received both the thing itself and its name from the Greeks (cf. Vitruvius x. chap. ii., *De machinis tractoriis*). The three-sheaved tackle they called τρίσπαστος, the five-sheaved πεντάσπαστος, the multi-sheaved generally πολύσπαστος. These names were certainly better than ours, for we have seen (§ 43) that the characteristic part of the tackle is the stretched rope or cord, and not the revolving pulley or sheave. A mere fixed guide pulley the Greeks called αρτέμον.

[44] (P. 237.) If we arrange the different forms of toothed wheels according to the increasing complexity of their theoretical treatment, we should have to adopt the order : spur-wheels, bevel-wheels, screw-wheels, hyperboloidal-wheels. It would, however, be a mistake to assume, without further inquiry, that this was the order of their natural development. As a matter of fact toothed wheels with crossed axes, and therefore with hyperboloidal axoids, appear to be the oldest, and to have led up to the conception of the simpler toothed wheels.

For we find wheels of the simplest possible form, consisting, namely, of nothing but a nave and radial spokes, in primitive water-lifting wheels, where the horizontal wheel-shaft was driven from a vertical one (cf. Fig. 50 of Ewbank's *Hydraulic and Other Machines*, 16th Edition, New York, 1870). The screw-wheels for parallel axes, the invention of which has been ascribed to an Englishman, White, are to be found in primitive Indian cotton ginning rollers (see a drawing in Leigh's *Modern Cotton Spinning*, London, 1873, as well as several complete machines in the Indian Museum, London). We may note also that toothed-wheels having intersecting axes, in the form of crown-wheel and pinion, have received more extensive application and attention in mill work, almost down to our own time, than spur-wheels. The latter, indeed, appear to come last in order, so that the real sequence of historical development is exactly the reverse of what we might expect, a hint that we must never confound what actually and practically lies nearest to us with what is geometrically simplest.

[45] (P. 238.) From the Arabic *sakai*, to water or supply water ; *sakkā*, a water-carrier in eastern countries.

[46] (P. 275.) It is not uninteresting to compare the different statements on this subject. We may give a few specimens of them :—

Poppe, *Maschinenkunde* (1821), p. 81 :—"The lever, the wheel and axle, the pulley, the inclined plane, the wedge, and the screw are included under the name simple machines, simple engines (*Rüstzeuge*) or mechanical powers. From these all machines, even the most complicated, are constructed. Since, however, the theory of the wheel and axle and of the pulley is based upon the law of the lever, and the theory of the wedge and the screw upon the law of the inclined plane, we may reduce the number of simple machines to two, the lever and the inclined plane."

Here it is clearly and emphatically stated that all machines, "even the most complicated," are formed from the simple machines, and that the latter may be reduced to two. Read, however, the following :—

Langsdorf, *Maschinenkunde* (1826), i. p. 277 :—"Even in the older text-books we find machines divided into simple and compound, the latter being those which are formed by a combination of several of the former. The simple machines are limited to the lever, the pulley, the inclined plane, the wedge, the screw, and the wheel and axle. The immoveable inclined plane should not, however, be included, it is no more a machine than is the slope of a mountain. We do indeed find a moveable inclined plane in the wedge ; inclined plane and wedge are not then machines of two different kinds. I put in their place the roller. . . . ." Here, then, it is said to be wrong to treat the inclined plane as a simple machine, while before it was treated as the foundation of several others.

Gerstner, *Handbuch der Mechanik* (1831), i. p. 73 :—"Machines are commonly divided into simple and compound. The simplest machine, which we shall first consider, is the lever. We shall then proceed to the wheel and axle, the pulley and the pulley-tackle (!), the inclined plane, the screw and the wedge. All these machines are simple machines ; compound machines consist always of a combination of several simple ones, after which, therefore, their treatment will come."

Kayser, *Handbuch der Statik* (1836), p. 460 :—" Machines are divided also into simple and compound. Strictly speaking only the cord (!), the lever and the inclined plane are simple machines. It is customary, however, to treat along with these also those into which compound machines can be resolved. These simple machines are seven in number, viz., the cord, the lever, the pulley, the wheel and axle, the inclined plane, the wedge, and the screw. They are also called machine organs or mechanical powers. Many writers do not reckon the cord among them."

Rühlman, *Mechanik* (1860), p. 231 :—" A machine of which no part is itself a machine is called simple, in the opposite case compound. The simple machines are the funicular machine, the lever, the pulley, the wheel and axle, the inclined plane and the wedge. Note : strictly speaking we need distinguish only three simple machines, the funicular machine, the lever, and the inclined plane, all the others may be resolved into these." This definition leaves something still wanting, and in itself it is a complete *petitio principii.* We have again the impossible derivation of the pulley from the lever.

Schrader, *Elemente der Mechanik und Maschinenlehre* (1860), p. 26 :—" The different kinds of simple machines. The originals of all simple machines are the lever and the inclined plane. From the lever are derived the pulley and the wheel and axle, and from the inclined plane the wedge and the screw. Note : in the lever the moving piece rotates, in the inclined plane it moves in a straight path." The pulley is, as usual, quite wrongly placed.

[I am sorry to say that the definitions of English authors have been no more satisfactory, as a rule, than those given above. Todhunter, for example, says :—" The most simple machines are called mechanical powers ; by combining these, all machines, however complicated, are constructed. These simple machines . . . . are usually considered to be seven in number; namely, the lever, the wheel and axle, the toothed-wheel, the pulley, the inclined plane, the wedge and the screw." We continually find, too, the loose expressions that the lever is a solid body "moveable about a fixed point," that it is "supported at one point" (as distinct from the wheel, which is supported at an axis), and so on.]

It is very remarkable that in all the examples we have given, with the exception of Langsdorf, the peculiarity of the screw as a simple machine is denied, although it is kinematically the general case of the three lower pairs, and ought therefore in every case to remain in the classification. The extraordinary confusion (for so we must call it) of ideas upon the subject arises from a peculiar misunderstanding which, so far as my experience goes, is very strongly rooted and may be only very slowly dislodged. It is that the similarity of the relations existing between the forces coming into action is mistaken for a similarity between the objects themselves. Because certain force-relations in the screw are conditioned similarly to those in the inclined plane, it does not follow that the two things are identical. Instead of examining the things themselves, people have concerned themselves with certain of their properties. The importance of the latter indeed cannot be disputed, but they ought logically to be kept apart from the actual nature of the combination of bodies to which they belong. When, on the other hand, the more recent writers apparently do away with the simple machines altogether, but in reality introduce them as " exercises," " examples," " applications " and so on, they have furthered a good result less than they believe.   For, as we have seen in the text, there is really s o m e truth at the bottom of these problems—no precautions have been able to expel the sense of this fact.   We read between the lines the sentence of Horace : *Naturam expellas furca, tamen usque recurret !*

It is not easy to say up to what limits it may be advisable for general Mechanics to follow the methods which we have worked out.   I believe, however, that it is certainly advisable that the simple machines should be treated in the way which our kinematic investigations have pointed out to us in elementary Mechanics.   This cannot but be of use in increasing the tangibility and definiteness of the ideas which the scholars receive upon the subject.

To the question how far generally Mechanics should concern itself with machines, we may say that this should be the case in decreasing degree from the lower to the higher Mechanics.   For those who want merely elementary notions of the subject, general Mechanics and the Mechanics of machinery are one and the same thing.   The higher the studies be pursued the more distinctly their differences make themselves felt.   Which of the many positions between the highest and the lowest should be adopted in the construction of a text-book must in every case be carefully considered.   But before everything in my opinion elementary text-books of Mechanics deserve much more careful treatment, especially as to the logical arrangement of their contents, than they have hitherto received.   They are too often deficient in that transparent clearness which we are entitled now to demand from Mechanics.   We have already noticed this in reference to friction.   How unconnected with everything else, also, the treatment of the strength of materials commonly is !   Apart from certain internal peculiarities in the way of new data, to which I have called attention in the preface to the last edition of my *Constructeur,* the general treatment of the matter appears to me defective, and it is never made sufficiently distinct that the " strength of materials " stands simply in the same relation to rigid bodies as hydrostatics and hydraulics occupy to liquids and aerostatics and aerodynamics to gaseous bodies.   (If we wished to include all under a general title we might use the words stereostatics and stereodynamics for that purpose.)   All three branches treat of the inner mechanical forces—our l a t e n t forces of § 1—which give the material its existence ; all three besides overlap each other in the limiting cases.   On the other hand just the same separation can be made with fluids as is done in the separate treatment of the " strength of materials" with solids ; the problems connected with their molecular condition can be separated from those concerning their relations as a whole to other bodies.   Very valuable analogies show themselves between the three departments if they only be looked for.   I believe that new life might be thrown into the whole study if its treatment were taken up afresh in the direction which I have pointed out.   [I may take this opportunity of pointing out that we have already in English an excellent word, introduced by Prof. Rankine, for for what Prof. Reuleaux calls a " latent force," namely s t r e s s.   I am not aware that it has any German equivalent.   It is very much to be wished that engineers—and physicists too, for that matter—would agree to use this word where now s t r a i n is often employed, and to keep the latter for its more obvious meaning of d e f o r m a t i o n, for which it is far better adapted.]

[47] (P. 290.) [The lines A B and D C (Fig. 208) cut 1, 4 and 2, 3 in such a way as to make the alternate angles 1 and 3 (and also of course 4 and 2) equal. Hence the name anti-parallel. See *e. g.* Reuleaux's *Constructeur*, 3rd ed., p. 71.]

[48] (P. 327.) It was Willis who first pointed out the nature of the conic crank-trains, and their analogy to the cylindric crank trains. (*Principles of Mechanism*, 2nd ed., 1870, p. 249, ff.) He called the trains "solid angular link work," and indicated several of their more important forms and characteristics. He had not, however, the idea of the kinematic chain, and missed therefore, some of their most essential properties; as a kinematist of the old school, too, the fourth (fixed) link altogether escaped him, as also the possibility of inversion, and with these some very remarkable practical applications of the chain, of which we shall have more to say further on.

[49] (P. 341.) The treatment of compound chains belongs to the more difficult problems of Kinematics. I refer to them again in Chapter XIII., particularly in § 160. The full advantage of the idea of chain reduction only makes itself felt in the study of these applications of Kinematics. I recommend the teacher to give his pupils exercises in the re-completion of reduced chains.

[50] (P. 384.) I have certainly not exhausted the list of chamber-trains which have been constructed from $(C_3'' \, P^\perp)^a$, although so many of them have been investigated. It is interesting to note that lately the crossed slider-crank chain (see § 73) has also been used in chamber-trains. Gibson's rotary steam-engine (*American Artisan*, Feb. 1874, p. 30) is an example of this; it is a combination of two trains of the form $(C_3'' \, P^+) \frac{a}{c} - b$.

[51] (P. 399.) For the determination of the axoids of the links b and d in $(C_3^\perp \, C^L)$ we have the following (see Fig. 448):—

$$\frac{w_1}{w} = \frac{r}{r_1} = \frac{\sin \gamma}{\sin \gamma_1} = \frac{\cos a}{1 - \sin^2 \omega \, \sin^2 a}$$

We require to find the values of γ and γ_1 corresponding to different values of ω. We have

$$\gamma_1 = 180° - (\gamma + a),$$

hence

$$\sin \gamma_1 = \sin (\gamma + a) = \sin \gamma \cos a + \cos \gamma \sin a,$$

and from this, putting $\dfrac{\cos a}{1 - \sin^2 \omega \, \sin^2 a} = A$, we obtain

$$\sin \gamma = A \, (\sin \gamma \cos a + \cos \gamma \sin a).$$

$$\frac{1}{A} = \cos a + \cot \gamma \sin a,$$

whence

$$\cot \gamma = \frac{\dfrac{1}{A} - \cos a}{\sin a},$$

or, again inserting the quantity represented by A :

$$cot\ \gamma = \frac{1 - cos^2\ a - sin^2\ \omega\ sin^2\ a}{cos\ a\ sin\ a} = \frac{sin^2\ a\ cos^2\ \omega}{cos\ a\ sin\ a},$$

that is,
$$cot\ \gamma = tan\ a\ cos^2\ \omega.$$

a relation which also may easily be determined graphically, as is shown, for example, in Fig. 452. It gives the ratio $\frac{cos^2\ \omega}{1} = \frac{x}{tan\ a}$, that is, $x\ cot\ \gamma$.

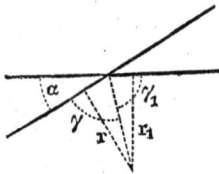

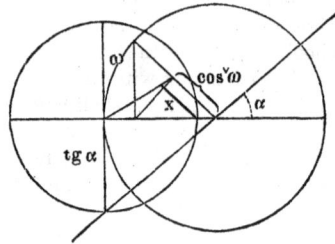

Fɪɢ 448.　　　　　　　　　Fɪɢ. 449.

[52] (P. 400.) It is not difficult to add new forms synthetically to the many old ones which we have mentioned. Indeed, the foregoing investigations permit such problems to be given directly as exercises in a course of machine instruction. I do not intend to urge here the use of such exercises, which would be suitable only for very advanced students ; they do not, however, essentially differ from those of modern chemistry, where the more advanced students in the laboratory are exercised in the synthetic development of new series of bodies. [I venture to go further in this matter than Prof. Reuleaux, and to hope that synthetic exercises may become both possible and popular among our students to a considerably greater extent than he suggests. The formation of chamber-trains alone gives immense scope for such exercises, and there are also other directions in which they could be worked without touching on problems of any serious difficulty. I know no kind of exercises which are likely to excite so much interest among the class of young men who in this country devote themselves to the profession of engineering.] I may give an example of this. It will be noticed that in the twelve conic chamber-trains formed from $(C_3^\perp\ C^L)^d$ and $(C_3^\perp\ C^L)^b$ the chambering $(V\mp)=c$, d is never used. This, might, however, be done as follows. Instead of making the element 4 of the link d into a diaphragm and guide for c (as in Fig. I., Pl. XXVIII.), it might be formed into a sector of a hollow cylinder with cylindrical openings at its ends, as in Fig. 450. In the chamber thus formed, the cross-section of which is similar to that of the former spheric-sector chambers, the element of c belonging to the pair 4 may be placed as a piston. This takes the form of a slice of the same cylinder, fitted with a shaft, the ends of which might project beyond the chamber. To these projecting ends an outer frame is fastened, which carries the cylinder, C+, of the pair 3 belonging to the link c. The link b has open cylinders both at 2 and at 3, and is simply a connecting-rod with its two open elements normal to each other. At 2 it is

paired with the crank $a = C+ \ldots \angle \ldots C+$, which (as before) finds its bearing 1 in the link d. In this way we obtain a chamber-train in which the piston c merely oscillates about its axis. Without discussing at all the usefulness of this form of engine, I may mention that something similar to it has already been made in Morton's disc-engine (*Deutsche Gewerbezeitung*, 1857,

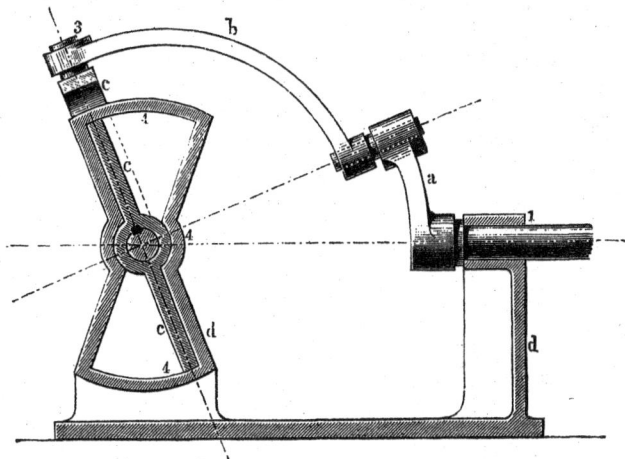

FIG. 450.

p. 31). Morton has, however, impressed by the form of the older disc-engines, made his piston c and chamber d unnecessarily as portions of spheres, and prides himself on the fact that the piston of his machine does not make the wobbling motions of those of the older engines, and that it has not the slot, nor the chamber the diaphragm and packing pieces formerly required. He has thought it necessary, however, to make the sides of his chamber with plane inner surfaces, parallel to the axis of 4, the piston lying upon these in its two extreme positions.

[53] (P. 413.) In the year before, 1858, a patent was taken out (dated 14th April, and taken out through Newton's Agency), for a steam-engine having such shoe-sole-shaped wheel pistons. Special packing-pieces were fitted into the ends of the teeth (*Propagation Industrielle*, iv., 1869, p. 179).

[54] (P. 437.) [This statement, I am sorry to say, does not apply to this country, where, so far as I know, the constructive elements as such have never yet received any systematic treatment. I have had to make a few small alterations in Chapter XI. on this account, omitting a few sentences which applied solely to the existing continental treatment of certain details. Professor Reuleaux's discussion as to the subdivision and classification of the constructive elements has therefore no direct bearing, as yet, upon English text-books, and hardly any upon English systems of instruction. I hope sincerely that such a state of matters may not long exist.]

[55] (P. 461.) [I think we might call the whole class c a m t r a i n s, under which the click-trains and slider-cam trains would come as special cases. The latter bear to the cam-trains the same relation that $(C_3'' P \perp)$ bears to (C_4'').

The higher pairing which occurs here receives further consideration in Chapter XIII. § 157.]

[56] (P. 498.) [The matter may also be looked at in a somewhat different way. The driver of the old engines formed an element of a sliding-pair, as t does still, for instance, in direct-acting pumping engines. The attempts at rotary engines seem to me attempts to replace this sliding-pair directly by a cylinder or turning-pair. Something equivalent to this is very frequently insisted on in descriptions and specifications, and has certainly, in a more or less indistinct form, been present in the minds of many inventors, who have persistently refused to see more than one moving part in their machines. It is one of the results most to be hoped from the acceptance of Professor Reuleaux's method of analysis, that the energies of these and other ingenious minds may be turned into worthier channels.]

[57] (P. 522.) [§ 137, as it appears here, is a summary of a much longer treatment of the subject given by Reuleaux, which I hope may be published at length in another form. He discusses in some detail the present position of workmen on the Continent, and the way in which they have been affected by the machine and machino-facture. The circumstances of the case, as he describes them, differ in some very important respects from the circumstances attending similar industries in this country.]

[58] (P. 533.) [It will be remembered that strictly the pair (C) is not closed, for it has not of itself the cross profiles necessary to prevent axial motion. This is a general difference between the pairs formed from S and those formed from $\bar{H}$; the former may be made completely constrained in themselves, while in the latter (as Figs. 363 to 367 for instance) the necessary constraint in one or more directions is obtained only by the use of pair-, chain-, or force-closure. This closure being provided, however, the form of these higher pairs determines motion as absolutely as the closed forms of the lower ones.]

[59] (P. 538.) [A comparison of this table with that given at p. 543 of the German edition will show some points of difference between the two. These appeared to me necessary to bring the table and the text into complete agreement,—especially as several corrections (see Preface) have been made in the latter,—but it is right to say that I have not been able to submit them to Prof. Reuleaux, who was on his way to Philadelphia when they were made. For my own purposes I have used a different classification, which I need not give here. There is one detail connected with the notation of the higher pairs which might be modified, I think, with advantage. Prof. Reuleaux uses, *e.g.* the symbols (C_z) and $(C_z;)$ for different pairings of C with Z. The first form appears to me much the better, so that I would suggest the use of (C_z), (C_z) and so on, instead of $(C_z;)$, $(C_z;)$, etc. It would then be understood that a small letter used simply as a suffix indicated some quality of the element denoted by the capital letter after which it was placed, and possessed in common by the two elements of the pair, while if a stop of any kind (comma, semicolon, etc.) were marked b e t w e e n the two letters the meaning would be that the pair consisted of two d i s s i m i l a r elements, paired in the manner pointed out by the stop. Thus (C_z), or more strictly $(C_{z;})$, would indicate a pair of elements each of the form C_z, while (C_z) would be a pairing of C with Z, etc.]

[60] (P. 552.) A model of the mechanism shown in Fig. 395, exhibited at Vienna, was placed beside a very nearly related piece, the so-called skew-disc (*schiefe Scheibe*). In the *Bairischen Industrie und Gewerbeblatt*, 1874, p. 100, Herr Schedlbauer gives a theory of the motion in this mechanism, and shows that the link *e* makes swinging motions which are given by the formula $(r\ tan\ a)\ sin\ \omega$,—in which *r* is the constant distance 1·6, *a* the angle between 1 and 2, and ω the angle of turning of the link *a* relatively to the fixed link *f*. According to this the link *e* would have a simple harmonic motion. The constructive conditions of Herr Schedlbauer's mechanism are, however, somewhat different from those of Fig. 395, and of my model. He assumes the link *b*, carrying an element of each of the pairs 2 and 3, to be of the form $C \dots \perp \dots P$, and also that the axis of the last-mentioned prism always intersects the axis of the prism *C* at a constant distance from 1. This would be a mechanism having for its formula,—beginning with the pair 1,— $(C{\llcorner}C{\llcorner}P{\perp}C{\llcorner}P''C'')_a$. In the train represented in Fig. 395 the link *e* makes motions which only approximate to simple harmonic oscillations, and which are given by the expression

$$y = \frac{(r\ sin\ a)\ sin\ \omega}{\sqrt{cos^2\omega\ cos^2 a + sin^2\omega}}$$

if *y* be the distance of a point of *e* from its middle position and *r* the length 2·4.

The difference between the two motions is very small with a small angle at *a*, and may be generally neglected in the cases which occur in machine practice. I mention the matter merely as another illustration of what I have already noticed in note 46, that the motion of a mechanism, or more correctly one of the motions occurring in it, has been often investigated without any examination having been made of the actual combination itself by which that motion was produced. The latter is, however, in the case before us the more important part of the problem, for we have already an immense number of trains in which an exact or approximate simple harmonic motion occurs, [I have called the latter *distorted* harmonic motion], while the various forms of the train (C_6^+) have never yet received investigation.

The steam-engine of Robertson, mentioned upon p. 551, contained one detail which deserves further mention. Robertson used, namely, according to the published description, the driving mechanism shown in Fig. 451, which has occasioned no little astonishment. Here *c* is a spur-wheel driven by the wheel *a* not by means of teeth but by the water *b* held in the hollow ring of *a ;* or, as we should say more rightly,

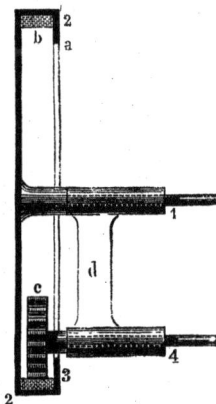

Fig. 451.

paired with *a*, for the water slides in *a* if the velocity of this wheel fluctuates. The water-ring *b* can therefore turn uniformly, although the wheel *a* moves with a variable speed received from the train (C_4^+). It is claimed that the transference of motion will be very "sweet" ; at starting only a little water will be spattered about, afterwards the centrifugal force is sufficient to keep it against the rim of the wheel *a*. We have here an application of the pair C_z, Q_λ or $(C_{z,\lambda})$ belonging to order XVI. (§ 149.) The vessel $V-$, with which Q_λ is paired on the other side, has, on account of the sliding of the water-ring, the form $C-$. We might write the whole train therefore :—

$$\overset{a}{C+\ldots\,|}\ldots C-,\,\overset{b}{Q\gamma\ldots\ldots}Q\gamma,\,\overset{c}{C_z^+\ldots\,|}\ldots\underline{C_{\underline{=}}^+C_{\underline{=}}^-\ldots\,\overset{d}{\|}}\ldots C_{\underline{=}}^-$$

We note also again here how practical machine construction has taken the road already pointed out by our synthesis. The pairing of Q_λ with C_z is, as we know, nothing new in itself; it exists, for instance, in the common water-wheel and other mechanisms; the mechanism of Robertson is interesting only as an attempt to use the pairing in a free manner in a driving-train intended for common factory work.

INDEX.